BASIC PRINCIPLES OF INDUCTION LOGGING

BASIC PRINCIPLES OF INDUCTION LOGGING

Electromagnetic Methods in Borehole Geophysics

ALEXANDER KAUFMAN

GREGORY ITSKOVICH

ELSEVIER

Elsevier
Radarweg 29, PO Box 211, 1000 AE Amsterdam, Netherlands
The Boulevard, Langford Lane, Kidlington, Oxford OX5 1GB, United Kingdom
50 Hampshire Street, 5th Floor, Cambridge, MA 02139, United States

© 2017 Elsevier Inc. All rights reserved.

No part of this publication may be reproduced or transmitted in any form or by any means, electronic or mechanical, including photocopying, recording, or any information storage and retrieval system, without permission in writing from the publisher. Details on how to seek permission, further information about the Publisher's permissions policies and our arrangements with organizations such as the Copyright Clearance Center and the Copyright Licensing Agency, can be found at our website: www.elsevier.com/permissions.

This book and the individual contributions contained in it are protected under copyright by the Publisher (other than as may be noted herein).

Notices
Knowledge and best practice in this field are constantly changing. As new research and experience broaden our understanding, changes in research methods, professional practices, or medical treatment may become necessary.

Practitioners and researchers must always rely on their own experience and knowledge in evaluating and using any information, methods, compounds, or experiments described herein. In using such information or methods they should be mindful of their own safety and the safety of others, including parties for whom they have a professional responsibility.

To the fullest extent of the law, neither the Publisher nor the authors, contributors, or editors, assume any liability for any injury and/or damage to persons or property as a matter of products liability, negligence or otherwise, or from any use or operation of any methods, products, instructions, or ideas contained in the material herein.

Library of Congress Cataloging-in-Publication Data
A catalog record for this book is available from the Library of Congress

British Library Cataloguing-in-Publication Data
A catalogue record for this book is available from the British Library

ISBN: 978-0-12-802583-3

For information on all Elsevier publications
visit our website at https://www.elsevier.com/books-and-journals

Publisher: Candice Janco
Acquisition Editor: Marisa LaFleur
Editorial Project Manager: Marisa LaFleur
Production Project Manager: Maria Bernard
Cover Designer: Mark Rogers

Typeset by SPi Global, India

DEDICATION

To the memory of my parents Basia and Boris Itskovich and grandma Golda.
To my dear wife Emma and children Semion and Galina.
To my faithful friend David Shulman.

CONTENTS

Introduction xi
Acknowledgments xv
List of Symbols xvii

1. System of Equations of the Stationary Electric and Magnetic Fields — 1

1.1 Equations of the Stationary Electric Field in a Conducting and Polarizable Medium — 2
1.2 Interaction of Currents, Biot-Savart Law, and Magnetic Field — 4
1.3 Vector Potential of the Magnetic Field — 13
1.4 System of Equations of the Stationary Magnetic Field — 18
1.5 Examples of Magnetic Field of Current-Carrying Objects — 21
1.6 System of Equations for the Stationary Fields — 37
References — 37
Further Reading — 37

2. Physical Laws and Maxwell's Equations — 39

2.1 Faraday's Law — 40
2.2 Principle of Charge Conservation — 46
2.3 Distribution of Electric Charges — 48
2.4 Displacement Currents — 59
2.5 Maxwell's Equations — 67
2.6 Equations for the Fields E and B — 73
2.7 Electromagnetic Potentials — 75
2.8 Maxwell's Equations for Sinusoidal Fields — 78
2.9 Electromagnetic Energy and Poynting Vector — 81
2.10 Uniqueness of the Forward Problem Solution — 86
Reference — 90
Further Reading — 90

3. Propagation of Electromagnetic Field in a Nonconducting Medium — 91

3.1 Plane Wave in a Uniform Medium — 91
3.2 Quasistationary Field in a Nonconducting Medium — 104
3.3 Induction Current in a Thin Conducting Ring Placed in a Time-Varying Field — 111
Further Reading — 131

4. Propagation and Diffusion in a Conducting Uniform Medium — 133

- 4.1 Sinusoidal Plane Wave in a Uniform Medium — 133
- 4.2 Field of the Magnetic Dipole in a Uniform Medium (Frequency Domain) — 141
- 4.3 Transient Field of the Magnetic Dipole in a Uniform Medium — 149
- 4.4 The Field in a Nonconducting Medium — 153
- 4.5 The Transient Field in a Conducting Medium — 157
- Further Reading — 161

5. Quasistationary Field of Magnetic Dipole in a Uniform Medium — 163

- 5.1 Expressions for the Field — 163
- 5.2 Low and High Frequency Asymptotic — 166
- 5.3 Expression for Induced Currents — 168
- Further Reading — 172

6. Geometrical Factor Theory of Induction Logging — 173

- 6.1 Two-Coil Probe — 174
- 6.2 The Vertical Responses of the Two-Coil Probe in the Media With the Horizontal Boundaries — 182
- 6.3 Radial Characteristics of Two-Coil Induction Probe — 195
- 6.4 Multicoil or "Focusing" Induction Probe — 204
- 6.5 Corrections of the Apparent Conductivity — 222
- References — 226
- Further Reading — 226

7. Integral Equations and Their Approximations — 227

- 7.1 Physical Principles of the Hybrid Method — 227
- 7.2 Derivation of the Equation for the Field — 228
- 7.3 A Volume Integral Equation and Its Linear Approximation — 235
- 7.4 A Surface Integral Equation for the Electric Field — 238
- References — 247
- Further Reading — 248

8. Electromagnetic Field of a Vertical Magnetic Dipole in Cylindrically Layered Formation — 249

- 8.1 The Boundary Value Problem for the Vector Potential — 249
- 8.2 Expressions for the Field Components — 254
- 8.3 The Magnetic Field in the Range of Small Induction Number — 257
- 8.4 Far Zone of Magnetic Field on the Axis of Borehole — 266

8.5 Displacement of the Probe from the Borehole Axis		284
Reference		288
Further Reading		288

9. Quasistationary Field of the Vertical Magnetic Dipole in a Bed of a Finite Thickness — 289

9.1 Vertical Component of the Field of a Magnetic Dipole		289
9.2 The Field of the Vertical Magnetic Dipole in the Presence of a Thin Conducting Plane		295
9.3 The Two-Coil Induction Probe in Beds With a Finite Thickness		297
9.4 Profiling Curves for a Two-Coil Probe in a Bed of Finite Thickness		306
Reference		310

10. Induction Logging Based on Transient EM Measurements — 311

10.1 Transient Field of the Magnetic Dipole in a Uniform Medium		312
10.2 Transient Field of the Magnetic Dipole in a Medium With Cylindrical Interfaces		321
10.3 Transient Field of the Vertical Magnetic Dipole in a Medium With Horizontal Boundaries		337
10.4 Transient Field in Application to Deep-Reading Measurements While Drilling		343
10.5 Inversion of Transient Data in the Task of Geo-Steering		365
References		383
Further Reading		383

11. Induction Logging Using Transversal Coils — 385

11.1 Electromagnetic Field of the Magnetic Dipole in a Uniform Isotropic Medium		385
11.2 Boundary Value Problem for the Horizontal Magnetic Dipole in the Cylindrically Layered Formation		388
11.3 Magnetic Field in the Range of Small Parameter		396
11.4 Magnetic Field in the Far Zone		406
11.5 Magnetic Field in a Medium With Two Cylindrical Interfaces		417
11.6 Magnetic Field in Medium With a Thin Resistive Cylindrical Layer		421
11.7 Magnetic Field in Medium With One Horizontal Interface		426
11.8 Magnetic Field of the Horizontal Dipole in the Formation With Two Horizontal Interfaces		432
11.9 Profiling With a Two-Coil Induction Probe in a Medium With Horizontal Interfaces		441
Further Reading		445

12. The Influence of Anisotropy on the Field — **447**

- 12.1 Anisotropy of a Layered Medium — 448
- 12.2 Electromagnetic Field of Magnetic Dipole in a Uniform and Anisotropic Medium — 452
- 12.3 Magnetic Field in an Anisotropic Formation of Finite Thickness — 459
- Further Reading — 465

Appendix: Electromagnetic Response of Eccentred Magnetic Dipole in Cylindrically Layered Media — *467*

Index — *493*

INTRODUCTION

Electromagnetic induction logging is the main method of evaluating water and hydrocarbon saturation in shaly sand and other formations, and has been successfully applied for more than 70 years by oil service companies around the world.

During the last two decades, this technology has undergone significant progress with respect to development of wireline array induction tools: the Schlumberger AIT[1] and the Baker Hughes HDIL™[2] systems, for example, permit increased depth of investigation of up to several feet while maintaining high vertical resolution down to 1 ft. These systems, comprised of coils, whose axes are aligned parallel to the borehole axis, became the standard tools for detecting and evaluating low-resistivity pay zones. If the formation dip is small, the induced currents flow mainly parallel to the bedding planes, thus enabling measurements that are sensitive to the horizontal resistivity of the formation. However, many geologic formations exhibit resistivity anisotropy (i.e., the resistivity varies with direction). For example, in thinly laminated sand/shale sequences, where the sand is hydrocarbon-bearing, the resistivity in the direction perpendicular to the bedding is larger than the horizontal resistivity. The conductive shales dominate the horizontal resistivity whereas the vertical resistivity is affected more by the low-conductivity sand layers. Induction tools with vertically oriented coils cannot accurately detect and delineate this type of reservoir because the measured resistivity will be biased toward the low-resistive shales. To resolve formation parameters in an electrically anisotropic formation and find the relative dip, all major service companies employ tools with transversal coils, e.g., the Baker Hughes 3DeX™ and the Schlumberger Rt Scanner.

Also in recent decades, exciting developments occurred in logging-while-drilling (LWD), in which resistivity logging (e.g., the Baker Hughes VisiTrak™, the Schlumberger PeriScope, the Halliburton ADR™,[3] and the Weatherford GuideWave®[4]) became part of the bottom hole assembly. LWD is now successfully used for geo-steering and formation evaluation

---

[1] AIT, Rt Scanner, and PeriScope are marks of Schlumberger Limited.
[2] HDIL, 3DeX, and VisiTrak are trademarks of Baker Hughes Incorporated.
[3] ADR is a trademark of Halliburton.
[4] GuideWave is a registered trademark of Weatherford.

(especially for real-time and high-angle wells). Advances in resistivity logging were accompanied by numerous publications describing modeling and interpretation. Although these publications focus on application of sophisticated numerical techniques, including integral equations, finite difference, and finite elements, we believe that the potential of classical approaches has not been exhausted yet. To a large degree, this book is dedicated to a semianalytical and asymptotic treatment of the corresponding boundary value problems of induction logging, leading to the ultra-fast and sufficiently accurate simulation of electromagnetic responses.

To some extent, our monograph can be considered as the second edition of the book by A. Kaufman and G. Keller, *Induction Logging*, published 25 years ago by Elsevier. The current edition includes numerous updates to the first edition, and new results describing the theory of induction logging.

The theory is governed by Maxwell equations, which include terms representing the conductivity of the medium. These terms lead to decay of the wave amplitude as the wave propagates through the medium. The rate of decay is characterized by skin depth, which depends on the conductivity of the medium. Understanding relationships between measured fields and properties of the medium plays a key role in research and development of induction logging. For this reason, the purpose of the first four chapters is to acquaint the reader with basic equations of field theory. The behavior of the field of magnetic dipole in a uniform conducting medium is discussed in Chapter 5. In spite of the simplicity of the medium, the study of the field leads to an understanding of such important concepts as quadrature and in-phase components and their fundamentally different dependence on conductivity.

Chapter 6 consists of two parts. The first describes Doll's theory of induction logging, including basic concepts of geometrical factor, radial and vertical responses of the probes, and the apparent resistivity concept. In the second part, we discuss the so-called focusing probes, their parameters, and radial and vertical responses. In particular, we give special attention to three-coil probes, which allow us to compensate for the primary electromotive force and reduce an influence of the borehole and an invasion zone.

Chapter 7 describes an approximate technique, or so-called hybrid method, for solving forward problems. We show that although the hybrid method might be quite useful for quick calculations, it is not as powerful as the Born approximation. In fact, we show that Doll's theory of induction logging and the hybrid method follow from the Born approximation.

The frequency responses of the vertical magnetic dipole in a medium with cylindrical boundaries are the subject of Chapter 8. The components, as well as their depth of investigation, are described in detail. We give special attention to the behavior of the amplitude and phase at the far zone by deriving asymptotic formulas and showing that the ratio of amplitudes and phase difference enables us in many cases to greatly reduce the influence of the borehole and invasion. Also in this chapter, we investigate the effect of the displacement of the two- and three-coil probes in the borehole, and show that the position of the probe has a different influence on the quadrature and in-phase components, as well as the ratio of amplitudes and phase difference.

In Chapter 9, we study the vertical responses of the induction probes in a medium with horizontal boundaries. We use a derived expression for the vertical component of the magnetic field, excited by the vertical magnetic dipole, to study the vertical responses of probes, located symmetrically with respect to boundaries. We give special attention to asymptotic behavior of the field at different frequency ranges.

The subject of Chapter 10 is the possibility of application of the transient field in borehole geophysics. First, we obtain expressions for the late stage in a medium with cylindrical boundaries and demonstrate how the depth of investigation increases with time in the case of wireline measurements—at the late stage, sensitivity of this field to the formation resistivity can be even higher than that of the quadrature component in the frequency domain. In the second part of the chapter, we discuss the potential of the transient measurements in while-drilling applications. Inasmuch as such measurements have to be performed in the presence of highly conductive pipe, we pay special attention to means of reducing the effect from the pipe. We first analyze behavior of the field in the case of an ideally conductive pipe, and then by making use of Leontovich conditions, proceed to the case of the pipe with the finite conductivity. We show how spacing, observation time, and different shields may help in addressing undesirable effects of the pipe.

In the last part of the chapter, some aspects of inversion of LWD transient data are discussed. Emphasis is placed on means of improving stability of the inversion.

Chapters 11 and 12 describe basic aspects of induction logging with transversal coils. Analysis of the field in the range of a small parameter leads to the important observation that the magnetic field is represented as a sum of two terms, each depending either on the conductivity of the borehole or the formation. This feature is favorable for application of the focusing probes, which permit a significant decrease of the influence of the borehole and invasion zone. Numerical examples are presented to confirm the expectations. In Chapter 12, we demonstrate sensitivity of the measurements to an anisotropy coefficient under a different scenario: an anisotropic layer surrounded by more conductive and less conductive shoulders. Presented data help identify the range in which the anisotropy coefficient can be reliably resolved.

ACKNOWLEDGMENTS

We acknowledge **our colleagues**: Samuel Akselrod, Sushant Dutta, Alex Kagansky, Marina Nikitenko, Michael Oristaglio, Michael Rabinovich, Lev Tabarovsky, and Sergey Terentyev, with whom we have had the privilege to work on different aspects of induction logging.

We are thankful to **Baker Hughes** for promoting innovative technologies in borehole geophysics.

We thank our **technical editor** Erika Guerra for her assistance in the preparation of the manuscript.

LIST OF SYMBOLS

a_1 borehole radius
a_2 invasion zone radius
A magnetic vector potential
B magnetic induction vector
D dielectric displacement vector
e charge
f frequency
E electric field vector
E_n normal component of electric field
E_0 primary electric field
H_0 primary magnetic field
I_ν, K_ν modified Bessel functions of the first and second kind
In in-phase part
I current
j current density vector
J_ν, Y_ν Bessel functions of the first and second kind
$k^2 = i\omega\mu_0(\gamma + i\omega\varepsilon)$ square of a complex wave number
M_T transmitter moment
M_R receiver moment
n unit vector normal to surface
n number of coil turns
L inductance
l linear dimension
$p = L/\delta$ length in units of skin depth
G geometric factor
$R = (r^2 + z^2)^{1/2}$ radius in spherical coordinates
Q quadrature part
s ratio of conductivities
t time (s)
t_τ ramp time
T period
U scalar potential
δ volume charge density
Δ difference
γ electrical conductivity
γ_a apparent conductivity
ε_0 dielectric permittivity of free space
Φ magnetic flux
λ anisotropy coefficient
μ magnetic permeability

xvii

μ_0 magnetic permeability of free space
$\tau_0 = \rho \varepsilon_0$ time constant
$\tau = (2\pi \rho t \times 10^7)^{1/2}$ scaled variable of time
ω circular frequency

CHAPTER ONE

System of Equations of the Stationary Electric and Magnetic Fields

Contents

1.1 Equations of the Stationary Electric Field in a Conducting and Polarizable Medium	2
1.2 Interaction of Currents, Biot-Savart Law, and Magnetic Field	4
1.2.1 Ampere's Law and Interaction of Currents	4
1.2.2 Magnetic Field and Biot-Savart Law	5
1.2.3 Lorentz Force and Electromotive Force Acting on the Moving Circuit	9
1.3 Vector Potential of the Magnetic Field	13
1.3.1 Relation Between Magnetic Field and Vector Potential	13
1.3.2 Divergence and Laplacian of Vector Potential	16
1.4 System of Equations of the Stationary Magnetic Field	18
1.5 Examples of Magnetic Field of Current-Carrying Objects	21
1.5.1 Example One: Magnetic Field of the Current Filament	22
1.5.2 Example Two: The Vector Potential A and the Magnetic Field B of a Current in a Circular Loop	24
1.5.3 Example Three: Magnetic Fields of the Magnetic Dipole	27
1.5.4 Example Four: Magnetic Field Due to a Current in a Cylindrical Conductor	30
1.5.5 Example Five: Magnetic Field of Infinitely Long Solenoid	32
1.5.6 Example Six: Magnetic Field of a Current Toroid	35
1.6 System of Equations for the Stationary Fields	37
References	37
Further Reading	37

Before describing time-varying electromagnetic fields we focus our attention first on stationary electric magnetic fields that do not vary in time. Coulomb's and Biot–Savart laws governing these fields, also play fundamental roles in the understanding of the quasistationary fields used in most electromagnetic methods of borehole geophysics. We begin with studying the main features of the stationary electric field.

1.1 EQUATIONS OF THE STATIONARY ELECTRIC FIELD IN A CONDUCTING AND POLARIZABLE MEDIUM

As shown in Ref. [2], Maxwell's equations have three forms for the stationary electric field at regular points:

$$curl\, \mathbf{E} = 0, \quad div\, \mathbf{E} = \delta/\varepsilon_0 \qquad (1.1)$$

or

$$curl\, \mathbf{E} = 0, \quad div\, \mathbf{D} = \delta_0 \qquad (1.2)$$

or

$$curl\, \mathbf{E} = 0, \quad div\, \mathbf{j} = 0 \qquad (1.3)$$

Here $\mathbf{E}$ is the electric field, $\mathbf{D}$ is the vector of electric induction, $\mathbf{D} = \varepsilon \mathbf{E}$, and ε is the dielectric constant of a medium. In accordance with Ohm's law,

$$\mathbf{j} = \gamma \mathbf{E} \qquad (1.4)$$

where $\mathbf{j}$ is the vector of current density characterizing an ordered movement of free charges in space and γ is a conductivity. The vector $\mathbf{E}$ is

$$\mathbf{E} = \mathbf{E}^c + \mathbf{E}^{ext}$$

where $\mathbf{E}^c$ and $\mathbf{E}^{ext}$ are Coulomb and external (nonCoulomb) electric fields, respectively. The total charge density δ is a sum of the densities of free δ_0 and bound δ_b charges:

$$\delta = \delta_0 + \delta_b \qquad (1.5)$$

Eqs. (1.1)–(1.3) are written at regular points where the field's derivatives exist. By definition of divergence for any vector $\mathbf{M}$ we have

$$div\, \mathbf{M} = \lim \frac{\oint \mathbf{M} \cdot d\mathbf{S}}{\Delta V}, \quad \text{as } \Delta V \to 0 \qquad (1.6)$$

which is the divergence of the field that characterizes the flux of the field through a closed surface, surrounding an elementary volume. This equation is valid everywhere, although it is not convenient for calculations because it requires computation of a surface integral. Taking into account that the surface S and distance between opposite sides are small, it is possible to replace integration by differentiation, which is much simpler to perform. Because this form of divergence contains derivatives, it is valid only at regular points. Also the curl of a field M can be defined as

$$curl\,\mathbf{M} = \frac{\oint_L \mathbf{M} \cdot d\mathbf{l}}{dS}\mathbf{n}, \quad \text{as } \Delta S \to 0 \qquad (1.7)$$

Here ΔS is the elementary area, and L is a closed path surrounding the area. If $\mathbf{n}$ is the unit vector perpendicular to ΔS and $d\mathbf{l}$ is the linear element of L, vectors $d\mathbf{l}$ and $\mathbf{n}$ obey the right-hand thumb rule. It is essential that an area ΔS in Eq. (1.7) is oriented in such a way that the numerator has a maximal value. Again, similar to Eq. (1.6), as the contour L becomes small, it is possible to replace integration by differentiation. The replacement can be performed only at regular points where derivatives exist. At interfaces between media with different electric parameters in place of Eqs. (1.1)–(1.3), we have the surface analog of the equations:

$$E_{2t} - E_{1t} = 0, \quad E_{2n} - E_{1n} = \sigma/\varepsilon_0$$

or

$$E_{2t} - E_{1t} = 0, \qquad D_{2n} - D_{1n} = \sigma_0 \qquad (1.8)$$

or

$$E_{2t} - E_{1t} = 0, \quad \gamma_2 E_{2n} - \gamma_1 E_{1n} = 0$$

where E_{1t}, E_{1n} and E_{2t}, E_{2n} are tangential and normal components of the electric field at the back and front sides of an interface, respectively, and the normal n is directed toward the front side into the medium with index "2." The conductivity of a medium can be expressed as

$$\gamma = \delta_0^+ u^+ + |\delta_0^-| u^- \qquad (1.9)$$

Here u^+ and u^- are the mobility of the positive and negative charges, respectively, which are extremely small numbers in a medium. Thus, the velocity of free charges engaged in an orderly motion in a conductor is usually very small and does not exceed 10^{-6} m/s. Nevertheless, these barely moving charges may create a strong magnetic field. Free charge in a medium is a charge that can move through distances exceeding the molecule size; bound charges move only within a fixed molecule.

As follows from Eq. (1.3), the stationary electric field in a conducting and polarizable medium is independent of the dielectric constant, and distribution of bound charges does not influence the electric field. Such a remarkable feature of the field is also observed in quasistationary fields: the density of the total charge coincides with that of free charges δ_0 in a non-polarizable medium where $\varepsilon = \varepsilon_0$.

The volume and surface densities of charge are related to the field and conductivity by [2]

$$\delta(p) = -\varepsilon_0 \frac{\mathbf{E} \cdot grad\,\gamma}{\gamma}, \quad \sigma(p) = 2\varepsilon_0 K_{12} E_n^{av} \qquad (1.10)$$

Here

$$K_{12} = \frac{\rho_2 - \rho_1}{\rho_2 + \rho_1}$$

and E_n^{av} is the mean value of the normal component of electric field at point p located at the boundary between media with resistivity ρ_1 and ρ_2; the normal n is directed from medium 1 to medium 2. Besides, it is assumed that an external force is absent in the vicinity of point p. The physical meaning of $E_n^{av}(p)$ is simple: it is the normal component of the field caused by all charges in the medium except those at the point p. These charges, placed at the boundary between ρ_1 and ρ_2, do not participate in the current flow. The second equation of the system for the electric field at regular points is

$$div\,\mathbf{E} = \delta/\varepsilon_0 \quad \text{or} \quad div\,\mathbf{D} = \delta_0$$

and, along with its surface analog, remains valid for the time-varying fields. Either one represents the third of Maxwell's equation.

Note that this equation can be derived from Coulomb's law by taking into account polarization and bound charges.

1.2 INTERACTION OF CURRENTS, BIOT-SAVART LAW, AND MAGNETIC FIELD

1.2.1 Ampere's Law and Interaction of Currents

Numerous experiments performed two centuries ago demonstrated that currents in two circuits interact with each other; that is, mechanical forces act at every element of a current circuit. This force depends on the magnitude of the current, the direction of charge movement, the shape and dimension of the current circuit, as well as the distance and mutual orientation of the circuits with respect to each other. This list of factors indicates that the mathematical formulation of this phenomenon should be a much

Stationary Electric and Magnetic Fields

more complicated task than that for the stationary electric field. Nevertheless, Ampere was able to find an expression for the force of interaction between two elementary currents in a relatively simple form:

$$d\mathbf{F}(p) = \frac{\mu_0}{4\pi} I_1 I_2 \frac{d\mathbf{l}_1(p) \times [d\mathbf{l}_2(q) \times \mathbf{L}_{qp}]}{L_{qp}^3} \tag{1.11}$$

where I_1 and I_2 are magnitudes of currents in the linear elements $d\mathbf{l}_1$ and $d\mathbf{l}_2$, respectively, and their direction coincides with that of the current density; L_{qp} is the distance between these elements; and $\mathbf{L}_{qp}$ is directed from point q to point p, which are located at the center of the current elements. Finally, μ_0 is a constant equal to

$$\mu_0 = 4\pi \times 10^{-7}\, H/m$$

which is called the magnetic permeability of free space, despite the fact that the term "free space" implies medium without physical properties. The distance between current elements L_{qp} is much greater than their lengths:

$$L_{qp} \gg dl_1, \quad L_{qp} \gg dl_2$$

Examples, illustrating an interaction of elementary currents, are given in Fig. 1.1.

Making use of the superposition principle, the force of interaction between two arbitrary closed current circuits is defined as

$$\mathbf{F} = \frac{\mu_0}{4\pi} I_1 I_2 \oint_{L_1} \oint_{L_2} \frac{d\mathbf{l}_1 \times (d\mathbf{l}_2 \times \mathbf{L}_{qp})}{L_{qp}^3} \tag{1.12}$$

where integration is performed along current lines L_1 and L_2, $p \neq q$. The resultant force $\mathbf{F}$ is a sum of forces acting on different elements of the contour and is measured in the SI newtons (N) if the lengths are in meters (m) and the currents are in units of amperes (A).

1.2.2 Magnetic Field and Biot-Savart Law

The interaction between currents suggests that current in a contour creates a field, and the existence of this field causes other currents to experience the action of the force $\mathbf{F}$. This field is called the magnetic field, and it is introduced from Ampere's law as

$$d\mathbf{F}(p) = I(p)\, d\mathbf{l}(p) \times d\mathbf{B}(p) \tag{1.13}$$

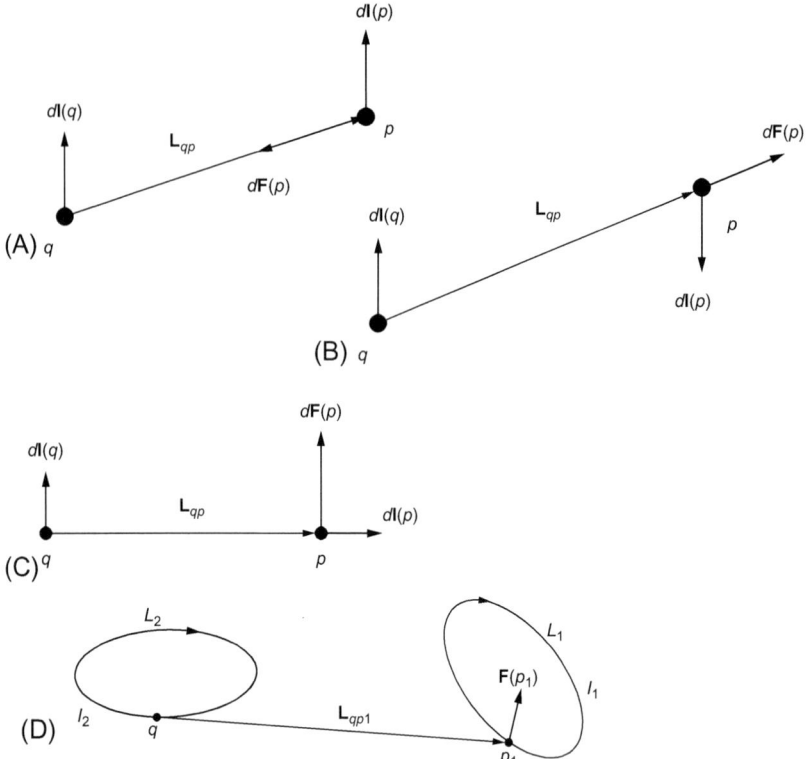

Fig. 1.1 (A) Interaction of currents having the same direction; (B) interaction of currents having opposite directions; (C) interaction of current elements perpendicular to each other; and (D) interaction of two current loops.

where

$$d\mathbf{B}(p) = \frac{\mu_0}{4\pi} I(q) \frac{d\mathbf{l}(q) \times \mathbf{L}_{qp}}{L_{qp}^3} \tag{1.14}$$

and $I(q)$ is the current of the element $d\mathbf{l}(q)$. Eq. (1.14), called the Biot-Savart law, describes the relationship between the elementary linear current and the magnetic field $d\mathbf{B}$. By definition the magnitude of the magnetic field caused by the elementary current is

$$dB(p) = \frac{\mu_0}{4\pi} I(q) \frac{dl}{L_{qp}^2} \sin\left(\mathbf{L}_{qp}, d\mathbf{l}\right) \tag{1.15}$$

Here $(\mathbf{L}_{qp}, d\mathbf{l})$ is the angle between the vectors $\mathbf{L}_{qp}$ and the element $d\mathbf{l}$; the vector $d\mathbf{B}$ is perpendicular to these vectors, as shown in Fig. 1.2A, and these three vectors obey the right-hand thumb rule. The unit vector $\mathbf{b}_0$, characterizing the direction of the field, is defined as

Stationary Electric and Magnetic Fields

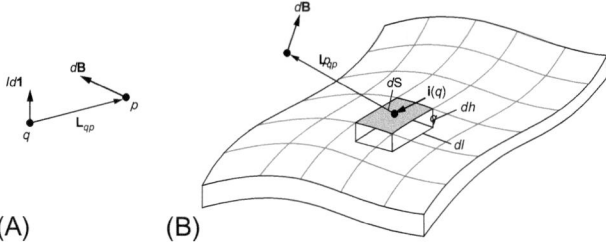

Fig. 1.2 (A) Magnetic field of a current element. (B) Magnetic field of the surface current.

$$\mathbf{b}_0 = \frac{d\mathbf{l} \times \mathbf{L}_{qp}}{|d\mathbf{l} \times \mathbf{L}_{qp}|}$$

In SI units, the magnetic field is measured in tesla (T) and is related to other common units, such as the gauss and the gamma as

$$1\,\text{tesla} = 10^4\,\text{gauss} = 10^9\,\text{gamma}$$

A nanotesla (nT) is **equivalent** to one **gamma.** Now we generalize Eq. (1.14), assuming that, along with linear currents, there are also the volume and surface currents. First let us represent the product $Id\mathbf{l}$ as

$$Id\mathbf{l} = jdSd\mathbf{l} = \mathbf{j}dSdl = \mathbf{j}dV \quad (1.16)$$

where dS is the cross section of the elementary current tube, $d\mathbf{l}$ is oriented along this tube, and $\mathbf{j}$ is the volume current density. If the current is concentrated in a relatively thin layer with thickness dh, which is small enough with respect to the distance to an observation point, it is convenient to replace this layer by a surface current. As seen in Fig. 1.2B, the product $Id\mathbf{l}$ can be modified as follows:

$$Id\mathbf{l} = \mathbf{j}dV = \mathbf{j}dhdS = \mathbf{i}dS \quad (1.17)$$

Here dS is the surface element, and

$$\mathbf{i} = \mathbf{j}dh$$

is the density of the surface current. The resultant force $\mathbf{F}$ is a sum of forces caused by different elementary currents. Applying the principle of superposition for all three types of currents (volume, surface, and linear) and making use of Eqs. (1.14), (1.16), (1.17), we obtain the generalized form of the Biot-Savart law:

$$\mathbf{B}(p) = \frac{\mu_0}{4\pi}\left[\int_V \frac{\mathbf{j} \times \mathbf{L}_{qp}}{L_{qp}^3}dV + \int_S \frac{\mathbf{i} \times \mathbf{L}_{qp}}{L_{qp}^3}dS + \sum_n I_n \oint \frac{d\mathbf{l} \times \mathbf{L}_{qp}}{L_{qp}^3}\right] \quad (1.18)$$

Eq. (1.18) allows us to calculate the magnetic field everywhere inside and outside of volume currents. In general, the currents arise from the motion of free charges and magnetization of a magnetic medium, which can be related to magnetization currents. Correspondingly, the current density is a sum

$$\mathbf{j} = \mathbf{j}_c + \mathbf{j}_m$$

where $\mathbf{j}_c$ and $\mathbf{j}_m$ are the volume density of the conduction and magnetization currents, respectively. The corresponding magnetic fields of these currents obey the Biot-Savart law. In most applications of borehole electrical methods, it is assumed that magnetization is absent. According to Eq. (1.18), the magnetic field caused by a given distribution of currents depends on location of observation point p only and is independent of the presence of other currents. The right-hand side of Eq. (1.18) does not contain any terms characterizing physical properties of a medium. Therefore, the field **B** at point p, generated by a specific distribution of currents, remains the same if free space is replaced by a nonuniform conducting and polarizable medium. For instance, if the current circuit is placed in a magnetic medium, the field **B** caused by this current is the same as if it were in free space. Of course, the presence of such magnetic medium results in a change of the magnetic field **B**, indicating the presence inside a medium of some other currents (magnetization currents) and producing a magnetic field. This observation directly follows from Eq. (1.18), which states that any change of the field **B** is caused by a change in distribution of current. Unlike the volume density currents, their linear and surface analogies are mathematical idealizations of the real current distributions. Normally, they are introduced to simplify calculations of the field and study its behavior. For this reason, the equation

$$\mathbf{B}(p) = \frac{\mu_0}{4\pi} \int_V \frac{\mathbf{j}(q) \times \mathbf{L}_{qp}}{L_{qp}^3} dV \tag{1.19}$$

is applicable for calculation of a magnetic field for all possible distributions of the current.

As will be shown later, the Biot-Savart law (Eq. 1.18) is also valid for a time-varying magnetic field when it is possible to neglect by so-called displacement currents.

The experiments, which allowed Ampere to derive Eq. (1.11), were carried out with closed circuits. At the same time Eq. (1.11), as well as Eq. (1.14), is written for the element $d\mathbf{l}$, where a current cannot exist if this element does not constitute a part of the closed circuit. In other words, Eqs. (1.11), (1.14) cannot be proved experimentally, but the interaction between closed current circuits takes place in such manner, as if the magnetic field **B**, caused by the current element *Idl*, were described by Eq. (1.14). In accordance with the Biot-Savart law, current is the sole source of a stationary magnetic field, and the distribution of this source is characterized by the magnitude and direction of the current density vector **j** whose vector lines are always closed. Magnetic field **B** is also, unlike the Coulomb's electric field, of the vortex type.

1.2.3 Lorentz Force and Electromotive Force Acting on the Moving Circuit

As follows from Eqs. (1.13), (1.16) the current in the elementary volume, placed in the magnetic field **B**, is subjected to the action of a force:

$$\mathbf{F} = (\mathbf{j} \times \mathbf{B}) dV \tag{1.20}$$

The latter allows us to find force acting on a single electron or ion moving with velocity **v**. By definition, the current density **j** can be represented as

$$\mathbf{j} = ne\mathbf{v}$$

where n is the number of particles in the unit volume, and e is the charge of electron or ion. Therefore, the force of the magnetic field **B** acting on all particles is

$$\mathbf{F}_B = ne(\mathbf{v} \times \mathbf{B}) dV$$

and, correspondingly, every moving particle, for example, the electron, is subjected to a force equal to

$$\mathbf{F}_B = e(\mathbf{v} \times \mathbf{B}) \tag{1.21}$$

Thus, this elementary charge is subjected to the total force equal to

$$\mathbf{F} = \mathbf{F}_e + \mathbf{F}_m = e\mathbf{E} + e(\mathbf{v} \times \mathbf{B}) \tag{1.22}$$

which is called the Lorentz force. Here

$$\mathbf{F}_e = e\mathbf{E}_c \quad \text{and} \quad \mathbf{F}_m = e(\mathbf{v} \times \mathbf{B})$$

are forces caused by the electric and magnetic fields, respectively. By analogy with Coulomb's law, let us introduce this nonCoulomb electric field as

$$\mathbf{E}_m = \mathbf{v} \times \mathbf{B} \tag{1.23}$$

which, in the presence of the magnetic field, acts on moving charge. By definition, this field is perpendicular to the velocity and the magnetic field, and it reaches a maximum when the angle between these two vectors is equal to $\pi/2$. As in the case of Coulomb's electric field, the voltage of this electric field along an elementary and arbitrary path is

$$\Delta V = \mathbf{E}_m \cdot d\mathbf{l} = (\mathbf{v} \times \mathbf{B}) d\mathbf{l} \quad \text{and} \quad V = \int (\mathbf{v} \times \mathbf{B}) d\mathbf{l}. \qquad (1.24)$$

In particular, the electromotive force caused by field $\mathbf{E}_m$ is

$$\Xi = \int (\mathbf{v} \times \mathbf{B}) d\mathbf{l}$$

Unlike the voltage of the Coulomb's electric field, the second equation in the set (1.24) is path dependent; in general, the electromotive force due to this field does not vanish.

The existence of this non-Coulomb electric field directly follows from Ampere's law, originally derived for the direct current. Let us consider several examples.

Example One

Suppose that the current circuit does not move and is placed in a magnetic field $\mathbf{B}$ (Fig. 1.3A). The moving electrons along the circuit are subjected to the action of the field $\mathbf{E}_m$, which is usually very small, because the electron

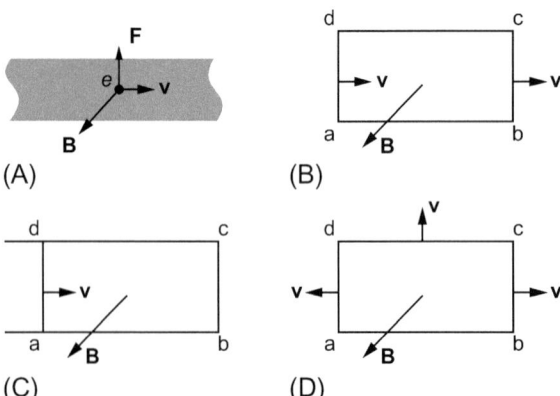

Fig. 1.3 (A) Magnetic force acting on a charge, moving with velocity **v**; (B) rectangular circuit moving with velocity **v** in the magnetic field; (C) movement of one side of the circuit with velocity **v**; and (D) movement and deformation of a contour in the magnetic field.

velocity is of the order of 10^{-6} m/s. By definition, this field is perpendicular to the Coulomb's field and may cause only an insignificant shift of charges toward the surface of the circuit, where the positive and negative charges would tend to appear. But their electric field prevents further movement of charges; eventually, there is no current flow caused by $\mathbf{E}_m$ field.

Example Two
Consider the rectangular and conducting circuit *abcd* (Fig. 1.3B), which moves at the constant velocity **v** along the *x* axis. The uniform magnetic field **B** is perpendicular to the circuit. Taking into account that direction of currents along paths *ad* and *cb* are opposite to each other, the voltages

$$\Delta V_{ad}, \quad \Delta V_{cb}$$

differ only by sign. The voltages along lines *ab* and *cd* are equal to zero. Indeed, according to vector algebra for the voltage along an arbitrary element *d***l** of the line, we have

$$(\mathbf{v} \times \mathbf{B}) \cdot d\mathbf{l} = (d\mathbf{l} \times \mathbf{v}) \cdot \mathbf{B} \tag{1.25}$$

Because in case of lines *ab* and *cd* vectors *d***l** and ***v*** have the same or opposite direction, we conclude that the voltage along these elements is zero; therefore, the electromotive force is also zero. As is seen in Fig. 1.3B, the flux of the magnetic field Φ through the area, surrounded by the path, remains constant. Thus, we have

$$\Xi = 0 \quad \text{and} \quad \frac{d\Phi}{dt} = 0$$

Example Three
Now suppose that only the side *ad* slides at velocity **v**, while the other part of the circuit is at rest (Fig. 1.3C). Then, the electromotive force coincides with the voltage ΔV_{ad}:

$$\Xi = \pm v_x B_z ad \tag{1.26}$$

where the sign depends on the orientation of the current in this line. The product $v_x ad$ represents the rate of a change of the area, enclosed by the circuit; therefore Eq. (1.26) can be rewritten as

$$\Xi = \pm \frac{d\Phi}{dt}$$

that is, the electromotive force arising in the circuit is defined by the rate of a change of the flux of the magnetic field through the area surrounded by the circuit. By definition, the flux is equal to

$$\Phi = \int_S \mathbf{B} \cdot d\mathbf{S}$$

If the direction along the circuit and the vector $d\mathbf{S}$ obey the right-hand thumb rule, we have

$$\Xi = -\frac{d\Phi}{dt} \qquad (1.27)$$

Example Four

In this case, the magnetic field is aligned in the plane of a circuit that moves with velocity v. Then, the voltage along an arbitrary element of the circuit is equal to

$$(\mathbf{v} \times \mathbf{B}) \cdot d\mathbf{l} = 0$$

therefore, the electromotive force is absent despite a motion of the circuit and the presence of the magnetic field. Inasmuch as the field $\mathbf{B}$ is tangential to the circuit, its flux is also equal to zero, and the electromotive force is

$$\Xi = -\frac{d\Phi}{dt} = 0$$

Thus, only the normal component of the magnetic field has an influence on the moving charge.

Example Five

Suppose that an arbitrary conducting circuit is located in some plane, and each element moves with velocity $\mathbf{v}$, which may vary from point to point (Fig. 1.3D). In this case, the circuit experiences a motion and deformation. The component of the magnetic field normal to this plane also may vary from point to point. Consider again the elementary voltage along the element $d\mathbf{l}$:

$$\Delta V = (\mathbf{v} \times \mathbf{B}) \cdot d\mathbf{l} = (d\mathbf{l} \times \mathbf{v}) \cdot \mathbf{B}$$

The magnitude of the vector product $d\mathbf{l} \times \mathbf{v}$ is equal to the area covered by the element $d\mathbf{l}$ during the unit of time; correspondingly, ΔV is equal to the rate of a change of the elementary flux of the field **B**. Integrating along the circuit and using the right-hand thumb rule, we obtain the electromotive force:

$$\Xi = -\frac{d\Phi}{dt} \quad (1.28)$$

Later we will describe Faraday's law, which has exactly the same form. However, it has one fundamental difference: it shows that an electromotive force may arise not only because of a movement and deformation of the circuit but also due to a rate of change of the magnetic field with time when the circuit does not move. Moreover, Faraday's law is applied to any closed path, which can be, for example, an insulator.

1.3 VECTOR POTENTIAL OF THE MAGNETIC FIELD

1.3.1 Relation Between Magnetic Field and Vector Potential

Although calculation of the magnetic field using the Biot-Savart law is not a complicated procedure, it is still useful to find a more convenient way of determining the field. To proceed, by analogy with the scalar potential of the electric field, we introduce a new function. In addition, this function allows one to derive a system of equations for the magnetic field. Let us start from the Biot-Savart law:

$$\mathbf{B}(p) = \frac{\mu_0}{4\pi} \int_V \frac{\mathbf{j}(q) \times \mathbf{L}_{qp}}{L_{qp}^3} dV \quad (1.29)$$

Taking into account that

$$\frac{\mathbf{L}_{qp}}{L_{qp}^3} = \overset{q}{\nabla} \frac{1}{L_{qp}} = -\overset{p}{\nabla} \frac{1}{L_{qp}} \quad (1.30)$$

and substituting Eq. (1.30) into Eq. (1.29), we obtain

$$\mathbf{B}(p) = \frac{\mu_0}{4\pi} \int_V \mathbf{j}(q) \times \overset{q}{\nabla} \frac{1}{L_{qp}} dV = \frac{\mu_0}{4\pi} \int_V \overset{p}{\nabla} \frac{1}{L_{qp}} \times \mathbf{j}(q) dV \quad (1.31)$$

Here q and p indicate that derivatives are taken with respect to coordinates of the point q and p. For instance, in the Cartesian system of coordinates, we have:

$$\overset{q}{\nabla}\frac{1}{L_{qp}} = \overset{q}{grad}\frac{1}{L_{qp}} = \frac{\partial}{\partial x_q}\frac{1}{L_{qp}}\mathbf{i}_x + \frac{\partial}{\partial y_q}\frac{1}{L_{qp}}\mathbf{i}_y + \frac{\partial}{\partial z_q}\frac{1}{L_{qp}}\mathbf{i}_z$$

where $\mathbf{i}_x, \mathbf{i}_y,$ and $\mathbf{i}_z$ are orthogonal unit vectors. Now we make use of the equality

$$\overset{p}{\nabla} \times \frac{\mathbf{j}}{L_{qp}} = \overset{p}{\nabla}\frac{1}{L_{qp}} \times \mathbf{j} + \frac{\overset{p}{\nabla} \times \mathbf{j}}{L_{qp}}$$

which follows from the vector identity:

$$\nabla \times (\phi\mathbf{a}) = \nabla\phi \times \mathbf{a} + \phi\nabla \times \mathbf{a} \qquad (1.32)$$

From Eqs. (1.31), (1.32), we have

$$\mathbf{B}(p) = \frac{\mu_0}{4\pi}\int_V \overset{p}{\nabla} \times \frac{\mathbf{j}}{L_{qp}}dV - \frac{\mu_0}{4\pi}\int_V \frac{\overset{p}{\nabla} \times \mathbf{j}}{L_{qp}}dV \qquad (1.33)$$

The current density is a function of the coordinates of q and does not depend on the location of the observation point p. Therefore, the integrand of the second integral is zero and

$$\mathbf{B}(p) = \frac{\mu_0}{4\pi}\int_V \overset{p}{curl}\,\frac{\mathbf{j}(q)}{L_{qp}}dV \qquad (1.34)$$

Inasmuch as the integration and differentiation in Eq. (1.34) are carried out with respect to different points q and p, we can interchange the order of operations that gives

$$\mathbf{B}(p) = \overset{p}{curl}\frac{\mu_0}{4\pi}\int_V \frac{\mathbf{j}(q)}{L_{qp}}dV$$

or

$$\mathbf{B}(p) = curl\,\mathbf{A} \qquad (1.35)$$

where

$$\mathbf{A}(p) = \frac{\mu_0}{4\pi}\int_V \frac{\mathbf{j}(q)}{L_{qp}}dV \qquad (1.36)$$

Thus, the magnetic field $\mathbf{B}$ caused by direct currents can be expressed through the vector potential $\mathbf{A}$ defined by Eq. (1.36). Comparing

Eqs. (1.29) and (1.36), we see that function **A** is related to the distribution of currents in a much simpler way than the magnetic field. One reason for introducing this function is thus already demonstrated. According to Eq. (1.36), **A**, unlike the potential of an electric field, is a vector, and its magnitude and direction depend essentially on the current distribution. Now let us derive expressions for the vector potential **A**, caused by surface and linear currents. Making use of Eq. (1.17):

$$\mathbf{j}dV = \mathbf{i}dS$$

and we have

$$\mathbf{A} = \frac{\mu_0}{4\pi}\int_S \frac{\mathbf{i}dS}{L_{qp}} \quad \text{and} \quad \mathbf{A} = \frac{\mu_0 I}{4\pi}\oint_L \frac{d\mathbf{l}}{L_{qp}} \tag{1.37}$$

In the general case when we have the volume, surface, and linear currents, we obtain

$$\mathbf{A} = \frac{\mu_0}{4\pi}\left[\int_V \frac{\mathbf{j}dV}{L_{qp}} + \int_S \frac{\mathbf{i}dS}{L_{qp}} + \sum_{i=1}^{n} I_i \oint_L \frac{d\mathbf{l}}{L_{qp}}\right] \tag{1.38}$$

The components of the vector potential can be derived directly from this equation. For instance, in the Cartesian coordinates, we obtain

$$A_x = \frac{\mu_0}{4\pi}\left[\int_V \frac{j_x dV}{L_{qp}} + \int_S \frac{i_x dS}{L_{qp}} + \sum_{i=1}^{n} I_i \oint \frac{dl_x}{L_{qp}}\right]$$

$$A_y = \frac{\mu_0}{4\pi}\left[\int_V \frac{j_y dV}{L_{qp}} + \int_S \frac{i_y dS}{L_{qp}} + \sum_{i=1}^{n} I_i \oint \frac{dl_y}{L_{qp}}\right] \tag{1.39}$$

$$A_z = \frac{\mu_0}{4\pi}\left[\int_V \frac{j_z dV}{L_{qp}} + \int_S \frac{i_z dS}{L_{qp}} + \sum_{i=1}^{n} I_i \oint \frac{dl_z}{L_{qp}}\right]$$

Similar expressions can be written for the vector potential components in other systems of coordinates. Eq. (1.38) implies that, if a current flows along a single straight line, the vector potential has only one component parallel to this line. Similarly, if currents are situated in a single plane, then the vector potential **A** at every point is parallel to this plane. Later we consider several examples illustrating the behavior of the vector potential and magnetic field.

1.3.2 Divergence and Laplacian of Vector Potential

Let us derive two useful relations for the function **A**, which simplify to a great extent the derivation of magnetic field equations.

First, we determine the divergence of the vector potential **A**. As follows from Eq. (1.36), we have

$$\overset{p}{div} \mathbf{A} = \overset{p}{div} \frac{\mu_0}{4\pi} \int_V \frac{\mathbf{j}(q)}{L_{qp}} dV$$

Inasmuch as differentiation and integration in this expression are performed with respect to different points, we can change the order of operations, which gives

$$\overset{p}{div} \mathbf{A} = \frac{\mu_0}{4\pi} \int_V \overset{p}{div} \frac{\mathbf{j}(q)}{L_{qp}} dV \qquad (1.40)$$

The volume of integration includes all currents; therefore, outside a surface S currents are absent. Correspondingly, the normal component of the current density at this surface equals zero:

$$j_n = 0 \qquad (1.41)$$

The integrand in Eq. (1.40) can be represented as

$$\overset{p}{\nabla} \frac{\mathbf{j}}{L_{qp}} = \frac{\overset{p}{\nabla} \mathbf{j}}{L_{qp}} + \mathbf{j} \cdot \overset{p}{\nabla} \frac{1}{L_{qp}} = \mathbf{j} \cdot \overset{p}{\nabla} \frac{1}{L_{qp}}$$

because the current density does not depend on the observation point and

$$\overset{p}{div} \mathbf{j}(q) = 0$$

Then, we have

$$\mathbf{j} \cdot \overset{p}{\nabla} \frac{1}{L_{qp}} = -\mathbf{j} \cdot \overset{q}{\nabla} \frac{1}{L_{qp}} = -\overset{q}{\nabla} \frac{\mathbf{j}}{L_{qp}} + \frac{\overset{q}{\nabla} \mathbf{j}}{L_{qp}}$$

As follows from the principle of charge conservation for direct currents

$$\overset{q}{div} \mathbf{j} = 0$$

therefore,

$$\mathbf{j} \cdot \overset{p}{\nabla} \frac{1}{L_{qp}} = -\overset{q}{div} \frac{\mathbf{j}}{L_{qp}}$$

Correspondingly, Eq. (1.40) can be written as

$$div\,\mathbf{A} = -\frac{\mu_0}{4\pi} \int_V \overset{q}{div} \frac{\mathbf{j}}{L_{qp}} dV$$

Unlike Eq. (1.40) on the right-hand side of this equation, both integration and differentiation are performed with respect to the same point q. By applying Gauss's theorem, we have

$$\int_V div\,\mathbf{M}\,dV = \oint_S \mathbf{M} \cdot d\mathbf{S}$$

Thus,

$$div\,\mathbf{A} = -\frac{\mu_0}{4\pi} \int_V \overset{q}{div} \frac{\mathbf{j}}{L_{qp}} dV = -\frac{\mu_0}{4\pi} \oint_S \frac{\mathbf{j} \cdot d\mathbf{S}}{L_{qp}} = -\frac{\mu_0}{4\pi} \oint_S \frac{j_n dS}{L_{qp}}.$$

Taking into account that the normal component of the current density j_n vanishes at the surface S, surrounding all currents (Eq. 1.41), we obtain

$$div\,\mathbf{A} = 0 \qquad (1.42)$$

This is the first relation that is useful for deriving the system of field equations. Let us note that, in accordance with Eq. (1.42), the vector lines of the field **A** are always closed. Next we obtain one more equation describing this function. As is well known [1], the potential of the electric field U satisfies Poisson's equation

$$\nabla^2 U = -\frac{\delta}{\varepsilon_0}$$

and its solution is

$$U = \frac{1}{4\pi\varepsilon_0} \int_V \frac{\delta dV}{L_{qp}}$$

As follows from Eq. (1.36), every component of the vector potential has the same form as the potential U; therefore, by analogy it also satisfies Poisson's equation:

$$\nabla^2 A_x = -\mu_0 j_x, \quad \nabla^2 A_y = -\mu_0 j_y, \quad \nabla^2 A_z = -\mu_0 j_z$$

Multiplying each of these equations by the corresponding unit vector $\mathbf{i}_x, \mathbf{i}_y, \mathbf{i}_z$ and performing the summation, we arrive at the Poisson's equation for the vector potential:

$$\nabla^2 \mathbf{A} = -\mu_0 \mathbf{j} \qquad (1.43)$$

1.4 SYSTEM OF EQUATIONS OF THE STATIONARY MAGNETIC FIELD

Now we are ready to derive the system of equations of the stationary magnetic field. First, making use of Eq. (1.35), we discover that divergence of the field $\mathbf{B}$ vanishes. In fact, we have

$$div\,\mathbf{B} = div(curl\,\mathbf{A}) \qquad (1.44)$$

From vector analysis, the right-hand term of Eq. (1.44) is identically zero. Therefore,

$$div\,\mathbf{B} = 0 \qquad (1.45)$$

This means that the magnetic field does not have sources, like charges, and, correspondingly, the vector lines of the magnetic field $\mathbf{B}$ are always closed. Applying Gauss's theorem, we obtain the integral form of this equation:

$$\oint_S \mathbf{B} \cdot d\mathbf{S} = 0 \qquad (1.46)$$

that is, the total flux of the field $\mathbf{B}$ through any closed surface is always equal to zero. Next we derive the surface analogy of Eq. (1.45) and, with this purpose in mind, consider the flux through an elementary cylindrical surface (Fig. 1.4A). It is equal to

$$\mathbf{B}^{(2)} \cdot d\mathbf{S}_2 + \mathbf{B}^{(1)} \cdot d\mathbf{S}_1 + \mathbf{B} \cdot d\mathbf{S}_* = 0 \qquad (1.47)$$

Here $d\mathbf{S}_2 = dS\mathbf{n}$, $d\mathbf{S}_1 = -dS\mathbf{n}$, and dS^* is the lateral surface of the cylinder. Then, reducing the height of the cylinder to zero in place of Eq. (1.47), we obtain

$$B_n^{(2)} dS - B_n^{(1)} dS = 0 \quad \text{or} \quad B_n^{(2)} = B_n^{(1)} \qquad (1.48)$$

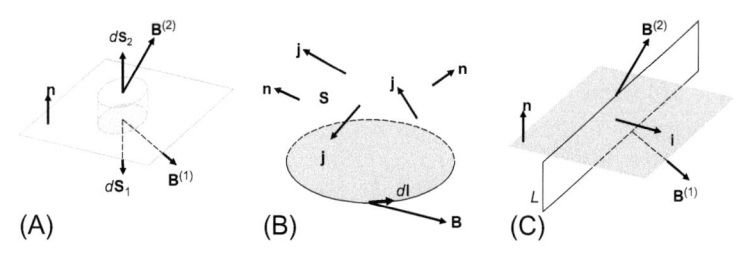

Fig. 1.4 (A) Surface analogy of Eq. (1.45); (B) illustration of Eq. (1.52); and (C) surface analogy of Eq. (1.50).

Thus, the normal component of the magnetic field **B** is always a continuous function of the spatial variables. We have three forms of the equation that describe the magnetic field caused by direct currents:

$$\oint_S \mathbf{B} \cdot d\mathbf{S} = 0, \quad div\,\mathbf{B} = 0, \quad B_n^{(2)} - B_n^{(1)} = 0 \quad (1.49)$$

Each of them expresses the same fact, namely, the absence of magnetic charges. Eq. (1.49) have been derived assuming that the field **B** is caused by conduction currents. However, they remain valid in the presence of the magnetic medium when the field also is generated by magnetization currents. The equations were obtained from the Biot-Savart law for direct currents, but actually they are still valid for the time-varying magnetic fields and, in effect, represent Maxwell's fourth equation.

Next we derive the second equation for the magnetic field. Making use of Eq. (1.35) and the identity

$$curl\,curl\,\mathbf{M} = grad\,div\,\mathbf{M} - \nabla^2 \mathbf{M}$$

we have:

$$curl\,\mathbf{B} = grad\,div\,\mathbf{A} - \nabla^2 \mathbf{A}$$

Considering

$$div\,\mathbf{A} = 0$$

and taking into account Eq. (1.43), we obtain

$$curl\,\mathbf{B} = -\nabla^2 \mathbf{A} = \mu_0 \mathbf{j}$$

Thus, the second equation for the magnetic field at regular points is

$$curl\,\mathbf{B} = \mu_0 \mathbf{j} \quad (1.50)$$

Consequently, outside of currents we have

$$curl\,\mathbf{B} = 0 \quad (1.51)$$

Eq. (1.50) states that currents are vortex-type sources, capable of generating a magnetic field. Applying Stokes theorem,

$$\oint_L \mathbf{M} \cdot d\mathbf{l} = \int_S curl\,\mathbf{M} \cdot d\mathbf{S}$$

where S is the surface bounded by the contour L, we obtain the integral form of the second equation:

$$\oint_L \mathbf{B} \cdot d\mathbf{l} = \int_S curl\,\mathbf{B} \cdot d\mathbf{S} = \mu_0 \int_S \mathbf{j} \cdot d\mathbf{S}$$

or

$$\oint_L \mathbf{B} \cdot d\mathbf{l} = \mu_0 I \quad (1.52)$$

Here I is the current flowing through the surface S bounded by the path L (Fig. 1.4B). It is proper to notice that the mutual orientation of vectors $d\mathbf{l}$ and $d\mathbf{S}$ is not arbitrary but obeys the right-hand thumb rule. Thus, the circulation of the magnetic field is defined by the value of current I piercing the surface surrounded by the contour L, and it does not depend on currents located outside the perimeter of this area. Of course, this path L can go through media with different physical properties. To derive the surface analogy of Eq. (1.52), consider a closed contour surrounding an element of surface current with density $\mathbf{i}(p)$ (Fig. 1.4C). Applying Eq. (1.52) to such a path and neglecting contribution from elements perpendicular to the surface current, we obtain

$$B_t^{(2)} - B_t^{(1)} = \mu_0 i_l \quad \text{or} \quad \mathbf{n} \times \left(\mathbf{B}^{(2)} - \mathbf{B}^{(1)}\right) = \mu_0 \mathbf{i} \quad (1.53)$$

where t and l represent two mutually perpendicular directions tangential to the surface. Thus, the tangential component of the magnetic field is a discontinuous function at points where the density of surface current differs from zero. We have derived three forms of the second equation of the field $\mathbf{B}$:

$$\oint_L \mathbf{B} \cdot d\mathbf{l} = \mu_0 I, \quad curl\,\mathbf{B} = \mu_0 \mathbf{j}, \quad \mathbf{n} \times \left(\mathbf{B}^{(2)} - \mathbf{B}^{(1)}\right) = \mu_0 \mathbf{i} \quad (1.54)$$

Here $\mathbf{i}$ is the vector of density of surface currents. It is interesting to note that the last of these equations is valid for any time-varying magnetic field, and it is usually regarded as the surface analogy of Maxwell's second equation. In addition, as pointed out earlier, the first two equations of the set (1.54) remain valid for quasistationary fields, which are widely used in the most electromagnetic methods of the borehole geophysics. Now let us summarize these results and present the system of equations of the magnetic field caused by conduction currents in differential form:

1. The system, shown below, has been derived from the Biot-Savart law in the same way that the system of equations for the electric field was derived from Coulomb's law.
2. The Biot-Savart law and Eq. (1.55) contain the same information about the magnetic field. This field is a classical example of the vortex field, which is caused by current density vector **j**.

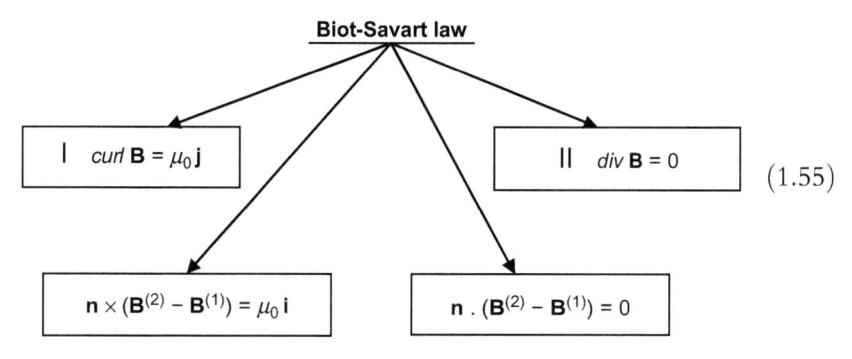

$$
\begin{array}{c}
\text{Biot-Savart law} \\
\\
\text{I} \quad \text{curl } \mathbf{B} = \mu_0 \mathbf{j} \qquad\qquad \text{II} \quad \text{div } \mathbf{B} = 0 \\
\\
\mathbf{n} \times (\mathbf{B}^{(2)} - \mathbf{B}^{(1)}) = \mu_0 \mathbf{i} \qquad \mathbf{n} \cdot (\mathbf{B}^{(2)} - \mathbf{B}^{(1)}) = 0
\end{array}
\tag{1.55}
$$

3. At surfaces where the current density **i** equals zero, both the normal and tangential components of the magnetic field are continuous functions.
4. The system (1.55) describes the field in free space as well as in any non-magnetic conducting medium. Moreover, it turns out that Eq. (1.55) are still valid in the presence of a medium that has an influence on the field (magnetic material), provided that the right-hand side of the first equation

$$\text{curl } \mathbf{B} = \mu_0 \mathbf{j}$$

includes also the magnetization currents.

5. As will be shown later, this system correctly defines the time-varying magnetic field, assuming that propagation effect is disregarded.

1.5 EXAMPLES OF MAGNETIC FIELD OF CURRENT-CARRYING OBJECTS

Now we consider several examples illustrating the behavior of both magnetic field and vector potential.

1.5.1 Example One: Magnetic Field of the Current Filament

Consider the case of current flowing along a straight current filament. Taking into account the axial symmetry of the problem (Fig. 1.5A), let us choose a cylindrical system of coordinates (r, ϕ, z) with its origin situated on the current-carrying line. Starting from the Biot-Savart law, one can see that the magnetic field has only the component B_ϕ, which is independent of the coordinate ϕ. From the principle of superposition, it follows that the total field is the sum of fields contributed by the current elements $I d\mathbf{z}$. Then we have

$$\mathbf{B}_\phi = \frac{\mu_0 I}{4\pi} \int_{z_1}^{z_2} \frac{d\mathbf{z} \times \mathbf{L}_{qp}}{L_{qp}^3} \tag{1.56}$$

where $L_{qp} = (r^2 + z^2)^{1/2}$, and z is the coordinate of the element dz. The coordinates of the observation point are r and $z=0$, while z_1, z_2 are terminal points of the current line. It is clear that the absolute value of the cross product is

$$\left| d\mathbf{z} \times \mathbf{L}_{qp} \right| = dz L_{qp} \sin\left(d\mathbf{z}, \mathbf{L}_{qp}\right) = dz L_{qp} \sin\beta = dz L_{qp} \cos\alpha$$

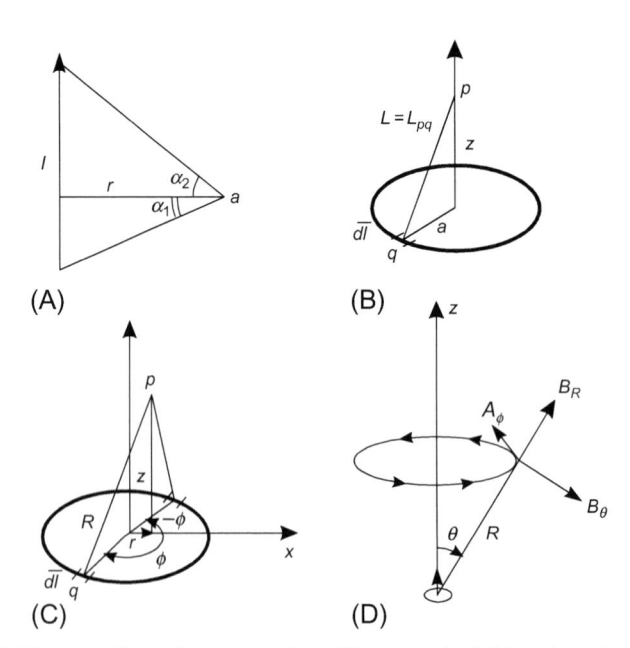

Fig. 1.5 (A) Magnetic field of a current line; (B) magnetic field at the axis of a current loop; (C) magnetic field of the current loop at an arbitrary point; and (D) magnetic field of magnetic dipole in spherical and cylindrical coordinate systems.

Thus

$$B_\phi = \frac{\mu_0 I}{4\pi} \int_{z_1}^{z_2} \frac{dz}{L_{qp}^2} \cos\alpha. \tag{1.57}$$

Inasmuch as $z = r\tan\alpha$, we have

$$dz = r\sec^2\alpha\, d\alpha \quad \text{and} \quad L_{qp}^2 = r^2(1 + \tan^2\alpha) = r^2\sec^2\alpha$$

Substituting these expressions into Eq. (1.57), we obtain

$$B_\phi = \frac{\mu_0 I}{4\pi r} \int_{\alpha_1}^{\alpha_2} \cos\alpha\, d\alpha$$

Thus, the expression for the magnetic field caused by the current flowing along a straight line is

$$B_\phi(p) = \frac{\mu_0 I}{4\pi r}(\sin\alpha_2 - \sin\alpha_1) \tag{1.58}$$

Here α_2 and α_1 are the angles, as shown in Fig. 1.5A. First, suppose that the current-carrying line is infinitely long so that the two angles α_2 and α_1 are $\pi/2$ and $-\pi/2$, respectively. Then

$$B_\phi(p) = \frac{\mu_0 I}{2\pi r} \tag{1.59}$$

In the case of a semiinfinite line, $\alpha_1 = 0$ and $\alpha_2 = \pi/2$, we have

$$B_\phi(p) = \frac{\mu_0 I}{4\pi r} \tag{1.60}$$

Now we assume that $\alpha_2 = \alpha$ and $\alpha_1 = -\alpha$. Then, in accordance with Eq. (1.58), we obtain

$$B_\phi(p) = \frac{\mu_0 I}{2\pi r}\sin\alpha = \frac{\mu_0 I}{2\pi r}\frac{l}{(l^2 + r^2)^{1/2}} \tag{1.61}$$

where $2l$ is the length of the current-carrying line. If l is much greater than the distance r, the right-hand side of Eq. (1.61) can be expanded in a series in terms of parameter $(r/l)^2$. This gives

$$B_\phi = \frac{\mu_0 I}{2\pi r}(1 + r^2/l^2)^{-1/2} \approx \frac{\mu_0 I}{2\pi r}\left(1 - \frac{1}{2}\frac{r^2}{l^2} + \frac{3}{8}\frac{r^4}{l^4} - \cdots\right)$$

If length of the current line $2l$ is a few times greater than the separation r, the field is practically the same as in the case of an infinitely long current line.

1.5.2 Example Two: The Vector Potential A and the Magnetic Field B of a Current in a Circular Loop

Consider next a circular loop of current. First, assume that the observation point is situated on the axis of a loop with radius a, as is shown in Fig. 1.5B. Then, in accordance with Eq. (1.37), we have

$$\mathbf{A} = \frac{\mu_0 I}{4\pi} \oint_L \frac{d\mathbf{l}}{L_{qp}}$$

Because the distance L_{qp} is the same for all points of the loop, we have

$$\mathbf{A} = \frac{\mu_0 I}{4\pi L_{qp}} \oint_L d\mathbf{l}$$

By definition, the sum of the elementary vectors $d\mathbf{l}$ along any closed path is zero. Therefore, the vector potential $\mathbf{A}$ at the z-axis of a circular current loop vanishes. Now we calculate the magnetic field on the z-axis. Because we do not know derivatives of the vector potential on the axis, we cannot use Eq. (1.35) and have to proceed from the Biot-Savart law. As can be seen from Eq. (1.14), in a cylindrical system of coordinates, each current element $Id\mathbf{l}$ creates two field components dB_z and dB_r. However, it is always possible to find two current elements $Id\mathbf{l}$ that contribute at any point of the z-axis the same horizontal components of opposite signs. Therefore, the magnetic field along the z-axis has only a vertical component, which is (Fig. 1.5B)

$$dB_z = |d\mathbf{B}| \frac{a}{L_{qp}} = \frac{\mu_0 I}{4\pi} \frac{dl}{L_{qp}^2} \frac{a}{L_{qp}} = \frac{\mu_0 I a}{4\pi} \frac{dl}{L_{qp}^3}$$

because $|d\mathbf{l} \times \mathbf{L}_{qp}| = L_{qp} dl$. After integration along the loop, we obtain

$$B_z = \frac{\mu_0 I a 2\pi a}{4\pi (a^2 + z^2)^{3/2}} = \frac{\mu_0 I a^2}{2(a^2 + z^2)^{3/2}} = \frac{\mu_0 M}{2\pi (a^2 + z^2)^{3/2}} \qquad (1.62)$$

where

$$M = I\pi a^2 = IS$$

and S being the area of the loop. When the distance z is much greater than the radius of the loop a, we arrive at the following expression for the magnetic field:

$$B_z = \frac{\mu_0 M}{2\pi |z|^3} \quad \text{if } z \gg a \tag{1.63}$$

The last expression plays an important role in electromagnetic fields applied in the induction logging. The intensity of the field is defined by the product $M = IS$, which is called the magnetic moment of the loop. Thus, a small current loop with radius a creates the same magnetic field as a magnetic dipole having the magnitude of the moment equal to $M = \pi a^2 I$. When the distance z in Eq. (1.62) is at least four times greater than the radius a, the treatment of the loop as the magnetic dipole situated at the center of the loop results in an error of no more than 10%. Thus far the vector potential and the magnetic field were considered only along the z-axis. Now we study a general case and first calculate the vector potential at the arbitrary point p. Due to symmetry, the vector potential does not depend on coordinate ϕ. For simplicity let us choose the point p in the x-z plane, where $\phi = 0$. Every pair of current elements (Fig. 1.5C), equally distant from the point p and having coordinates ϕ and $-\phi$, creates a vector potential $d\mathbf{A}$ perpendicular to the x-z plane because each element $I d\mathbf{l}$ causes potential of the same orientation as $d\mathbf{l}$. Inasmuch as the whole loop can be represented as the sum of such pairs, we conclude that the vector potential $\mathbf{A}$ caused by the current-carrying loop has only the component A_ϕ. Therefore, from Eq. (1.36) it follows that

$$A_\phi = \frac{\mu_0 I}{4\pi} \oint \frac{dl_\phi}{R} = \frac{\mu_0 I}{2\pi} \int_0^\pi \frac{a \cos\phi \, d\phi}{(a^2 + r^2 - 2ar\cos\phi + z^2)^{1/2}} \tag{1.64}$$

where dl_ϕ is the component of $d\mathbf{l}$ along coordinate line ϕ:

$$dl_\phi = dl \cos\phi \quad \text{and} \quad L_{qp} = (a^2 + r^2 - 2ar\cos\phi + z^2)^{1/2}$$

Letting $\phi = \pi + 2\alpha$, we have

$$d\phi = 2d\alpha, \quad \cos\phi = 2\sin^2\alpha - 1$$

and, therefore,

$$A_\phi = \frac{aI\mu_0}{\pi} \int_0^{\pi/2} \frac{(2\sin^2\alpha - 1)d\alpha}{\left[(a+r)^2 + z^2 - 4ar\sin^2\alpha\right]^{1/2}}$$

Introducing new variable

$$k^2 = \frac{4ar}{(a+r)^2 + z^2}$$

and carrying out some algebraic operations, we obtain

$$A_\phi = \frac{kI\mu_0}{2\pi}\left(\frac{a}{r}\right)^{1/2}\left[\left(\frac{2}{k^2}-1\right)\int_0^{\pi/2}\frac{d\alpha}{(1-k^2\sin^2\alpha)^{1/2}} - \frac{2}{k^2}\int_0^{\pi/2}(1-k^2\sin^2\alpha)^{1/2}d\alpha\right]$$

$$= \frac{kI\mu_0}{2\pi}\left(\frac{a}{r}\right)^{1/2}\left[\left(1-\frac{k^2}{2}\right)K - E\right]$$

(1.65)

where K and E are complete elliptical integrals of the first and second kind:

$$K(k) = \int_0^{\pi/2}\frac{d\alpha}{(1-k^2\sin^2\alpha)^{1/2}}, \qquad E(k) = \int_0^{\pi/2}(1-k^2\sin^2\alpha)^{1/2}d\alpha \quad (1.66)$$

The functions $K(k)$ and $E(k)$ can be estimated using widely available computer subroutines.

Using the relationship (1.35) between the vector potential and magnetic field, we have, in cylindrical coordinates,

$$B_r = -\frac{\partial A_\phi}{\partial z}, \quad B_\phi = 0, \quad B_z = \frac{1}{r}\frac{\partial}{\partial r}(rA_\phi) \quad (1.67)$$

For elliptical integrals, we have

$$\frac{dK}{dk} = \frac{E}{k(1-k^2)} - \frac{K}{k}, \quad \frac{dE}{dk} = \frac{E}{k} - \frac{K}{k}$$

and

$$\frac{\partial k}{\partial z} = -\frac{zk^3}{4ar}, \quad \frac{\partial k}{\partial r} = \frac{k}{2r} - \frac{k^3}{4r} - \frac{k^3}{4a}$$

Therefore, for the magnetic field, we derive

Stationary Electric and Magnetic Fields

$$B_r = \frac{\mu_0 I}{2\pi} \frac{z}{r\left[(a+r)^2 + z^2\right]^{1/2}} \left[-K + \frac{a^2 + r^2 + z^2}{(a-r)^2 + z^2} E\right]$$

$$B_z = \frac{\mu_0 I}{2\pi} \frac{1}{\left[(a+r)^2 + z^2\right]^{1/2}} \left[K + \frac{a^2 - r^2 - z^2}{(a-r)^2 + z^2} E\right]$$

(1.68)

Thus, the magnetic field caused by a circular current loop is expressed in terms of elliptical integrals.

1.5.3 Example Three: Magnetic Fields of the Magnetic Dipole

Suppose that the distance from the center of the current-carrying loop to the observation point R is considerably greater than the loop radius, that is

$$R = \left(r^2 + z^2\right)^{1/2} \gg a$$

Then Eq. (1.64) can be simplified to

$$A_\phi \approx \frac{\mu_0 I a}{2\pi} \int_0^\pi \frac{\cos \phi \, d\phi}{(R^2 - 2ar \cos \phi)^{1/2}} = \frac{\mu_0 I a}{2\pi R} \int_0^\pi \frac{\cos \phi \, d\phi}{[1 - (2ar/R^2) \cos \phi]^{1/2}}$$

$$\approx \frac{\mu_0 I a}{2\pi R} \int_0^\pi \left(1 + \frac{ar}{R^2} \cos \phi\right) \cos \phi \, d\phi = \frac{\mu_0 I a}{2\pi R} \int_0^\pi \cos \phi \, d\phi + \frac{\mu_0 I a^2 r}{2\pi R^3} \int_0^\pi \cos^2 \phi \, d\phi$$

(1.69)

where approximation

$$(1+x)^{-n} \approx 1 - nx$$

has been used assuming $nx \ll 1$. The first integral in Eq. (1.69) vanishes, and we obtain

$$A_\phi = \frac{\mu_0 I a^2 r}{4R^3} \quad \text{or} \quad \mathbf{A}_\phi = A_\phi \mathbf{i}_\phi = \frac{\mu_0 I S r}{4\pi R^3} \mathbf{i}_\phi$$

(1.70)

where S is the area of the loop. Now we make use of the spherical system of coordinates, (R, θ, ϕ) with the origin at the center of the loop (Fig. 1.5D). Then, Eq. (1.70) can be written as

$$\mathbf{A} = \mathbf{i}_\phi \frac{\mu_0 I S}{4\pi R^2} \sin \theta$$

(1.71)

Next we introduce the moment of the small loop as a vector directed along the z-axis, whose magnitude is equal to the product of the current and area of the loop:

$$\mathbf{M} = IS\mathbf{i}_z = M\mathbf{i}_z \qquad (1.72)$$

where $M = IS$. The moment $\mathbf{M}$ and direction of the current form the right-hand side system. Thus, instead of Eq. (1.71) we have:

$$\mathbf{A} = \frac{\mu_0 \mathbf{M} \times \mathbf{R}}{4\pi R^3} \qquad (1.73)$$

since

$$\mathbf{M} \times \mathbf{R} = \mathbf{i}_\phi MR \sin\theta$$

Now proceeding from Eqs. (1.35), (1.73), and taking into account that

$$A_R = A_\theta = 0$$

we obtain the following expressions for the magnetic field in a spherical system of coordinates:

$$B_R = \frac{1}{R \sin\theta} \frac{\partial(\sin\theta A_\phi)}{\partial \theta}, \quad B_\theta = -\frac{1}{R} \frac{\partial(RA_\phi)}{\partial R}, \quad B_\phi = 0$$

Whence

$$B_R = \frac{2\mu_0 M}{4\pi R^3} \cos\theta, \quad B_\theta = \frac{\mu_0 M}{4\pi R^3} \sin\theta, \quad B_\phi = 0 \qquad (1.74)$$

These equations describe the magnetic field of a small current loop, assuming that its radius is much smaller than the distance between the center of the loop and the observation point. This is the most important condition for use of Eq. (1.74), while the values of the loop radius and the distance R are not essential. Eq. (1.74) describes the magnetic field of magnetic dipole with the moment M.

Some Comments

1. In the case of the electric field, a "dipole" means a combination of equal charges having opposite signs when the separation of the charges is much smaller than the distance to the point at which the field is determined. The notion "magnetic dipole" is the limit of a closed loop of electric current, as the radius of the loop is reduced to zero while keeping the magnetic moment constant.

2. The magnetic field of any current system, regardless of the shape and dimensions, is equivalent to that of the magnetic dipole at distances much greater than the size of this system. For instance, a distribution of conduction currents within the upper part of the earth's core is complicated and changes with time. However, at the earth's surface, relatively far away from the core, the magnetic field of these currents is close to that of the magnetic dipole.
3. In most cases of induction logging, current-carrying coils within the logging tool can be treated as magnetic dipoles. Expressions (1.74), derived for the direct current, are also valid for the case of quasistationary fields, which is the main reason why we considered this example in detail. The main features of the field of the magnetic dipole follow directly from Eq. (1.74):
   (a) At the z-axis the field of the magnetic dipole has only one component B_z directed along this z-axis, and it drops with z as

   $$B_z = \frac{\mu_0 M}{2\pi z^3} \qquad (1.75)$$

   (b) At the equatorial plane $\theta = \pi/2$, the radial component B_R vanishes, and the field has the direction opposite to that of the magnetic moment M:

   $$B_z = -\frac{\mu_0 M}{4\pi r^3} \qquad (1.76)$$

   here r is the distance from the dipole to an observation point.

   (c) Along any radius (θ = constant) both components, B_R and B_θ, decrease inversely proportional to R^3. At the same time, their ratio, as well as an orientation of the total vector $\mathbf{B}$ with respect to $\mathbf{R}$, does not change. In fact, in accordance with Eq. (1.74), we have

   $$\frac{B_\theta}{B_R} = \frac{1}{2}\tan\theta \qquad (1.77)$$

   It is also useful to consider the components of the field in the cylindrical system. As follows from Fig. 1.5D, for components B_r and B_z, we have

   $$B_r(r, z) = B_R \sin\theta + B_\theta \cos\theta \quad \text{and} \quad B_z(r, z) = B_R \cos\theta - B_\theta \sin\theta$$

   where

   $$R = (r^2 + z^2)^{1/2}$$

Taking into account (1.74), we obtain

$$B_r(r,z) = \frac{3\mu_0 M}{4\pi R^3}\sin\theta\cos\theta, \quad B_z(r,z) = \frac{\mu_0 M}{4\pi R^3}\left(2\cos^2\theta - \sin^2\theta\right)$$

or

$$B_r(r,z) = \frac{3\mu_0 Mrz}{4\pi(r^2+z^2)^{5/2}}, \quad B_z = \frac{\mu_0 M}{4\pi(r^2+z^2)^{5/2}}\left(2z^2 - r^2\right) \qquad (1.78)$$

Behavior of B_r and B_z components along vertical line parallel to the z-axis at fixed r is presented in (Fig. 1.6). It is clear that the radial component B_r is an odd function of z changing sign in the equatorial plane of the dipole. At the same time, the vertical component is an even function of z changing sign at points

$$z = \pm\frac{r}{\sqrt{2}}.$$

1.5.4 Example Four: Magnetic Field Due to a Current in a Cylindrical Conductor

Consider an infinitely long and homogeneous cylindrical conductor (Fig. 1.7A) with the radius a and current I. In this case, the current density **j** is uniformly distributed over the cross section S and has only z component:

$$j = j_z = \text{constant} \qquad (1.79)$$

In the cylindrical system of coordinates r, ϕ, z where the z-axis is directed along the conductor, the magnetic field can be characterized by three components: B_r, B_ϕ, B_z. However, two of these components are equal to zero.

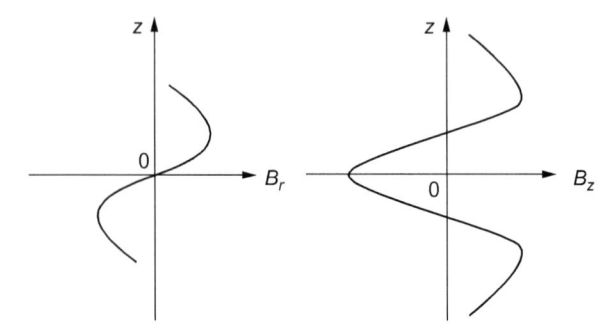

Fig. 1.6 Components B_r and B_z as functions of z.

Stationary Electric and Magnetic Fields

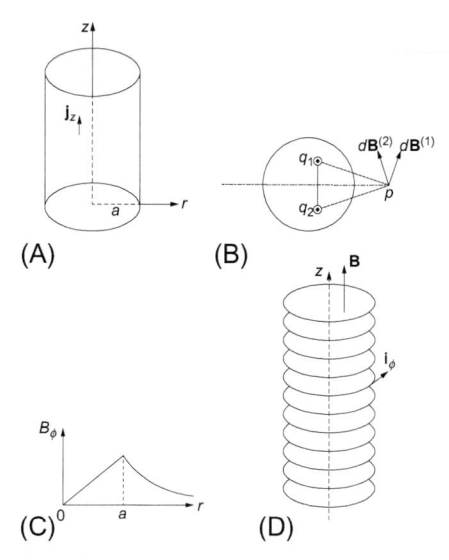

Fig. 1.7 (A) Cylindrical conductor with current; (B) summation of radial components of the magnetic field of cylindrical conductor; (C) behavior of the magnetic field; and (D) infinitely long solenoid.

As follows from the Biot-Savart law, the magnetic field caused by the current element is perpendicular to the current density **j**; therefore $B_z = 0$. Next consider two current elements located symmetrically with respect to the half-plane $\phi =$ constant (Fig. 1.7B).

Obviously, the sum of radial components of the field is equal to zero. Because the entire conductor can be presented as a combination of such pairs of current elements, the total magnetic field does not have the radial component, B_r either. Thus, we demonstrated that $\mathbf{B} = (0, B_\phi, 0)$. Taking into account the symmetry in distribution of the currents, the vector lines of the magnetic field are circles located in horizontal planes with centers on the z-axis. In order to determine the component B_ϕ, we apply the first equation of the system (1.54):

$$\oint_L \mathbf{B} \cdot d\mathbf{l} = \oint_L B_\phi dl = B_\phi \oint_L dl = 2\pi r B_\phi = \mu_0 I_S$$

Here I_S is the current passing through any area bounded by the magnetic line. In the derivation, we take advantage of the axial symmetry of the field and parallel character of vectors **B** and d**l**. Thus, the field outside and inside the current is

$$B_\phi^e = \frac{\mu_0 I}{2\pi r} \qquad \text{if } r \geq a \qquad (1.80)$$

and

$$B_\phi^i = \frac{\mu_0 j}{2} r, \qquad \text{if } r \leq a \qquad (1.81)$$

because $I_S = \pi r^2 j$. In accordance with Eqs. (1.80), (1.81), the magnetic field is equal to zero at the z-axis and increases linearly inside. At the surface of the conductor, it reaches maximum, equal to

$$B_\phi(a) = \frac{\mu_0 j}{2} a \qquad (1.82)$$

and then the field decreases inversely proportional to the distance r (Fig. 1.7C). Considering the magnetic field of the linear current (Eq. 1.58), we have shown that the field tends to infinity when an observation point approaches the surface of the current line ($r \to 0$). Obviously, this is a result of replacement of real distribution of currents by its fictitious model. At the surface of the conductor, the field has a finite value defined by Eq. (1.82).

1.5.5 Example Five: Magnetic Field of Infinitely Long Solenoid

Suppose that, at each point of the cylindrical surface S, the current density has only one component $\mathbf{i}_\phi$ (Fig. 1.7D). In this case, we have $B_\phi = 0$. The radial component also vanishes. Indeed, consider two elementary current circuits located symmetrically with respect to the plane which includes an observation point (Fig. 1.8A). It is seen that the sum of radial components

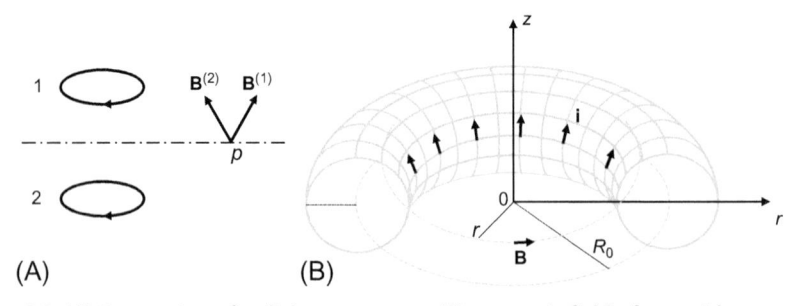

Fig. 1.8 (A) Summation of radial components; (B) magnetic field of a toroid.

is equal to zero. Taking into account the fact that the solenoid is infinitely long, one can always find such a pair of current loops that provide the resultant radial component equal to zero. Thus, the total field has only a z component:

$$\mathbf{B} = (0, 0, B_z) \tag{1.83}$$

In general, the field can be evaluated by the Biot-Savart law and integration of the fields caused by elementary currents with the same radius a. But we can simplify calculations by using Poisson's equation for the vector potential

$$\Delta \mathbf{A} = -\mu_0 \mathbf{j} \tag{1.84}$$

Taking into account the symmetry of the problem and the fact that the vector potential has the same component as the current density, we have

$$\mathbf{A} = A_\phi \mathbf{i}_\phi \tag{1.85}$$

Outside the currents, the vector potential satisfies Laplace's equation:

$$\Delta \mathbf{A} = \Delta(A_\phi \mathbf{i}_\phi) = 0$$

According to vector calculus, we have

$$\mathbf{i}_\phi = -\mathbf{i}_x \sin\phi + \mathbf{i}_y \cos\phi$$

Here $\mathbf{i}_x$ and $\mathbf{i}_y$ are independent of the coordinate unit vectors of the Cartesian system. Thus

$$\Delta \mathbf{A} = -\mathbf{i}_x \Delta(A_\phi \cos\phi) + \mathbf{i}_y \Delta(A_\phi \sin\phi)$$

For two arbitrary scalar functions u and v, we have

$$\Delta(uv) = v\Delta u + u\Delta v + 2(\text{grad}\, u \cdot \text{grad}\, v)$$

In our case, A_ϕ depends only on r, while the second function is either $\cos\phi$ or $\sin\phi$.

The term $2(\text{grad}\, u \cdot \text{grad}\, v)$ vanishes because gradients are orthogonal to each other, and we obtain

$$\Delta \mathbf{A} = -\mathbf{i}_x \left[\cos\phi \cdot \Delta A_\phi + A_\phi \Delta \cos\phi\right] + \mathbf{i}_y \left[\sin\phi \cdot \Delta A_\phi + A_\phi \Delta \sin\phi\right]$$

By definition in the cylindrical system of coordinates

$$\Delta \cos\phi = -\frac{1}{r^2}\cos\phi \quad \text{and} \quad \Delta \sin\phi = -\frac{1}{r^2}\sin\phi$$

Whence

$$\Delta\mathbf{A} = -\mathbf{i}_x \cos\phi \left[\Delta A_\phi - \frac{1}{r^2}A_\phi\right] + \mathbf{i}_y \sin\phi \left[\Delta A_\phi - \frac{1}{r^2}A_\phi\right] = 0$$

which holds when

$$\Delta A_\phi - \frac{1}{r^2}A_\phi = 0 \tag{1.86}$$

The operator Δ is

$$\Delta A_\phi = \frac{1}{r}\frac{\partial}{\partial r}\left(r\frac{\partial A_\phi}{\partial r}\right) \tag{1.87}$$

Substitution of the latter into Eq. (1.86) gives Laplace's equation for a scalar component A_ϕ:

$$\frac{d}{dr}\left[r\frac{dA_\phi(r)}{dr}\right] - \frac{A_\phi(r)}{r} = 0 \tag{1.88}$$

The solution to this ordinary differential equation of the second order is

$$A_\phi(r) = Cr + Dr^{-1} \tag{1.89}$$

Because magnetic field has a finite value everywhere and tends to zero at infinity, we represent the vector potential inside and outside of the solenoid as

$$A_\phi^{(i)} = Cr, \qquad A_\phi^{(e)} = Dr^{-1} \tag{1.90}$$

where C and D are unknown coefficients. By definition

$$\mathbf{B} = \operatorname{curl} \mathbf{A} = \frac{1}{r}\begin{vmatrix} \mathbf{1}_r & r\mathbf{1}_\phi & \mathbf{1}_z \\ \frac{\partial}{\partial r} & \frac{\partial}{\partial \phi} & \frac{\partial}{\partial z} \\ 0 & rA_\phi & 0 \end{vmatrix} \tag{1.91}$$

Whence

$$B_r = 0, \quad B_\phi = 0, \qquad B_z = \frac{1}{r}\frac{\partial}{\partial r}(rA_\phi) \tag{1.92}$$

Substitution of $A_\phi^{(e)}$ into Eq. (1.92) yields

$$B_z^{(e)} = 0, \quad \text{if } r > a$$

proving that the surface currents of the solenoid do not create a magnetic field outside the solenoid. In the same manner for the field inside of the solenoid, we obtain

$$B_z^{(i)} = 2C, \quad \text{if } r \leq a \qquad (1.93)$$

In order to determine C, we recall that the difference of tangential components at both sides of the solenoid is

$$2C = \mu_0 i_\phi \quad \text{or} \quad B_z^{(i)} = \mu_0 i_\phi$$

Thus, for the field **B**, we have

$$B_z^{(i)} = \mu_0 i_\phi, \text{ if } r \leq a \quad \text{and} \quad B_z^{(e)} = 0, \quad \text{if } r > a \qquad (1.94)$$

Such behavior of the field is not obvious. First, because the field caused by a single current loop varies greatly along the radius it is difficult to predict that the field inside, $B_z^{(i)}$, is uniform over the cross section. Also, it is not obvious that the field outside a solenoid is zero; that is, the elementary fields caused by all current loops compensate each other. However, imagine a plane $z = $ const where an observation point outside a solenoid is situated. Current circuits located relatively close to this plane generate a negative component along the z-axis, while current loops situated far away provide at the same observation point a positive contribution, $r > a$. Correspondingly, the field outside solenoid is an algebraic sum of elementary fields, and it turns out that, in the case of infinitely long solenoid, this sum is equal to zero. Note that inside the solenoid all elementary fields are positive. Of course, if a solenoid has a finite length along the z-axis, the field outside is not zero and has two components B_r, B_z.

1.5.6 Example Six: Magnetic Field of a Current Toroid

Let us introduce a cylindrical system of coordinates with the z-axis perpendicular to the toroid, having radius R_0 (Fig. 1.8B). The current density in toroid is **i**. Due to axial symmetry, the magnetic field and vector potential are independent of the coordinate ϕ.

Also imagine two current loops of the toroid located symmetrically with respect to the vertical plane, where a point of observation is located. As can be seen, the sum of vector potentials due to these elementary currents does not have the ϕ-component. Thus, for the vector potential, we have

$$\mathbf{A} = (A_r, 0, A_z)$$

Taking into account that

$$\mathbf{B} = \frac{1}{r}\begin{vmatrix} \mathbf{i}_r & r\mathbf{i}_\phi & \mathbf{i}_z \\ \dfrac{\partial}{\partial r} & \dfrac{\partial}{\partial \phi} & \dfrac{\partial}{\partial z} \\ A_r & 0 & A_z \end{vmatrix}$$

we obtain

$$B_r = 0, \qquad B_\phi = \frac{\partial A_r}{\partial z} - \frac{\partial A_z}{\partial r}, \qquad B_z = 0 \qquad (1.95)$$

Thus, the magnetic field has only one component B_ϕ but cannot be calculated from Eq. (1.95) unless the vector potential is known. The problem can be solved by using the Biot-Savart law in the integral form Eq. (1.52):

$$\oint_L \mathbf{B} \cdot d\mathbf{l} = \mu_0 I_S \qquad (1.96)$$

where L is a circular path of radius r located in the horizontal plane with the center situated at the toroid axis, and I_S is the current passing through a surface S surrounded by this path L. Taking into account an axial symmetry and the same directionality of $\mathbf{B}$ and $d\mathbf{l}$, we have

$$B_\phi 2\pi r = \mu_0 I_S \qquad (1.97)$$

First, consider a point p, located outside the toroid. In such a case the current either does not intersect the surface S, or its net value passing through the surface is equal to zero (equal current is passing in opposite directions through the surface). This means that $B_\phi = 0$; therefore, the magnetic field is zero outside the toroid, as in the case of the solenoid:

$$B_\phi^{(e)} = 0 \qquad (1.98)$$

Next consider the magnetic field inside of the toroid. As follows from Eq. (1.97) the field $B_\phi^{(i)}$ is not uniform and equals

$$B_\phi^{(i)} = \frac{\mu_0 I_S}{2\pi r} \qquad (1.99)$$

In this case, the path of integration is inside the toroid. Suppose that it is located in the plane $z = 0$; then, a change of its radius does not change the flux of the current density. Therefore, within the range,

$$R_0 - r_0 < r < R_0 + r_0$$

an increase of r results in a decrease of the field inversely proportional to r. If we consider circular paths in planes with $z \neq 0$, then the current I_S becomes smaller as z increases. Thus, we observe a nonuniform magnetic field inside the toroid. It is natural to expect that, with an increase of the ratio of the radius R_0 to the radius of its cross section r_0, the field inside becomes more uniform. Note that if the toroid has an arbitrary but constant cross section and current density is independent on the coordinate ϕ, we can still apply Eq. (1.99) and conclude that the field $\mathbf{B}$ is equal to zero outside the toroid. Of course, if the current density is not constant in the last two examples, the magnetic field appears outside, too, $\mathbf{B}^{(e)} \neq 0$.

1.6 SYSTEM OF EQUATIONS FOR THE STATIONARY FIELDS

Let us summarize the results that follow from Coulomb's law, the Biot-Savart law, Ohm's law, and the principle of charge conservation. As shown above, we have the following equations at regular points:

$$\begin{aligned} curl\,\mathbf{E} &= 0, & div\,\mathbf{D} &= \delta_0 \\ curl\,\mathbf{B} &= \mu_0 \mathbf{j}_c, & div\,\mathbf{B} &= 0 \end{aligned} \quad (1.100)$$

and boundary conditions at interfaces:

$$\begin{aligned} \mathbf{n} \times (\mathbf{E}_2 - \mathbf{E}_1) &= 0, & \mathbf{n} \cdot (\mathbf{D}_2 - \mathbf{D}_1) &= \sigma_0 \\ \mathbf{n} \times (\mathbf{B}_2 - \mathbf{B}_1) &= \mathbf{i}_c, & \mathbf{n} \cdot (\mathbf{B}_2 - \mathbf{B}_1) &= 0 \end{aligned} \quad (1.101)$$

Here

$$\mathbf{D} = \varepsilon \mathbf{E} \quad \text{and} \quad \mathbf{j}_c = \gamma \mathbf{E}$$

REFERENCES

[1] Kaufman AA, Anderson BI. Principles of electric methods in surface and borehole geophysics. Amsterdam: Elsevier; 2010.
[2] Alpin LM. Field theory. Moscow: Nedra; 1966.

FURTHER READING

[1] Kaufman A, Kleinberg R, Hansen R. Principles of the magnetic methods in geophysics. Amsterdam: Elsevier; 2009.

CHAPTER TWO

Physical Laws and Maxwell's Equations

Contents

2.1 Faraday's Law	40
2.2 Principle of Charge Conservation	46
2.3 Distribution of Electric Charges	48
2.3.1 Equation for the Volume Charge Density	49
2.3.2 Uniform Medium	49
2.3.3 Nonuniform Medium	50
2.3.4 Quasi-Stationary Field	52
2.3.5 Behavior of Charge Density δ_{02}	53
2.3.6 Surface Distribution of Charges	54
2.3.7 Slowly Varying (Quasi-Stationary) Field	57
2.4 Displacement Currents	59
2.4.1 Second Source of the Magnetic Field	59
2.4.2 Total Current and the Charge Conservation Principle	61
2.4.3 Currents in the Circuit With a Capacitor	63
2.5 Maxwell's Equations	67
2.5.1 Introduction	67
2.5.2 Maxwell's Equations	68
2.5.3 Second Form of Maxwell's Equations	70
2.5.4 Maxwell's Equations in a Piecewise Uniform Medium	72
2.6 Equations for the Fields E and B	73
2.7 Electromagnetic Potentials	75
2.8 Maxwell's Equations for Sinusoidal Fields	78
2.9 Electromagnetic Energy and Poynting Vector	81
2.9.1 Principle of Energy Conservation and Joule's Law	81
2.9.2 Energy Density and Poynting Vector	83
2.9.3 Current Circuit and Transmission Line	85
2.10 Uniqueness of the Forward Problem Solution	86
2.10.1 Uniqueness Theorem	86
2.10.2 Formulation of the Boundary Value Problem	89
Reference	90
Further Reading	90

Proceeding from Faraday's, Coulomb's, Biot-Savart's, and Ohm's laws, governing static and time-varying electric and magnetic fields, we arrive at Maxwell's equations. From a historical perspective, such approach is natural because most of these laws, as well as Faraday's field concept, were known before Maxwell derived his system.

2.1 FARADAY'S LAW

Experiments performed by Faraday demonstrated that time-varying flux Φ of the magnetic field **B** through any surface S, bounded by a closed contour L (Fig. 2.1A), creates an electromotive force Ξ along this contour:

$$\Xi = -\frac{\partial \Phi}{\partial t} \tag{2.1}$$

where

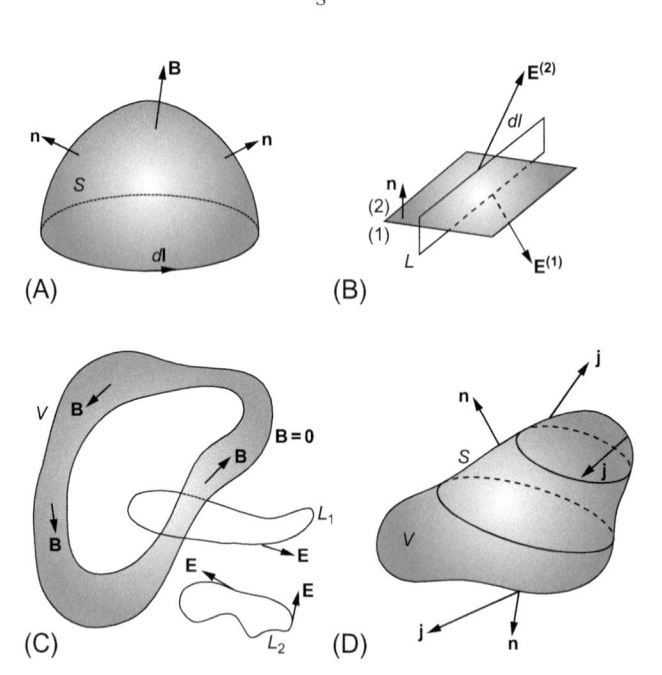

$$\Phi = \int_S \mathbf{B} \cdot d\mathbf{S}$$

Fig. 2.1 (A) Flux of the field $\dot{\mathbf{B}}$. (B) Electric field near an interface. (C) Faraday's law. (D) Flux of the current density.

is the flux of the magnetic field and $\partial\Phi/\partial t$ is its derivative with respect to time. The contour L is a closed path that can have any form and can pass in general through media with different electric properties, including insulators. Of course, such a path L usually does not coincide with the actual current lines. By definition, the electromotive force is

$$\Xi = \oint_L \mathbf{E} \cdot d\mathbf{l} \tag{2.2}$$

where E is the electric field at each point of the contour L. Thus an electromotive force may exist only if there is an electric field. Consequently, in place of Eq. (2.1) we have

$$\oint_L \mathbf{E} \cdot d\mathbf{l} = -\frac{\partial \Phi}{\partial t} \tag{2.3}$$

A change of the magnetic flux Φ with time gives rise to an electric field. In other words, a time-varying magnetic field B is a source of an electric field in the same sense that electric charges are the source of a stationary electric field. This phenomenon, first observed by Faraday, is called electromagnetic induction. The relationship between the electric field and the rate of the change of the magnetic flux, as described by Eq. (2.3), is one of the most fundamental laws of nature. By convention, the electric field that appears due to the electromagnetic induction is called the inductive electric field $\mathbf{E}^{ind}$ to emphasize its origin. Thus we can rewrite Eq. (2.3) in the form

$$\oint_L \mathbf{E}^{ind} \cdot d\mathbf{l} = -\frac{\partial \Phi}{\partial t} \tag{2.4}$$

Because electric field $\mathbf{E}^{ind}$ appears in the integrand, its determination from Eq. (2.3) requires additional information, even for known function $\partial\Phi/\partial t$. In general, there are two sources of the electric field, namely, charges and a change of the magnetic field with time, as shown later.

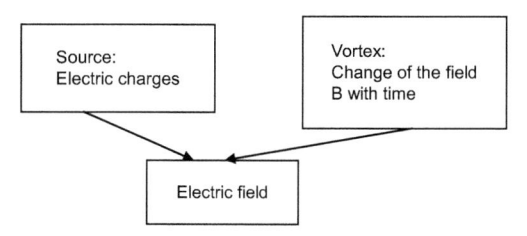

```
┌──────────────────┐      ┌──────────────────────┐
│ Source:          │      │ Vortex:              │
│ Electric charges │      │ Change of the field  │
│                  │      │ B with time          │
└────────┬─────────┘      └──────────┬───────────┘
         │                           │
         └───────────┐   ┌───────────┘
                    ┌▼───▼┐
                    │ Electric field │
                    └────────────────┘
```

At the same time, we can readily think of particular cases in which only one of these sources exists, such as:
1. A static field in which the derivative with respect to time is zero, and the electric field arises only due to the presence of the electric charges.
2. An alternating electromagnetic field in which the electric field has only tangential component at interfaces between media with different electrical conductivities. In such a case charges are absent, and the electric field has an inductive origin only:

$$\mathbf{E} = \mathbf{E}^c + \mathbf{E}^{ind} \quad (2.5)$$

where $\mathbf{E}^c$ is the electric field generated by charges and at every instant obeys Coulomb's law, while $\mathbf{E}^{ind}$ is the part of the electric field, which arises due to a change of the magnetic field with time. Combining Eqs. (2.4), (2.5), we have

$$\oint_L \mathbf{E} \cdot d\mathbf{l} - \oint_L \mathbf{E}^c \cdot d\mathbf{l} = -\frac{\partial \Phi}{\partial t}$$

Because the circulation of the Coulomb's electric field is equal to zero, we have

$$\oint_L \mathbf{E}^{ind} \cdot d\mathbf{l} = \oint_L \mathbf{E} \cdot d\mathbf{l} = -\frac{\partial \Phi}{\partial t} \quad (2.6)$$

This result may lead to confusion about the role played by charges in creating an electromagnetic field. Eq. (2.6) shows that the electromotive force caused by the Coulomb electric field vanishes. But this conclusion, as in the case of a stationary field, does not mean that the Coulomb field plays no role in the distribution of currents and electromagnetic fields. In fact, the Coulomb field does influence the distribution of currents in a conducting medium, and these currents in turn can create an alternating magnetic field. Therefore, in general, both parts of the field, $\mathbf{E}^c$ and $\mathbf{E}^{ind}$, are closely related to each other. Next, we describe different forms of Faraday's law. First, using the definition of the magnetic flux we have

$$\oint_L \mathbf{E} \cdot d\mathbf{l} = -\frac{\partial}{\partial t} \int_S \mathbf{B} \cdot d\mathbf{S}$$

As previously mentioned, a change of the flux may happen due to a change of the magnetic field with time, as well as a change of a position of the path L. The influence of the last factor was studied earlier (Lorenz force), but from now we assume that the path L along which the electromotive force is calculated does not move. Then, the last equation can be rewritten as

$$\oint_L \mathbf{E} \cdot d\mathbf{l} = -\int_S \dot{\mathbf{B}} \cdot d\mathbf{S} \qquad (2.7)$$

where

$$\dot{\mathbf{B}} = \frac{\partial \mathbf{B}}{\partial t}$$

Eq. (2.7) is a formulation of Faraday's law, and, as will be seen later, it is the first of Maxwell's equations in integral form. In this equation $d\mathbf{l}$ is elementary displacement along the contour and indicates the direction in which integration is carried along the contour L, while the vector $d\mathbf{S}$ has the direction of the unit vector $\mathbf{n}$ normal to the surface $\mathbf{S}$, bounded by the contour L (Fig. 2.1A). To retain the physical meaning of Faraday's law, the vectors $d\mathbf{l}$ and $d\mathbf{S}$ are chosen according to the right-hand rule. This means that an observer facing in the direction of the vector $-d\mathbf{S}$ sees that $d\mathbf{l}$ indicates a direction along the path L, which is counterclockwise. Only in this case Eq. (2.7) correctly describes the phenomenon of electromagnetic induction. Now, making use of Stokes theorem,

$$\oint \mathbf{M} \cdot d\mathbf{l} = \int curl\, \mathbf{M} \cdot d\mathbf{S}$$

we obtain the differential form of Eq. (2.7)

$$\oint_L \mathbf{E} \cdot d\mathbf{l} = \int_S curl\, \mathbf{E} \cdot d\mathbf{S} = -\int_S \frac{\partial \mathbf{B}}{\partial t} \cdot d\mathbf{S}$$

Because this equation is valid for any surface S, it follows that the integrands on either side are equal:

$$curl\, \mathbf{E} = -\frac{\partial \mathbf{B}}{\partial t} \qquad (2.8)$$

where $\mathbf{E}$ and $\mathbf{B}$ are considered in the same point. Both Eqs. (2.7), (2.8) describe the same physical phenomenon, but the differential form Eq. (2.8) applies only at regular points in which all components of the electric field are continuous functions of the spatial variables.

Considering that in most problems we must examine electromagnetic fields in media with discontinuous changes in physical properties (interfaces), it is useful to derive a surface analog of Eq. (2.8). For example, it is clear that Eq. (2.8) cannot be used for points of the interface between media having different values of dielectric permittivity and conductivity because the normal component of the electric field is a discontinuous function at such interfaces. For this reason, we proceed with Eq. (2.7) performing integration along the elementary path, as shown in Fig. 2.1B. Taking into account that the flux through the area surrounded by this contour tends to zero, we obtain

$$E_{2t} - E_{1t} = 0 \qquad (2.9)$$

where t indicates an arbitrary direction tangential to the interface. The vector form of this equation is

$$\mathbf{n} \times (\mathbf{E}_2 - \mathbf{E}_1) = 0$$

Here $\mathbf{E}_1$ and $\mathbf{E}_2$ are electric fields at the back and front sides of the surface, respectively.

In accordance with this equation, the tangential component of the time-varying field is a continuous function at the interface, as it would be in the case of the field caused by static electric charges. Thus we have derived three forms of Faraday's law:

$$\oint_L \mathbf{E} \cdot d\mathbf{l} = -\int_S \dot{\mathbf{B}} \cdot d\mathbf{S}, \quad \text{curl}\, \mathbf{E} = -\frac{\partial \mathbf{B}}{\partial t}, \quad \mathbf{n} \times (\mathbf{E}_2 - \mathbf{E}_1) = 0 \qquad (2.10)$$

and each of them describes the first Maxwell's equation. Later we consider numerous examples illustrating electromagnetic induction and application of Eq. (2.10).

Several comments:
1. Suppose that a change of magnetic field with time occurs within some volume V, but outside of V the field $\mathbf{B}$ is absent. As follows from Eq. (2.3), the electromotive force along the contour L_1, surrounding this volume (Fig. 2.1C), is nonzero, regardless of location of the volume V inside L_1. In other words, the time-varying field $\mathbf{B}$ in some region results in an appearance of the electric field $\mathbf{E}^{ind}$ everywhere in space. Here we see the direct analogy with Coulomb's electric field caused by charges. But neither Coulomb's law nor Faraday's law can explain how the field reaches an observation point. Later we will discuss this subject in detail.

2. Consider an arbitrary closed path L_2 that does not enclose a volume where vortices $\partial \mathbf{B}/\partial t$ are located. In general, an electric field caused by the sources may exist at each point of this contour and vary in both magnitude and direction (Fig. 2.1C). However, the electromotive force in this case is equal to zero.

3. This analysis shows that, at every point of a closed contour, the inductive electric field can be presented as the sum of two fields. The first is caused by vortices intersecting the area surrounded by the loop, while the second field is generated by vortices that do not cross the area of the loop. In accordance with Faraday's law, the electromotive force Ξ is given by only the first part of the electric field.

This fact allows us to measure the rate of change, $\partial \mathbf{B}/\partial t$, at any point in space with a relatively small loop around this point.

4. It is well known that the voltage of a Coulomb electric field between two points is path independent. In general, taking into account (2.4), the voltage of the inductive electric field

$$\int_a^b \mathbf{E}^{ind} \cdot d\mathbf{l}$$

depends on the path of integration. Taking into account Faraday's law and modifying the system of equations, derived for the static field, we obtain

$$\begin{array}{ll} \operatorname{curl} \mathbf{E} = -\dfrac{\partial \mathbf{B}}{\partial t}, & \operatorname{div} \mathbf{D} = \delta_0, \\ \operatorname{curl} \mathbf{B} = \mu_0 \mathbf{j}_c, & \operatorname{div} \mathbf{B} = 0 \end{array} \qquad (2.11)$$

At first glance, this set of equations fully describes the time-varying electromagnetic field because it takes into account electromagnetic induction as well as the Coulomb's and Biot-Savart's laws. In fact, as we will see later, it characterizes fields in the so-called quasi-stationary approximation, which plays a dominant role in the induction logging. However, the set in Eq. (2.11) also suggests that the field instantly appears at any point of space regardless of distance from the source. Suppose that the conduction current has changed at some instant t. Then, in accordance with the equation

$$\oint_L \mathbf{B} \cdot d\mathbf{l} = \mu_0 I$$

the magnetic field synchronously changes at any observation point. The same is true for the electric field. There is another fact, which implies that there is a problem with Eq. (2.11). Indeed, from the equation

$$curl\, \mathbf{B} = \mu_0 \mathbf{j}_c$$

we have

$$div\mathbf{j}_c = 0 \qquad (2.12)$$

because from vector calculus $div curl\, \mathbf{B} = 0$. Eq. (2.12) describes charge conservation law for the static electric field (Chapter 1). But we will see it later that it contradicts the principle of charge conservation law for time-varying fields. Understanding this discrepancy leads to the discovery of electromagnetic field propagation and formulation of complete system of Maxwell's equations. For this reason, it is proper to describe in detail the principle of charge conservation.

2.2 PRINCIPLE OF CHARGE CONSERVATION

In general, the principle of charge conservation is written in the form

$$\oint_S \mathbf{j} \cdot d\mathbf{S} = -\frac{\partial e_0}{\partial t} \qquad (2.13)$$

where S is an arbitrary closed surface, e_0 is a free charge in the volume V, surrounded by the surface S, and $d\mathbf{S}$ is directed outside the volume (Fig. 2.1D). Here

$$\mathbf{j} = \mathbf{j}_c$$

is the conduction current only. In accordance with Eq. (2.13), a flux of the current density through S defines the rate of a change of charge over time inside the volume. If, for instance, the flux is positive, the charge e_0 decreases; by contrast, when the flux is negative more charges arrive than leave the volume. At the same time, experiments show that, in the absence of electrical current, it is impossible to have the appearance or disappearance of charges from any volume. Thus any change of the charge inside the volume V may occur only due to the flux of free charges through the surface S. Now, applying Gauss's theorem for regular points of a medium, we obtain

a
Physical Laws and Maxwell's Equations

$$\oint_S \mathbf{j} \cdot d\mathbf{S} = \int_V div\mathbf{j}\, dV = -\frac{\partial}{\partial t}\int_V \delta_0\, dV \qquad (2.14)$$

where δ_0 is the volume density of free charges. Assuming that the volume does not change with time, we have

$$\int_V div\mathbf{j}\, dV = -\int_V \frac{\partial \delta_0}{\partial t} dV \quad \text{or} \quad div\mathbf{j} = -\frac{\partial \delta_0}{\partial t} = -\dot{\delta}_0 \qquad (2.15)$$

Next, consider the surface analogy of Eq. (2.15). With this purpose in mind, let us determine the flux of the current density through the surface of an elementary cylinder. Making use of Eq. (2.14), we have

$$\mathbf{j}_2 \cdot d\mathbf{S}_2 + \mathbf{j}_1 \cdot d\mathbf{S}_1 + \int_{S_l} \mathbf{j} \cdot d\mathbf{S} = -\dot{\sigma}_0 dS$$

where $\mathbf{j}_2$ and $\mathbf{j}_1$ are the current density at the front and back sides of the surface, respectively, and S_l is the lateral surface of the cylinder,

$$d\mathbf{S}_2 = \mathbf{n}\, dS, \quad d\mathbf{S}_1 = -\mathbf{n}\, dS$$

In the limit when the cylinder height tends to zero, we have

$$j_{2n} - j_{1n} = -\dot{\sigma}_0 \qquad (2.16)$$

that is, the difference of normal components of the current density defines the rate at which a surface density of charges, σ_0, changes. Thus we have derived three forms of equations describing the principle of charge conservation:

$$\oint_S \mathbf{j} \cdot d\mathbf{S} = -\frac{\partial e_0}{\partial t}, \quad div\mathbf{j} = -\dot{\delta}_0, \quad j_{2n} - j_{1n} = -\dot{\sigma}_0 \qquad (2.17)$$

These equations are always valid for any electromagnetic field, and they show that a change of a charge in one place is always accompanied by such a change of charge in other places so that the total charge remains the same. This is the reason why the phenomenon is called the principle of charge conservation. At the same time, one can imagine at least two cases when it is possible to neglect a change of the charge with time.

Case One: The Stationary Field

By definition, the right-hand side of Eq. (2.13) vanishes, and the flux of the current density is equal to zero through any closed surface. This means that charges may exist, but they do not change with time.

Case Two: Quasi-Stationary Electromagnetic Field

Suppose that a medium is conductive and a time-varying field changes relatively slowly. Then, it turns out that, under certain conditions, the right-hand side of Eq. (2.17) can be neglected. In these cases, at each instant of time the amount of charge entering any volume is equal to the amount leaving the volume. As will be shown later, such an approximation gives a sufficiently correct result when displacement currents are much smaller than the conduction ones. As mentioned earlier, this scenario is of great importance for the induction logging. Thus in these special cases in place of Eq. (2.17) we have

$$\oint_S \mathbf{j} \cdot d\mathbf{S} \approx 0, \quad div\,\mathbf{j} \approx 0, \quad j_{2n} \approx j_{1n} \qquad (2.18)$$

Returning back to the principle of charge conservation for the time-varying field, it is natural to expect that the set (2.17) should follow from the system of equations (2.11). However, it turns out that, in general, these equations do not follow from the set (2.11) because the third equation of this set does not take into account one more sources of the magnetic field.

2.3 DISTRIBUTION OF ELECTRIC CHARGES

Now proceeding from Eq. (2.17) and the second equation of the set (2.11):

$$\oint_S \varepsilon \mathbf{E} \cdot d\mathbf{S} = e_0 \quad \text{or} \quad div\,\varepsilon \mathbf{E} = \delta_0 \qquad (2.19)$$

we study a distribution of charges in a conducting and polarizable medium with conductivity γ and dielectric permittivity ε. In this light it is proper to note that Eq. (2.19) was derived from Coulomb's law. This may create impression that our results are applied only for the static field. However, as will be shown later, these equations are applied for the time-varying electromagnetic fields, too.

2.3.1 Equation for the Volume Charge Density

First consider some points of a medium where equations

$$div\,\mathbf{j} = -\frac{\partial \delta_0}{\partial t} \quad \text{and} \quad div\,\varepsilon\mathbf{E} = \delta_0 \qquad (2.20)$$

are valid. Taking into account Ohm's law, we have

$$div\,\mathbf{j} = div\,\gamma\mathbf{E} = div\frac{\gamma}{\varepsilon}\varepsilon\mathbf{E} = \frac{\gamma}{\varepsilon}div\,\varepsilon\mathbf{E} + \varepsilon\mathbf{E}\cdot grad\,\frac{\gamma}{\varepsilon} = -\frac{\partial \delta_0}{\partial t}$$

or

$$\frac{\gamma}{\varepsilon}\delta_0 + \varepsilon\mathbf{E}\cdot grad\,\frac{\gamma}{\varepsilon} = -\frac{\partial \delta_0}{\partial t}$$

Thus we have arrived at the following differential equation for the volume density of free charges:

$$\frac{\partial \delta_0}{\partial t} + \frac{1}{\tau_0}\delta_0 = -\varepsilon\mathbf{E}\cdot grad\,\frac{1}{\tau_0} \qquad (2.21)$$

where

$$\tau_0 = \frac{\varepsilon}{\gamma} = \varepsilon\rho \qquad (2.22)$$

is often called the time constant of the medium.

2.3.2 Uniform Medium

Suppose that, in the vicinity of some point, the parameter τ_0 does not change or the field $\mathbf{E}$ and $grad(1/\tau_0)$ are perpendicular to each other:

$$\mathbf{E}\cdot\nabla\frac{1}{\tau_0} = 0 \qquad (2.23)$$

Then, Eq. (2.21) is simplified, and we have

$$\frac{\partial \delta_0}{\partial t} + \frac{1}{\tau_0}\delta_0 = 0 \qquad (2.24)$$

The solution of this equation is

$$\delta_0(t) = C\exp(-t/\tau_0) \qquad (2.25)$$

where C is the density of the free charge at the initial instant. In a conducting and polarizable medium, the parameter τ_0 is usually small. For example, if $\rho = 100\,\text{ohm}\,\text{m}$ and $\varepsilon = 10\varepsilon_0$ then

$$\tau_0 = 100 \cdot 10 \cdot (36\pi)^{-1} 10^{-9} \text{s} < 10^{-8} \text{s}$$

Thus a free charge placed inside a conducting medium quickly disappears. If we are concerned only with charges that exist at times greater than τ_0 ($t \gg \tau_0$) and described by Eq. (2.24), we can assume that they are, in practice, absent. In addition, the initial volume charge is usually equal to zero inside the conducting medium, that is $C=0$. Therefore, we conclude that at points where the medium is uniform with respect to τ_0 or condition (2.23) is met, there are no electric charges and

$$\text{div}\,\varepsilon \mathbf{E} = 0 \qquad (2.26)$$

Earlier (Chapter 1) it was mentioned that free charges are accompanied by bound ones:

$$\text{div}\,\mathbf{E} = \frac{\delta_0 + \delta_b}{\varepsilon_0} \quad \text{and} \quad \text{div}\,\mathbf{E} = \frac{\delta_0}{\varepsilon}$$

because in our case $\text{grad}\,\varepsilon = 0$. Whence

$$\frac{\delta_0 + \delta_b}{\varepsilon_0} = \frac{\delta_0}{\varepsilon} \quad \text{and} \quad \delta_b = (\varepsilon_0/\varepsilon - 1)\delta_0 \qquad (2.27)$$

Therefore, both the bound and free charges, located in the vicinity of some point where $\nabla \tau_0 = 0$, decay in the same manner and

$$\text{div}\,\mathbf{E} = 0 \quad \text{if} \quad t \gg \tau_0 \qquad (2.28)$$

Similarly, with the case of the static field, the total density of the decaying charge is smaller than the free charge by the factor ε_r.

$$\delta = \frac{\delta_0(t)}{\varepsilon_r} \qquad (2.29)$$

where $\varepsilon_r = \varepsilon/\varepsilon_0$.

2.3.3 Nonuniform Medium

Thus far we have studied the behavior of the charge in the vicinity of points where either the medium is uniform or the condition (2.23) is met. It was established that charge decays rapidly, and such behavior is observed regardless of the presence of the electromagnetic field caused by the source located at some place of the medium. A different situation occurs when the medium is not uniform and either $\mathbf{E} \cdot \nabla \gamma \neq 0$ or $\mathbf{E} \cdot \nabla \varepsilon \neq 0$. In this case the right-hand side of Eq. (2.21) does not vanish, and we have an inhomogeneous differential equation of the first order:

$$\frac{dy}{dt} + \frac{1}{\tau_0}y = f(t) \tag{2.30}$$

where $y = \delta_0(t)$ and $f(t) = -\varepsilon \mathbf{E} \cdot \nabla \frac{1}{\tau_0}$. The general solution of Eq. (2.30) has the form

$$y(t) = y_0 \exp(-t/\tau_0) + \exp(-t/\tau_0) \int_0^t \exp(x/\tau_0) f(x) dx \tag{2.31}$$

where y_0 is the value of the function $y(t)$ at the instant $t=0$. In accordance with Eq. (2.31),

$$\delta_0(t) = C \exp(-t/\tau_0) - \exp(-t/\tau_0)\varepsilon \int_0^t \exp\left(\frac{x}{\tau_0}\right) \mathbf{E} \cdot \nabla \frac{1}{\tau_0} dx \tag{2.32}$$

If the direction of the electric field does not change with time, the last equation can be rewritten as

$$\delta_0(t) = C \exp(-t/\tau_0)$$
$$- \exp(-t/\tau_0)\varepsilon \int_0^t \exp\left(\frac{x}{\tau_0}\right) E(x) dx \left(\mathbf{e}_0 \cdot \nabla \frac{1}{\tau_0}\right) \tag{2.33}$$

Here

$$\mathbf{E}(t) = E(t)\mathbf{e}_0$$

We can recognize two types of charges whose behavior is quite different as a function of time:

$$\delta_0(t) = \delta_{01}(t) + \delta_{02}(t) \tag{2.34}$$

where

$$\delta_{01}(t) = C \exp(-t/\tau_0)$$

and

$$\delta_{02}(t) = -\exp(-t/\tau_0)\varepsilon \int_0^t \exp(x/\tau_0)E(x)dx(\mathbf{e}_0 \cdot \nabla(1/\tau_0)). \qquad (2.35)$$

The behavior of the function $\delta_{01}(t)$ is the same as in the case of a uniform medium. According to Eq. (2.35), a free charge $\delta_{02}(t)$ arises in the neighborhood of any point where a medium is not uniform, provided that the field $E(t)$ is not perpendicular to the direction of the gradient of τ_0. In general, the density $\delta_{02}(t)$ depends on the resistivity and dielectric permittivity of the medium as well as on the magnitude and direction of the electric field.

2.3.4 Quasi-Stationary Field

Now we consider a special case, which is of practical interest for induction logging. Suppose the following inequality holds:

$$\frac{\partial \delta_0}{\partial t} \ll \frac{\delta_0}{\tau_0} \qquad (2.36)$$

Then, instead of Eq. (2.21), we obtain an approximate equation

$$\gamma \frac{\delta_0(t)}{\varepsilon} + \varepsilon \mathbf{E}(t) \cdot \nabla \frac{1}{\tau_0} = 0 \qquad (2.37)$$

Correspondingly, the density of free charge is

$$\delta_0(t) = -\varepsilon^2 \rho \mathbf{E}(t) \cdot \nabla \frac{1}{\tau_0}$$

or $\qquad (2.38)$

$$\delta_0(t) = \varepsilon \mathbf{E}(t) \cdot \frac{\nabla \rho}{\rho} + \mathbf{E}(t) \cdot \nabla \varepsilon$$

since

$$\nabla \frac{1}{\tau_0} = \nabla \frac{1}{\varepsilon \rho} = -\frac{\nabla \varepsilon}{\varepsilon^2} \frac{1}{\rho} - \frac{\nabla \rho}{\rho^2} \frac{1}{\varepsilon}$$

Therefore, free charges arise in the vicinity of points where either conductivity or dielectric permittivity changes. Of course, this happens only if the electric field is not perpendicular to the direction of the maximal rate of change of these parameters. Note that, in the frequency domain, the inequality (2.36) is equivalent to the following:

$$\frac{\omega \varepsilon}{\gamma} \ll 1$$

Physical Laws and Maxwell's Equations

As we already know, the free charges are usually accompanied by bound charges, and their density is

$$\delta_b = \delta - \delta_0 = div[(\varepsilon_0 - \varepsilon)\mathbf{E}] = div[(\varepsilon_0 - \varepsilon)\mathbf{j}\rho]$$

Making use of Eq. (2.38) and equation

$$div\,\mathbf{E} = \frac{\delta_0 + \delta_b}{\varepsilon_0}$$

it can be shown that

$$\delta_b(t) = -\mathbf{E}(t) \cdot \nabla\varepsilon + (\varepsilon_0 - \varepsilon)\mathbf{E}(t) \cdot \frac{\nabla\rho}{\rho} \tag{2.39}$$

From Eqs. (2.38), (2.39) it follows that the total charge is

$$\delta(t) = \varepsilon_0 \mathbf{E}(t) \cdot \frac{\nabla\rho}{\rho} \tag{2.40}$$

This means that, at points where only ε varies, the total charge is equal to zero. In such places the free and bound charges compensate each other. At the same time at points where both parameters change, the total charge is the same as if polarization were absent.

2.3.5 Behavior of Charge Density δ_{02}

Now we return to the general case (Eq. 2.35) and consider two examples that illustrate the behavior of the charge $\delta_{02}(t)$ when variation in time is taken into account (2.21).

Example One
Let us assume that the electric field varies with time as

$$\mathbf{E}(t) = E_0 \exp(-t/\tau)\mathbf{e}_0 \tag{2.41}$$

and τ is the parameter characterizing the rate of the field decay. Correspondingly, Eq. (2.35) becomes

$$\delta_{02}(t) = -\varepsilon \exp\left(-\frac{t}{\tau}\right)\left(\mathbf{e}_0 \cdot \nabla\frac{1}{\tau_0}\right)\int_0^t \exp\left[\left(\frac{1}{\tau_0} - \frac{1}{\tau}\right)x\right]dx E_0$$

Carrying out integration, we obtain

$$\delta_{02}(t) = -\frac{\varepsilon\tau_0 E_0 \exp(1 - t/\tau_0)}{1 - \tau_0/\tau}\{\exp[t(1/\tau_0 - 1/\tau)] - 1\}\mathbf{e}_0 \cdot \nabla\frac{1}{\tau_0} \tag{2.42}$$

As follows from this equation, the charge is absent at the instant $t = 0$. Then it increases, reaches a maximum and, afterward, decays exponentially. Thus, in general, the dependence of this charge density and the electric field $\mathbf{E}(t)$ on time differs from each other. Assuming that the electric field decays relatively slowly $(\tau \gg \tau_0)$ and measurements are performed at sufficiently large times $(t \gg \tau_0)$, in place of Eq. (2.42) we obtain

$$\delta_{02}(t) = -\varepsilon\tau_0 \exp(-t/\tau) E_0 \left(\mathbf{e}_0 \cdot \nabla \frac{1}{\tau_0}\right) \quad (2.43)$$

Thus the volume density of free charge and the electric field decay in the same manner. For instance, when the time constant of the field τ is 1 s, the function $\delta_{02}(t)$ also decreases with a time constant 1 s regardless of the conductivity and dielectric permittivity of the medium.

Example Two
Now suppose that the electric field varies as a sinusoidal function:

$$\mathbf{E}(t) = E_0 \mathbf{e}_0 \sin \omega t.$$

Substituting this expression into Eq. (2.35) and integrating, we have

$$\delta_{02}(t) = -\frac{\varepsilon E_0 \tau_0}{1+\omega^2\tau_0^2}[\omega\tau_0 \exp(-t/\tau_0) + (\sin \omega t - \omega\tau_0 \cos \omega t)]\left(\mathbf{e}_0 \cdot \nabla \frac{1}{\tau_0}\right) \quad (2.44)$$

In particular, assuming that

$$t \gg \tau_0 \quad \text{and} \quad T \gg \tau_0 \quad (2.45)$$

where T is the period of oscillations, we have

$$\delta_{02}(t) = -\varepsilon\tau_0 E_0 \sin \omega t \left(\mathbf{e}_0 \cdot \nabla \frac{1}{\tau_0}\right) \quad (2.46)$$

Notice that conditions Eqs. (2.36), (2.45) have the same meaning.

2.3.6 Surface Distribution of Charges

So far we have studied the distribution of volume charge density. Now consider time-varying free charges that arise at interfaces between media with different electric properties. Applying equations

$$j_{2n} - j_{1n} = -\frac{\partial \sigma_0}{\partial t}, \quad D_{2n} - D_{1n} = \sigma_0$$

where

$$D_{in} = \varepsilon_i E_{in}, \quad i = 1, 2$$

we have

$$\frac{\gamma_2}{\varepsilon_2} D_{2n} - \frac{\gamma_1}{\varepsilon_1} D_{1n} = \frac{1}{2}\left[\left(\frac{\gamma_2}{\varepsilon_2} + \frac{\gamma_1}{\varepsilon_1}\right)(D_{2n} - D_{1n}) + \left(\frac{\gamma_2}{\varepsilon_2} - \frac{\gamma_1}{\varepsilon_1}\right)(D_{2n} + D_{1n})\right] = -\frac{\partial \sigma_0}{\partial t}$$

or

$$\alpha^{av} \sigma_0 + (\alpha_2 - \alpha_1) D_n^{av} = -\frac{\partial \sigma_0}{\partial t}$$

Whence

$$\frac{\partial \sigma_0(t)}{\partial t} + \frac{1}{\tau_{0s}} \sigma_0(t) = (\alpha_1 - \alpha_2) D_n^{av} \qquad (2.47)$$

where

$$\tau_{0s} = \frac{1}{\alpha^{av}} = \frac{2}{1/\tau_{01} + 1/\tau_{02}}$$

is the relaxation time for surface charges, and

$$\alpha_1 - \alpha_2 = \frac{1}{\tau_{01}} - \frac{1}{\tau_{02}}, \quad D_n^{av} = \frac{\varepsilon_1 E_{1n} + \varepsilon_2 E_{2n}}{2},$$

$$\tau_{01} = \varepsilon_1 \rho_1, \quad \tau_{02} = \varepsilon_2 \rho_2, \quad \alpha_1 = \frac{1}{\tau_{01}}, \quad \alpha_2 = \frac{1}{\tau_{02}}$$

Thus the equation for the surface density of free charges is a differential equation of the first order similar to that for the volume density. In accordance with Eq. (2.35), the solution of Eq. (2.47) is

$$\sigma_0(t) = C \exp(-t/\tau_{0s}) + \exp(-t/\tau_{0s})(\alpha_1 - \alpha_2) \int_0^t D_n^{av}(x) \exp(x/\tau_{0s}) dx$$

(2.48)

that is

$$\sigma_0 = \sigma_{01} + \sigma_{02}$$

where

$$\sigma_{01}(t) = C\exp(-t/\tau_{0s}), \quad \sigma_{02}$$
$$= (\alpha_1 - \alpha_2)\exp(-t/\tau_{0s})\int_0^t D_n^{av}(x)\exp(x/\tau_{0s})dx \quad (2.49)$$

Respectively, there are two types of surface charges. The first type σ_{01} corresponds to the case of free charge with density C placed at the interface. As follows from Eq. (2.49), it decays exponentially with time constant τ_{0s}. The decay is controlled by the conductivity and dielectric permittivity of the media on both sides of the interface, and it is independent of the electric field caused by other sources. Inasmuch as the relaxation time τ_{0s} is usually small and measurements are performed at times much greater than τ_{0s}, in most cases one can ignore the presence of this charge. Correspondingly, let us concentrate on the charges of the second type arising on a boundary. Of course, as in the case of volume density, the surface charges consist of the free and bound charges, and they are related to each other. In fact, from the equations

$$E_{2n} - E_{1n} = \frac{\sigma}{\varepsilon_0} \quad \text{and} \quad \varepsilon_2 E_{2n} - \varepsilon_1 E_{1n} = \sigma_0$$

we have

$$\sigma_0 = \frac{1}{2}[(\varepsilon_2 + \varepsilon_1)(E_{2n} - E_{1n}) + (\varepsilon_2 - \varepsilon_1)(E_{2n} + E_{1n})]$$

or

$$\sigma_0 = \frac{\varepsilon^{av}}{\varepsilon_0}\sigma + (\varepsilon_2 - \varepsilon_1)E_n^{av}$$

Here

$$\sigma = \sigma_0 + \sigma_b, \quad \varepsilon^{av} = \frac{\varepsilon_1 + \varepsilon_2}{2}, \quad E_n^{av} = \frac{E_{2n} + E_{1n}}{2}$$

Hence

$$\sigma_b = \frac{\varepsilon_0 - \varepsilon^{av}}{\varepsilon^{av}}\sigma_0 - \frac{\varepsilon_2 - \varepsilon_1}{\varepsilon^{av}}\varepsilon_0 E_n^{av} \quad (2.50)$$

2.3.7 Slowly Varying (Quasi-Stationary) Field

Now let us assume again that the time constant τ_{0s} is small with respect to time of measurements,

$$t \gg \tau_{0s}$$

or the period of sinusoidal oscillations of the field T is much greater that $\tau_{0s}: T \gg \tau_{0s}$. Then it is appropriate to replace the right-hand side of Eq. (2.49) by a series in the parameter τ_{0s}. Carrying out this expansion using integration by parts and discarding all terms except the first one, we obtain

$$\sigma_{02}(t) = (\alpha_1 - \alpha_2)\tau_{0s} D_n^{av}(t) \tag{2.51}$$

It is obvious that the same result follows from Eq. (2.47) if we neglect by the derivative $\partial \sigma_{02}(t)/\partial t$ in comparison with the term σ_{02}/τ_{0s}:

$$\frac{\partial \sigma_{02}}{\partial t} \ll \frac{\sigma_{0s}}{\tau_{0s}} \tag{2.52}$$

The free charges are accompanied by bound ones; however, it turns out that the density of the total charge σ does not depend on the dielectric permittivity. In other words, the total charge σ coincides with that of free charges, if the medium is not polarizable, provided that the condition (Eq. 2.52) is met. Correspondingly, letting $\varepsilon_1 = \varepsilon_2 = \varepsilon_0$, Eq. (2.51) can be written as

$$(\gamma_1 + \gamma_2)\frac{\sigma}{2\varepsilon_0} + (\gamma_2 - \gamma_1)E_n^{av} = 0 \tag{2.53}$$

where E_n^{av} is the average magnitude of the normal component of the electric field at point p, located at the interface. Therefore, we arrive at the following expression for the surface density of the total charge:

$$\sigma(p, t) = 2\varepsilon_0 K_{12} E_n^{av}(p, t) \tag{2.54}$$

Here

$$K_{12} = \frac{\rho_2 - \rho_1}{\rho_2 + \rho_1} \tag{2.55}$$

As we already know from (1.10), the same equation describes the density of charges when the field is time invariant. Thus Eq. (2.54) shows that, if the condition (Eq. 2.52) holds, the density of time-varying charges is related to the electric field and resistivity of the medium as the density of stationary charges. Eq. (2.54) plays a fundamental role for understanding the so-called galvanic part of the field. It is useful to represent the normal component of the field E on two sides of the interface as follows:

$$E_{1n}(p, t) = E_n(p, t) - \frac{\sigma(p, t)}{2\varepsilon_0}$$

and (2.56)

$$E_{2n}(p, t) = E_n(p, t) + \frac{\sigma(p, t)}{2\varepsilon_0}$$

Here $E_n(p, t)$ is the normal component of the field at the point p contributed by all sources except the charge at this point. As is shown Ref. [1], this surface charge creates in its vicinity the field

$$\pm \frac{\sigma(p, t)}{2\varepsilon_0}$$

and, in accordance with Eq. (2.56), we have to conclude that

$$E_n^{av}(p, t) = E_n(p, t) \qquad (2.57)$$

where the normal n is directed from the back side "1" to the front side, "2" of the interface. Therefore, the function $E_n^{av}(p, t)$ describes the normal component of the field caused by all sources except the field produced by the charge in the vicinity of point p. For this reason, the second term of Eq. (2.53) can be interpreted as the flux of the current density j through a closed surface of an elementary cylinder with a unit cross-section and an infinitely small height caused by all sources, located outside this surface. In other words, this flux characterizes the difference between the amount of charge that arrives and leaves this volume during each time interval, and this motion of charges is caused by external sources only. The term

$$(\gamma_1 + \gamma_2)\frac{\sigma_0}{2\varepsilon_0}$$

defines the flux of the current density through the same closed surface caused by the electric field of the charge inside the elementary cylinder. Thus, under the approximation (Eq. 2.52) the flux of the current density due to the external sources, such as charges and a change of the magnetic field with time, is compensated by the flux caused by the charge in the vicinity of the point p.

2.4 DISPLACEMENT CURRENTS
2.4.1 Second Source of the Magnetic Field

Next we demonstrate that the system (2.11) is not in agreement with the principle of charge conservation for time-varying fields. To proceed, let us demonstrate that the second equation of this system derived from the Biot-Savart's law

$$\text{curl } \mathbf{B} = \mu_0 \mathbf{j}_c \tag{2.58}$$

in general, contradicts the principle of charge conservation when an electromagnetic field changes with time. In fact, taking the divergence of both sides of Eq. (2.58), we have

$$\text{div curl } \mathbf{B} = \mu_0 \, \text{div} \mathbf{j}_c \quad \text{or} \quad \text{div} \mathbf{j}_c = 0$$

while, as follows from Eq. (2.15), $\mathbf{j}_c$ should be equal to the rate of decrease with time of the charge density. To remove this contradiction, we first assume that, on the right-hand side of Eq. (2.58), there is an additional term X, which disappears in the case of a stationary field. Then Eq. (2.58) becomes

$$\text{curl } \mathbf{B} = \mu_0 (\mathbf{j}_c + \mathbf{X}) \tag{2.59}$$

Now we choose the vector X in such a way that the principle of charge conservation is satisfied. Forming the divergence on both sides of Eq. (2.59), we obtain

$$0 = \text{div} \mathbf{j}_c + \text{div} \mathbf{X}$$

or, in accordance with Eq. (2.15),

$$\text{div} \mathbf{X} = \dot{\delta}_0 \tag{2.60}$$

It is a partial differential equation with respect to unknown vector **X**, and it is not clear how to solve it. However, the problem is greatly simplified if we take into account the third equation of the set (Eq. 2.11):

$$\text{div} \mathbf{D} = \delta_0$$

Assuming that this equation is valid for time-varying fields and taking the derivative with respect to time, we have

$$\text{div} \dot{\mathbf{D}} = \dot{\delta}_0 \tag{2.61}$$

Comparison of Eqs. (2.60) and (2.61) gives

$$div\,\dot{\mathbf{D}} = div\,\mathbf{X} \quad \text{or} \quad div(\dot{\mathbf{D}} - \mathbf{X}) = 0.$$

An infinite number of vectors X satisfy this last equation, and they may differ from each other by $curl\,\mathbf{M}$, where M is an arbitrary vector because

$$div\,curl\,\mathbf{M} = 0$$

However, Maxwell assumed the simplest solution of this equation and let

$$\mathbf{X} = \dot{\mathbf{D}} = \frac{\partial \varepsilon \mathbf{E}}{\partial t} \tag{2.62}$$

Numerous experimental studies performed during almost two centuries have shown the validity of this assumption, and the vector $\partial \mathbf{D}/\partial t$ is called the density of displacement current:

$$\mathbf{j}_d = \frac{\partial \mathbf{D}}{\partial t} \tag{2.63}$$

or

$$\mathbf{j}_d = \varepsilon \frac{\partial \mathbf{E}}{\partial t} \tag{2.64}$$

if we assume that the dielectric permittivity does not change with time. Consequently, instead of Eq. (2.59), we have

$$curl\,\mathbf{B} = \mu_0 \left(\mathbf{j}_c + \varepsilon \frac{\partial \mathbf{E}}{\partial t}\right) \tag{2.65}$$

Thus the time-varying magnetic field is caused by two types of sources in a nonmagnetic medium, namely, the conduction and displacement currents as illustrated later.

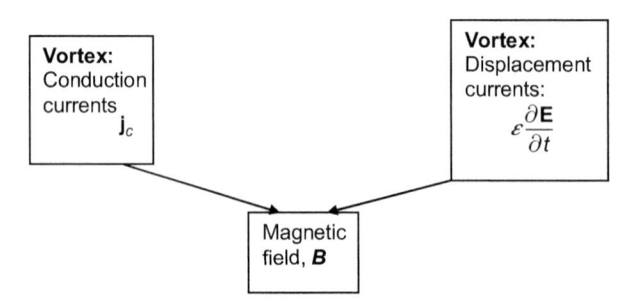

```
┌─────────────┐            ┌─────────────┐
│ Vortex:     │            │ Vortex:     │
│ Conduction  │            │ Displacement│
│ currents    │            │ currents:   │
│      j_c    │            │    ε ∂E/∂t  │
└─────────────┘            └─────────────┘
         \                  /
          \                /
           ┌──────────────┐
           │  Magnetic    │
           │  field, B    │
           └──────────────┘
```

Applying Stoke's theorem, we obtain the integral form of Eq. (2.65):

Physical Laws and Maxwell's Equations

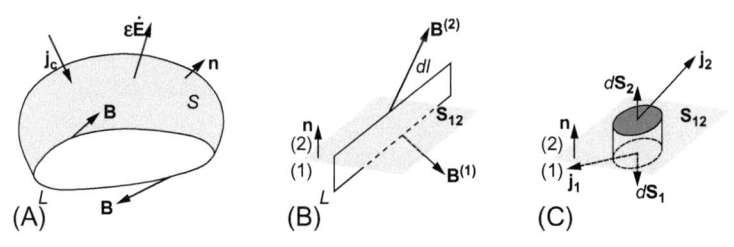

Fig. 2.2 (A) Flux of conduction and displacement currents. (B) Field **B** near interface. (C) Continuity of normal component of the total current near interface.

$$\oint_L \mathbf{B} \cdot d\mathbf{l} = \mu_0 \int_S \left(\mathbf{j}_c + \varepsilon \frac{\partial \mathbf{E}}{\partial t} \right) \cdot d\mathbf{S} \qquad (2.66)$$

which shows that the circulation of the magnetic field along any contour L is determined by the total current passing through any surface S bound by this contour (Fig. 2.2A). Now suppose that the path of integration L is an elementary contour, as shown in Fig. 2.2B. Then, taking into account the fact that in the limit when the area surrounded by the path L tends to zero, the flux of both the conduction and displacement currents vanishes, we obtain

$$\mathbf{n} \times (\mathbf{B}_2 - \mathbf{B}_1) = 0 \qquad (2.67)$$

Therefore, the tangential component of the magnetic field, as in the case of the static field, is a continuous function at the interface, if the surface density of conduction currents $\mathbf{i}_c$ is absent. However, sometimes it is convenient to assume that $\mathbf{i}_c \neq 0$; then, in place of Eq. (2.67), we have

$$\mathbf{n} \times (\mathbf{B}_2 - \mathbf{B}_1) = \mu_0 \mathbf{i}_c \qquad (2.68)$$

2.4.2 Total Current and the Charge Conservation Principle

Having introduced the displacement currents let us represent the charge conservation principle in a different form. Because

$$\mathrm{div}\,\mathbf{j}_c = -\dot{\delta}_0 \quad \text{and} \quad \mathrm{div}\,\mathbf{D} = \delta_0$$

we have

$$\mathrm{div}\,\mathbf{j}_c + \mathrm{div}\,\dot{\mathbf{D}} = 0 \quad \text{or} \quad \mathrm{div}\,\mathbf{j} = 0 \qquad (2.69)$$

where

$$\mathbf{j} = \mathbf{j}_c + \varepsilon \frac{\partial \mathbf{E}}{\partial t} \qquad (2.70)$$

is the density of the total current. In accordance with Eq. (2.69), the current lines of the field **j** are always closed; therefore, **j** is the vortex field. Applying Gauss's theorem, we obtain the integral form of Eq. (2.69):

$$\oint_S \mathbf{j} \cdot d\mathbf{S} = 0 \qquad (2.71)$$

which is the flux of the vector of the total current density through any closed surface and is always equal to zero. Considering again an elementary cylinder (Fig. 2.2C) and calculating the flux of **j** through this closed surface, we have

$$j_{1nc} + \frac{\partial D_{1n}}{\partial t} = j_{2nc} + \frac{\partial D_{2n}}{\partial t} \qquad (2.72)$$

Thus the normal component of the vector **j** is a continuous function at an interface. Let us write down equations for the total current density, describing the principle of charge conservation:

$$\oint_S \mathbf{j} \cdot d\mathbf{S} = 0, \quad div\,\mathbf{j} = 0, \quad j_{1n} = j_{2n} \qquad (2.73)$$

Comments:
1. Eq. (2.65) can be rewritten as

$$curl\,\mathbf{B} = \mu_0 \mathbf{j} \qquad (2.74)$$

where **j** is the vector of the total current density. The similarity of Eqs. (2.58), (2.74) is obvious. However, it does not mean that, in general, a time-varying magnetic field obeys the Biot-Savart's law. Nevertheless, if the influence of displacement currents is negligible, the magnetic field **B**(t) behaves practically in accordance with this law, and the field is a quasi-stationary one.
2. Displacement currents depend on the dielectric permittivity and electric field. In particular, in a nonpolarizable medium ($\varepsilon = \varepsilon_0$) displacement current is caused only by the rate of change of the electric field with time.
3. In an isolative medium there are only displacement currents, while in a conducting medium conduction currents usually prevail. Of course, with an increase of the frequency, the field is varying faster and the relative contribution of displacement currents becomes stronger.

4. Unlike electromagnetic induction, the introduction of displacement currents, which was discovered experimentally, was a bold assumption made by Maxwell and only later confirmed by experiments.
5. The quantity $\varepsilon \dfrac{\partial \mathbf{E}}{\partial t}$ is called the displacement currents density, though it is not related to a motion of free charges as in the case of conduction currents. In spite of this fundamental difference between the conduction and displacement currents, the latter is also called the current in order to emphasize that both can generate a magnetic field.
6. The charge conservation principle has two forms:

$$div\mathbf{j}_c + \frac{\partial \delta_0}{\partial t} = 0 \quad \text{and} \quad div\mathbf{j} = 0$$

In the case of the quasi-stationary field, we disregard with $\dfrac{\partial \delta_0}{\partial t}$ term assuming that an influence of displacement currents is negligible.

7. Among numerous phenomena based on the existence of displacement currents, we note only two:
   a. Propagation of electromagnetic waves with a finite velocity.
   b. Presence of the alternating current in a circuit with a capacitor.

In fact, Eq. (2.58) was derived from the Biot-Savart's law, which implies that the magnetic field B instantly appears at any point regardless of its distance from conduction currents. In other words, the velocity of propagation of the field is infinitely high. However, this conclusion contradicts all experimental observations that show that the field propagates with a finite velocity. For instance, in a nonpolarizable and nonmagnetic medium this velocity is equal to the speed of light:

$$c = 3 \times 10^8 \, \text{m/s}$$

Later we demonstrate that propagation of the electromagnetic field at a finite speed is impossible without displacement currents. Now consider the first example, illustrating the effect of displacement currents.

2.4.3 Currents in the Circuit With a Capacitor

Suppose that the circuit consists of a conducting part (wire) and an insulator bounded by two conducting plates, parallel to each other (Fig. 2.3), comprising a capacitor. At the beginning suppose that there is only a conducting current I_c in the wire, while displacement currents are absent. In accordance with such assumption there is a magnetic field around this circuit. Applying the equation

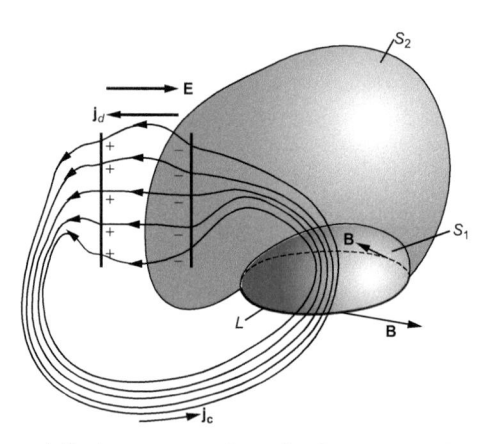

Fig. 2.3 Distribution of displacement and conduction currents in a circuit.

$$\oint_L \mathbf{B} \cdot d\mathbf{l} = \mu_0 I_C$$

to the closed contour L (Fig. 2.3), we discover a paradox. Indeed, if the surface S_l bounding the contour L intersects the conducting part of the circuit, the circulation of the magnetic field remains the same, and it is equal to $\mu_0 I_c$. However, if the surface S_l passes through the capacitor, this circulation becomes equal to zero because there is no conduction current inside the capacitor. This ambiguity indicates that our assumption was incorrect; in reality, there is a displacement current inside the capacitor. Moreover, this current has to be equal to the conduction current: $I_d = I_c$. Then, applying Eq. (2.66), we see that the circulation of the magnetic field is independent of the place where the surface S_l intersects the circuit. Later, we consider several examples illustrating the role of displacement currents, but now let us study the current in the circuit with a capacitor (Fig. 2.3). First, assume that, at some instant, two charges with equal magnitude and opposite sign are placed on the capacitor plates. To facilitate this analysis, we make several assumptions:

1. The inductive electric field, caused by a change of the magnetic field with time, can be disregarded. Therefore, the electric field $\mathbf{E}(t)$ is caused by charges only, and it obeys Coulomb's law. In particular, the field $\mathbf{E}(t)$ inside the capacitor is mainly caused by surface charges located on the conducting plates.
2. The distance between capacitor plates is small compared with their dimensions.
3. At any given instant of time, the current density has the same value at all points of the circuit.

Charges, located on plates, create an electric field everywhere including in the conducting part of the circuit. As a result, charges appear on the lateral surface of the wire. Due to the electric field of these charges, a conduction current with density **j** arises, and, correspondingly, a decrease of the plate charges is observed. Note that inside the capacitor there can be bound charges. The electric field **E**(t) in the capacitor is directed from the positive to negative charges, as shown in Fig. 2.3. Taking into account the fact that the field **E**(t) decreases, the displacement current has a direction that is opposite that of the electric field. Thus the conduction current in the wire and the displacement current in the capacitor have the same direction. Displacement currents appear inside the conducting part of the circuit, but they are relatively small ($j_c \gg j_d$). In addition, displacement currents exist around the circuit, but we assume that their influence is negligible. In this approximation Eq. (2.72) can be rewritten as

$$j_c = \varepsilon \frac{\partial E}{\partial t} = j \qquad (2.75)$$

and under our approximation charges on the plates are located only at points where wire is connected to the capacitor. Thus we demonstrated that the displacement current in the capacitor represents a continuation of the conduction current in the wire, and, in accordance with the charge conservation principle, the vector lines of the current density j are closed. Now we consider both types of currents in some detail. Suppose that at some instant t charges with density $\sigma_0(t)$ and $-\sigma_0(t)$ are located on the capacitor plates. Then, as follows from Eq. (2.16), the normal component of density of conduction current is related to the free charge on the plate as

$$j_c = -\frac{\partial \sigma_0}{\partial t} \qquad (2.76)$$

Respectively, the current in the wire is equal to

$$I = -\frac{\partial Q_0(t)}{\partial t} \qquad (2.77)$$

where $Q_0(t)$ is the amount of free charge at each plate. By definition the voltage of the electric field caused by charges is

$$\int_+^- \mathbf{E} \cdot d\mathbf{l} = U_+ - U_- = IR \qquad (2.78)$$

Here U_+ and U_- are potentials of the plates with positive and negative charges, respectively, and R is the resistance of the wire. The usage of the potential $U(t)$ is justified because the vortex part of the electric field is neglected. Next we find an expression for the same potential difference in terms of capacitor parameters. The electric field between plates is directly proportional to density σ_0 and

$$E(t) = \frac{\sigma_0(t)}{\varepsilon} \tag{2.79}$$

At the same time, the free charges on the plate and difference of potentials are related as

$$Q_0(t) = C[U_+(t) - U_-(t)] \tag{2.80}$$

where C is called the capacitance and is equal to amount of the charge on the plate when the difference of potentials equals unity. In particular in the [SI] units the capacitance is measured in farads:

$$1\,\text{F} = 1\,\text{Coulomb}/1\,\text{V} = 10^9\ \text{pF}$$

Assuming that the influence of plate edges is small and the medium between plates is uniform, it is easy to determine the capacitance C. In fact, from Eq. (2.79) we have

$$U_+ - U_- = \frac{\sigma_0}{\varepsilon} d = \frac{Q_0 d}{\varepsilon S} = \frac{Q_0}{C}$$

where S is the plate area and d is the distance between plates. Thus the capacitance in this case is

$$C = \frac{\varepsilon S}{d} \tag{2.81}$$

Now we derive the differential equation describing the behavior of the charge Q_0 and currents. From Eqs. (2.77), (2.78), (2.80), we have

$$\frac{dQ_0(t)}{dt} + \frac{Q_0}{CR} = 0 \tag{2.82}$$

Therefore, the charge decays exponentially with time:

$$Q_0(t) = Q_0^0 \exp(-t/CR) \tag{2.83}$$

Correspondingly, for the conduction and displacement currents, we have

$$I_C(t) = I_d(t) = \frac{Q_0^0}{CR} \exp(-t/CR) \tag{2.84}$$

With an increase of the resistance and capacitance the currents decay slower. This example illustrates how a displacement current passes through the capacitor.

2.5 MAXWELL'S EQUATIONS
2.5.1 Introduction

In the previous sections we have introduced two sources of the electromagnetic field, namely, the rate of a change of the electric and magnetic fields with time:

$$\varepsilon \frac{\partial \mathbf{E}}{\partial t} \quad \text{and} \quad \frac{\partial \mathbf{B}}{\partial t}$$

Together with charges and conduction currents, they form the complete set of sources of the electromagnetic field, as shown in Table 2.1.

Let us point out several facts concerning the relationship between the electromagnetic field and its sources:

1. In general, the electric field is caused by both charges δ and vortices $\partial \mathbf{B}/\partial t$.

   However, the magnetic field does not have sources; it is generated in a nonmagnetic medium by two types of vortices: conduction and displacement currents.

2. As is seen from Table 2.1, generators of the magnetic field are defined by the electric field, while one of generators of the electric field is caused by a change of the magnetic field B with time. Usually, electric and magnetic fields depend on each other, and it is impossible to determine them separately.

3. Behavior of the static electric and magnetic fields is governed by Coulomb's and Biot-Savart's laws. These laws require knowledge about distribution of charges and currents, causing the fields. But in the presence of conductive media, there are some additional secondary sources that affect the fields as well. In order to quantify these secondary sources,

Table 2.1 Sources and Vortexes of the Electromagnetic Field

Generators	Electric Field	Magnetic Field
Sources	δ	—
Vortexes	$\dfrac{\partial \mathbf{B}}{\partial t}$	$\mathbf{j}_c, \varepsilon \dfrac{\partial \mathbf{E}}{\partial t}$

we will derive system of Maxwell's equations that, along with the boundary conditions, uniquely defines these sources and permits quantitative description of the fields in space and time.
4. In the case of the time-varying fields, we also have to proceed from the system of field equations because (a) the medium affects the field; and (b) existence of interaction between electric and magnetic fields.

In the case of the static field Coulomb's law:

$$\mathbf{E}(p) = \frac{1}{4\pi\varepsilon_0} \int_V \frac{\delta(q)\mathbf{L}_{qp}}{L_{qp}^3} dV$$

allows one to determine the field $\mathbf{E}(p)$ at any point if the charge distribution is known. The same is valid for the Biot-Savart's law, and the magnetic field can be calculated as soon as the conduction current $\mathbf{j}_c$ is fully specified. At the same time Faraday's law

$$\oint \mathbf{E} \cdot d\mathbf{l} = -\int \dot{\mathbf{B}} \cdot d\mathbf{S}$$

establishes only the linkage between the flux of the vector $\dot{\mathbf{B}}$ and the circulation of the electric field along some line L, where $\mathbf{E}(t)$ usually changes from point to point. This implies that, even for the known magnetic field, the field $\mathbf{E}(t)$ cannot be determined at a specific location without additional information. Of course, this statement is also applied to the relationship between magnetic field and conduction and displacement currents. Therefore, in order to determine the electromagnetic field, we have to proceed from a system of field equations.

As is well known, the system of equations for any vector field $\mathbf{M}(p)$ at regular points consists of two equations:

$$\text{curl}\,\mathbf{M}(p) = \mathbf{W}(p), \quad \text{div}\,\mathbf{M}(p) = \omega(p)$$

where functions $\mathbf{W}(p)$ and $\omega(p)$ describe the distribution of vortices and sources, respectively. Thus existence of interconnected electric and magnetic fields leads to the system of four equations.

2.5.2 Maxwell's Equations

Before presenting this system, it is appropriate to remind readers that we restrict ourselves to the study of fields in a piecewise uniform and non-magnetic medium because, in most cases, this model properly describes a distribution of conductivity and dielectric permittivity of geologic media.

This model is widely used in the theory of the induction logging. Taking into account (2.8), (2.65) and assuming that equations

$$div\,\varepsilon\mathbf{E} = \delta_0, \quad div\,\mathbf{B} = 0$$

remain valid for time-varying fields, we obtain the following system of equations:

$$\begin{array}{ll} \text{I} \quad curl\,\mathbf{E} = -\dfrac{\partial \mathbf{B}}{\partial t} & \text{III} \quad div\,\varepsilon\mathbf{E} = \delta_0 \\ \text{II} \quad curl\,\mathbf{B} = \mu_0\left(\mathbf{j}_c + \varepsilon\dfrac{\partial \mathbf{E}}{\partial t}\right) & \text{IV} \quad div\,\mathbf{B} = 0 \end{array} \quad (2.85)$$

and their surface analogs:

$$\begin{array}{ll} \text{I} \quad \mathbf{n} \times (\mathbf{E}_2 - \mathbf{E}_1) = 0 & \text{III} \quad \mathbf{n} \cdot (\varepsilon_2 \mathbf{E}_2 - \varepsilon_1 \mathbf{E}_1) = \sigma_0 \\ \text{II} \quad \mathbf{n} \times (\mathbf{B}_2 - \mathbf{B}_1) = \mu_0 \mathbf{i}_c & \text{IV} \quad \mathbf{n} \cdot (\mathbf{B}_2 - \mathbf{B}_1) = 0 \end{array} \quad (2.86)$$

In Eq. (2.86) $\mathbf{E}_1, \mathbf{B}_1$ and $\mathbf{E}_2, \mathbf{B}_2$ are the electric and magnetic fields at the back and front sides of the interface, respectively. Eqs. (2.85), (2.86) are Maxwell's equations in differential form. The first Maxwell's equation describes Faraday's law, while the second equation is the result of generalization of the Biot-Savart's law, which takes into account the conduction and displacement current. The third equation was derived from Coulomb's law, and it is based on the assumption that it is valid for the time-varying fields. Finally, the last equation follows from the Biot-Savart's law, and it implies the magnetic field does not have sources in a form of magnetic charges. Maxwell derived the system by proceeding from the experimental laws, and his main assumption was that the magnetic field is also caused by displacement currents. Each equation of this system describes some specific features of the field. However, only a combination of all four equations describes such fundamental phenomenon as the propagation of electromagnetic waves. It is useful to represent Maxwell's equations in integral form, which are valid everywhere in space, including regular points and interfaces. Applying Stokes and Gauss's theorems, we have from Eq. (2.85)

$$\oint_L \mathbf{E} \cdot d\mathbf{l} = -\oint_S \frac{\partial \mathbf{B}}{\partial t} \cdot d\mathbf{S}, \quad \oint_S \varepsilon \mathbf{E} \cdot d\mathbf{S} = e_0,$$

$$\oint_L \mathbf{B} \cdot d\mathbf{l} = \mu_0 \int_S (\mathbf{j}_c + \varepsilon\frac{\partial \mathbf{E}}{\partial t}) \cdot d\mathbf{S}, \quad \oint_S \mathbf{B} \cdot d\mathbf{S} = 0 \quad (2.87)$$

As follows from Eq. (2.86) tangential components of the electric field are continuous functions. At the same time, there are cases when it is convenient to introduce the presence of a double layer. Then, the tangential component of the electric field $\mathbf{E}(t)$ might have a discontinuity at the layer surface. The tangential component of the magnetic field also has a discontinuity in cases when the actual distribution of currents near interfaces is replaced by that of a surface current. Finally, due to the absence of magnetic charges, the normal component of the field $\mathbf{B}(t)$ is always a continuous function, while the discontinuity of the normal component of the electric field is defined by the density of surface charges.

2.5.3 Second Form of Maxwell's Equations

The equations of the set (Eq. 2.85), which characterize the divergence of the fields $\mathbf{E}$ and $\mathbf{B}$, can be derived from the first two equations of this system and the charge conservation principle. In fact, taking the divergence of both sides of the equations:

$$\operatorname{curl} \mathbf{E} = -\frac{\partial \mathbf{B}}{\partial t}, \quad \operatorname{curl} \mathbf{B} = \mu_0 \left(\mathbf{j}_c + \varepsilon \frac{\partial \mathbf{E}}{\partial t} \right), \qquad (2.88)$$

we obtain

$$\frac{\partial}{\partial t} \operatorname{div} \mathbf{B} = 0 \quad \text{and} \quad \frac{\partial}{\partial t}(-\delta_0 + \varepsilon \operatorname{div} \mathbf{E}) = 0$$

because

$$\operatorname{div} \mathbf{j}_c = -\frac{\partial \delta_0}{\partial t}$$

Therefore,

$$\operatorname{div} \mathbf{B} = C_1 \quad \text{and} \quad \operatorname{div} \mathbf{D} = \delta_0 + C_2$$

where C_1 and C_2 are independent of time. It is natural to assume that, at some time in the past, the fields $\mathbf{E}$ and $\mathbf{B}$, as well as charges, were absent; therefore, constants C_1 and C_2 should be equal to zero. Thus we again obtain the second pair of Maxwell's equations at regular points:

$$\operatorname{div} \varepsilon \mathbf{E} = \delta_0, \quad \operatorname{div} \mathbf{B} = 0 \qquad (2.89)$$

Next, let us show that the surface analog of Eq. (2.89) also follows from the first two equations of the set (Eq. 2.87). For simplicity assume that the

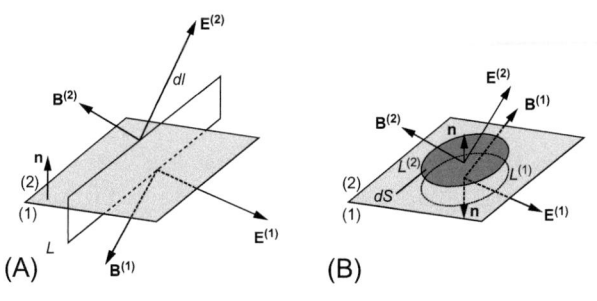

Fig. 2.4 (A) Illustration of the proof of the continuity of tangential components of the electric and magnetic fields across the interface. (B) Illustration of the continuity of the normal components of the magnetic field and current density across the interface.

surface density of currents is equal to zero. Then, applying these equations to any elementary contour (Fig. 2.4A) intersecting an interface, we can see that tangential components of the electric and magnetic fields are continuous functions. Next, we apply the following equation

$$\oint_L \mathbf{E} \cdot d\mathbf{l} = -\int_S \dot{\mathbf{B}} \cdot d\mathbf{S} = -\int_S \dot{B}_n dS$$

to both sides around elementary closed paths of the interface (Fig. 2.4B). Because the tangential component of the electric field is continuous across the interface, the left-hand side of this equation has the same value for both elementary paths. Therefore, we have

$$\dot{B}_{1n} dS - \dot{B}_{2n} dS = 0$$

from which follows continuity of the normal components of the field **B**. Finally, applying the equation

$$\oint_L \mathbf{B} \cdot d\mathbf{l} = \mu_0 \int_S \left(\mathbf{j}_c + \varepsilon \dot{\mathbf{E}} \right) \cdot d\mathbf{S}$$

to the same closed paths, we find that the normal component of the total current density consists of continuous functions:

$$j_{1nc} + \varepsilon_1 \dot{E}_{1n} = j_{2nc} + \varepsilon_2 \dot{E}_{2n}$$

where

$$j_{1nc} = \gamma_1 E_{1n}, \quad j_{2nc} = \gamma_2 E_{2n}$$

Integrating both sides of the last equality over time and taking into account that

$$j_{2nc} - j_{1nc} = -\dot{\sigma}_0$$

we again obtain the third equation of the set (Eq. 2.86). In other words, equations (2.89) and their integral form are valid for any time-varying electromagnetic field. This analysis allows us to represent the system of Maxwell's equations in a different form:

$$\operatorname{curl} \mathbf{E} = -\frac{\partial \mathbf{B}}{\partial t}, \quad \operatorname{curl} \mathbf{B} = \mu_0 \left(\mathbf{j}_c + \varepsilon \frac{\partial \mathbf{E}}{\partial t} \right)$$

and

$$\mathbf{n} \times (\mathbf{E}_2 - \mathbf{E}_1) = 0, \quad \mathbf{n} \times (\mathbf{B}_2 - \mathbf{B}_1) = 0, \quad \text{if } \mathbf{i}_c = 0 \qquad (2.90)$$

Also we can write down the integral form of these equations:

$$\oint_L \mathbf{E} \cdot d\mathbf{l} = -\int_S \frac{\partial \mathbf{B}}{\partial t} d\mathbf{S}, \quad \oint_L \mathbf{B} \cdot d\mathbf{l} = \int_S \left(\mathbf{j}_c + \varepsilon \frac{\partial \mathbf{E}}{\partial t} \right) \cdot d\mathbf{S} \qquad (2.91)$$

where $\mathbf{j}_c = \gamma \mathbf{E}$.

Let us emphasize again that, in deriving Maxwell's equations, we proceeded from the following physical laws:
1. Coulomb's law;
2. Biot-Savart's law;
3. Faraday's law;
4. Charge conservation principle;
5. Ohm's law; and
6. Maxwell's concept of displacement currents.

2.5.4 Maxwell's Equations in a Piecewise Uniform Medium

The theory of electromagnetic methods in geophysics is mainly based on the assumption that the Earth is a piecewise uniform medium. Then, as previously shown, the density of volume charges is equal to zero and in place of Eq. (2.85), we have at regular points

$$\begin{aligned}
&\text{I} \quad \operatorname{curl} \mathbf{E} = -\frac{\partial \mathbf{B}}{\partial t} & &\text{III} \quad \operatorname{div} \mathbf{E} = 0 \\
&\text{II} \quad \operatorname{curl} \mathbf{B} = \mu_0 \left(\mathbf{j}_c + \varepsilon \frac{\partial \mathbf{E}}{\partial t} \right) & &\text{IV} \quad \operatorname{div} \mathbf{B} = 0
\end{aligned} \qquad (2.92)$$

Here, let us make several comments:
1. As follows from the third equation of this set, the volume charges are absent, but they can be present at interfaces between media with different electric properties.
2. By definition, Eq. (2.92) represents the system of eight scalar partial differential equations of the first order with six unknown components of the electric and magnetic fields. In general, it is a complicated system, and it is difficult to identify important features of the field directly using this set. In fact, it is possible to reduce the system (Eq. 2.92) to a simpler system, which is the subject of the next section.

2.6 EQUATIONS FOR THE FIELDS E AND B

Now we replace Maxwell's equations by two equations that contain either the field **E** or the field **B**. Taking the curl of both sides of the first equation of the set (Eq. 2.92), we have

$$curl\ curl\ \mathbf{E} = -curl\ \dot{\mathbf{B}}$$

or

$$grad\ div\mathbf{E} - \nabla^2 \mathbf{E} = -\frac{\partial}{\partial t} curl\ \mathbf{B}$$

Making use of the second and third equations of the same set:

$$curl\ \mathbf{B} = \mu_0 (\mathbf{j}_c + \varepsilon \dot{\mathbf{E}}) \quad \text{and} \quad div\mathbf{E} = 0$$

we obtain

$$-\nabla^2 \mathbf{E} = -\frac{\partial}{\partial t}\left(\gamma \mu_0 \mathbf{E} + \varepsilon \mu_0 \frac{\partial \mathbf{E}}{\partial t}\right)$$

or

$$\nabla^2 \mathbf{E} - \gamma \mu_0 \frac{\partial \mathbf{E}}{\partial t} - \varepsilon \mu_0 \frac{\partial^2 \mathbf{E}}{\partial t^2} = 0$$

By analogy, taking the curl of the second of Maxwell's equations and using the first and fourth equations of the system (Eq. 2.92), we have

$$\text{curl curl } \mathbf{B} = \text{curl}\left(\gamma\mu_o \mathbf{E} + \varepsilon\mu_0 \frac{\partial \mathbf{E}}{\partial t}\right)$$

or

$$\text{grad div} \mathbf{B} - \nabla^2 \mathbf{B} = -\gamma\mu_0 \frac{\partial \mathbf{B}}{\partial t} - \varepsilon\mu_0 \frac{\partial^2 \mathbf{B}}{\partial t^2}$$

and

$$\nabla^2 \mathbf{B} - \gamma\mu_0 \frac{\partial \mathbf{B}}{\partial t} - \varepsilon\mu_0 \frac{\partial^2 \mathbf{B}}{\partial t^2} = 0$$

Thus, instead of the system of differential equations of the first order with respect to two fields, we have derived one differential equation of the second order for fields $\mathbf{E}$ and $\mathbf{B}$. These equations are valid at regular points of the conducting and polarizable medium:

$$\nabla^2 \mathbf{E} - \gamma\mu_0 \frac{\partial \mathbf{E}}{\partial t} - \varepsilon\mu_0 \frac{\partial^2 \mathbf{E}}{\partial t^2} = 0$$
$$\nabla^2 \mathbf{B} - \gamma\mu_0 \frac{\partial \mathbf{B}}{\partial t} - \varepsilon\mu_0 \frac{\partial^2 \mathbf{B}}{\partial t^2} = 0$$
(2.93)

Then, the electromagnetic fields can be described at regular points and at interfaces by groups of equations. For the electric field, we have

$$\nabla^2 \mathbf{E} - \gamma\mu_0 \frac{\partial \mathbf{E}}{\partial t} - \varepsilon\mu_0 \frac{\partial^2 \mathbf{E}}{\partial t^2} = 0$$

and (2.94)

$$\mathbf{n} \times (\mathbf{E}_2 - \mathbf{E}_1) = 0, \quad \gamma_1 E_{1n} + \varepsilon_1 \frac{\partial E_{1n}}{\partial t} = \gamma_2 E_{2n} + \varepsilon_2 \frac{\partial E_{2n}}{\partial t}$$

while for the magnetic field

$$\nabla^2 \mathbf{B} - \gamma\mu_0 \frac{\partial \mathbf{B}}{\partial t} - \varepsilon\mu_0 \frac{\partial^2 \mathbf{B}}{\partial t^2} = 0$$

and (2.95)

$$\mathbf{n} \times (\mathbf{B}_2 - \mathbf{B}_1) = 0, \quad \mathbf{n} \cdot (\mathbf{B}_2 - \mathbf{B}_1) = 0$$

Here, let us observe the following:
1. The electric and magnetic fields, defined from these equations, are interconnected because they obey the set (Eq. 2.92).
2. The differential equations for the fields $\mathbf{E}$ and $\mathbf{B}$ have a remarkable feature that is not obvious from the original set of Maxwell's equations: the individual equations for $\mathbf{E}$ and $\mathbf{B}$ discover two fundamental features

of all electromagnetic fields. Suppose that the last term in Eq. (2.93) is much greater than the second one; that is,

$$\varepsilon\mu_0 \frac{\partial^2 \mathbf{E}}{\partial t^2} \gg \gamma\mu_0 \frac{\partial \mathbf{E}}{\partial t} \quad \text{or} \quad \varepsilon\mu_0 \frac{\partial \mathbf{E}}{\partial t} \gg \gamma\mu_0 \mathbf{E}$$

and displacement currents greatly exceed the conduction currents:

$$\nabla^2 \mathbf{E} = \varepsilon\mu_0 \frac{\partial^2 \mathbf{E}}{\partial t^2} \quad \text{and} \quad \nabla^2 \mathbf{B} = \varepsilon\mu_0 \frac{\partial^2 \mathbf{B}}{\partial t^2} \tag{2.96}$$

These equations describe an important class of fields that propagate through a medium with the finite velocity

$$c = (\varepsilon\mu_0)^{-1/2}$$

Next, consider the opposite case when the conduction currents prevail; then, in place of Eq. (2.93), we obtain

$$\nabla^2 \mathbf{E} = \gamma\mu_0 \frac{\partial \mathbf{E}}{\partial t}, \quad \nabla^2 \mathbf{B} = \gamma\mu_0 \frac{\partial \mathbf{B}}{\partial t} \tag{2.97}$$

These two equations describe a process called "diffusion." Thus, in accordance with Eq. (2.93), the electromagnetic fields always display two fundamental features: propagation and diffusion. For instance, in a resistive medium the influence of diffusion may be insignificant, and mainly propagation is observed. By contrast, in a relatively conductive medium, the diffusion usually prevails, but propagation is always present. Later we will discuss this subject in detail.

2.7 ELECTROMAGNETIC POTENTIALS

Another useful approach in solving Maxwell's equations is based on the concept of vector potentials. In many cases, it is possible to describe the fields $\mathbf{E}$ and $\mathbf{B}$ with only two or even one component of the vector potential and, thus, greatly simplify the boundary value problem. To introduce potentials, we make use of two of Maxwell's equations:

$$div\,\mathbf{E} = 0 \quad \text{and} \quad div\,\mathbf{B} = 0. \tag{2.98}$$

When the divergence of a vector field is zero at regular points, the field can be represented as the curl of an auxiliary function. Thus,

$$\mathbf{E} = curl\,\mathbf{A}_m \quad \text{and} \quad \mathbf{B} = curl\,\mathbf{A}_e \tag{2.99}$$

where $\mathbf{A}_m$ and $\mathbf{A}_e$ are called the vector potentials of the magnetic and electric types, respectively. It is clear that an infinite number of vector potentials describe the same electromagnetic field. For instance, adding functions $grad\,\phi_m$ and $grad\,\phi_e$ to the vector potentials $\mathbf{A}_m$ and $\mathbf{A}_e$, new vector potentials

$$\mathbf{A}_{m2} = \mathbf{A}_{m1} + grad\,\phi_m \quad \text{and} \quad \mathbf{A}_{e2} = \mathbf{A}_{1e} + grad\,\phi_e$$

also describe the same field because

$$curl\,(grad\,\phi_m) = curl\,(grad\,\phi_e) = 0$$

Eq. (2.99) defines the vector potentials up to the gradient of some functions ϕ_m and ϕ_e, which are called scalar potentials of the electromagnetic field. This ambiguity in $\mathbf{A}_m$ and $\mathbf{A}_e$ can be used to our advantage in simplifying the equations. Let us start with the function $\mathbf{A}_m$. Substituting

$$\mathbf{E} = curl\,\mathbf{A}_m$$

into the second equation of Eq. (2.92), we have

$$curl\,\mathbf{B} = \gamma\mu_0\,curl\,\mathbf{A}_m + \varepsilon\mu_0\,curl\,\dot{\mathbf{A}}_m$$

or

$$curl\,(\mathbf{B} - \gamma\mu_0\mathbf{A}_m - \varepsilon\mu_0\dot{\mathbf{A}}_m) = 0$$

Whence

$$\mathbf{B} - \gamma\mu_0\mathbf{A} - \varepsilon\mu_0\dot{\mathbf{A}}_m = grad\,\phi_m. \tag{2.100}$$

Here ϕ_m is the scalar potential of the magnetic type and, as in the case of the vector potential, an infinite number of these functions describe the same electromagnetic field. Substituting expressions for the fields $\mathbf{E}$ and $\mathbf{B}$ in terms of potentials into the first Maxwell's equation, we obtain

$$curl\,curl\,\mathbf{A}_m = -\gamma\mu_0\dot{\mathbf{A}}_m - \varepsilon\mu_0\ddot{\mathbf{A}}_m - grad\,\dot{\phi}_m$$

or $\hspace{5cm}$ (2.101)

$$grad\,div\,\mathbf{A}_m - \nabla^2\mathbf{A}_m = -\gamma\mu_0\dot{\mathbf{A}}_m - \varepsilon\mu_0\ddot{\mathbf{A}}_m - grad\,\dot{\phi}_m,$$

where

$$\dot{\mathbf{A}}_m = \frac{\partial \mathbf{A}_m}{\partial t}, \quad \ddot{\mathbf{A}}_m = \frac{\partial^2 \mathbf{A}_m}{\partial t^2}, \quad \text{and} \quad \dot{\phi}_m = \frac{\partial \phi_m}{\partial t}$$

In Eq. (2.101) we can select the pair $\mathbf{A}_m$ and ϕ_m, which simplifies the system to the greatest extent, namely,

$$div\,\mathbf{A}_m = -\frac{\partial \phi_m}{\partial t} \qquad (2.102)$$

and we obtain for the vector potential $\mathbf{A}_m$ exactly the same equation as for the electromagnetic field:

$$\nabla^2 \mathbf{A}_m - \gamma\mu_0 \frac{\partial \mathbf{A}_m}{\partial t} - \varepsilon\mu_0 \frac{\partial^2 \mathbf{A}_m}{\partial t^2} = 0 \qquad (2.103)$$

Again, using the gage condition (Eq. 2.102), both fields $\mathbf{E}$ and $\mathbf{B}$ can be expressed in terms of the vector potential $\mathbf{A}_m$ only. In fact, from Eqs. (2.99)–(2.100) we have

$$\mathbf{E} = curl\,\mathbf{A}_m$$

and $\qquad (2.104)$

$$\dot{\mathbf{B}} = \gamma\mu_0 \dot{\mathbf{A}} + \varepsilon\mu_0 \ddot{\mathbf{A}}_m + grad\,div\,\mathbf{A}_m$$

Taking the divergence of both sides of Eq. (2.103) and integrating over time, we find that the scalar potential ϕ_m also satisfies the same equation as $\mathbf{A}_m$:

$$\nabla^2 \phi_m - \gamma\mu_0 \frac{\partial \phi_m}{\partial t} - \varepsilon\mu_0 \frac{\partial^2 \phi_m}{\partial t^2} = 0 \qquad (2.105)$$

Next we derive an equation for the vector potential of the electric type. Substituting the equation

$$\mathbf{B} = curl\,\mathbf{A}_e$$

into the first of Maxwell's equation, we obtain

$$curl\,\mathbf{E} = -curl\,\frac{\partial \mathbf{A}_e}{\partial t} \quad \text{or} \quad \mathbf{E} = -\frac{\partial \mathbf{A}_e}{\partial t} + grad\,\phi_e \qquad (2.106)$$

where ϕ_e is the scalar potential of the electric type. This equation suggests that the electric field is caused by a change of the magnetic field with time and electric charges. In other words, there are two parts of this field: the vortex and galvanic one. Replacing the fields $\mathbf{E}$ and $\mathbf{B}$ in the second equation of the set (Eq. 2.92), we have

$$\text{curl curl } \mathbf{A}_e = \mu_0 \left(-\gamma \frac{\partial \mathbf{A}_e}{\partial t} + \gamma \operatorname{grad} \phi_e - \varepsilon \frac{\partial^2 \mathbf{A}_e}{\partial t^2} + \varepsilon \operatorname{grad} \frac{\partial \phi_e}{\partial t} \right)$$

or

$$\operatorname{grad} \operatorname{div} \mathbf{A}_e - \nabla^2 \mathbf{A}_e = -\gamma \mu_0 \frac{\partial \mathbf{A}_e}{\partial t} - \varepsilon \mu_0 \frac{\partial^2 \mathbf{A}_e}{\partial t^2} + \operatorname{grad} \left(\gamma \mu_0 \phi_e + \varepsilon \mu_0 \frac{\partial \phi_e}{\partial t} \right)$$
(2.107)

Assuming that a pair of the vector and scalar potentials obeys the condition

$$\operatorname{div} \mathbf{A}_e = \gamma \mu_0 \phi_e + \varepsilon \mu_0 \frac{\partial \phi_e}{\partial t} \qquad (2.108)$$

we obtain for the vector potential $\mathbf{A}_e$ the same equation as that for the function $\mathbf{A}_m$:

$$\nabla^2 \mathbf{A}_e - \gamma \mu_0 \frac{\partial \mathbf{A}_e}{\partial t} - \varepsilon \mu_0 \frac{\partial^2 \mathbf{A}_e}{\partial t^2} = 0 \qquad (2.109)$$

In this case, the electromagnetic field cannot be expressed in terms of the vector potential only; thus, we have

$$\mathbf{B} = \operatorname{curl} \mathbf{A}_e, \quad \mathbf{E} = -\frac{\partial \mathbf{A}_e}{\partial t} + \operatorname{grad} \phi_e. \qquad (2.110)$$

At the same time, in the absence of electric charges the latter is greatly simplified:

$$\mathbf{E} = -\frac{\partial \mathbf{A}_e}{\partial t}$$

One should not be confused that equations for $\mathbf{A}_m$ (Eq. 2.103) and $\mathbf{A}_e$ (Eq. 2.109) are exactly the same. These equations do not describe the same fields because the corresponding boundary value problems apply different boundary (and initial) conditions leading to different solutions.

2.8 MAXWELL'S EQUATIONS FOR SINUSOIDAL FIELDS

Until now we have not made any assumptions about the dependence of the electromagnetic field on time. Let us examine important case of sinusoidal with time fields. This leads to significant simplifications. First, consider the scalar function

$$M = M_0 \sin(\omega t + \phi) \qquad (2.111)$$

where M_0 is the amplitude of the oscillation, ϕ is the initial phase, and ω is the angular frequency ($\omega = 2\pi f = 2\pi/T$) with T being the period of oscillation. Making use of Euler's formula,

$$e^{-i(\omega t + \phi)} = \cos(\omega t + \phi) - i\sin(\omega t + \phi)$$

we can present the right-hand side of Eq. (2.111) as the imaginary part of the exponential function:

$$M_0 \sin(\omega t + \phi) = -\operatorname{Im} M^* e^{-i\omega t} \qquad (2.112)$$

Here M^* is the complex amplitude given by

$$M^* = M_0 e^{-i\phi} \qquad (2.113)$$

Therefore, we have

$$M^* e^{-i\omega t} = M_0 e^{-i\phi} e^{-i\omega t} = M_0 e^{-i(\omega t + \phi)}$$

and

$$-\operatorname{Im} M^* e^{-i\omega t} = -\operatorname{Im}[M_0 \cos(\omega t + \phi) - iM_0 \sin(\omega t + \phi)] = M_0 \sin(\omega t + \phi)$$

Similarly, a cosine function can be presented by the real part of the complex function:

$$M_0 \cos(\omega t + \phi) = \operatorname{Re} M^* e^{-i\omega t}$$

where, as before, $M^* = M_0 e^{-i\phi}$. It is essential that the complex amplitude M^* is defined by the amplitude of oscillation M_0 and the initial phase ϕ:

$$|M^*| = M_0 \quad \text{and} \quad \operatorname{Arg} M^* = \phi$$

In other words, the complex amplitude contains all information about the corresponding sinusoidal function. Suppose that functions $M_0 \sin(\omega t + \phi)$ and $M_0 \cos(\omega t + \phi)$, describing any component of the electromagnetic field, are solutions of Maxwell's equations. Then, taking into account that these equations are linear, the sum of functions

$$M^* e^{-i\omega t} = M_0 \cos(\omega t + \phi) - iM_0 \sin(\omega t + \phi)$$

is also a solution of this system. Therefore, we can represent any component of the electric and magnetic fields as a complex quantity:

$$M^* e^{-i\omega t},$$

but, after solving the equations, only the imaginary or the real part of the solution should be considered. This form of a solution, $M^* \exp(-i\omega t)$, has one remarkable feature: namely, it is a product of two functions. One is the complex amplitude M^*, which depends on the geometry and the physical parameters of the medium, the position of an observation point and frequency. The second part, the function $\exp(-i\omega t)$, has a simple time dependence; after differentiation, it still remains an exponential function. This fact permits us to write equations in a form that does not contain the argument t. Because sinusoidal functions have infinite duration in time, there is no need to study the field at the initial moment when the electromagnetic fields arise. Now we generalize this result for the vector function. Suppose that

$$\mathbf{M} = M_{01} \sin(\omega t + \phi_1) \mathbf{1}_x + M_{02} \sin(\omega t + \phi_2) \mathbf{1}_y + M_{03} \sin(\omega t + \phi_3) \mathbf{1}_z$$

where $\mathbf{1}_x$, $\mathbf{1}_y$, and $\mathbf{1}_z$ are unit vectors along the coordinate axes. The latter can be rewritten as

$$\mathbf{M} = -\text{Im}\left[M_{01} e^{-i\phi_1} \mathbf{1}_x + M_{02} e^{-i\phi_2} \mathbf{1}_y + M_{03} e^{-i\phi_3} \mathbf{1}_z \right] e^{-i\omega t}$$

or (2.114)

$$\mathbf{M} = -\text{Im}\, \mathbf{M}^* e^{-i\omega t}$$

Here

$$\mathbf{M}^* = M_{01} e^{-i\phi_1} \mathbf{1}_x + M_{02} e^{-i\phi_2} \mathbf{1}_y + M_{03} e^{-i\phi_3} \mathbf{1}_z \qquad (2.115)$$

is the complex amplitude of the sinusoidal vector function $\mathbf{M}$, which is described by the complex vector. Then, representing the field in the form

$$\mathbf{E} = -\text{Im}\left(\mathbf{E}^* e^{-i\omega t}\right), \quad \mathbf{B} = -\text{Im}\left(\mathbf{B}^* e^{-i\omega t}\right)$$

and substituting these expressions into the first two Maxwell's equations, we obtain

$$\text{curl}\,\mathbf{E}^* = i\omega \mathbf{B}^*, \quad \text{curl}\,\mathbf{B}^* = \mu_0 (\gamma - i\omega\varepsilon) \mathbf{E}^* \qquad (2.116)$$

because

$$\frac{\partial}{\partial t} e^{-i\omega t} = -i\omega e^{-i\omega t}$$

The conditions at the interfaces for the complex amplitudes are the same as those for the field and

$$\mathbf{n} \times \left(\mathbf{E}_2^* - \mathbf{E}_1^*\right) = 0, \quad \mathbf{n} \times \left(\mathbf{B}_2^* - \mathbf{B}_1^*\right) = 0 \qquad (2.117)$$

Correspondingly, instead of Eq. (2.93) we have

$$\nabla^2 \mathbf{E}^* + k^2 \mathbf{E}^* = 0, \quad \nabla^2 \mathbf{B}^* + k^2 \mathbf{B}^* = 0 \qquad (2.118)$$

where

$$k^2 = i\gamma\mu_0\omega + \omega^2 \varepsilon\mu_0 \qquad (2.119)$$

The quantity k is called the wavenumber. It is obvious that complex amplitudes of potentials of the electromagnetic field also satisfy the same equations:

$$\nabla^2 \mathbf{A}_m^* + k^2 \mathbf{A}_m^* = 0, \quad \nabla^2 \mathbf{A}_e^* + k^2 \mathbf{A}_e^* = 0 \qquad (2.120)$$

The earlier equations allow us to determine only the complex amplitudes; to find vector potentials, we have to multiply these amplitudes by $\exp(-i\omega t)$ and take either the imaginary or real part of this product. This consideration shows at least two important merits of sinusoidal oscillations:

1. The system of Maxwell's equations for the complex amplitudes of the field, as well as Eqs. (2.118), (2.120), does not contain functions of time.
2. If parameters of medium are independent of time and the external field is a sinusoidal one, the electromagnetic field still remains a sinusoidal function of time of the same frequency.

This is an important feature of the sinusoidal field. In general, the primary and total fields might have been different on the sinusoidal dependence on time. Using Fourier's transform, the primary field (input) with arbitrary dependence on time can be represented as a combination of sinusoidal oscillations and, then, the field (output) is also described as a combination of sinusoids having different amplitudes and phases. Sinusoidal fields are of a great practical interest because they are used in induction logging.

2.9 ELECTROMAGNETIC ENERGY AND POYNTING VECTOR

2.9.1 Principle of Energy Conservation and Joule's Law

Until now we have focused on equations, describing the electric and magnetic fields as a function of space coordinates and time. It is also useful to describe the fields in terms of their energy. Suppose that a distribution of energy of an electromagnetic field is characterized by an energy density $u(p, t)$. Then the amount of energy inside some volume V is

$$W = \int_V u(p,t)dV \qquad (2.121)$$

The change of this energy is caused by several reasons. First is the presence of electromagnetic energy sources with density $P(p,t)$. Here, p is a point inside the volume. Second is a motion of charges (current) through a medium. To create the current, the electromagnetic field performs work on the charges; correspondingly, the electromagnetic energy decreases by some amount $Q(p,t)$, converting into heat. We already emphasized that the electromagnetic field could display propagation and diffusion phenomena.

The energy also moves through space, and there is the corresponding electromagnetic flux, causing the energy change. It is defined as

$$\oint_S \mathbf{Y} \cdot d\mathbf{S} \qquad (2.122)$$

Here, S is the surface surrounding the volume V. The scalar product $\mathbf{Y} \cdot d\mathbf{S}$ characterizes the flux of energy passing the surface dS during the unit of time. By definition, $\mathbf{Y}$ shows a direction of movement, and its magnitude is equal to the amount of energy passing during units of time through a unit area oriented perpendicular to the flux. From the principle of conservation of energy, we have

$$\frac{\partial W}{\partial t} = \int_V P(p,t)dV - \int_V Q(p,t)dV - \oint_S \mathbf{Y}(p,t) \cdot d\mathbf{S} \qquad (2.123)$$

or, making use of Gauss formula,

$$\oint_S \mathbf{Y} \cdot d\mathbf{S} = \int_V \mathrm{div}\,\mathbf{Y}\,dV$$

we obtain

$$\frac{\partial W}{\partial t} = \int_V P(p,t)dV - \int_V Q(p,t)dV - \int_V \mathrm{div}\,\mathbf{Y}\,dV. \qquad (2.124)$$

The last two terms have a negative sign because transformation into heat and the positive flux of energy cause a decrease of the energy in the volume V.

Next, using Ohm's law and Maxwell's equations, we express both sides of Eq. (2.124) in terms of the electromagnetic field. As pointed out earlier, the electromagnetic field causes a motion of charges (current), and the force acting on the moving charge δ in the unit volume is

$$\mathbf{F} = \delta(\mathbf{E}_t + \mathbf{v} \times \mathbf{B}) \tag{2.125}$$

so the work performed by this force per unit of time (the power) in a unit volume is

$$A = \delta(\mathbf{E}_t + \mathbf{v} \times \mathbf{B}) \cdot \mathbf{v} = \delta \mathbf{E}_t \cdot \mathbf{v} \tag{2.126}$$

because

$$(\mathbf{v} \times \mathbf{B}) \cdot \mathbf{v} = 0$$

Here $\mathbf{v}$ is charge velocity, $\mathbf{E}_t$ is the total electric field, comprised of fields caused by charges, a change of the magnetic field with time and field caused by external forces. By definition, δ is the charge density. Inasmuch as the product $\delta \mathbf{v}$ is equal to the current density $\mathbf{j}$, we have

$$A = \mathbf{E}_t \cdot \mathbf{j}$$

or (2.127)

$$A = \mathbf{j} \cdot \mathbf{E} + \mathbf{j} \cdot \mathbf{E}^{ext} = \mathbf{j} \cdot \mathbf{E} + P$$

Here $\mathbf{E}^{ext}$ is the external field caused by sources inside the elementary volume, and

$$\mathbf{j} \cdot \mathbf{E} = Q \tag{2.128}$$

is the work performed by the electromagnetic field in a conducting medium and converted into heat (Joule's law).

2.9.2 Energy Density and Poynting Vector

Taking into account (2.121) and (2.128), we have from Eq. (2.124)

$$\mathbf{E} \cdot \mathbf{j} = P - \frac{\partial u}{\partial t} - \nabla \cdot \mathbf{Y} \tag{2.129}$$

The latter describes a distribution of energy in the unit volume of a medium in the presence of external source and allows us to find formulas for the energy density u and vector $\mathbf{Y}$ in terms of the electric and magnetic fields. To proceed, we express the left-hand side in terms of the fields $\mathbf{E}$ and $\mathbf{B}$ only and at the beginning assume that an external source is absent. From the second Maxwell's equation, we have

$$\mathbf{E} \cdot \mathbf{j} = \frac{1}{\mu_0} \mathbf{E} \cdot (\nabla \times \mathbf{B}) - \varepsilon \mathbf{E} \cdot \frac{\partial \mathbf{E}}{\partial t} \tag{2.130}$$

As follows from vector analysis:

$$\nabla \cdot (\mathbf{E} \times \mathbf{B}) = \mathbf{B} \cdot (\nabla \times \mathbf{E}) - \mathbf{E} \cdot (\nabla \times \mathbf{B})$$

From Eq. (2.130), we have

$$\mathbf{E} \cdot \mathbf{j} = -\frac{1}{\mu_0} \nabla \cdot (\mathbf{E} \times \mathbf{B}) + \frac{1}{\mu_0} \mathbf{B} \cdot (\nabla \times \mathbf{E}) - \varepsilon \mathbf{E} \cdot \frac{\partial \mathbf{E}}{\partial t}$$

Applying the first Maxwell's equation, we obtain

$$\mathbf{E} \cdot \mathbf{j} = -\frac{1}{\mu_0} \nabla \cdot (\mathbf{E} \times \mathbf{B}) - \frac{1}{\mu_0} \mathbf{B} \cdot \frac{\partial \mathbf{B}}{\partial t} - \varepsilon \mathbf{E} \cdot \frac{\partial \mathbf{E}}{\partial t}$$

or (2.131)

$$\mathbf{E} \cdot \mathbf{j} = -\frac{1}{\mu_0} \nabla \cdot (\mathbf{E} \times \mathbf{B}) - \frac{\partial}{\partial t} \frac{1}{2} \left(\frac{1}{\mu_0} B^2 + \varepsilon E^2 \right)$$

Introducing notations for the energy (u) and flux density $\mathbf{Y}$,

$$u = \frac{1}{2} \left(\varepsilon E^2 + \frac{1}{\mu_0} B^2 \right) \quad \text{and} \quad (2.132)$$

$$\mathbf{Y} = \frac{1}{\mu_0} (\mathbf{E} \times \mathbf{B}) \quad (2.133)$$

we arrive at the conservation energy principle for the unit volume

$$\frac{\partial u}{\partial t} = -Q - \operatorname{div} \mathbf{Y} \quad (2.134)$$

or in more general case ($P \neq 0$):

$$\frac{\partial u}{\partial t} = P - Q - \operatorname{div} \mathbf{Y} \quad (2.135)$$

Performing integration over an arbitrary volume and using Gauss's formula, we obtain Eq. (2.123), which shows that the flux of energy through a closed surface is equal to

$$\oint \mathbf{Y} \cdot d\mathbf{S} \quad (2.136)$$

where $\mathbf{Y}$ is called the Poynting vector, describing the rate at which energy flows through a unit surface area perpendicular to the direction of wave propagation. The SI unit of the Poynting vector is the watt per square meter (W/m^2).

As follows from Eq. (2.123) when the energy does not change with time, we have

$$\int_V P \, dV = \int_V Q \, dV + \oint_S \mathbf{Y} \cdot d\mathbf{S} \qquad (2.137)$$

One part of the energy, created by the generators inside the volume, is transformed into heat, while the other part forms the flux, intersecting the surface S surrounding this volume. If the external force is absent, we have

$$\oint_S \mathbf{Y} \cdot d\mathbf{S} = -Q \qquad (2.138)$$

and electromagnetic energy arrives at the volume (the flux is negative), then it is converted into heat. Inasmuch as conversion of energy cannot take place without a propagation, even a static field is based on the propagation. To illustrate a movement of the energy and the usage of the Poynting vector, we consider two examples.

2.9.3 Current Circuit and Transmission Line

As shown in Fig. 2.5A, inside the internal part of the circuit, the electric field of the Coulomb's origin and current has opposite directions, while in the external part both vectors have the same direction. By definition, the Poynting vector is directed outside and inside within the internal and external parts of the circuit, respectively. Electromagnetic energy travels away from the internal part into the surrounding medium and then returns back into the circuit of the external part.

This description is rather approximate because the surface charges arise at the lateral surface of the contour and create the normal component of the

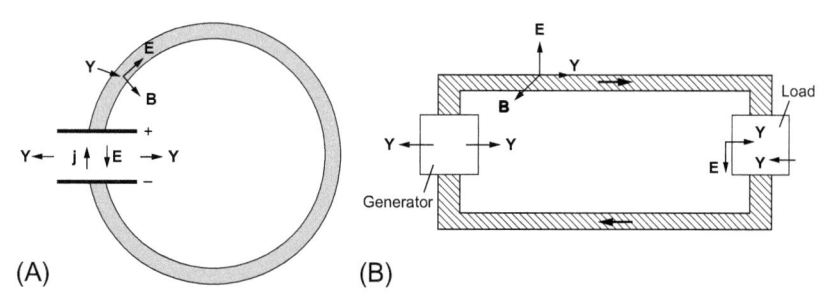

Fig. 2.5 (A) Flux energy around a current circuit. (B) Poynting vector in the vicinity of transmission line.

electric field. As a result, at these points of the surface the Poynting vector has both the tangential and normal components.

Suppose we have the system (Fig. 2.5B) consisting of three parts: the internal part where external forces produce the work that results in electromagnetic energy; the long transmission line; and the relatively resistive load.

As previously demonstrated, the electromagnetic energy leaves the internal part of the circuit and travels through the surrounding medium. Now we consider the field and Poynting vector in the vicinity of the transmission line and the load. Inasmuch as the line has low resistivity, the tangential component of the electric field **E** is small inside the line. In fact, from Ohm's law we have

$$E_t = \rho j$$

Due to continuity of the tangential component, the field **E** is also small on the external side of the conductor. At the same time surface charges create outside the line a normal component of the field E_n, which is much greater than the tangential component E_t: $E_n \gg E_t$. Then, as shown in Fig. 2.5B, the Poynting vector is practically tangential to the transmission line, and electromagnetic energy travels along this line, which plays the role of a guide, determining direction of movement of the energy toward the load; otherwise, the energy would travel in all directions from the external source. Due to the presence of the tangential component of the electric field, a small amount of the electromagnetic energy moves into the transmission line and converts into a heat. This is a pure loss, which reduces the amount of energy arriving to the load. Unlike the transmission line, the load is relatively resistive; correspondingly, the tangential component of the electric field prevails over the normal component, $E_t \gg E_n$. Therefore, the Poynting vector is mainly directed inward.

2.10 UNIQUENESS OF THE FORWARD PROBLEM SOLUTION

2.10.1 Uniqueness Theorem

The theory of induction logging is based on an analysis of the forward problems that allow one to establish a relationship between the field and the electric and geometric parameters of a medium. In general, Maxwell's equations have an infinite number of solutions. To determine the field uniquely, it is necessary to impose additional conditions. Unlike the static field, these

Case One

Suppose that at the instant $t = t_0$ the field is considered in some volume V, where all points are regular, and this volume is surrounded by the surface S. At each point, the fields obey the equations

$$\nabla^2 \mathbf{E} - \gamma \mu_0 \frac{\partial \mathbf{E}}{\partial t} - \varepsilon \mu_0 \frac{\partial^2 \mathbf{E}}{\partial t^2} = 0 \quad \text{and} \quad \nabla^2 \mathbf{B} - \gamma \mu_0 \frac{\partial \mathbf{B}}{\partial t} - \varepsilon \mu_0 \frac{\partial^2 \mathbf{E}}{\partial t^2} = 0$$

Also, at the initial instant ($t = t_0$) the electric and magnetic fields are known at every point of the volume:

$$\mathbf{E}(p, t_0) = \mathbf{N}(p), \quad \mathbf{B}(p, t_0) = \mathbf{M}(p) \qquad (2.139)$$

In addition, we have to formulate boundary conditions at the surface S. In the static case, Gauss's formula serves as the "bridge" between values of the field inside of the volume and the surface S. In our case, the conservation energy principle (Eq. 2.123) plays a similar role. In the beginning, it is assumed that the external sources of the field are absent: $P(p, t) = 0$. Consider two solutions of equations for both fields

$$\mathbf{E}_1, \mathbf{B}_1, \quad \mathbf{E}_2, \mathbf{B}_2$$

which, at the initial moment, have the same values inside the volume. Taking into account linearity, the differences of these solutions

$$\mathbf{E}_3 = \mathbf{E}_2 - \mathbf{E}_1 \quad \text{and} \quad \mathbf{B}_3 = \mathbf{B}_2 - \mathbf{B}_1 \qquad (2.140)$$

also obey the same equations, while the initial condition has the form

$$\mathbf{E}_3(p, t_0) = 0 \quad \text{and} \quad \mathbf{B}_3(p, t_0) = 0 \qquad (2.141)$$

To establish the boundary conditions, we apply the principle of energy conservation in the following form:

$$\frac{\partial}{\partial t} \int_V \left(\frac{\varepsilon}{2} \mathbf{E}_3^2 + \frac{\mathbf{B}_3^2}{2\mu_0} \right) dV = -\int_V \rho \mathbf{j}_0^2 dV - \frac{1}{\mu_0} \oint_S (\mathbf{E}_3 \times \mathbf{B}_3) \cdot d\mathbf{S} \qquad (2.142)$$

Suppose that either tangential components of $\mathbf{E}_1$ and $\mathbf{E}_2$ or tangential components of $\mathbf{B}_1$ and $\mathbf{B}_2$ coincide with each other at points of the surface S for $t > t_0$. For instance, in the case of the electric field, this corresponds to the tangential component of $\mathbf{E}_3$ equal to zero:

$$\mathbf{n} \times \mathbf{E}_3 = 0 \qquad (2.143)$$

where **n** is the unit vector normal to the surface S. Because

$$\oint_S (\mathbf{E}_3 \times \mathbf{B}_3) \cdot d\mathbf{S} = \oint_S (\mathbf{E}_3 \times \mathbf{B}_3)\mathbf{n} dS = \oint_S (\mathbf{n} \times \mathbf{E}_3) \cdot \mathbf{B}_3 dS$$

the surface integral in Eq. (2.142) vanishes, and we have

$$\frac{\partial}{\partial t} \int_V \left(\frac{\varepsilon}{2} \mathbf{E}_3^2 + \frac{1}{2\mu_0} \mathbf{B}_3^2 \right) dV = -\int_V \rho \mathbf{j}_3^2 dV \qquad (2.144)$$

Here $\mathbf{j}_3$ is the vector of current density caused by the field $\mathbf{E}_3$. The right-hand side of Eq. (2.144) can be either negative or equal to zero. At the same time, the left-hand side is either equal to zero or positive. In general, the derivative of energy with respect to time can be either positive or negative. However, in our case, when at the initial moment the energy is zero (Eq. 2.141), the amount of energy must either remain zero or become positive when $t \geq t_0$; otherwise, the energy would be negative. Thus the equality (2.144) holds only when both right and left sides are equal to zero, and for $t > t_0$ we have

$$\mathbf{E}_3 = \mathbf{E}_2 - \mathbf{E}_1 = 0, \quad \mathbf{B}_3 = \mathbf{B}_2 - \mathbf{B}_1 = 0$$

Therefore, electromagnetic field within the volume V is uniquely defined for $t > t_0$ by the initial values of the electric and magnetic fields inside the volume V and by the tangential component of either the electric or magnetic field at the surface S, surrounding V.

1. Proof of the uniqueness theorem remains the same if the volume V is surrounded by several surfaces.
2. If the surface S tends to infinity, we can assume that a medium has a finite conductivity; due to conversion of energy into heat, the surface integral in Eq. (2.142) still tends to zero.
3. We have assumed that external forces are absent inside the volume V. At the same time, if fields $\mathbf{E}_1, \mathbf{B}_1$ and $\mathbf{E}_2, \mathbf{B}_2$ are caused by the same sources, the initial and boundary conditions remain sufficient to provide uniqueness of the solution.

Case Two

Consider a more complicated case when inside the volume there is an interface S_{12} separating media with different electrical properties. Inasmuch as the differential equations for the field do not apply at the interface, we surround

S_{12} by the surface S^*. Correspondingly, at the right-hand side of Eq. (2.142), we have an additional integral:

$$-\frac{1}{\mu_0}\oint_{S^*}(\mathbf{E}_3 \times \mathbf{B}_3)\cdot d\mathbf{S}$$

In approaching S^* to S_{12} we integrate at both sides of the interfaces:

$$\frac{1}{\mu_0}\oint_{S_{12}}\left[(\mathbf{E}'_3 \times \mathbf{B}'_3) - (\mathbf{E}''_3 \times \mathbf{B}''_3)\right]d\mathbf{S} \qquad (2.145)$$

where $\mathbf{E}_3', \mathbf{B}_3'$ and $\mathbf{E}_3'', \mathbf{B}_3''$ are the electric and magnetic fields at the back and front sides of the interface, respectively. In accordance with Maxwell's equations, the tangential components of the electric field $\mathbf{E}$ and magnetic field $\mathbf{B}$ are continuous functions at the interface. Thus the integrand in Eq. (2.145) can be represented as

$$\left[(\mathbf{E}'_3 - \mathbf{E}''_3) \times \mathbf{B}_3\right]\cdot d\mathbf{S} = \left(E'_{3n} - E''_{3n}\right)(\mathbf{n}\times\mathbf{B}_3)\cdot\mathbf{n}dS$$
$$= \left(E'_{3n} - E''\right)(\mathbf{n}\times\mathbf{n})\mathbf{B}_3 dS = 0$$

Therefore, the integral over the interface vanishes if the solutions at the back and front sides of the interface satisfy Maxwell's equations.

2.10.2 Formulation of the Boundary Value Problem

Let us summarize conditions that uniquely define the electromagnetic field in a general boundary value problem:
1. At regular points of the volume V, the field should obey equations

$$\nabla^2\mathbf{E} - \gamma\mu_0\frac{\partial\mathbf{E}}{\partial t} - \varepsilon\mu_0\frac{\partial^2\mathbf{E}}{\partial t^2} = 0, \quad \nabla^2\mathbf{B} - \gamma\mu_0\frac{\partial\mathbf{B}}{\partial t} - \varepsilon\mu_0\frac{\partial^2\mathbf{E}}{\partial t^2} = 0$$

2. At the initial moment $t = t_0$, the field

$$\mathbf{E}(p, t_0) \text{ and } \mathbf{B}(p, t_0)$$

should be given at each regular point of the volume.
3. At the surface S, surrounding the volume V, either the tangential component of the electric or magnetic field

$$\mathbf{n}\times\mathbf{E} \text{ or } \mathbf{n}\times\mathbf{B}$$

should be given at all instances $t \geq t_0$.

4. At interfaces inside the volume, a solution should obey the surface analog of Maxwell's equations.

As we have shown, these conditions uniquely define electromagnetic fields. Because the field also might be expressed in terms of the vector potentials, the corresponding boundary value problem can be formulated in terms of these functions as well.

REFERENCE
[1] Kaufman AA, Anderson BI. Principles of electric methods in surface and borehole geophysics. Amsterdam: Elsevier; 2010.

FURTHER READING
[1] Alpin LM. Field theory. Moscow: Nedra; 1966.
[2] Bursian VR. Theory of electromagnetic fields applied in electrical methods. Moscow: Nedra; 1972.

CHAPTER THREE

Propagation of Electromagnetic Field in a Nonconducting Medium

Contents

3.1 Plane Wave in a Uniform Medium	91
3.1.1 Solution to the Wave Equation	92
3.1.2 Velocity of Propagation of Plane Wave	93
3.1.3 Propagation of the Plane Wave	95
3.1.4 Note on the Plane Wave Model	97
3.2 Quasistationary Field in a Nonconducting Medium	104
3.3 Induction Current in a Thin Conducting Ring Placed in a Time-Varying Field	111
3.3.1 Equation of Induced Current in the Conductive Ring	111
3.3.2 Transient Responses of Induced Current	113
3.3.3 A Step-Function Varying Primary Magnetic Field	115
3.3.4 Sinusoidal Primary Magnetic Field	117
3.3.5 Two Inductively Connected Rings Excited by an External Source	122
3.3.6 Notes on Measurements of Induced Electric and Magnetic Fields	128
Further Reading	131

Proceeding from Maxwell equations, we study the main features of propagating electromagnetic fields by considering several simple and educational examples. Special attention is paid to the quasistationary approximation, describing fields of conductive objects in nonconductive environment.

3.1 PLANE WAVE IN A UNIFORM MEDIUM

Suppose that a nonconducting medium with parameters ε and μ_0 is uniform and that the electric and magnetic fields depend on the z coordinate only; that is, the field is constant on any plane perpendicular to the z axis. Also assume that the electric field has a single vector component along the x axis:

$$\mathbf{E} = E_{0x} e(z, t) \mathbf{i}_x \tag{3.1}$$

where E_{0x} is a constant, $e(z, t)$ is a function that depends on the coordinate z and time t, and $\mathbf{i}_x$ is the unit vector directed along the x axis. Because the field is independent of the x and y coordinates, the first equation of the set (Eq. 2.93) in a noncoducting medium is greatly simplified, and we have

$$\frac{\partial^2 e}{\partial z^2} - \epsilon\mu_0 \frac{\partial^2 e}{\partial t^2} = 0 \tag{3.2}$$

3.1.1 Solution to the Wave Equation

Eq. (3.2) is the well-known partial differential equation of the second order describing wave propagation. Applying the trial and error method, D'Alembert found its solution in the following form:

$$e(z, t) = Af\left[a\left(t - \sqrt{\epsilon\mu_0}z\right)\right] + Bg\left[a\left(t + \sqrt{\epsilon\mu_0}z\right)\right] \tag{3.3}$$

Here A and B are some constants, and f and g are functions having the first and second derivatives. The constant a must have dimensions s^{-1}, because the argument of any function should be dimensionless. It is a simple matter to show that function $e(z, t)$ obeys (Eq. 3.2). In fact, introducing the variable

$$u = a\left(t - \sqrt{\epsilon\mu_0}z\right)$$

we have for derivatives of the function $f(z, t)$:

$$\frac{\partial f}{\partial z} = \frac{\partial f}{\partial u}\frac{\partial u}{\partial z} = -a\sqrt{\epsilon\mu_0}f'_u$$

where f'_u is the first derivative with respect to the argument u. Therefore,

$$\frac{\partial^2 f}{\partial z^2} = a^2 \epsilon\mu_0 f''_{uu} \tag{3.4}$$

Also,

$$\frac{\partial f}{\partial t} = \frac{\partial f}{\partial u}\frac{\partial u}{\partial t} = af'_u \quad \text{and} \quad \frac{\partial^2 f}{\partial t^2} = a^2 f''_{uu} \tag{3.5}$$

The last two equations show that the function

$$f\left[a\left(t - \sqrt{\epsilon\mu_0}z\right)\right]$$

satisfies Eq. (3.2). Of course, this is also true for the function

$$g\left[a\left(t + \sqrt{\epsilon\mu_0}z\right)\right]$$

This clearly indicates that any function can be a solution of Eq. (3.2), provided the arguments z and t are comprised in the following combination u:

$$u = a\left(t \pm \sqrt{\varepsilon\mu_0}\,z\right) \tag{3.6}$$

It is useful to recognize that, if the argument of a function has the slightly different form

$$b\left(z \pm t/\sqrt{\varepsilon\mu_0}\right)$$

then it is also a solution of Eq. (3.2). D'Alembert found such a relationship between arguments z and t that a corresponding function obeys Eq. (3.2), and this is the essence of his solution.

Next let us assume that the electric field is described by only the function

$$f\left[a\left(t - \sqrt{\varepsilon\mu_0}\right)z\right]$$

Thus

$$\mathbf{E} = E_{0x} f\left\{a\left[t - (\varepsilon\mu)^{1/2} z\right]\right\} \mathbf{i}_x \tag{3.7}$$

and consider its physical meaning.

3.1.2 Velocity of Propagation of Plane Wave

Analyzing the argument u (Eq. 3.6), we may notice that:
1. At any point with coordinate z the field $\mathbf{E}$, in general, changes with time, while at any given instant t, it can have different values at different coordinates z.
2. The electric field $\mathbf{E}$ has the same value at different points and different time if the argument

$$u = a\left(t - \sqrt{\varepsilon\mu_0}\,z\right)$$

remains the same. As follows from the definition of this argument, with an increase of the distance z, the same value of the field is observed at greater times. Imagine a system of parallel planes, $z = \text{const}$, which are perpendicular to the z axis and think of these planes as the surfaces where the field $\mathbf{E}$ has the same value as the time changes (Fig. 3.1A). Each of these planes corresponds to some instant of time, and the relationship between z and t is

$$t_2 - t_1 = \sqrt{\varepsilon\mu_0}\,(z_2 - z_1), \quad t_3 - t_1 = \sqrt{\varepsilon\mu_0}\,(z_3 - z_1)$$

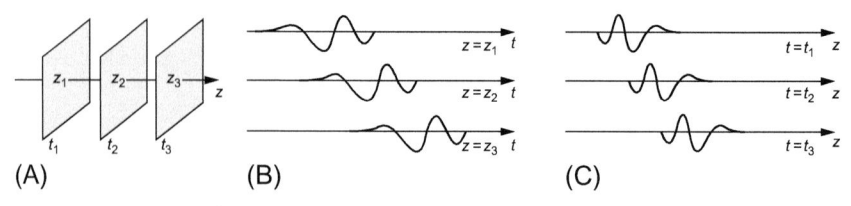

Fig. 3.1 (A) Motion of wave. (B) Wave at different distances. (C) Change of wave with time.

and

$$t_n - t_i = \sqrt{\varepsilon\mu_0}(z_n - z_i)$$

because

$$u_i = t_i - \sqrt{\varepsilon\mu_0}\, z_i = \text{constant}$$

We can interpret this infinite series of parallel planes as a movement of only one plane, characterized by the same argument, with velocity

$$v = \frac{1}{\sqrt{\varepsilon\mu_0}} = \frac{c}{\varepsilon_r^{1/2}} \tag{3.8}$$

where c is the speed of light

$$c = 3 \times 10^8 \text{ m/s}$$

In particular, in a free space $v = c$, but, for example, in water

$$v \approx 0.33 \times 10^8 \text{ m/s}$$

In other words, we observe a motion of the field **E** along the z axis, and this phenomenon is called propagation or wave motion or, even simpler, a wave. For this reason, Eq. (3.2) is called the wave equation. It is proper to remind that, at all points of each plane $z = \text{const}$, the electric field has the same value independent of coordinates x and y, and it is natural that such motion of the field is called the plane wave. For illustration, the distribution of the wave as a function of t and z is given in Fig. 3.1, which represents the wave distribution along the z axis, Fig. 3.1, and the change of the field with time, Fig. 3.1C, respectively.

3.1.3 Propagation of the Plane Wave

Now consider the magnetic field that accompanies the field **E**. To proceed we represent Maxwell's equations (assuming that conduction currents are absent) in Cartesian coordinates. Then, using the expression for *curl*, we obtain

$$\begin{vmatrix} \mathbf{i}_x & \mathbf{i}_y & \mathbf{i}_z \\ \frac{\partial}{\partial x} & \frac{\partial}{\partial y} & \frac{\partial}{\partial z} \\ E_x & 0 & 0 \end{vmatrix} = -\frac{\partial \mathbf{B}}{\partial t}, \quad \begin{vmatrix} \mathbf{i}_x & \mathbf{i}_y & \mathbf{i}_z \\ \frac{\partial}{\partial x} & \frac{\partial}{\partial y} & \frac{\partial}{\partial z} \\ B_x & B_y & B_z \end{vmatrix} = \varepsilon \mu_0 \frac{\partial \mathbf{E}}{\partial t}$$

Equating corresponding components of the fields from both sides of these equations, we have

$$\frac{\partial B_x}{\partial t} = 0, \quad \frac{\partial B_y}{\partial t} = -\frac{\partial E_x}{\partial z}, \quad \frac{\partial B_z}{\partial t} = 0, \\ \frac{\partial B_y}{\partial z} = -\varepsilon \mu_0 \frac{\partial E_x}{\partial t}, \quad \frac{\partial B_x}{\partial z} = 0 \quad (3.9)$$

In deriving these equations, we used the fact that the electric field has only the component E_x, and both the electric and magnetic fields are independent of the x and y coordinates.

The equation for the magnetic field component B_y derived in Chapter 2 directly follows from Eq. (3.9), which gives

$$\frac{\partial^2 B_y}{\partial z^2} - \varepsilon \mu_0 \frac{\partial^2 B_y}{\partial t^2} = 0 \quad (3.10)$$

Substituting Eq. (3.7) into Eq. (3.9) and taking into account Eq. (3.8), we obtain

$$\frac{\partial B_y}{\partial t} = \frac{1}{v} E_{ox} a f''_u[a(t-z/v)] \quad \text{and} \quad \frac{\partial B_y}{\partial z} = -\varepsilon \mu_0 E_{ox} a f''_u[a(t-z/v)]$$

It is obvious that the function

$$B_y(z, t) = B_{0y} f[a(t - z/v)] \quad (3.11)$$

satisfies both equations of the set (Eq. 3.9), provided that

$$B_{0y} = \sqrt{\varepsilon \mu_0} E_{0x} \quad \text{or} \quad E_{0x} = v B_{0y} \quad (3.12)$$

Thus, we have demonstrated that the electromagnetic field propagates along the z axis with the velocity v, and it is described by two vectors:

$$\mathbf{E}_x(z,t) = E_{0x}e(z,t)\mathbf{i}_x, \quad \mathbf{B}_y = B_{0y}b(z,t)\mathbf{i}_y \quad (3.13)$$

where the coefficients E_{0x} and B_{0y} are related to each other by Eq. (3.12) and

$$e = b = f[a(t - z/v)] \quad (3.14)$$

From Eq. (3.13), we may note that the electric and magnetic fields are perpendicular to each other and the direction of propagation. Such an electromagnetic field is called the transverse plane wave. By definition, the Poynting vector, representing the directional energy flux density, is

$$\mathbf{Y} = \frac{1}{\mu_0}(\mathbf{E} \times \mathbf{B})$$

Taking into account (Eqs. 3.12–3.14), we have

$$\mathbf{Y}(z,t) = \left(\frac{\varepsilon}{\mu_0}\right)^{1/2} E_{0x}^2 e^2 \mathbf{i}_z \quad (3.15)$$

which shows the direction of the wave motion. As follows from Eq. (3.9), the electric and magnetic fields support each other at every point of space. In fact, when the magnetic field changes with time it generates an electric field (Faraday's law),

$$\frac{\partial E_x}{\partial z} = -\frac{\partial B_y}{\partial t}$$

while a variation of the field $\mathbf{E}$ with time (displacement currents) creates a magnetic field:

$$\frac{\partial B_y}{\partial z} = -\varepsilon\mu_0 \frac{\partial E_x}{\partial t}$$

Supporting each other, the magnetic and electric fields form an electromagnetic wave that propagates through a nonconducting medium with velocity v. Note that these two generators of the field,

$$\frac{\partial \mathbf{B}}{\partial t} \quad \text{and} \quad \varepsilon \frac{\partial \mathbf{E}}{\partial t}$$

are vital in forming electromagnetic waves: if one of them is disregarded the effect of propagation disappears. For instance, suppose that displacement currents do not have any influence on the field. From the mathematical point of view, this corresponds to the case of the dielectric permittivity ε equal to zero. Therefore, in accordance with Eq. (3.8), the velocity of propagation becomes infinitely large, and, respectively, the fields **E** and **B** arrive instantly at all points of a medium. This contradicts the concept of propagation of the field with a finite velocity. Also, in Chapter 2 we described the behavior of displacement currents inside a capacitor, assuming that the change of the magnetic fields of these currents with time is negligible. In other words, the analysis did not take into account the inductive electric field. Correspondingly, the inductive effect vanishes and the field **E** between the capacitor plates behaves as a Coulomb's field.

3.1.4 Note on the Plane Wave Model

In many cases, it is convenient to use a model of a plane wave; however, this model is not quite realistic, because existence of plane wave requires an infinitely large energy. We may think of such source as two plates of infinite dimension, located at the plane XOY, with charges of equal magnitude but opposite sign uniformly distributed over the plates. At some instant $t=0$ one of the plates starts to move in the x direction, forming a current with the surface density $\mathbf{i}_x$, which is independent of the x and y coordinates (Fig. 3.2A):

$$i_x = \begin{cases} 0 & t<0 \\ I_0 \psi(at) & t \geq 0 \end{cases} \quad (3.16)$$

Here $\psi(at)$ is an arbitrary function of time. The current Eq. (3.16) causes a magnetic field, arising in the vicinity of the plane $z=0$. Applying the integral form of the second Maxwell's equation to the elementary path around this surface current and located in the plane perpendicular to the x axis, we obtain

$$B_{2y} - B_{1y} = -\mu_0 i_x, \quad \text{if } z=0$$

where B_{2y} and B_{1y} are components of the magnetic field at the front and back sides of the plate, respectively. Because the magnetic field B_y is an odd function with respect to the coordinate z, we have

$$B_{2y} = -B_{1y} = -\frac{\mu_0 i_x}{2}, \quad \text{if } z=0 \quad (3.17)$$

Thus, in accordance with Eq. (3.16), the magnetic field at the front side of the current plate behaves as

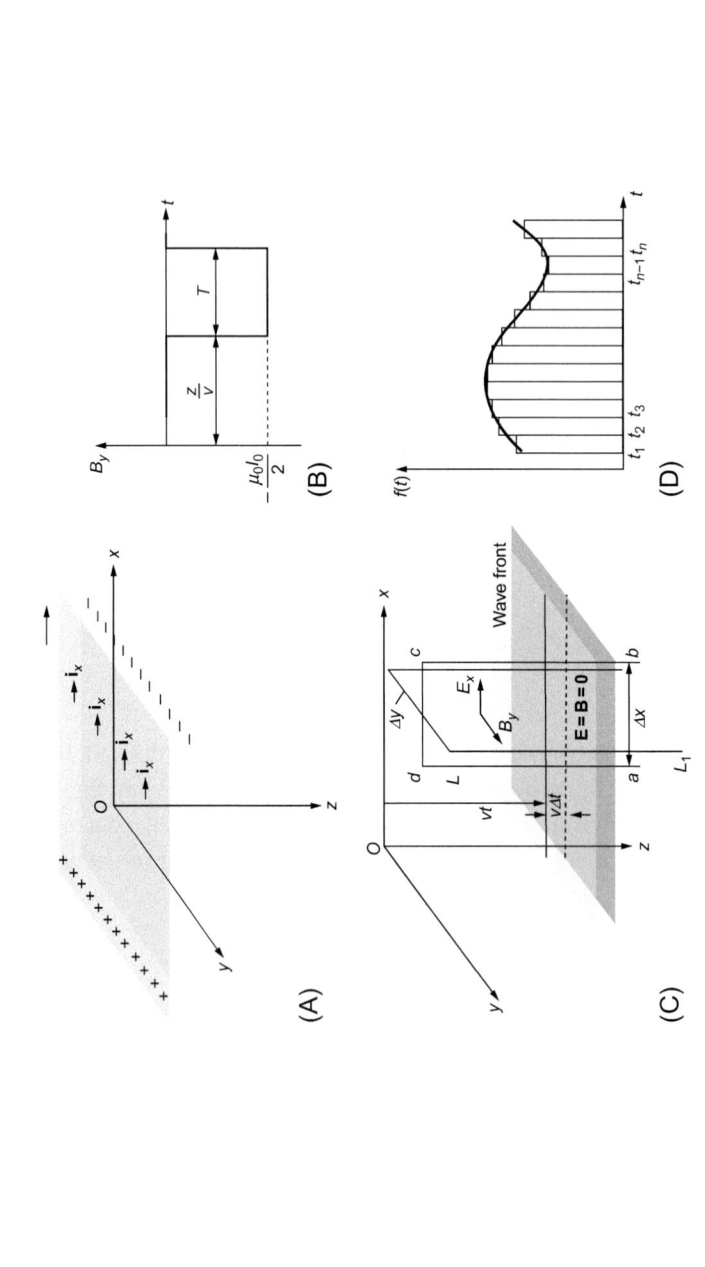

Fig. 3.2 (A) Moving plate with surface current density i_x. (B) Propagation of impulse. (C) Illustration of relation between fields **E** and **B**. (D) Wave propagation in the case of the current source being arbitrary function of time.

$$B_y(0, t) = \begin{cases} 0 & t < 0 \\ -\dfrac{\mu_0 I_0}{2}\Psi(at) & t \geq 0 \end{cases} \qquad (3.18)$$

At the same time, as follows from Eq. (3.11), the dependence of the field on the coordinate z and time t has to be defined by the same function for all points. It is also essential (Eq. 3.18) that this function is equal to zero if the argument is negative. Therefore, we can represent the magnetic field regardless of the distance from the current plate, in the form

$$B_y(z, t) = \begin{cases} 0 & z > \nu t \\ -\dfrac{\mu_0 I_0}{2}\Psi[a(t - z/\nu)] & z \leq \nu t \end{cases} \qquad (3.19)$$

Thus, in the vicinity of the plate the magnetic field varies almost synchronously with the current density of the source, but at some distance z the field is observed with time delay z/v.

In accordance with Eq. (3.12), the electric field is

$$E_x(z, t) = \begin{cases} 0 & z > \nu t \\ -\dfrac{I_0}{2}\left(\dfrac{\mu_0}{\varepsilon}\right)^{1/2}\Psi[a(t - z/\nu)] & z \leq \nu t \end{cases} \qquad (3.20)$$

Let us notice that the coefficient $(\mu_0/\varepsilon)^{1/2}$ can be presented as

$$\sqrt{\dfrac{\mu_0}{\varepsilon}} = \dfrac{120\pi}{(\varepsilon_r)^{1/2}}\ \text{Ohm} \qquad (3.21)$$

It is an appropriate to notice:
1. In deriving Eq. (3.17) we took into account that the flux of displacement currents through an infinitely small area surrounded by the elementary path is equal to zero. This fact allowed us to establish a relation between the magnetic field in the vicinity of the source and its current.
2. In essence, Eq. (3.17) is a boundary condition near the source; later, it will be derived in a similar manner for more complicated cases.
3. The electric and magnetic fields of a plane wave have the same dependence on distance and time. Such behavior is an idealization, which is not observed in the case of real sources.
4. Electromagnetic field of the plane wave is described by the same function, regardless of the distance from the current source. This is another idealization, which does not take place in a realistic electromagnetic field.

5. The argument $a(t-z/v)$ is called the phase of the wave, while the planes on which the phase has a constant value are called phase surfaces. Thus, the propagation of the plane wave can be treated as a movement of the phase surface with the velocity v, and the Poynting vector being perpendicular to this plane. To illustrate the effect of propagation we consider several examples.

Example One

Suppose that the current in the plate differs from zero only during some time interval T:

$$i_x = \begin{cases} I_0 \Psi(at) & 0 \leq t \leq T \\ 0 & t < 0,\ t > T \end{cases}$$

Then at the point of observation, located at the distance z, the field exists when

$$\frac{z}{v} \leq t \leq \frac{z}{v} + T$$

Hence if the distance z is such that

$$T < \frac{z}{v}$$

the field will be observed at times after the conduction current on the plate has stopped. This fact vividly demonstrates that the electromagnetic field propagates due to the interaction of the electric and magnetic fields; in this sense, the wave does not require the current source to remain active.

Example Two

Now consider propagation of a square waveform shown in Fig. 3.2B. In accordance with (3.19), we have

$$B_y(z, t) = \begin{cases} 0 & t < \frac{z}{v} \\ -\dfrac{\mu_0 I_0}{2} & \dfrac{z}{v} \leq t \leq \dfrac{z}{v} + T \\ 0 & t \geq \dfrac{z}{v} + T \end{cases} \quad (3.22)$$

and

$$E_x = v B_y$$

Taking into account the simplicity of the form (Eq. 3.22), let us study the relationship between the electric and magnetic fields proceeding directly from the integral form of Maxwell's equations. We can imagine a closed rectangular contour L, situated in any plane that is parallel to the plane XOZ (Fig. 3.2C) and assume that the wave front is located somewhere inside the contour, while the back side of the plane wave has not yet reached its side $cd = \Delta x$. Then, at any moment, t, the flux of the magnetic field Φ, intersecting the contour is equal to

$$\Phi(t) = z^* \Delta x B_y$$

where $z^* \Delta x = vt\Delta x$ characterizes the area of the loop where the field B is not zero. Inasmuch as the electromagnetic field moves along the z axis, this area as well as the flux increases. In particular, at the instant $t + \Delta t$ the flux is

$$\Phi(t + \Delta t) = \Delta x (z^* + v\Delta t) B_y$$

Therefore,

$$\frac{d\Phi}{dt} = B_y v \Delta x \qquad (3.23)$$

and, in accordance with Faraday's law

$$\oint_L \mathbf{E} \cdot d\mathbf{l} = -\frac{d\Phi}{dt}$$

an electromotive force appears in the contour. Integrals along paths, which are parallel to the z axis, vanish because the dot product of two perpendicular vectors

$$\mathbf{E} = E_x \mathbf{i}_x, \quad \mathbf{dl} = dl\mathbf{i}_z$$

is zero. At the same time, the integral along the path ab is also zero because the field has not yet arrived at this side of the contour. Respectively, the electromotive force is defined by the voltage along the path cd and is equal to $-E_x \Delta x$; that is,

$$\oint_L \mathbf{E} \cdot d\mathbf{l} = -E_x \Delta x \qquad (3.24)$$

Therefore, due to a movement of the magnetic field an inductive electric field arises and, as follows from Eqs. (3.23), (3.24)

$$E_x = vB_y$$

Next we apply the second of Maxwell's equation,

$$\oint \mathbf{B} \cdot d\mathbf{l} = \varepsilon \mu_0 \int \dot{\mathbf{E}} \cdot d\mathbf{S}$$

to the rectangular contour L_1 located in the plane YOZ. The same approach as before shows that the rate of change of the flux of displacement currents through an area enclosed by the contour L_1 is

$$\varepsilon \mu_0 E_x v \Delta y$$

At the same time, the circulation of the magnetic field along L_1 is

$$B_y \Delta y$$

Thus, displacement currents generate a field $\mathbf{B}$, which is equal to

$$B_y = \varepsilon \mu_0 v E_x = \frac{1}{v} E_x$$

that, of course, coincides with the relationship between the vectors $\mathbf{E}$ and $\mathbf{B}$ derived before.

Example Three

Now consider a more general case when the current in the source $I_0 \psi(at)$ is an arbitrary function of time. This function can be represented as a system of rectangular pulses with different magnitudes arising at different times (Fig. 3.2D). With decrease of the width of each pulse, this approximation becomes more accurate. Therefore, at the instant t, the observed field is caused by the current impulse, which appears earlier at the instant

$$t_1 = t - z/v$$

For example, if the current in the source I_0 remains constant when $t > 0$, then the time-invariant field is observed at any point with coordinate z, provided $z < vt$. This occurs because the current pulses are identical, and they follow each other continuously. In other words, the front and back of neighbor pulses arise at the same time. Because they are characterized by opposite directions, we observe the constant field. Thus, the time-invariant electric and magnetic fields arise due to wave propagation.

Example Four

Another example of the function $\psi(z,t)$ is a sinusoidal oscillation that also can be treated as a system of pulses with different magnitudes and signs. In accordance with Eqs. (3.19), (3.20), if

$$i_x = I_0 \sin \omega t, \quad \text{if} \quad t \geq 0$$

Then

$$B_y(z,t) = \begin{cases} -\dfrac{\mu_0 I_0}{2} \sin \omega(t - z/v) & z < vt \\ 0 & z > vt \end{cases}$$

$$E_x(z,t) = \begin{cases} -\dfrac{I_0}{2} \left(\dfrac{\mu_0}{\varepsilon}\right)^{1/2} \sin \omega(t - z/v) & z < vt \\ 0 & z > vt \end{cases} \quad (3.25)$$

Taking into account that $\omega = 2\pi/T$, we can rewrite Eq. (3.25):

$$B_y = -\dfrac{\mu_0 I_0}{2} \sin(\omega t - \phi) \quad \text{and} \quad E_x = v B_y$$

Here,

$$\phi = \dfrac{2\pi z}{\lambda} \quad \text{and} \quad \lambda = vT \quad (3.26)$$

The parameter λ is called the wavelength, and it characterizes the distance passed by every elementary pulse during one period. The quantity ϕ is the phase shift between the electromagnetic field and function describing the current source $i_x(0,t)$, and it is defined by the distance from this source expressed in units of the wavelength λ. It is not easy to visualize the propagation of sinusoidal waves, or any periodical function, because there is no front or back of the wave. At the same time, the wave nature of the field can be established considering the behavior of the phase at different times and distances from the source. In a nonconducting medium the field is generated by two types of vortices at regular points:

$$\dfrac{\partial \mathbf{B}}{\partial t}, \varepsilon \dfrac{\partial \mathbf{E}}{\partial t}$$

However, when the field has a discontinuity, there is also a surface distribution of vortices. For instance, if the wave is represented by the rectangular pulse, the vortices are located at the front and back sides of the pulse. The simplest example of a plane wave in a uniform and nonconducting medium can be treated as an introduction to the propagation of a wave through a conducting medium. Now we describe one important approximation for the field that is widely used in the induction logging.

3.2 QUASISTATIONARY FIELD IN A NONCONDUCTING MEDIUM

An electromagnetic field arising somewhere in a space cannot reach any place instantly but rather always requires some time that is defined by two parameters, namely, the distance between the two points and the velocity of propagation. This phenomenon occurs in any medium, regardless of its conductivity and dielectric permittivity, and both electromagnetic induction and displacement currents are vital for field propagation. For instance, letting the parameter ε equal zero, that is, neglecting the displacement currents,

$$\mathbf{j}_d = \varepsilon \frac{\partial \mathbf{E}}{\partial t} \to 0$$

we arrive at an infinite velocity of propagation of the electromagnetic field. Of course, in reality, there is no propagation at an infinite velocity. Quite opposite, it always has a finite value even though it is very large; that is, the propagation effect without exception takes place. However, there are conditions when, with a given accuracy of measurements, it is practically impossible to observe the wave phenomenon. In such cases, the field is called the quasistationary one. We first study quasistationary fields in a nonconducting medium; later, the influence of conductivity will be investigated in detail. Suppose that the field is caused by conduction currents, and they are distributed uniformly in the plane as shown in Fig. 3.2A. Then we have for the magnetic field,

$$B_y(z, t) = B_0 \psi[a(t - z/v)] \tag{3.27}$$

where $\psi(0, t)$ is defined by conduction currents of the source, and the ratio z/v characterizes the time needed for the field to travel from the source to an observation point. Disregarding displacement currents and assuming that

$$z/v = 0$$

we come to conclusion that at every point the magnetic field and the current in the source vary synchronously. From a mathematical point of view, this corresponds to the propagation with infinite velocity, and this fact can be interpreted in the following way. Suppose that in some area the field is observed at an instant t, which is much greater than the delay time z/v:

$$t \gg z/v \qquad (3.28)$$

Thus, we can say that both the current in the source and the quasistationary field **B** are practically described by the same argument of the function $\psi(at)$. From the physical point of view, this is an indication that the magnetic field is, in fact, a quasistationary one, and it obeys the Biot-Savart law. It is also useful to represent the inequality (Eq. 3.28) in a form that corresponds to a sinusoidal electromagnetic field. Multiplying both sides of this relationship by the frequency ω, we obtain

$$\omega t \gg \frac{\omega z}{v} = \frac{2\pi z}{\lambda} \qquad (3.29)$$

Thus, the field caused by the sinusoidal current source is quasistationary if the distance between this source and the observation point is much smaller than the wavelength λ:

$$z/\lambda \ll 1 \qquad (3.30)$$

Now we consider several examples that illustrate the behavior of the quasistationary field in a nonconducting medium.

Example One: Inductive Electric Field of the Solenoid

Suppose that a magnetic field arises due to an alternating current flowing in an infinitely long cylindrical solenoid, as shown in Fig. 3.3A. In the quasistationary approximation the magnetic field satisfies Biot-Savart law and coincides in phase with the current flowing in the solenoid. Using results of the Chapter 2, we can say that, inside the solenoid, the field is uniform and directed along its axis, while outside the field **B** vanishes. Because the magnetic field changes with time, an inductive electric field arises. Taking into account that both vectors **B** and $\partial \mathbf{B}/\partial t$ have the vertical component only, the electrical field is tangential to the horizontal planes (Fig. 3.3A). Moreover, due to axial symmetry the vector lines of **E** are circles with centers located on the solenoid axis.

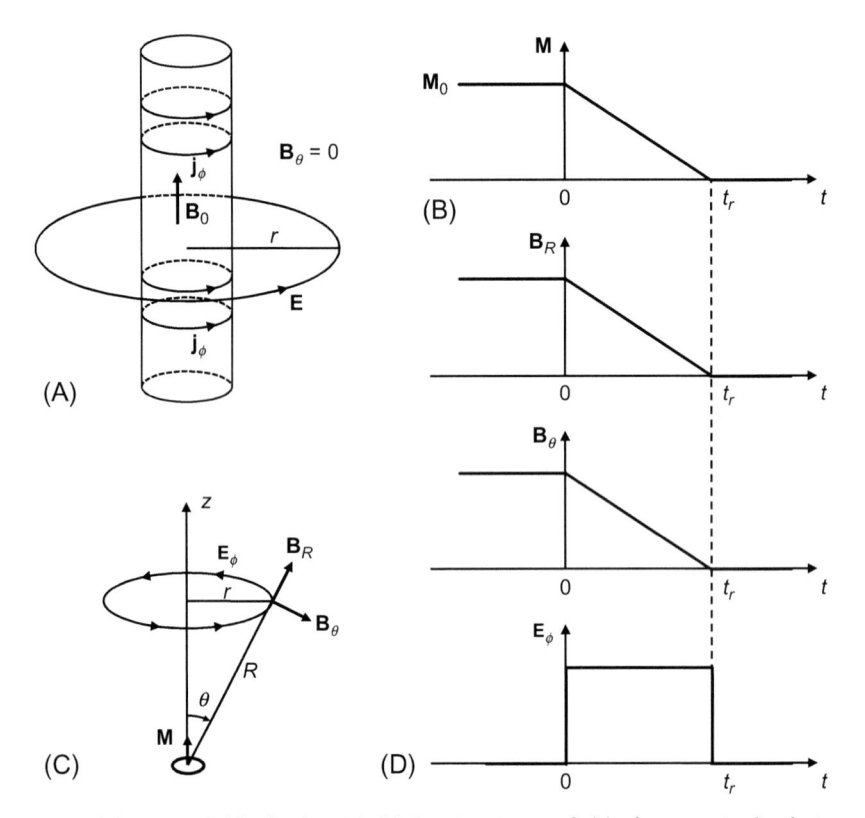

Fig. 3.3 (A) Vortex field of solenoid. (B) Quasistationary field of magnetic dipole in a nonconducting medium. (C) Time-variable dipole moment. (D) Time-variable magnetic and electric field.

Therefore, the electric field has only component E_ϕ, which is a function of distance r. Making use of Faraday's law,

$$\Xi = -\frac{\partial \Phi}{\partial t}$$

as well as the axial symmetry, for any circle located in a horizontal plane, we obtain

$$\oint \mathbf{E} \cdot d\mathbf{l} = 2\pi r E_\phi = -\frac{\partial \Phi}{\partial t}$$

or

$$E_\phi = -\frac{1}{2\pi r}\frac{\partial \Phi}{\partial t} \qquad (3.31)$$

where $\partial \Phi/\partial t$ is the rate of change of the magnetic flux within the area bounded by the circle with radius r. Bearing in mind that the magnetic field inside the solenoid is uniform:

$$B = B_0 f(t)$$

we have for the electric field inside the solenoid

$$E_\phi^i = -\frac{\pi r^2}{2\pi r} B_0 f'(t) = -\frac{B_0 r}{2} f'(t) \quad \text{if } r \le a \qquad (3.32)$$

where a is the radius of the solenoid. Thus, the electric field inside the solenoid increases linearly with the distance from solenoid axis. For all horizontal circles with radii r exceeding the solenoid radius a the flux Φ, as well as the derivative $\partial \Phi/\partial t$, remains the same at any given instant of time, and it is equal to

$$\Phi = \pi a^2 B_0 f(t) \quad \text{and} \quad \dot{\Phi} = B_0 \pi a^2 f'(t)$$

Therefore, the voltage (electromotive force) along any of these circles does not change with further increase of their radius and, in accordance with Eq. (3.31), we have

$$E_\phi^e = -\frac{B_0}{2\pi r}\pi a^2 f'(t) = -\frac{B_0 a^2}{2r} f'(t), \quad \text{if } r \ge a \qquad (3.33)$$

As follows from this equation, the vortex electric field outside the solenoid is inversely proportional to the radius r. This example vividly demonstrates a case when a vortex electric field in the quasistationary approximation is nonzero at points where the magnetic field is absent. In reality, due to a change of the electric field with time, displacement currents arise everywhere, and they also generate the magnetic field. In our approximation, this effect is negligible, but it provides a propagation of the field no matter how small the rate of a change of the current in the solenoid is. Taking into account that, outside the solenoid,

$$\text{curl } \mathbf{E} = 0$$

the inductive electric field can be expressed in terms of the potential.

Example Two: The Quasistationary Field of a Magnetic Dipole in a Nonconducting Medium

Consider a magnetic dipole with the moment **M**(*t*) directed along the *z* axis and situated at the origin of a spherical system of coordinates (Fig. 3.3B). We again assume that, regardless of the distance, the magnetic field at any instant *t* is defined by the magnitude of the dipole current at the same moment (quasistationary approximation). Then, making use of the expressions for the magnetic dipole with steady current (chapter one), we obtain

$$B_R(t) = \frac{2\mu_0 M(t)}{4\pi R^3} \cos\theta, \quad B_\theta(t) = \frac{\mu_0 M(t)}{4\pi R^3} \sin\theta, \quad B_\phi = 0 \quad (3.34)$$

The magnetic field is located in longitudinal planes of the spherical system of coordinates, and it possesses the axial symmetry. In this case, as follows from Maxwell's equations, the inductive electric field arising due to a change of the field **B** with time has only a single component $E_\phi(t)$. Therefore, vector lines of the electric field are circles, and their centers are located at the *z* axis. We can write

$$E_\phi = -\frac{1}{2\pi r}\dot{\Phi} \quad (3.35)$$

where Φ is the flux piercing the area bounded by a circle with radius *r* (Fig. 3.3B). Taking into account that vector d**S** is parallel to the *z* axis, we have the following expression for the flux Φ:

$$\Phi = \int_S \mathbf{B} \cdot d\mathbf{S} = 2\pi \int_0^r r B_z dr \quad (3.36)$$

where $dS = 2\pi r dr$ and B_z is the vertical component of the magnetic field. As is shown in Fig. 3.3B,

$$B_z(t) = B_R \cos\theta - B_\theta \sin\theta$$

and, considering Eq. (3.34), we obtain

$$B_z(t) = \frac{\mu_0 M(t)}{4\pi R^3}(3\cos^2\theta - 1) \quad (3.37)$$

Substituting this expression into Eq. (3.36) and integrating, we have

$$\dot{\Phi} = \frac{d\Phi}{dt} = \frac{\mu_0}{2}\frac{\dot{M}}{R^3}r^2 \qquad (3.38)$$

where

$$R = (r^2 + z^2)^{1/2} \quad \text{and} \quad \dot{M} = dM/dt$$

Therefore, the vortex electric field is

$$E_\phi = -\frac{\dot{M}(t)}{4\pi R^2}\sin\theta \qquad (3.39)$$

Thus, in the quasistationary approximation, when the instantaneous magnitude of the dipole moment defines the magnetic field at the same instant, the expressions for the electromagnetic fields are

$$B_R(t) = \frac{2\mu_0 M(t)}{4\pi R^3}\cos\theta, \quad B_\theta(t) = \frac{\mu_0 M(t)}{4\pi R^3}\sin\theta, \quad E_\phi = -\frac{\dot{M}(t)}{4\pi R^2}\sin\theta \qquad (3.40)$$

It should be expected that the electric field is zero on the z axis ($\theta = 0, \pi$), because the flux through a surface bounded by a circle of radius r tends to zero when the radius decreases. At the same time, as the radius increases, the magnetic vector lines begin to intersect the surface twice. In other words, the component B_z could have an opposite sign at different points of the surface. For this reason, if r is sufficiently large, the flux Φ gradually decreases in spite of the unlimited increase of the surface. Thus, the flux Φ as a function of r has a maximum whose position depends on the coordinate z and with its increase, the maximum is observed at greater distances from the z axis. As follows from Eq. (3.40), at every point of a medium, the magnetic field is accompanied by an inductive electric field. If a medium is conductive, this electric field gives rise to a current. The field described by Eq. (3.40) is generated by the current of the magnetic dipole only, and is called the primary electromagnetic field. Now let us consider this field when the dipole moment varies with time in a relatively simple manner.

Case One

Suppose that the current in the dipole changes as a sinusoidal function, that is,

$$M = M_0 \sin\omega t \qquad (3.41)$$

where M_0 is the moment amplitude and $\omega = 2\pi f$ is the angular frequency with $T = 1/f$ being the period of oscillations.

Then, in accordance with Eqs. (3.40), (3.41), for the quasistationary field, we have

$$B_R(t) = \frac{2\mu_0 M_0}{4\pi R^3}\cos\theta\sin\omega t, \quad B_\theta(t) = \frac{\mu_0 M_0}{4\pi R^3}\sin\theta\sin\omega t,$$
$$E_\phi(t) = \frac{\omega\mu_0 M_0}{4\pi R^2}\sin\theta\sin(\omega t - \pi/2) \tag{3.42}$$

Thus, one can say that the primary electric field exhibits a phase shift of 90 degrees with respect to the current flowing in the dipole or to the primary magnetic field. Eq. (3.42) is useful for understanding of induction logging utilizing a magnetic dipole as the primary source.

Case Two

Next assume that the dipole moment varies with time, as shown in Fig. 3.3C:

$$M(t) = \begin{cases} M_0 & \text{if } t \leq 0 \\ M_0 - at, & \text{if } 0 \leq t \leq t_r \\ 0 & \text{if } t \geq t_r \end{cases} \tag{3.43}$$

where $a = M_0/t_r$. As follows from Eq. (3.40), the primary magnetic field is constant if $t < 0$, then it decreases linearly within the interval $0 < t < t_r$ and equals zero when $t > t_r$. Respectively, the primary electric field of vortex origin exists only within the time interval where the magnetic field changes ($0 \leq t \leq t_r$), and in view of its linear dependence on time, the electric field is constant. Thus, we have

$$B_R = \frac{2\mu_0 M_0}{4\pi R^3}\cos\theta, \quad B_\theta = \frac{\mu_0 M_0}{4\pi R^3}\sin\theta, \quad E_\phi = 0, \quad \text{if } t \leq 0$$
$$B_R(t) = \frac{2\mu_0 M(t)}{4\pi R^3}\cos\theta, \quad B_\theta(t) = \frac{\mu_0 M(t)}{4\pi R^3}\sin\theta, \quad E_\phi(t) = \frac{\mu_0 M_0}{4\pi R^2 t_r}\sin\theta, \quad \text{if } 0 < t \leq t_r$$
$$B_R = B_\theta = 0, \quad E_\phi = 0, \quad \text{if } t > t_r$$
$$\tag{3.44}$$

The curves shown in Fig. 3.3D illustrate the behavior of the magnetic and electric fields as functions of time. Of course, per our considerations we do not take into account the propagation of the field, and in this approximation the electric field exists only within the time interval where the dipole moment changes with time.

3.3 INDUCTION CURRENT IN A THIN CONDUCTING RING PLACED IN A TIME-VARYING FIELD

3.3.1 Equation of Induced Current in the Conductive Ring

Consider an example that will be later used for explanation of the skin effect in a conducting medium. Assume that the quasistationary approximation is accurate enough and that a thin conducting ring with a radius r is placed into the primary field $\mathbf{B}_0$ (Fig. 3.4A). The appearance of currents in a conducting ring can be described as follows. The time-varying primary magnetic field is accompanied by the inductive electric field. For simplicity, we assume that this electric field has a simple component $E_{0\phi}$ only, which is tangential to the ring surface. This field is the primary cause of the conduction current in the ring. In turn, these currents generate a secondary electromagnetic field. The induced current in the ring depends on both the primary and secondary electric fields. According to Ohm's law, we have

$$j_\phi = \gamma \left(E_{0\phi} + E_{s\phi} \right) \qquad (3.45)$$

where j_ϕ is the current density, γ is the ring conductivity, and $E_{0\phi}$ and $E_{s\phi}$ are the primary and secondary electric fields, respectively. To determine the current in the ring, we use Faraday's law

$$\Xi = -\frac{d\Phi}{dt} \qquad (3.46)$$

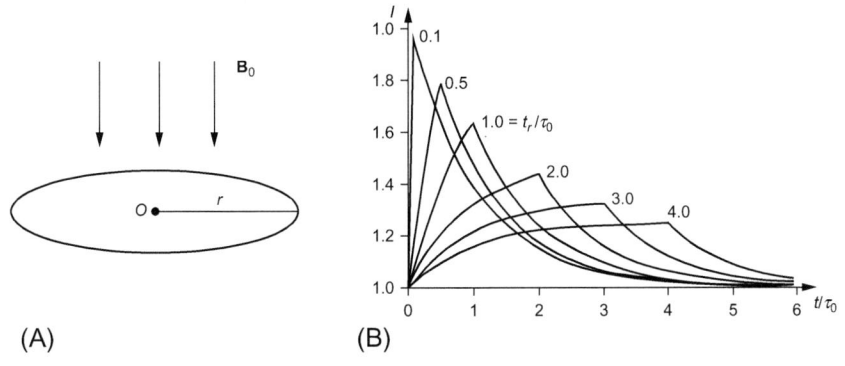

Fig. 3.4 (A) Conducting ring in magnetic field. (B) Transient responses of the current. Index of curves is t_r/τ_0.

The flux Φ through the area bounded by the ring is

$$\Phi = \Phi_0 + \Phi_s \qquad (3.47)$$

Here Φ_0 is the flux of the primary magnetic field caused by a current source, while Φ_s is the flux of the magnetic field generated by the induction current in the ring. Correspondingly, Eq. (3.46) can be written as

$$\Xi = -\frac{d\Phi_0}{dt} - \frac{d\Phi_s}{dt}. \qquad (3.48)$$

In this equation, only the term $d\Phi_0/dt$ is known, whereas the electromotive force Ξ and the rate of a change of the secondary flux $d\Phi_s/dt$ are unknown. Our objective is to determine the current I flowing in the ring and express both unknowns in terms of this function. Applying Ohm's law we have

$$\Xi = IR \qquad (3.49)$$

where R is the ring resistance given by

$$R = \rho \frac{l}{S}, \quad \text{if } r \gg a_0 \qquad (3.50)$$

Here, ρ is the resistivity of the ring, l is its circumference, and the area of the ring cross-section is $S = \pi a_0^2$, where a_0 is the radius of the ring cross-section. As follows from the Biot-Savart law, the magnetic flux Φ_s caused by the ring's current is directly proportional to I, and it can be represented as

$$\Phi_s = LI \qquad (3.51)$$

Here, L is a coefficient of proportionality known as the inductance of the ring. According to Eq. (3.51), the ring inductance is the ratio of the secondary magnetic flux through the ring and the current creating the flux:

$$L = \frac{\Phi_s}{I}$$

In other words, numerically the inductance is the flux caused by the unit current. It is defined only by the geometrical parameters of the ring. In general, determination of inductance involves rather complicated calculations based on the Biot-Savart law. But, in a special case of a thin circular ring, the self-inductance is defined by the well-known formula:

$$L = \mu_0 r \left(\ln \frac{8r}{a_0} - 1.75 \right) \tag{3.52}$$

Inductance is measured in henrys in SI units. If we have a coil (solenoid) with n rings per unit length, the inductance increases as the square of number of turns n of the solenoid:

$$L = \mu_0 n^2 r \left(\ln \frac{8r}{a_0} - 1.75 \right) \tag{3.53}$$

Thus, the simple form of the conductor and the assumption about uniform distribution of the current density over the cross-section of the ring, have allowed us to find the coefficient of proportionality between the secondary flux Φ_s and the induced current in the ring. Substituting Eqs. (3.49), (3.51) into Eq. (3.48), we arrive at a differential equation for the current I:

$$L \frac{dI}{dt} + RI = -\frac{d\Phi_0}{dt} \quad \text{or} \quad \frac{dI}{dt} + \frac{1}{\tau_0} I = f(t) \tag{3.54}$$

here,

$$\tau_0 = \frac{L}{R} \quad \text{and} \quad f(t) = -\frac{1}{L} \frac{d\Phi_0}{dt} \tag{3.55}$$

are given. The solution to this ordinary differential equation of the first order is

$$I(t) = I_0 \exp(-t/\tau_0) - \exp(-t/\tau_0) \frac{1}{L} \int_0^t \exp(x/\tau_0) \frac{d\Phi_0(x)}{dx} dx \tag{3.56}$$

where I_0 is the current at the instant $t = 0$. Now we study the behavior of induced currents in two cases.

3.3.2 Transient Responses of Induced Current

First, suppose that the primary magnetic field varies with time in a similar way, as shown in Fig. 3.3B:

$$\frac{d\Phi_0}{dt} = \begin{cases} 0 & \text{if } t < 0 \\ -\dfrac{\Phi_0}{t_r} & \text{if } t \leq 0 < t_r \\ 0 & \text{if } t \geq t_r \end{cases} \tag{3.57}$$

During the time interval ($t < 0$), there are no induced currents in the ring; that is,

$$I(t) = 0, \quad \text{if } t < 0$$

Within the ramp time, the primary flux Φ_0 varies linearly with time; therefore, an induced current arises. Its magnitude is defined by the rate of change of the primary magnetic field as well as two parameters of the ring R and L. When the primary field disappears ($t > t_r$), the behavior of the induced current is controlled by the time constant τ_0 only. In fact, in this time range Eq. (3.54) is simplified, and we have

$$\frac{dI}{dt} + \frac{1}{\tau_0} I = 0, \quad \text{if } t \geq \tau_r \tag{3.58}$$

and its solution is

$$I(t) = C \exp(-t/\tau_0), \quad \text{if } t \geq \tau_r. \tag{3.59}$$

In order to determine the constant C, we look at the behavior of the induced currents during the ramp time. In accordance with Eqs. (3.56), (3.57), we obtain

$$I(t) = I_0 \exp(-t/\tau_0) + \frac{\tau_0 \Phi_0}{t_r L}[1 - \exp(-t/\tau_0)], \quad \text{if } 0 \leq t \leq t_r \tag{3.60}$$

Because the induced current is absent at the instant $t = 0$, that is $I_0 = 0$, we have

$$I(t) = \frac{\tau_0 \Phi_0}{t_r L}[1 - \exp(-t/\tau_0)], \quad \text{if } 0 \leq t \leq t_r \tag{3.61}$$

The constant C is readily found from Eqs. (3.59), (3.61). In fact, letting $t = t_r$ in both equations, we obtain

$$I(t_r) = C \exp(-t_r/\tau_0) = \frac{\tau_0 \Phi_0}{t_r L}[1 - \exp(-t_r/\tau_0)]$$

Thus

$$C = \frac{\tau_0 \Phi_0}{t_r L}[\exp(t_r/\tau_0) - 1] \tag{3.62}$$

Correspondingly, the expressions describing the induced current in the ring are

$$I(t) = \begin{cases} 0 & t < 0 \\ \dfrac{\tau_0 \Phi_0}{t_r L}[1 - \exp(-t/\tau_0)] & t \leq 0 < t_r \\ \dfrac{\tau_0 \Phi_0}{t_r L}[\exp(t_r/\tau_0) - 1]\exp(-t/\tau_0) & t \geq t_r \end{cases} \quad (3.63)$$

As follows from Eq. (3.63), the induced current gradually increases during the ramp time, reaches a maximum at the moment $t = t_r$, and then decreases exponentially. Suppose that the ramp time t_r is much less than the time constant τ_0: $t_r \ll \tau_0$. Then, expanding the exponential terms in Eq. (3.63) in power series and discarding all terms but those of the first and second order, we obtain

$$I(t) = \begin{cases} 0 & t < 0 \\ \dfrac{t \Phi_0}{t_r L} & t \leq 0 < t_r \\ \dfrac{\Phi_0}{L} \exp(-t/\tau_0) & t \geq t_r \end{cases} \quad (3.64)$$

In this case the induced current increases linearly during the ramp time, and outside this range ($t \geq t_r$) the current magnitude is independent on the parameter t_r. It is obvious that the magnetic field caused by this current has the same features. As will be shown later, a similar behavior is observed in a more general case of induced currents in volume conductors. In the opposite case of $t_r \gg \tau_0$, the current I increases linearly at the beginning ($t \ll t_r$) and then slowly approaches a maximum equal to

$$\frac{\tau_0 \Phi_0}{t_r L} \ll \frac{\Phi_0}{L}, \quad \text{if } t = t_r$$

Of course, at late time the current decays exponentially. Curves, illustrating the behavior of induced current at different parameters t_r/τ_0, are shown in Fig. 3.4B.

3.3.3 A Step-Function Varying Primary Magnetic Field

When the primary magnetic flux changes as a step function the current in the ring I is described by the last equation of the set (Eq. 3.64) when the ramp time approaches zero. Thus, we have

$$I(t) = \frac{\Phi_0}{L}\exp(-t/\tau_0), \quad \text{if } t > 0 \text{ and } t_r \to 0 \quad (3.65)$$

and the initial value of the induced current does not depend on the ring resistance but rather is determined by the primary flux Φ_0 and the inductance L. Inasmuch as under real conditions there is always a nonzero ramp time, the initial value of the current Φ_0/L is the value at the instant $t = t_r$, provided that t_r is much less than τ_0. At the same time, the current at the initial moment ($t = 0$) is equal to zero. For better understanding of the skin effect, it is useful to derive the same result directly from Eq. (3.54). Integrating both parts of this equation within the ramp time interval, we have

$$R\int_0^{t_r} I(t)\,dt + L\int_0^{t_r} \frac{dI(t)}{dt}\,dt = -\int_0^{t_r} \frac{d\Phi_0}{dt}\,dt$$

Whence

$$R\int_0^{t_r} I(t)\,dt + L[I(t_r) - I(0)] = \Phi_0(0) - \Phi_0(t_r) \qquad (3.66)$$

Inasmuch as at the initial moment

$$\Phi_0(0) = \Phi_0, \quad I(0) = 0$$

and at the instant $t = t_r$ the primary flux disappears, Eq. (3.66) can be written as

$$R\int_0^{t_r} I(t)\,dt + LI(t_r) = \Phi_0 \qquad (3.67)$$

By definition, the integrand $I(t)\,dt$ indicates the amount of charge passing through the ring cross-section during the time interval dt. With decrease of the ramp time, the total amount of charge tends to zero. Therefore, in the limit when the primary flux varies as a step function, we have

$$LI(0) = \Phi_0, \quad \text{if } t_r = 0 \qquad (3.68)$$

that is, the initial current is

$$I(0) = \frac{\Phi_0}{L} \qquad (3.69)$$

It is natural that Eqs. (3.65), (3.68) give the same magnitude for the initial current. As follows from Eqs. (3.63), errors caused by discarding the integral in Eq. (3.67) become smaller as the ratio t_r/τ_0 decreases. In other words, with an increase of the inductance L or a decrease of the resistance R, Eq. (3.68) defines the initial current more accurately. Eq. (3.68), characterizing initial distribution of the current, constitutes the essential feature of the electromagnetic induction and later will be generalized and applied to more complicated medium. In fact, the left-hand side of Eq. (3.68) defines the magnetic flux through the area of the ring caused by induced current at the instant $t=0$ when the primary flux disappears. Thus, the induced current arises at the ring of such magnitude $I(0)$ that, at the first instant magnetic flux, $LI(0)$ is exactly equal to the primary flux Φ_0. This induced current is trying to preserve the flux due to the primary field. If, for example, the primary magnetic field instantly arises at the moment $t=0$, then the induced current has such direction and magnitude that the total flux Φ through the area, bounded by the ring, is equal to zero at $t=0$. In essence we observe the fundamental phenomenon of the inertia of magnetic flux. This study clearly shows that there are two factors governing the behavior of induced current. One is the inertia of the magnetic flux Φ, which tends to keep the current unchanged. The second is a conversion of the electromagnetic energy into heat, which results in a decrease of the current with time. The larger the resistivity R the faster is the decay.

3.3.4 Sinusoidal Primary Magnetic Field

Suppose that the primary magnetic field varies as a sinusoidal function $A \sin \omega t$. To determine the induced current, we use Eq. (3.56). Because the primary flux is presented as $\Phi_0 \sin \omega t$, for the current $I(t)$, we have

$$I(t) = I_0 \exp(-t/\tau_0) - \frac{\omega \Phi_0}{L} \exp(-t/\tau_0) \int_0^t \exp\left(\frac{x}{\tau_0}\right) \cos \omega x\, dx$$

Taking into account that

$$\int \exp(\alpha x) \cos \beta x\, dx = \frac{\exp \alpha x}{\alpha^2 + \beta^2}(\alpha \cos \beta x + \beta \sin \beta x)$$

for the induced current in the ring, we have

$$I(t) = I_0 \exp(-t/\tau_0) - \frac{\Phi_0}{L} \frac{\omega\tau_0 \cos\omega t}{1+(\omega\tau_0)^2} - \frac{\Phi_0}{L} \frac{(\omega\tau_0)^2 \sin\omega t}{1+(\omega\tau_0)^2}$$
$$+ \frac{\Phi_0}{L} \frac{\omega\tau_0}{1+(\omega\tau_0)^2} \exp(-t/\tau_0)$$

Inasmuch as the initial value of the current is equal to zero ($I_0 = 0$), for the sinusoidal currents at ($t \gg \tau_0$), we have

$$I(t) = -\frac{\Phi_0}{L} \frac{\omega\tau_0 \cos\omega t}{1+(\omega\tau_0)^2} - \frac{\Phi_0}{L} \frac{(\omega\tau_0)^2 \sin\omega t}{1+(\omega\tau_0)^2} \quad (3.70)$$

Let us introduce notations

$$a(\omega) = -\frac{\Phi_0}{L} \frac{(\omega\tau_0)^2}{1+(\omega\tau_0)^2}, \quad b(\omega) = -\frac{\Phi_0}{L} \frac{\omega\tau_0}{1+(\omega\tau_0)^2} \quad (3.71)$$

This gives

$$I(t) = a\sin\omega t + b\cos\omega t \quad (3.72)$$

and the induced current is presented as a sum of two oscillations. One is $a\sin\omega t$, which changes synchronously with the primary magnetic field and is called the in-phase component:

$$InI = a\sin\omega t$$

The second oscillation $b\cos\omega t$ is shifted in phase by 90 degrees with respect to the primary magnetic field and is called the quadrature component:

$$QI = b\cos\omega t$$

There is another interpretation of Eq. (3.72). In fact, let us represent magnitudes of these components in the form

$$a = A\cos\phi, \quad b = A\sin\phi \quad (3.73)$$

Then the induced current is written as

$$I(t) = A(\cos\phi\sin\omega t + \sin\phi\cos\omega t) = A\sin(\omega t + \phi) \quad (3.74)$$

Therefore, we can say that induced current $I(t)$ is the single sinusoidal oscillations with the same frequency ω as the primary field $\mathbf{B}_0(t)$ and phase shift ϕ. As follows from Eqs. (3.71), (3.73), for the amplitude and phase of the current, we have

Propagation of Electromagnetic Field

$$A = (a^2 + b^2)^{1/2} = \frac{\Phi_0}{L} \frac{\omega\tau_0}{[1+(\omega\tau_0)^2]^{1/2}}, \quad \phi = \tan^{-1}\frac{1}{\omega\tau_0} \quad (3.75)$$

Frequency responses of the quadrature and in-phase components, as well as the amplitude and phase, are shown in Fig. 3.5. Again, we can interpret Eq. (3.73) as two currents shifted by phase in 90 degrees with respect to each other or a single current with an amplitude and phase defined by Eq. (3.75). Both interpretations are equivalent and widely used by engineers. In spite of the apparent simplicity of the analyzed thin ring object, the considered frequency responses contain general features typical for much more complicated conducting objects.

The Range of Small and Large Parameters of $\omega\tau_0$

Assuming $\omega\tau_0 < 1$, we can expand the right hand side of Eq. (3.71) in a series. This gives

$$a(\omega) = \frac{\Phi_0}{L}\left[-(\omega\tau_0)^2 + (\omega\tau_0)^4 - (\omega\tau_0)^6 + (\omega\tau_0)^8 - \cdots\right]$$
$$b(\omega) = \frac{\Phi_0}{L}\left[-\omega\tau_0 + (\omega\tau_0)^3 - (\omega\tau_0)^5 + (\omega\tau_0)^7 - \right] \quad (3.76)$$

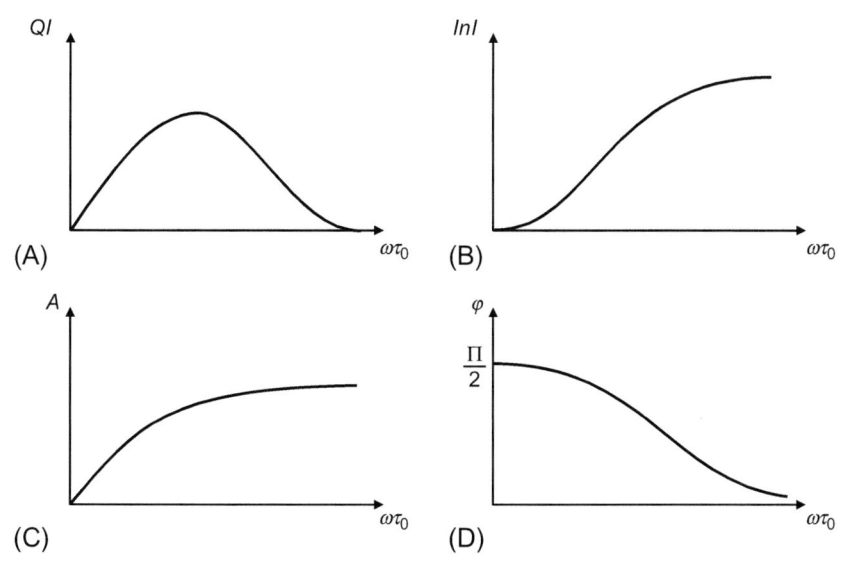

Fig. 3.5 (A) Quadrature component of the current. (B) In-phase component of the current. (C) Amplitude of the current. (D) Phase of the current.

Thus in the low-frequency limit the quadrature and in-phase components of the induced current can be represented as a series containing either odd or even powers of ω. It is interesting that this feature remains valid for induced currents arising in any confined conductor surrounded by an insulator and even in some special cases of a medium unbounded dimension. Both series converge if:

$$\omega \tau_0 < 1 \tag{3.77}$$

In other words, the radius of convergence of these series is

$$\omega = \frac{1}{\tau_0} \tag{3.78}$$

As follows from the theory of complex variables, the radius of convergence of the power series is the distance from the origin ($\omega = 0$) to the nearest singularity of the functions $a(\omega)$ and $b(\omega)$. To determine the location of this singularity, we have to treat these functions as functions of complex variable ω and consider the denominator in Eq. (3.71). It becomes equal to zero when $\omega = \pm \dfrac{i}{\tau_0}$. That is, the spectrum has two poles located on the imaginary axis of ω. It is essential that the radius of convergence of the series, describing the low-frequency part of the spectrum, is expressed in terms of the time constant of the ring. This fact reflects an important relationship between the low-frequency part of the spectrum and the late stage of the transient response observed in confined conductors. Now suppose that the frequency is so low that we can consider only the first term in series. Then, we have

$$\begin{aligned} a(\omega) &\approx -\frac{\Phi_0}{L}(\omega \tau_0)^2 \text{ and } b(\omega) = -\frac{\Phi_0}{L}\omega \tau_0, \\ QI(\omega) &\approx -\frac{\Phi_0}{L}(\omega \tau_0)\cos \omega t, \quad InI(\omega) \approx -\frac{\Phi_0}{L}(\omega \tau_0)^2 \sin \omega t, \text{ if } \omega \tau_0 \ll 1 \end{aligned} \tag{3.79}$$

In this frequency range, the quadrature component is dominant and directly proportional to the conductivity of the ring and frequency. Also, it does not depend on the inductance L because $\tau_0 = L/R$. Such behavior can be explained as follows. If we disregard the flux caused by induced current, $\Phi_s \ll \Phi_0$, then the total flux is practically equal to the primary one:

$$\Phi \approx \Phi_0 \sin \omega t$$

Respectively, the electromotive force in the ring is

$$\Xi = -\frac{d\Phi}{dt} = -\omega\Phi_0 \cos\omega t$$

Applying Ohm's law, we obtain for the quadrature component of induced current

$$QI(\omega) = -\frac{\omega\Phi_0}{R}\cos\omega t = -\frac{\Phi_0}{L}(\omega\tau_0)\cos\omega t$$

Thus, the first term of the series of the quadrature component describes the current that arises due to the primary flux only. This feature is essential for understanding the signals that are recorded in induction logging. In contrast, the in-phase component is caused by a secondary flux. In our approximation, $\omega\tau_0 \ll 1$, the flux generated by the quadrature component of the current is

$$\Phi_1 = LQI(\omega) = -\frac{\omega\Phi_0}{R}L\cos\omega t = -\omega\tau_0\Phi_0 \cos\omega t$$

Therefore,

$$\frac{d\Phi_1}{dt} = \omega^2\tau_0\Phi_0 \sin\omega t$$

and for the in-phase component of the current induced in the ring, we have

$$InI(\omega) = -\frac{\omega^2\tau_0\Phi_0}{R}\sin\omega t = -\frac{\Phi_0}{L}(\omega\tau_0)^2 \sin\omega t$$

which is identical to the first term of the series of the in-phase component (Eq. 3.76). Applying the same approach, we can obtain subsequent terms of the series. Note that the term "the low frequency part of the spectrum" is sometimes confusing. In fact, it does not mean that the equations are valid only when frequencies are small. In fact, validity of equations and frequency is also defined by resistivity and geometry of the object. For instance, if the parameter τ_0 is small, the upper limit of "the low frequency spectrum" can be large and increase with reduction of the parameter $\tau_0 = L/R$.

In the high frequency limit, we have

$$\omega\tau_0 \gg 1. \tag{3.80}$$

As follows from Eq. (3.71), in this range the in-phase component dominates, and, with an increase of the frequency, it approaches a constant value determined by the primary magnetic flux and geometric parameters of the ring:

$$b(\omega) \to 0, \quad a(\omega) \to -\frac{\Phi_0}{L} \quad \text{if } \omega\tau_0 \to \infty \qquad (3.81)$$

Comparing Eqs. (3.69), (3.81), we see that the magnitude of the induced current at the early stage of the transient response ($t \ll \tau_0$) coincides with that at the high frequency part of the spectrum. This is not accidental and is valid for an arbitrary conductive medium. The behavior of the frequency and transient responses, given by Eqs. (3.81), (3.69), is analogous to induced currents in confined conductors with an arbitrary shape of cross-section.

As follows from the Biot-Savart law, the quadrature and in-phase components of the secondary magnetic field are generated by the corresponding components of the induced current. Therefore, the frequency and transient responses of magnetic field and induced currents are similar.

3.3.5 Two Inductively Connected Rings Excited by an External Source

This example illustrates how inductive coupling affects induced currents in the neighboring object. The objects are circular rings with R_1, L_1 and R_2, L_2 that are the resistor and inductance of the first ring and the second ring, correspondingly. Mutual inductance between rings is M. The rings are excited by another source-ring carrying current I_0 and having the radius r_0. Induced currents in the rings are denoted as I_1 and I_2. The centers of the rings are placed on the z axis, as in Fig. 3.6.

We analyze two regimes of excitation: the harmonic excitation when current source is sinusoidal function of time and transient regime when current is abruptly changing from finite value to zero.

Harmonic Excitation

Variable with time current source $I_0(t)$ induces current $I_1(t)$ in the first and $I_2(t)$ in the second ring. According to Kirchhoff law, the sum of all the voltages around the ring is equal to zero. This leads to the following system of equations with respect to $I_1(t)$ and $I_2(t)$:

$$\begin{cases} E_{10}(t) = I_1(t)R_1 + L_1\dfrac{dI_1(t)}{dt} + M\dfrac{dI_2(t)}{dt} \\ E_{20}(t) = I_2(t)R_2 + L_1\dfrac{dI_2(t)}{dt} + M\dfrac{dI_1(t)}{dt} \end{cases} \qquad (3.82)$$

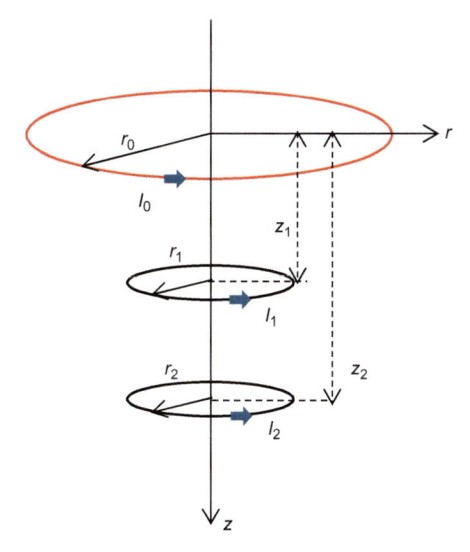

Fig. 3.6 Two inductively connected rings excited by an external source.

where $E_{10}(t)$ and $E_{20}(t)$ are external electromotive forces induced by the current ring in the first and the second ring, M_{10} and M_{20} are the mutual inductances between current ring r_0 and ring r_1 and r_2, correspondingly. In the case of harmonic excitation

$$I(t) = I_0 \sin \omega t, \quad E_{10}(t) = -M_{10} I_0 \omega \cos \omega t, \quad E_{20}(t) = -M_{20} I_0 \omega \cos \omega t$$

and we have the following system with respect to the complex amplitudes I_1^* and I_2^*:

$$\begin{cases} E_{10} = I_1^* R_1 - I_1^* j\omega L_1 - I_2^* j\omega M = I_1^* Z_1 - I_2^* j\omega M \\ E_{20} - I_1^* j\omega M + I_2^* R_2 - I_2^* j\omega L_2 = -I_1^* j\omega M + I_2^* Z_2 \end{cases} \quad (3.83)$$

where Z_1 and Z_2 are the impedances of the rings:

$$Z_1 = R_1 - j\omega L_1$$
$$Z_2 = R_2 - j\omega L_2$$

Solving Eq. (3.83), we receive

$$I_1^* = E_{10} \frac{(Z_2 - j\omega M)}{Z_1 Z_2 + (\omega M)^2}$$

$$I_2^* = E_{20} \frac{(Z_1 - j\omega M)}{Z_1 Z_2 + (\omega M)^2} \quad (3.84)$$

Equating the denominator in Eq. (3.84) to zero, we find poles that are related to characteristic decay τ_1, τ_2 of the rings:

$$(\tau_1\tau_2 - \tau^2)\omega^2 + i(\tau_1 + \tau_2)\omega - 1 = 0$$

where $\tau_1 = L_1/R_1$ and $\tau_2 = L_2/R_2$, $\tau^2 = M^2/R_1R_2$.
The latter gives

$$\omega = \frac{-i(\tau_1 + \tau_2) \pm \left[-(\tau_1 + \tau_2)^2 + 4(\tau_1\tau_2 - \tau^2)\right]^{1/2}}{2(\tau_1\tau_2 - \tau^2)} \qquad (3.85)$$

or introducing $\omega = j\omega^*$, we find two poles:

$$\omega^* = \frac{-(\tau_1 + \tau_2) \pm \left[(\tau_1 + \tau_2)^2 - 4(\tau_1\tau_2 - \tau^2)\right]^{1/2}}{2(\tau_1\tau_2 - \tau^2)} \qquad (3.86)$$

Let us consider several scenarios corresponding to the different couplings between rings.

a. *Case one*: No interaction between rings, $M = 0$.

When mutual inductance $M = 0$, we receive well-known expressions for characteristic decays of two independent circuits:

$$\omega^* = \frac{-(\tau_1 + \tau_2) \pm (\tau_1 - \tau_2)}{2\tau_1\tau_2} \quad \text{and} \quad \omega_1^* = -\frac{1}{\tau_1}, \quad \omega_2^* = -\frac{1}{\tau_2}$$

b. *Case two*: Weak interaction, $M^2 \ll L_1 L_2$.

In this case $\tau_1\tau_2 \gg \tau^2$, $(\tau_1 \neq \tau_2)$, and we have

$$\omega^* \approx \frac{-(\tau_1 + \tau_2) \pm (\tau_1 - \tau_2) + 2\tau^2(\tau_1 - \tau_2)^{-1}}{2\tau_1\tau_2}$$

For the poles ω_1^* and ω_2^*, we obtain

$$\omega_1^* \approx \frac{-\tau_2 + \tau^2(\tau_1 - \tau_2)^{-1}}{\tau_1\tau_2} = -\frac{1}{\tau_1} + \frac{\tau^2}{\tau_1\tau_2(\tau_1 - \tau_2)} \qquad (3.87)$$

$$\omega_2^* \approx \frac{-\tau_1 + \tau^2(\tau_1 - \tau_2)^{-1}}{\tau_1\tau_2} = -\frac{1}{\tau_2} + \frac{\tau^2}{\tau_1\tau_2(\tau_1 - \tau_2)} \qquad (3.88)$$

If interaction is small, the position of poles is close to the limiting case of $M = 0$, but the presence of the second term of an opposite sign in Eqs. (3.87), (3.88) indicates the shift of the poles toward a smaller value of $1/\tau_1$ and $1/\tau_2$. The consequence of this shift has important physical implications that will be better understood when we analyze transient regime.

c. *Case three*: Strong interaction, $M^2/L_1L_2 \approx 1$.

This scenario takes place when the distance between rings $(z_2 - z_1)$ is smaller than the radius of rings. For simplicity, we assume that $\tau_1 = \tau_2 = \tau_0 \approx \tau$. In this case, two rings behave as one ring with characteristic decay $2\tau_0$ and pole, located at

$$\omega^* \approx \frac{-2\tau_0 \pm \left[(4\tau_0^2 - 4(\tau_0^2 - \tau^2)\right]^{1/2}}{2(\tau_0^2 - \tau^2)} = -\frac{1}{(\tau_0 + \tau)} \approx -\frac{1}{2\tau_0} \quad (3.89)$$

Step-Function (Transient) Excitation

Now assume that, at $t=0$, the current in the source is abruptly changing from I_0 to 0.

Taking into account that $\frac{d}{dt}(1(t)) = \delta(t)$, the transient process in the rings is described by the following system:

$$\begin{cases} E_{10}(t) = -M_{10}\dfrac{dI_0(t)}{dt} = -I_0 M_{10}\delta(t) = I_1(t)R_1 + L_1\dfrac{dI_1(t)}{dt} + M\dfrac{dI_2(t)}{dt} \\ E_{20}(t) = -M_{20}\dfrac{dI_0(t)}{dt} = -I_0 M_{20}\delta(t) = I_2(t)R_2 + L_2\dfrac{dI_2(t)}{dt} + M\dfrac{dI_1(t)}{dt} \end{cases}$$
(3.90)

The matrix form of the system (Eq. 3.90) is

$$\begin{bmatrix} \dfrac{dI_1}{dt} \\ \dfrac{dI_2}{dt} \end{bmatrix} = \begin{bmatrix} L_{11} & M \\ M & L_{22} \end{bmatrix}^{-1} \cdot \begin{bmatrix} -I_1 R_1 - I_0 M_{01}\delta(t) \\ -I_2 R_2 - I_0 M_{02}\delta(t) \end{bmatrix} \quad (3.91)$$

where the inverse matrix $\widehat{L}$ is calculated as

$$\widehat{L} = \begin{bmatrix} L_1 & M \\ M & L_2 \end{bmatrix}^{-1} = \frac{1}{\det}\begin{bmatrix} L_2 & -M \\ -M & L_1 \end{bmatrix}$$

and $\det = L_1 L_2 - M^2$. This system of two ordinary differential equations (3.91) along with initial conditions $I_1(0) = I_2(0) = 0$ can be solved numerically using the Runge-Kutta method. The delta function can be approximated as a rectangular pulse with the height $1/h_t$ and width h_t.

(In the numerical implementation, it is important to maintain an integration time step several times smaller than h_t). Resistor R and inductance are defined by Eqs. (3.50), (3.52). The mutual inductance between two rings, having radius r_i and r_j, and separated vertically by the distance z, is calculated as

$$L_{ij} = \mu_0 \sqrt{r_i r_j} \left[\left(\frac{2}{k} - k \right) K(k) - \frac{2}{k} E(k) \right]$$

where

$$k^2 = \frac{4 r_i r_j}{(r_i + r_j)^2 + z^2}$$

and $K(k)$, $E(k)$ are full elliptic integrals of the first and second kind, correspondingly.

In the following example, we select the radius of the source ring $r_0 = 1$ m, while $r_1 = r_2 = 0.1$ m and $a = 0.01$ m. Resistivity of the first ring ρ_1 is fixed and equal to 1 ohm, resistivity of the second ring ρ_2 is varying.

a. *Case one*: Weak interaction, $M^2 \ll L_1 L_2$.

To simulate weak interaction, we separate rings vertically by $z = z_1 - z_2 = 0.75$ m $\gg 0.1$ m, where z_1 and z_2 are distances between source ring and two other rings, correspondingly. The solutions of the system (3.91), $I_1(t)$ and $I_2(t)$, are presented in Fig. 3.7.

The transient decay in the first ring (Fig. 3.7A) is solely defined by the time decay τ_1, and only at the very late stage, $t/\tau_1 \geq 10$, the second ring manifests itself: the smaller the resistivity of the second ring the earlier it manifests itself by slowing down the time decay. At the same time, when resistivity approaches $\rho_2 = 1$ ohmm or above, the effect of the second ring becomes negligible. In other words, in the case of weak interaction, only the more conductive second ring, $\rho_2/\rho_1 < 1$, may slow down the time decay in the first ring.

The transient currents in the second ring, $I_2(t)$, are presented in Fig. 3.7B (solid lines). Also, there is the set of curves (dashed lines), corresponding to the case when interaction is absent ($M = 0$). When resistivity is relatively small ($\rho_2/\rho_1 < 1$), the dashed and solid lines coincide, demonstrating no effect of the first ring on the transient current $I_2(t)$: the transient process takes place with characteristic decay τ_2. At the same time when $\rho_2/\rho_1 \geq 1.5$ the current $I_2(t)$ is affected by the transient process in the first ring. This is especially pronounced at

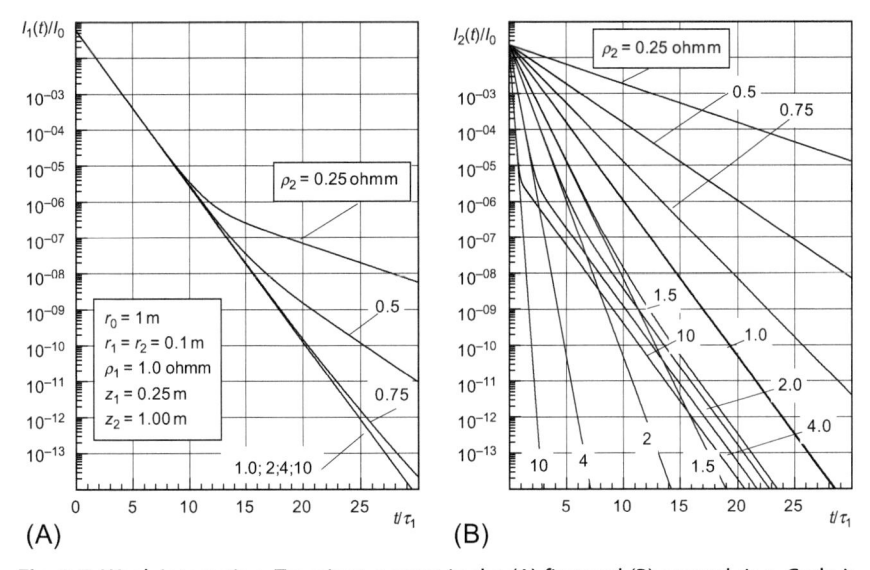

Fig. 3.7 Weak interaction. Transient current in the (A) first and (B) second ring. Code is the resistivity of the second ring.

$\rho_2/\rho_1 = 10$, when only an initial stage, $t/\tau_1 \leq 1$, is defined by the characteristic decay τ_2. At $t/\tau_1 \geq 1$, the transient process is in the second ring is driven by the characteristic time τ_1 of the first ring. The time range $\Delta t/\tau_1$ when decay of $I_2(t)$ is defined by τ_2 is expanded with increase of the conductivity of the second ring $1/\rho_2$. For example, at $\rho_2/\rho_1 = 2$ it is about $\Delta t/\tau_1 \approx 5$ (Fig. 3.7B).
 b. *Case two*: Strong interaction, $M^2 \ll L_1 L_2$.

Strong interaction is simulated by reducing the distance between rings to $\Delta z = z_2 - z_1 = 0.03$ m (Fig 3.8).
Again, the left subplot (Fig. 3.8A) corresponds to the transient process $I_1(t)$ in the first ring, and the subplot on the right (Fig. 3.8B) depicts $I_2(t)$. Due to the closeness of the rings, the mutual coupling is much more pronounced compared with that in the previous case. Particularly, when conductivity of the second ring is high ($\rho_2 = 0.25$ ohmm), the transient decay in the first ring is driven by parameter τ_2 if $t/\tau_1 \geq 3$. When $\rho_2 = \rho_1 = 1$ ohmm, in accordance with (3.89), the characteristic decay is equal to $2\tau_1$. The influence of the second ring practically disappears if $\rho_2/\rho_1 > 4$. Similarly, a strong influence of the first ring is observed on $I_2(t)$, (Fig. 3.8B). Although at $\rho_2 = 0.25$ ohmm, the presence of the first ring with $\rho_1 = 1$ ohmm is practically negligible (compare solid and dashed lines), it becomes significant at all resistivity values

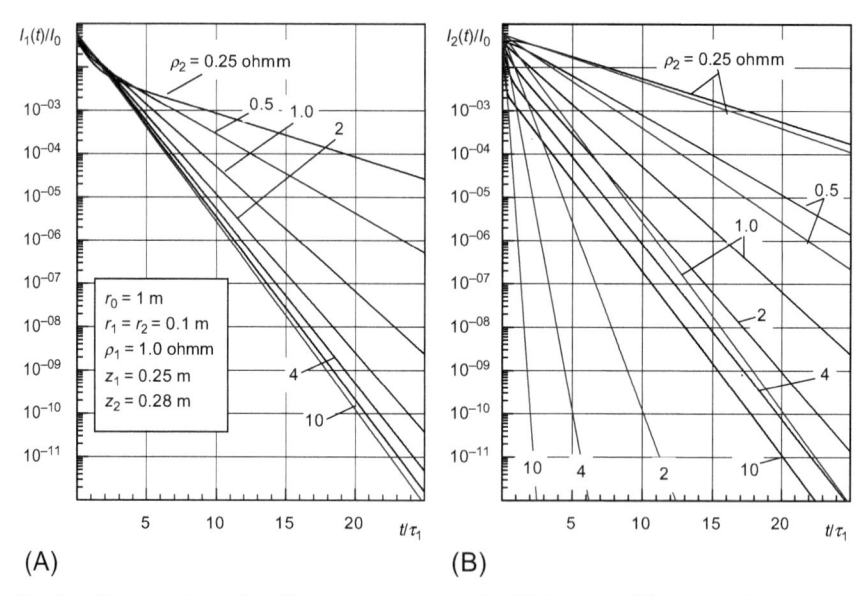

Fig. 3.8 Strong interaction. Transient current in the (A) first and (B) second ring. Code is the resistivity of the second ring.

above 0.5 ohm. (In the case of the weak interaction, the influence of the first ring was not visible at $\rho_2/\rho_1 < 1.0$). Thus, we see that a better conductor may affect a transient process in the second object. When coupling is strong, the influence is observed at earlier times and at smaller conductivity contrast compared with the case of the weak interaction. Moreover, when coupling is strong, even less conductive object slows down a transient process in the more conductive one.

3.3.6 Notes on Measurements of Induced Electric and Magnetic Fields

Coils are often used for measuring time-varying magnetic fields. Suppose that a conducting loop, as shown in Fig. 3.9, is placed in the magnetic field $\mathbf{B}(t)$. In general, the field $\mathbf{E}$ is arbitrary oriented with respect to the loop, and the voltmeter connected in series with the receiver measures the voltage along the path L between terminal points, b and c:

$$V = \int_c^b \mathbf{E} \cdot d\mathbf{l} = \int_c^b E_l dl \tag{3.92}$$

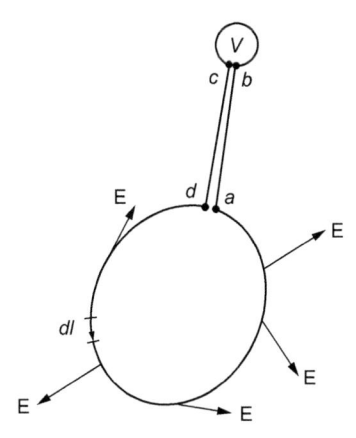

Fig. 3.9 Electric field along receiving loop.

where E_l is the tangential to the loop component dl. When the radius of the loop cross-section is much smaller than its length, the voltage is practically independent of the position of the path L inside the loop. If the receiver consists of n loops, then the voltage is

$$V = n \int_c^b E_l dl$$

When the circuit intervals ab and cd are close to each other, we can write

$$\int_a^b \mathbf{E} \cdot d\mathbf{l} = -\int_c^d \mathbf{E} \cdot d\mathbf{l} \quad \text{or} \quad \int_a^b \mathbf{E} \cdot d\mathbf{l} + \int_c^d \mathbf{E} \cdot d\mathbf{l} = 0 \qquad (3.93)$$

Taking into account Eq. (3.93) and almost coincident positions of points a and d, Eq. (3.92) can be rewritten as

$$V = \int_d^a \mathbf{E} \cdot d\mathbf{l} = \oint_L \mathbf{E} \cdot d\mathbf{l} = \oint_L E_l dl = \Xi. \qquad (3.94)$$

Thus, the voltmeter measures, in essence, the electromotive force along the receiver loop. As a rule, the internal resistance of the voltmeter is high; therefore, the current in the loop is extremely small. For this reason, we can disregard the influence of its magnetic field; correspondingly, the electromotive force is defined by only the external electric field $\mathbf{E}(t)$.

Inasmuch as the voltage is path dependent, a change of the position, size, and shape of the loop results in a change of the electromotive force, even though terminal points of the voltmeter remain in the same place. As is seen in Fig. 3.9, in general, regardless of how small the loop is, it is impossible to determine the electric field from the electromotive force Ξ. However, there is one exception in which measurements with the loop allow us to calculate $\mathbf{E}(t)$. In fact, suppose that this field is tangential to the loop surface, and its magnitude is constant. Then, in accordance with Eq. (3.94), we have

$$E_l = \frac{\Xi}{l}$$

where l is the loop length. A Coulomb electric field $\mathbf{E}^c$ caused by charges has no influence on the electromotive force because

$$\oint \mathbf{E}^c \cdot d\mathbf{l} = 0$$

In particularly, these charges are often located on the surface of the loop. Moreover, in the quasistationary approximation the field of charges is described by Coulomb's law, and, due to the electrostatic induction, the Coulomb's electric field is equal to zero inside the receiver loop.

By measuring electromotive force Ξ, we also can estimate the rate of change of the magnetic flux through the loop. Indeed, according to Faraday's law, we have

$$\frac{d\Phi}{dt} = \int_S \dot{\mathbf{B}} \cdot d\mathbf{S} = \int_S \dot{B}_n dS = -\Xi \tag{3.95}$$

If the size of the loop is small enough, the normal component of the field $\dot{B}_n$ is uniform within the loop and equal to

$$\dot{B}_n = -\frac{\Xi}{S}$$

where S is the area of the loop.

The magnetic field also can be determined:

$$B_n(t) = B_n(t_*) - \frac{1}{S}\int_{t_*}^{t} \Xi(x)\,dx \tag{3.96}$$

Here, t_* is the time at which the field is known. In particular, if $t_* \to \infty$ and $B_n(t_*) = 0$, then

$$B_n(t) = -\frac{1}{S}\int_{t_*}^{t} \Xi(x)\,dx$$

FURTHER READING
[1] Kaufman AA. Geophysical field theory and methods, Part B. New York: Academic Press; 1994.
[2] Tamm IE. Principles of electromagnetic theory. Moscow: Nauka; 1987.

CHAPTER FOUR

Propagation and Diffusion in a Conducting Uniform Medium

Contents

4.1 Sinusoidal Plane Wave in a Uniform Medium	133
4.1.1 Expressions for the Field	133
4.1.2 The Plane Wave as a Function of Time and Distance	136
4.1.3 The High and Low Frequency Limits	138
4.2 Field of the Magnetic Dipole in a Uniform Medium (Frequency Domain)	141
4.2.1 Solution of Helmholtz Equation	141
4.2.2 Dependence of the Field on the Frequency and Observation Point	146
4.3 Transient Field of the Magnetic Dipole in a Uniform Medium	149
4.3.1 Expression for the Vector Potential and Field Components	149
4.4 The Field in a Nonconducting Medium	153
4.4.1 Expressions for the Field	153
4.4.2 Duhamel's Integral	153
4.5 The Transient Field in a Conducting Medium	157
Further Reading	161

Before we describe the theory of induction logging, it is useful to study the propagation of electromagnetic fields in a conducting and polarizable medium and formulate conditions when quasi-stationary approximation is valid. We start from the simplest case of a plane wave whose surfaces of constant phase form a plane surface normal to the direction of propagation.

4.1 SINUSOIDAL PLANE WAVE IN A UNIFORM MEDIUM

4.1.1 Expressions for the Field

Suppose that in the plane XOY there is a current source with the density

$$i_x = i_0 f(at)$$

that is independent of the coordinate y. As in the case of a nonconducting medium, we assume that the electromagnetic field is independent of coordinates x and y and has only two nonzero components: E_x and B_y. From Eq. (2.93) it follows that

$$\frac{\partial^2 E_x}{\partial z^2} - \gamma\mu_0 \frac{\partial E_x}{\partial z} - \varepsilon\mu_0 \frac{\partial^2 E_x}{\partial t^2} = 0$$
$$\frac{\partial^2 B_y}{\partial z^2} - \gamma\mu_0 \frac{\partial B_y}{\partial z} - \varepsilon\mu_0 \frac{\partial^2 B_y}{\partial t^2} = 0$$
(4.1)

When the second term is zero ($\gamma = 0$), we arrive at the wave equation derived earlier. Applying d'Alembert's method, the wave equation can be solved for the arbitrary function $f(at)$, characterizing the primary source.

The essential feature of this method is that an argument of the solution must have the form $t \pm z/v$ or $z \pm vt$, where v is the velocity of propagation. But this approach cannot be applied to Eq. (4.1) due to the presence of the term, proportional to the conductivity. As a result, in general, there is no closed-form solution of Eq. (4.1), except for a special case of a sinusoidal time-varying source. In this case, the field is also a sinusoidal function of the same frequency. This an important fact allows one to apply a Fourier integral and obtain a general solution when the primary current is an arbitrary function of time. Suppose that the current i_x and electromagnetic field vary with time as

$$i_x = i_0 \sin \omega t$$
$$E_x(z,t) = E_{0x}(z)\cos(\omega t - \phi), \quad B_y(z,t) = B_{0y}(z)\cos(\omega t - \psi)$$
(4.2)

As shown in Chapter 2, the last expressions can also be presented as

$$E_x(z,t) = \mathrm{Re}\, E^*_{0x} \exp(-i\omega t), \quad B_y(z,t) = \mathrm{Re}\, B^*_{0y} \exp(-i\omega t) \quad (4.3)$$

through the complex amplitudes

$$E^*_{0x} = E_{0x} \exp(i\phi), \quad B^*_{0y} = B_{0y} \exp(i\psi) \quad (4.4)$$

containing information about the amplitude and phase of the field. As was demonstrated earlier, the form Eq. (4.3) greatly simplifies a solution of Eq. (4.1). Substituting Eq. (4.3) into Eq. (4.1), we obtain equations for the complex amplitudes:

$$\mathrm{Re}\left\{ \frac{\partial^2 E^*_{0x}}{\partial z^2} + i\gamma\mu_0\omega E^*_{0x} + \omega^2 \varepsilon\mu_0 E^*_{0x} \right\} = 0$$

$$\text{Re}\left\{\frac{\partial^2 B_{0y}^*}{\partial z^2} + i\gamma\mu_0\omega B_{0y}^* + \omega^2\varepsilon\mu_0 B_{0y}^*\right\} = 0$$

These equations are satisfied if the complex amplitudes are solutions of the one-dimensional Helmholtz equation:

$$\frac{\partial^2 E_{0x}^*}{\partial z^2} + k^2 E_{0x}^* = 0 \tag{4.5}$$

$$\frac{\partial^2 B_{0y}^*}{\partial z^2} + k^2 B_{0y}^* = 0 \tag{4.6}$$

where $k^2 = i\gamma\mu_0\omega + \omega^2\varepsilon\mu_0$ is the square of the wavenumber k. The solutions of Eqs. (4.5), (4.6) are well known:

$$\begin{aligned} E_{0x}^* &= C_1 \exp(ikz) + C_2 \exp(-ikz) \\ B_{0y}^* &= D_1 \exp(ikz) + D_2 \exp(-ikz) \end{aligned} \tag{4.7}$$

where C and D are constants. The wavenumber is a complex value

$$k = a + ib \tag{4.8}$$

and, correspondingly, in place of Eq. (4.7) we have

$$\begin{aligned} E_{0x}^* &= C_1 \exp(-bz)\exp(iaz) + C_2 \exp(bz)\exp(-iaz) \\ B_{0y}^* &= D_1 \exp(-bz)\exp(iaz) + D_2 \exp(bz)\exp(-iaz) \end{aligned} \tag{4.9}$$

In addition, it is assumed that $a > 0$ and $b > 0$. A plane wave has to decay with the distance, because its amplitude is attenuated by conducting medium. The second terms in Eq. (4.9) do not satisfy this requirement and have to be discarded. This gives

$$E_{0x}^* = C_1 \exp(-bz)\exp(iaz), \quad B_{0y}^* = D_1 \exp(-bz)\exp(iaz) \tag{4.10}$$

Next, we find relationship between constants C_1 and D_1. We can substitute Eq. (4.3) into the first of Maxwell's equations

$$\text{curl } \mathbf{E} = -\frac{\partial \mathbf{B}}{\partial t}$$

Taking into account independence of the field from coordinates x and y, we obtain:

$$\frac{\partial E_{0x}^*}{\partial z} = i\omega B_{0y}^*$$

This gives

$$C_1 = \frac{\omega}{k} D_1 \qquad (4.11)$$

Since k is a complex number, we conclude that there is a phase shift between the electric and magnetic fields, which depends on conductivity, dielectric constant, and frequency. In particular, in a nonconducting medium the phase shift is zero. In contrast, when conduction currents are dominant, the phase shift is equal to $-\pi/4$. Applying Biot-Savart's law near the source, as in the case of a nonconducting medium, it is easy to express the real constant D_1 in terms of the current source i_x. As long as the electric and magnetic fields satisfy all conditions of the theorem of uniqueness, the expressions in Eq. (4.10) are the solution of the boundary value problem.

4.1.2 The Plane Wave as a Function of Time and Distance

To analyze the behavior of the plane wave consider the function $B_y(z,t)$

$$B_y = D_1 \exp(-bz) \cos\left[\omega\left(t - \frac{a}{\omega}z\right)\right] \qquad (4.12)$$

At each observation point the magnetic field is a sinusoidal function of time, but the spatial dependence on the distance z is described by the product of sinusoidal and exponential functions. Of course, the electric field has a similar form. From Eq. (4.12) it follows that the velocity of propagation of the sinusoidal wave is

$$v = \frac{\omega}{a} \qquad (4.13)$$

By definition, during the period T, the phase plane moves a distance equal to the wavelength λ:

$$\lambda = vT \quad \text{or} \quad \lambda = \frac{v}{f}$$

Thus,

$$\lambda = \frac{2\pi}{a} \qquad (4.14)$$

and the wavelength is inversely proportional to the real part of the wavenumber. Let us introduce a parameter

$$b\lambda = 2\pi\frac{b}{a} \tag{4.15}$$

which determines the attenuation of the wave within the distance equal to the wavelength. If this parameter is large, the field strongly decays over the distance of one wavelength, and the sinusoidal character is hardly observable. For instance, when $a = b$ the field decays by a factor of $\exp(-2\pi)$ over every following wavelength. It would require a great deal of imagination to see a propagation of such wave.

$$b\lambda > 1 \tag{4.16}$$

If we observe diffusion rather than propagation (Fig. 4.1A, solid line). Correspondingly, in place of Eq. (4.1) we have the following diffusion equations:

$$\frac{\partial^2 E_x}{\partial z^2} - \gamma\mu_0 \frac{\partial E_x}{\partial t} = 0, \quad \frac{\partial^2 B_y}{\partial z^2} - \gamma\mu_0 \frac{\partial B_y}{\partial t} = 0$$

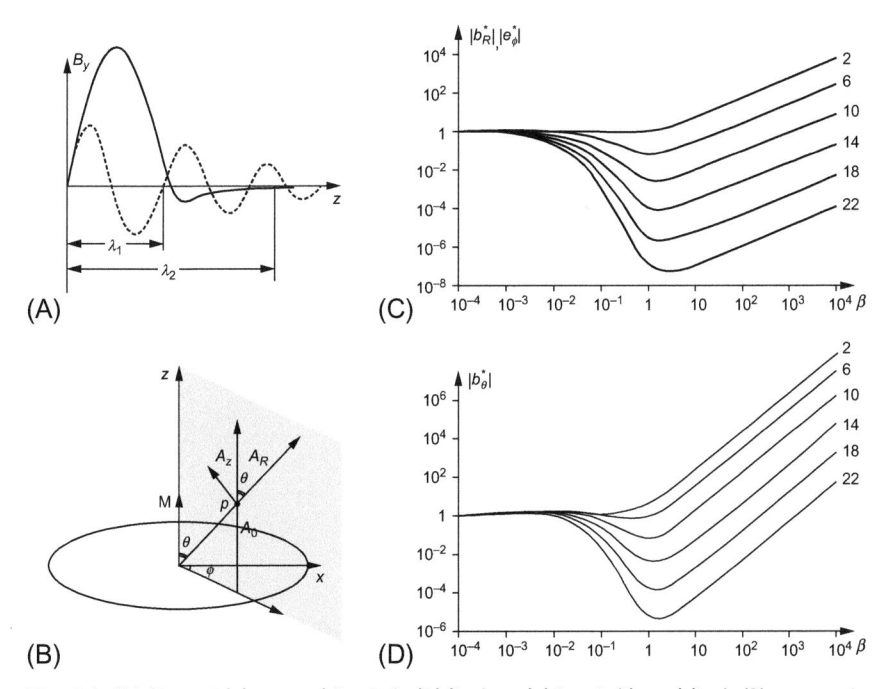

Fig. 4.1 (A) Sinusoidal wave, $b\lambda > 1$ *(solid line)* and $b\lambda \ll 1$ *(dotted line)*; (B) magnetic dipole and vector potential in spherical coordinates; (C) frequency responses of the field amplitudes b_R^* and e_ϕ^*; (D) b_θ^* curve index is the parameter $X = R/R_0$.

which describe the quasi-stationary fields. In contrast, when the product is small

$$b\lambda \ll 1 \qquad (4.17)$$

propagation becomes visible, and we can observe the sinusoidal character of the field. It propagates over several wavelengths before attenuation becomes noticeable (Fig. 4.1A, dotted line). This consideration allows us to qualitatively distinguish three possible scenarios where the field exhibits either features of propagation, diffusion, or both. Regardless of how small the frequency ω is, the field reaches any observation point by propagation as a wave. Indeed, a sinusoidal source current can always be presented as a system of pulses following one after another. Due to conversion of electromagnetic energy into heat in a conducting medium, there is always attenuation, which becomes stronger with an increase of a distance from the primary source.

Now we can study the dependence of attenuation, velocity, and wavelength of a plane wave in a uniform medium on the frequency and electric parameters of the medium. Let us determine the real and imaginary parts of the wavenumber:

$$k = a + ib = \left(i\gamma\mu_0\omega + \omega^2\varepsilon\mu_0\right)^{1/2} \qquad (4.18)$$

As follows from Eq. (4.12), the imaginary part of k defines the decrease of the field amplitude, while the real part affects the phase. Taking the square of both sides of Eq. (4.18), we obtain the system of equations with respect to a and b:

$$\omega^2\varepsilon\mu_0 = a^2 - b^2 \quad \text{and} \quad \gamma\mu_0\omega = 2ab \qquad (4.19)$$

After solving a system (4.19) we get:

$$a = k_0 \left[\frac{\left(1+\beta^{-2}\right)^{1/2}+1}{2}\right]^{1/2}, \quad b = k_0 \left[\frac{\left(1+\beta^{-2}\right)^{1/2}-1}{2}\right]^{1/2} \qquad (4.20)$$

Here $k_0 = \omega(\varepsilon\mu_0)^{1/2}$ and $\beta = \omega\varepsilon/\gamma$ is the ratio of the real and imaginary parts of the wavenumber.

4.1.3 The High and Low Frequency Limits

Suppose that parameter β is large; that is, displacement currents prevail. Expanding the right side of Eq. (4.20) in a series, we obtain expressions for the real and imaginary parts of the wavenumber:

Propagation and Diffusion in a Conducting Uniform Medium

$$a \approx a_\varepsilon \left(1 + \frac{1}{8}\beta^{-2}\right) \quad \text{and} \quad b \approx b_\varepsilon \left(1 - \frac{1}{8}\beta^{-2}\right) \tag{4.21}$$

where

$$a_\varepsilon = \omega(\varepsilon\mu_0)^{1/2}, \quad b_\varepsilon = \frac{\gamma}{2}\left(\frac{\mu_0}{\varepsilon}\right)^{1/2}. \tag{4.22}$$

Therefore, at the high frequency limit, $\beta \geq 1$, the real part of the wavenumber is practically independent of conductivity and is directly proportional to the frequency.

As follows from Eq. (4.13), in this frequency range the velocity of propagation is defined mainly by the dielectric constant that is, almost independent of frequency and conductivity. It is nearly the same as in a non-conducting medium:

$$v = \frac{1}{(\varepsilon\mu_0)^{1/2}} = \frac{c}{(\varepsilon_r)^{1/2}}, \quad \text{if } \beta \geq 1 \tag{4.23}$$

where $c = 3 \times 10^8$ m/s. The imaginary part of the wavenumber b, characterizing decrease of the field with distance, is governed by the term b_ε, which is directly proportional to conductivity and practically independent of the frequency. As follows from Eq. (4.22),

$$b_\varepsilon \lambda \approx \frac{\pi}{\beta} \ll 1$$

Correspondingly, with an increase of the frequency the effect of the field decay over one wavelength decreases, and the wave phenomena becomes noticeable. At the same time, the field does not propagate far away from the source because of attenuation. The ratio of a_ε and b_ε is

$$\frac{b_\varepsilon}{a_\varepsilon} = \frac{1}{2\beta} < 1 \tag{4.24}$$

As is seen from Eq. (4.21), even at β being close to unity, the real and imaginary parts of the wavenumber k differ only slightly from the limiting values a_ε and b_ε. Thus, at the high frequency limit sinusoidal waves decay almost at the same rate, and have practically equal velocity $(\varepsilon\mu_0)^{-1/2}$, which slightly increases with the frequency.

In contrast, when the conduction current prevails and $\beta < 1$, Eq. (4.20) gives

$$a \approx a_\gamma \left(1 + \frac{\beta}{2}\right) \quad \text{and} \quad b = b_\gamma \left(1 - \frac{\beta}{2}\right) \quad \text{if } \beta < 1 \qquad (4.25)$$

Here

$$a_\gamma = b_\gamma = \left(\frac{\gamma \mu_0 \omega}{2}\right)^{1/2} = \frac{1}{\delta} \qquad (4.26)$$

The parameter δ is called the skin depth and is expressed as:

$$\delta = \left(\frac{2}{\gamma \mu_0 \omega}\right)^{1/2} \qquad (4.27)$$

Thus, within the range of $\beta < 1$ the attenuation is described by Eq. (4.26), and becomes smaller with a decrease of conductivity and frequency. As follows from Eq. (4.12) the skin depth is equal to the distance where the magnitude of the sinusoidal plane wave decreases by a factor $e \approx 2.718$. It should be noted that under real conditions, when the field is generated by a finite size source and depends on coordinates y and z, the field decay is even stronger. The velocity of propagation is given by

$$v = \frac{\omega}{a} = \left(\frac{2\omega\rho}{\mu_0}\right)^{1/2} \quad \text{or} \quad v = (10\rho f)^{1/2} \text{ km/s} \qquad (4.28)$$

and it becomes smaller with a decrease of the frequency and resistivity. Also, the latter can be presented as

$$v = (2\beta)^{1/2} \frac{c}{(\varepsilon_r)^{1/2}}$$

The wavelength is determined from Eq. (4.28):

$$\lambda = 2\pi\delta \quad \text{or} \quad \lambda = (10\rho T)^{1/2} \text{ km} \qquad (4.29)$$

and increases as the frequency decreases. As follows from Eqs. (4.26), (4.29), if

$$b_\gamma \lambda = 2\pi$$

the decay over a distance of one wavelength is extremely strong. Within this frequency range the wave phenomena is practically invisible. Functions $a(\beta)$ and $b(\beta)$, normalized by their limiting values, as well as the product $b\lambda$, are shown in Fig. 4.2.

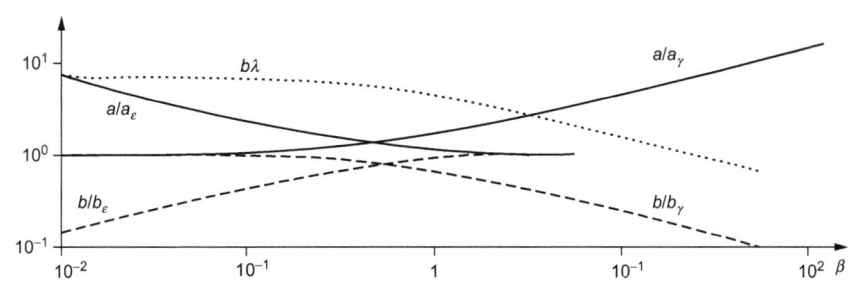

Fig. 4.2 Functions $b/b_\varepsilon(\beta)$, $a/a_\varepsilon(\beta)$, $b\lambda(\beta)$, $b/b_\gamma(\beta)$.

In the analysis we assumed that conductivity γ and dielectric permittivity ε are independent of frequency. In fact, some rocks exhibit dispersive behavior, thus parameters γ and ε do depend on the frequency. The dispersive behavior of the formation is utilized in the dielectric logging, where multifrequency measurements in combination with an advanced petrophysical interpretation permit a unique information on rock properties and fluid distribution. Also note that regardless of how low the frequency is, there is always propagation of waves through a conducting medium. Otherwise, the field would instantly appear at any point of a medium regardless of the distance between the source and observation point. The influence of conductivity is expressed in two ways. First, the field decays with distance from the source due to attenuation of the field by conducting media. Second, there is, in general, a frequency dispersion of the velocity caused by the dispersive conductivity.

4.2 FIELD OF THE MAGNETIC DIPOLE IN A UNIFORM MEDIUM (FREQUENCY DOMAIN)

4.2.1 Solution of Helmholtz Equation

Next we study the frequency responses of the magnetic and electric fields caused by the magnetic dipole (a small current loop) in a uniform conducting medium. The dipole moment is

$$\mathbf{M} = \operatorname{Re} M_0 \exp(-i\omega t)\mathbf{i}_z \tag{4.30}$$

Here $M_0 = I_0 nS$ is the magnitude of the moment, I_0 is the amplitude of the sinusoidal current, n is the number of turns in the loop, and S is the area enclosed by a single turn of the loop. The dipole moment $\mathbf{M}$ is directed along the z-axis (Fig. 4.1C) and $\mathbf{i}_z$ is the corresponding unit vector. Before we formulate the boundary value problem, it is useful to recall the behavior of the quasi-stationary electric field of the magnetic dipole in a non-conducting medium (Chapter 3). In the spherical system of coordinates (Fig. 4.1C) this field has only one component E_ϕ. For $M = M_0 \cos \omega t$ from Eq. (3.40) we obtain:

$$E_\phi(t) = \frac{\mu_0 \omega M_0 \sin \omega t}{4\pi R^2} \cos \theta \quad \text{or} \quad E_\phi^*(\omega) = \frac{i\omega \mu_0 M_0}{4\pi R^2} \cos \theta \qquad (4.31)$$

Here E_ϕ^* is the complex amplitude of the electric field caused by the primary time-variable magnetic dipole. Inasmuch as the electric field has only one component confined to horizontal planes, arising conduction and displacement currents also have only an azimuthal component j_ϕ^*. An inductive electric field can be presented as

$$\mathbf{E}^* = \operatorname{curl} \mathbf{A}^* \quad \text{or} \quad \mathbf{E}^* = \nabla \times \mathbf{A}^* \qquad (4.32)$$

where $\mathbf{A}^*$ is the complex amplitude of the vector potential of the magnetic type. As was demonstrated in Chapter 2, the function $\mathbf{A}^*$ satisfies the Helmholtz equation and fully describes electromagnetic field components $\mathbf{E}^*$ and $\mathbf{B}^*$.

We assume that the vector potential has a single z-component, which in the spherical coordinates depends on the coordinate R:

$$\mathbf{A}^* = A_z^*(k, R)\mathbf{i}_z \qquad (4.33)$$

Then, Helmholtz's equation is simplified to:

$$\frac{1}{R^2} \frac{d}{dR}\left(R^2 \frac{dA_z^*}{dR}\right) + k^2 A_z^* = 0 \qquad (4.34)$$

Introducing a new function

$$W = A_z^* R$$

and performing differentiation, we obtain:

$$\frac{dA_z^*}{dR} = -R^{-2} W + R^{-1} \frac{dW}{dR} \quad \text{and} \quad R^2 \frac{dA_z^*}{dR} = -W + R \frac{dW}{dR}$$

Whence

$$\frac{d}{dR}\left(R^2 \frac{dA_z^*}{dR}\right) = R \frac{d^2 W}{dR^2}$$

Therefore, Eq. (4.34) becomes

$$\frac{d^2 W}{dR^2} + k^2 W = 0$$

whose solutions are exponential functions $\exp(\pm ikR)$. Thus, the expression for the z-component of the vector potential is

$$A_z^* = C\frac{\exp(ikR)}{R} + D\frac{\exp(-ikR)}{R} \tag{4.35}$$

Inasmuch as

$$-ikR = -i(a+ib)R = -iaR + bR$$

an increase of R leads to an unlimited increase of the second term of Eq. (4.35), To meet conditions at infinity, we have to discard this term and thus reduce Eq. (4.35) to

$$A_z^*(k, R) = C\frac{\exp(ikR)}{R} \tag{4.36}$$

To satisfy the condition near the source, we have to determine the unknown C. Since the electric field has only the ϕ-component, Eq. (4.32) gives

$$E_\phi^* = \frac{1}{R}\left[\frac{\partial}{\partial R}(RA_\theta^*) - \frac{\partial A_R^*}{\partial \theta}\right] \tag{4.37}$$

As is seen from Fig. 4.1C,

$$A_R^* = A_z^* \cos\theta \quad \text{and} \quad A_\theta^* = -A_z^* \sin\theta$$

Substituting these expressions into Eq. (4.37) and performing simple algebraic operations, we obtain

$$E_\phi^* = \frac{C}{R^2}(1 - ikR)\exp(ikR)\sin\theta \tag{4.38}$$

Near the dipole, the electric field is approaching the value

$$E_\phi^* = \frac{C}{R^2}\sin\theta \tag{4.39}$$

which tends to that caused by the dipole source only (Eq. 4.31). In essence, the boundary condition near the source allows us to determine unknown constant C. Comparing Eqs. (4.31) and (4.39), we obtain

$$C = \frac{i\omega\mu_0 M_0}{4\pi} \tag{4.40}$$

Finally, the solution for the complex amplitude of the vector potential is

$$\mathbf{A}^* = \frac{i\omega\mu_0 M_0}{4\pi R} \exp(ikR)\mathbf{i}_z \tag{4.41}$$

From Eqs. (4.38), (4.40) for the complex amplitude of the electric field we have:

$$E_\phi^* = \frac{i\omega\mu_0 M_0}{4\pi R^2}(1 - ikR)\exp(ikR)\sin\theta \tag{4.42}$$

To determine the complex amplitudes of the magnetic field we use the equation derived in Chapter 2 (Eq. 2.104):

$$i\omega \mathbf{B}^* = k^2 \mathbf{A}^* + \mathrm{grad\,div}\,\mathbf{A}^*. \tag{4.43}$$

Inasmuch as the vector potential is directed along the z-axis there are only B_R and B_θ nonzero components of magnetic field. First, let us calculate $\mathrm{div}\,\mathbf{A}^*$. Taking into account that divergence is independent of the system of coordinates, it is convenient to perform derivation in Cartesian coordinates:

$$\mathrm{div}\,\mathbf{A}^* = \frac{dA_z^*}{dz}$$

Carrying out differentiation, we obtain

$$\mathrm{div}\,\mathbf{A}^* = -\frac{i\omega\mu_0 M_0}{4\pi R^2}\exp(ikR)(1 - ikR)\cos\theta \tag{4.44}$$

since

$$\frac{dR}{dz} = \cos\theta$$

Expressing *grad* in spherical coordinates, we have

$$i\omega B_R^* = k^2 A_R^* + \frac{\partial}{\partial R}\mathrm{div}\,\mathbf{A}^* \quad \text{and} \quad i\omega B_\theta^* = k^2 A_\theta^* + \frac{1}{R}\frac{\partial}{\partial \theta}\mathrm{div}\,\mathbf{A}^*$$

Differentiating and bearing in mind that

$$A_R^* = A_z^*\cos\theta \quad \text{and} \quad A_\theta^* = -A_z^*\sin\theta$$

we obtain

$$B_R^* = \frac{2\mu_0 M_0}{4\pi R^3} \exp(ikR)(1-ikR)\cos\theta$$
$$B_\theta^* = \frac{\mu_0 M_0}{4\pi R^3} \exp(1-ikR-k^2R^2)\sin\theta \tag{4.45}$$

Thus, Eqs. (4.42), (4.45) describe the complex amplitudes of the field of the magnetic dipole at any point in a uniform medium. The field is comprised of the primary dipole source, as well as the field caused by conduction and displacement currents arising in a medium. Correspondingly, it depends on several factors, such as the dipole moment, the frequency, the product kR, and the coordinates of the observation point. It is convenient to normalize the field by the primary field of the dipole in free space. Then, we have:

$$e_\phi^* = \frac{E_\phi^*}{E_{\phi 0}^*} = (1-ikR)\exp(ikR), \quad b_R^* = \frac{B_R^*}{B_{R0}^*} = (1-ikR)\exp(ikR)$$
$$b_\theta^* = \frac{B_\theta^*}{B_{\theta 0}^*} = (1-ikR-k^2R^2)\exp(ikR) \tag{4.46}$$

These expressions depend only on the parameter kR, which simplifies the field analysis. The primary field components $E_{\phi 0}^*$, B_{R0}^*, and $B_{\theta 0}^*$ are described by simple formulas given earlier. For the amplitude and phase, as well as the quadrature and in-phase components we have:

$$\left|e_\phi^*\right| = \exp(-bR)\left[(1+bR)^2 + a^2R^2\right]^{1/2}, \quad \psi_\phi = aR - \tan^{-1}\frac{aR}{1+bR} \tag{4.47}$$

$$Ine_\phi = Inb_R = \left|e_\phi^*\right|\cos\psi_\phi \quad \text{and} \quad Qe_\phi = Qb_R = \left|e_\phi^*\right|\sin\psi_\phi \tag{4.48}$$

Similarly, for the azimuthal component of the magnetic field we obtain

$$\left|b_\theta^*\right| = \exp(-bR)\left[\left(1+bR+b^2R^2-a^2R^2\right)^2 + \left(aR+2abR^2\right)^2\right]^{1/2} \tag{4.49}$$

$$\psi_\theta = aR - \frac{aR+2abR^2}{1+bR+b^2R^2-a^2R^2}$$

For the in-phase and quadrature components we have:

$$Inb_\theta = \left|b_\theta^*\right|\cos\psi_\theta, \quad Inb_\theta = \left|b_\theta^*\right|\sin\psi_\theta \tag{4.50}$$

In Eqs. (4.47), (4.49) a and b are the real and imaginary parts of the complex number k.

4.2.2 Dependence of the Field on the Frequency and Observation Point

To study the frequency responses of the field amplitudes it is convenient to introduce two parameters, namely a characteristic length R_0 and characteristic frequency ω_0. The parameter R_0 is defined from the relation $b_\varepsilon R_0 = 1$ or

$$R_0 = \frac{2}{\gamma(\mu_0/\varepsilon)^{1/2}} = \frac{\rho(\varepsilon_r)^{1/2}}{188.5} \qquad (4.51)$$

The characteristic frequency is the frequency at which the displacement and conduction currents are equal

$$\frac{\omega_0 \varepsilon}{\gamma} = 1 \qquad (4.52)$$

The complex amplitudes b_R^*, b_θ^*, and e_ϕ^* can be treated as functions of the dimensionless parameters $\beta = \omega/\omega_0$ and $X = R/R_0$, where R is the distance from the dipole to the observation point. A set of typical curves illustrating the dependence of the functions $|b_R^*|$ and $|b_\theta^*|$ on the parameter β is given in Fig. 4.1C and D, correspondingly. With an increase of frequency the electromagnetic field first decreases; near the characteristic frequency the amplitude of the field passes through a minimum and then grows. With a decrease of the characteristic length $R_0 = \frac{\rho(\varepsilon_r)^{1/2}}{188.5}$, the minimum becomes smaller and shifts toward higher frequencies. If the frequencies are lower than ω_0, the field is practically independent of dielectric permittivity (Eq. 4.25) and becomes quasi-stationary. For frequencies higher than the characteristic frequency, the field is greater and might be orders of magnitude larger than the primary field. In this part of the spectrum the field depends on both conductivity and dielectric permittivity, but the imaginary part of the wavenumber b is independent of the frequency. Measuring the field magnitude and phase we may obtain information about both conductivity and dielectric permittivity. Consider a field behavior as a function of the separation between dipole and observation point R. As follows from Eq. (4.46), and, taking into account that

$$k = a + ib = |k|\exp(i\xi) \quad \text{and} \quad B_R = \text{Re}\left[B_R^* \exp(-i\omega t)\right]$$

we obtain

$$B_R = \frac{2\mu_0 M_0}{4\pi} \exp(-bR) \left[\frac{\cos(\omega t - aR)}{R^3} - \frac{|k|}{R^2} \sin(\omega t - aR - \xi) \right] \cos\theta$$

or

$$B_R = \frac{2\mu_0 M_0}{4\pi R^3} \exp(-bR) [\cos(\omega t - aR) - |k|R \sin(\omega t - aR - \xi)] \cos\theta$$

Also,

$$B_\theta = \frac{\mu_0 M_0}{4\pi R^3} \exp(-bR) [\cos(\omega t - aR) - |k|R \sin(\omega t - aR - \xi) \\ - |k^2|R^2 \cos(\omega t - aR - 2\xi)] \sin\theta \quad (4.53)$$

and

$$E_\phi = \frac{\mu_0 \omega M_0}{4\pi} \exp(-bR) \left[\frac{\sin(\omega t - aR)}{R^2} - \frac{|k|}{R} \cos(\omega t - aR - \xi) \right] \sin\theta$$

or

$$E_\phi = \frac{\mu_0 M_0}{4\pi R^2} \omega \exp(-bR) [\sin(\omega t - aR) - |k|R \cos(\omega t - aR - \xi)] \sin\theta$$

As we see, the field is presented as a combination of sinusoidal waves decaying with the distance from the dipole. There are two factors which result in a decrease of the field: one is attenuation caused by conversion of electromagnetic energy into heat; the term $\exp(-bR)$ appears due to this factor. The second factor is geometry of the dipole source: the wave moves in all directions and the energy density decreases with the distance even in a nonconducting medium. As follows from Eq. (4.53) at relatively small separations, when $|k|R \ll 1$, the magnetic field almost synchronously varies with the dipole current and rapidly decreases with distance as $1/R^3$. This range, $|k|R \ll 1$, is called the near zone. For instance, if conduction currents prevail ($\omega < \omega_0$), we have:

$$k(i\omega\mu_0\gamma)^{1/2} = \frac{(2i)^{1/2}}{\delta} = \frac{1+i}{\delta}$$

$$\delta = \left(\frac{2}{\gamma\mu_0\omega}\right)^{1/2} \quad \text{or} \quad \delta = \frac{10^3}{2\pi} \left(\frac{10\rho}{f}\right)^{1/2} m \quad (4.54)$$

Correspondingly, the near zone is defined by the condition

$$R/\lambda \ll 1 \quad (4.55)$$

where $\lambda = 2\pi\delta$. For illustration, consider two examples. First, suppose that $\rho = 10$ ohmm and $f = 10$ kHz. Then $\delta = 16$ m and this distance defines, approximately, the boundary of the near zone. If the frequency is increased to $f = 10^6$ Hz, the first term of Eq. (4.53) describes the field at distances which are smaller than 1 m. In general, when terms aR and ξ are disregarded, the quasi-stationary field is observed; that is, both the current and magnetic field change synchronously. In the beginning of the *intermediate zone* the field decays as in the near zone and then starts to decrease slower. The phase shift between the field and the dipole current also changes with distance (Eq. 4.53). This change is caused by superposition of waves, which differently depend on distance R (Eq. 4.53). Finally, when $R > \lambda$ we observe the wave zone where

$$B_R = -\frac{2\mu_0 M_0}{4\pi} \exp(-bR) \frac{|k|}{R^2} \sin(\omega t - aR - \xi) \cos\theta$$

$$B_\theta = -\frac{\mu_0 M_0}{4\pi} \exp(-bR) \frac{|k^2|}{R} \cos(\omega t - aR - 2\xi) \sin\theta \qquad (4.56)$$

$$E_\phi = -\frac{\mu_0 \omega M_0}{4\pi} \exp(-bR) \frac{|k|}{R} \cos(\omega t - aR - \xi) \sin\theta$$

In the wave zone the field decays relatively slower and $B_\theta > B_R$, provided that the observation point is not placed in the vicinity of the z-axis. In summary, we may note:
1. Regardless of the frequency, there are always three zones of the different field behavior.
2. In each zone there is propagation of sinusoidal waves.
3. Their attenuation is caused by the geometric spreading and conversion of electromagnetic energy into heat.
4. At high frequency, $\omega > \omega_0$, the electromagnetic wave propagates with almost the same velocity as in a nonconducting medium and attenuation is directly proportional to the conductivity and, practically, independent of frequency. In contrast, at low frequency $\omega < \omega_0$, the velocity is mainly defined by conductivity and frequency:

$$v = \frac{\omega}{a} = \left(\frac{2\omega}{\gamma\mu_0}\right)^{1/2}$$

and the field is governed by the diffusion equation. The latter is equivalent to the assumption that displacement currents are disregarded and the field instantly appears at all points of a medium regardless of distance from the source.

4.3 TRANSIENT FIELD OF THE MAGNETIC DIPOLE IN A UNIFORM MEDIUM

4.3.1 Expression for the Vector Potential and Field Components

For better understanding of propagation and diffusion, we also consider a transient field caused by the magnetic dipole in a uniform medium. Suppose that the dipole current arises instantaneously, so that its magnetic moment M_0 is described as a step-function of time:

$$M_0 = \begin{cases} 0 & t < 0 \\ M_0 & t \geq 0 \end{cases} \quad (4.57)$$

To derive the transient field we proceed from the Fourier integrals:

$$F(t) = \frac{1}{2\pi} \int_{-\infty}^{\infty} S(\omega) \exp(-i\omega t) d\omega \quad \text{and} \quad S(\omega) = \int_{-\infty}^{\infty} F(t) \exp(i\omega t) dt \quad (4.58)$$

The first equation shows that the function $F(t)$ can be presented as a sum of an infinite number of sinusoids (harmonics), and their amplitudes and phases are characterized by the spectrum (complex amplitude) $S(\omega)$, given by the second equation. In accordance with Eq. (4.58) the spectrum of the step-function is

$$S(\omega) = M_0 \int_0^{\infty} \exp(i\omega t) dt$$

because the current is zero when $t < 0$. To calculate this integral, we start from a slightly different convergent integral:

$$\int_0^{\infty} \exp[(i\omega - p)t] dt$$

where p is a small positive number and then find the limit when $p \to 0$. Performing integration, we obtain the spectrum of the step function:

$$S(\omega) = -\frac{M_0}{i\omega} \quad (4.59)$$

Thus, the step-function is a sum of sinusoids of the same phase, but their amplitudes decrease with an increase of frequency. In other words, the maximum energy of the dipole is concentrated at the low frequencies. The field generated at these frequencies decays relatively slowly with the distance from the source and, correspondingly, has greater depth of penetration. As we already know, the vertical component of the complex amplitude A_z^* is

$$A_z^* = \frac{i\omega\mu_0 M_0}{4\pi}\frac{\exp(ikR)}{R}$$

If the moment is described as $-M_0/i\omega$, then for the vector potential at each frequency we have:

$$A_z^*(\omega) = -\frac{\mu_0 M_0}{4\pi}\frac{\exp(ikR)}{R}$$

Applying Fourier's integral to the last expression, we obtain the formula for the vector potential in the time domain

$$A_z(t) = -\frac{\mu_0 M_0}{8\pi^2 R}\int_{-\infty}^{\infty}\exp i(kR-\omega t)d\omega \qquad (4.60)$$

Here $k = (i\gamma\omega\mu_0 + \varepsilon\mu_0\omega^2)^{1/2}$, and $A_z(t)$ is the vector potential of magnetic type. Integration in Eq. (4.60), using a table of integrals, gives the following expression for $A_z(t)$:

$$A_z = -\begin{cases} 0 & t < \tau_0 \\ \dfrac{\mu_0 M_0}{4\pi R}\left\{\exp(-q\tau_0)\delta(t-\tau_0) + q\tau_0\exp(-qt)\dfrac{I_1\left[q(t^2-\tau_0^2)\right]}{(t^2-\tau_0^2)^{1/2}}\right\} & t \geq \tau_0 \end{cases}$$

(4.61)

Here

$$q = \frac{1}{2}\frac{\gamma}{\varepsilon}, \quad \tau_0 = (\varepsilon\mu_0)^{1/2}R \qquad (4.62)$$

$I_1\left[\left(q(t^2-\tau_0^2)^{1/2}\right)\right]$ is a modified Bessel's function of the first order, and $\delta(t-\tau_0)$ is the Dirac delta function defined as

$$\int_{b_1}^{b_2}f(x')\delta^{(n)}(x'-x)dx' = \begin{cases} (-1)^n f^{(n)}(x) & \text{if } b_1 \leq x \leq b_2 \\ 0 & \text{if } x < b_1; x > b_2 \end{cases} \qquad (4.63)$$

As follows from Eq. (4.61), the field at some point of a medium can be observed only after the instant

$$\tau_0 = (\varepsilon \mu_0)^{1/2} R \tag{4.64}$$

With increase of the distance R, the signal appears at later times. Thus, the wave front propagates with velocity defined by high frequency harmonics:

$$v = \frac{1}{(\varepsilon \mu_0)^{1/2}} = \frac{c}{(\varepsilon_r)^{1/2}} \tag{4.65}$$

Using Eq. (4.61), it is easy to derive expressions for the components of the electromagnetic field. By analogy with the frequency domain, we have:

$$E_\phi = -\frac{\partial A_z}{\partial R} \sin \theta$$

and, omitting some simple algebraic operations, we obtain

$$\begin{aligned} E_\phi &= 0 \text{ if } t < \tau_0 \\ E_\phi &= E_\phi^{(1)} + E_\phi^{(2)} \text{ if } t \geq \tau_0 \end{aligned} \tag{4.66}$$

where

$$E_\phi^{(1)} = -\frac{\mu_0 M_0}{4\pi R^2} \left[\left(1 + q\tau_0 + \frac{q^2 \tau_0^2}{2} \right) \delta(t - \tau_0) + \tau_0 \delta'(t - \tau_0) \right] \exp(-q\tau_0) \sin \theta \text{ if } t \geq \tau_0$$

$$E_\phi^{(2)} = -\frac{\mu_0 M_0}{4\pi R^2} q^2 \tau_0^3 \exp(-qt) \frac{I_2 \left[q(t^2 - \tau_0^2)^{1/2} \right]}{t^2 - \tau_0^2} \sin \theta \quad \text{if } t > \tau_0$$

$$\tag{4.67}$$

because

$$I_2(x) = I_0(x) - \frac{2I_1(x)}{x}$$

To determine the magnetic field we use the first Maxwell equation

$$\text{curl } \mathbf{E} = -\frac{\partial \mathbf{B}}{\partial t}$$

which gives

$$\dot{B}_R = \frac{1}{R\sin\theta} \frac{\partial}{\partial\theta}(E_\phi \sin\theta) \quad \text{and} \quad \dot{B}_\theta = \frac{1}{R} \frac{\partial}{\partial R}(RE_\phi)$$

The expression for the $\dot{B}_\theta$ turns out to be fairly complicated, so in the subsequent analysis we consider $\dot{B}_R$ and E_ϕ components only. For the time derivative of the radial component of the magnetic field we have:

$$\dot{B}_R = \dot{B}_R^{(1)} + \dot{B}_R^{(2)}$$

where each term is equal to zero if $t < \tau_0$. Taking into account (4.67), we have:

$$\dot{B}_R^{(1)} = \frac{\mu_0 M_0}{2\pi R^3}\left[\left(1 + q\tau_0 + \frac{q^2\tau_0^2}{2}\right)\delta(t-\tau_0) + \tau_0 \delta'(t-\tau_0)\right]\exp(-q\tau_0)]\cos\theta$$

$$\dot{B}_R^{(2)} = \frac{\mu_0 M_0}{2\pi R^3} q^2\tau_0^3 \exp(-qt)\frac{I_2\left[q(t^2-\tau_0^2)^{1/2}\right]}{t^2-\tau_0^2}\cos\theta$$

(4.68)

Integration of the last expressions with respect to time gives the magnetic fields:

$$B_R = 0 \quad \text{and} \quad B_\theta = 0 \quad \text{if } t < \tau_0 \qquad (4.69)$$

$$B_R^{(1)} = \frac{\mu_0 M_0}{2\pi R^3}[(1+q\tau_0)h(t-\tau_0) + \tau_0\delta(t-\tau_0)]\exp(-q\tau_0)\cos\theta \qquad (4.70)$$

$$B_R^{(2)} = \frac{\mu_0 M_0}{2\pi R^3} q^2\tau_0^3 \int_{\tau_0}^{t} \exp(-qx)\frac{I_2\left[q(x^2-\tau_0^2)^{1/2}\right]}{x^2-\tau_0^2}dx\cos\theta, \quad \text{if } t \geq \tau_0 \quad (4.71)$$

where I_2 is the modified Bessel functions of the second order, while $h(t-\tau_0)$ is the step function

$$h(t-\tau_0) = \begin{cases} 0 & t<\tau_0 \\ 1 & t>\tau_0 \end{cases}$$

As follows from Eqs. (4.69)–(4.71), the electromagnetic field depends on:
1. Time t, distance from the dipole R and angle θ.
2. The velocity of the high frequency propagating waves is defined as

$$v = \frac{1}{(\epsilon\mu_0)^{1/2}} = \frac{R}{\tau_0}$$

3. The parameter q is equal to $\gamma/2\varepsilon$, which has dimension of t^{-1} and characterizes the decay of high frequency waves in a medium.

Similar to the frequency domain, we present the magnetic fields in units of field of the static dipole and functions b_R, b_θ:

$$B_R = \frac{2\mu_0 M_0}{4\pi R^3} b_R \cos\theta, \quad B_\theta = \frac{\mu_0 M_0}{4\pi R^3} b_\theta \sin\theta$$

4.4 THE FIELD IN A NONCONDUCTING MEDIUM

4.4.1 Expressions for the Field

First, we look at the field in a nonconducting medium. In accordance with Eq. (4.70) when $q=0$ and $t>\tau_0$ we have:

$$\begin{aligned}b_R^{(1)} &= h(t-\tau_0) + \tau_0\delta(t-\tau_0) \\ e_\phi^{(1)} &= -\delta(t-\tau_0) - \tau_0\delta'(t-\tau_0)\end{aligned} \quad (4.72)$$

Also from Eq. (4.67)

$$E_\phi = \frac{\mu_0 M_0}{4\pi R^2} e_\phi \sin\theta \quad (4.73)$$

At the observation point R it is natural to distinguish three successive stages. If $t<\tau_0$ the field is absent. Then at the instant $t=\tau_0$ the wave front arrives and after it, $t>\tau_0$, the magnetic field instantly becomes a constant while the electric field vanishes. The sensor measures the mean value of the electric field within some small time interval:

$$E(\tau_0) = \frac{1}{\Delta t}\int_{\tau_0-\Delta t/2}^{\tau_0+\Delta t/2} E(t)\,dt \quad (4.74)$$

Taking into account (4.72)–(4.73), we obtain

$$E(\tau_0) = -\frac{\mu_0 M_0}{4\pi R^2}\sin\theta \quad (4.75)$$

4.4.2 Duhamel's Integral

Prior to this we have considered fields produced by a step-function source. To handle sources of an arbitrary shape

$$M(t) = \begin{cases} 0 & t < 0 \\ M(t) & t \geq 0 \end{cases}$$

it is convenient to use Duhamel's integral. The idea of the method is as follows. The input signal $M(t)$ is represented as the sum of standard signals for which the system response $H(t)$, called the response function is known. Normally, the Heaviside step-function $h(t)$ serves as the standard signal. This is illustrated in Fig. 4.3, where the dipole moment $M(t)$ is represented as a sum of subsequent elementary step-functions $h(t-\tau)$ with the amplitudes $M'(\tau)d\tau$. At the limit of $\Delta \tau \to 0$ the sum is presented as the integral

$$M(t) = M(0) + \int_0^t \frac{dM}{d\tau} h(t-\tau) d\tau$$

The output response of the system is expressed as the integral of the product of the delayed $H(t-\tau)$ and derivative of the input signal $M'(t)$. Thus, knowing the response of the system to the impact of the Heaviside Function, it is possible to predict the system response to an arbitrary input $M(t)$. In the case of magnetic field, assuming that the response function $H(t) = A_B$ is known, the magnetic field $B(t)$, corresponding to the $M(t)$ pulse excitation is calculated as

$$B(t) = M(0)A_B(t) + \int_0^t \frac{dM(\tau)}{d\tau} A_B(t-\tau) d\tau \qquad (4.76)$$

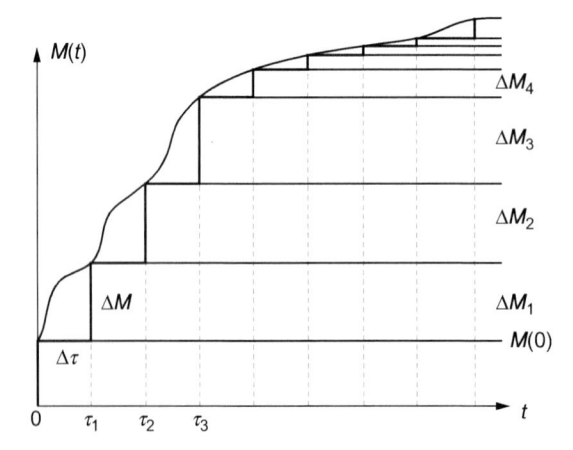

Fig. 4.3 Representation of an arbitrary function as a sum of step-functions.

Propagation and Diffusion in a Conducting Uniform Medium

For instance, in accordance with Eq. (4.72),

$$b_R^{(1)} = \int_0^t \frac{dM}{d\tau}[h(t-\tau_0-\tau) + \tau_0\delta(t-\tau_0-\tau)]d\tau$$

Taking into account that the step-function is equal to unity for positive values of argument $(0 < \tau < t - \tau_0)$, we have

$$\int_0^t \frac{dM}{d\tau} h(t-\tau_0-\tau)d\tau = \int_0^{t-\tau_0} M'(t-\tau_0-\tau)d\tau = M(t-\tau_0)$$

Also, applying Eq. (4.63), we obtain

$$b_R^{(1)}(t) = M(t-\tau_0) + \tau_0 M'(t-\tau_0) \qquad (4.77)$$

By analogy,

$$e_\phi^{(1)}(t) = -M'(t-\tau_0) - \tau_0 M''(t-\tau_0) \qquad (4.78)$$

Thus, the time domain electromagnetic field in a nonconducting medium is expressed in terms of the moment $M(t)$ and its first and second derivatives.

For illustration, consider two examples.

Example One
Suppose that

$$M(t) = \begin{cases} 0 & t < 0 \\ kt & 0 \leq t \leq T, \ k = \frac{1}{T} \\ 1 & t > T \end{cases} \qquad (4.79)$$

Applying Eq. (4.77), we have for the radial component of the field (Fig. 4.4)

$$b_R^{(1)}(t) = \begin{cases} 0 & t < \tau_0 \\ kt & \tau_0 \leq t \leq \tau_0 + T \\ 1 & t > \tau_0 + T \end{cases}$$

The azimuthal component behaves similarly. When the moment is a linear function of time, the magnetic field changes in a similar way, except at

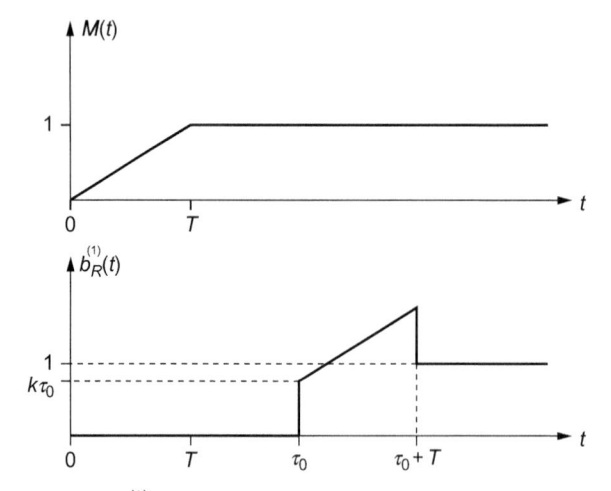

Fig. 4.4 Magnetic field $b_R^{(1)}(t)$ in the case of linear magnetic moment.

the instants $t = \tau_0$ and $t = \tau_0 + T$, where it changes abruptly. At the same time, the electric field is given by

$$e_\phi^{(1)} = \begin{cases} 0 & t < \tau_0 \\ 1/T & \tau_0 \leq t \leq \tau_0 + T \\ 0 & t > \tau_0 + T \end{cases}$$

which is zero except over the interval T where it is a constant.

Example Two

Consider the case when the moment varies as

$$M(t) = \begin{cases} 0 & t < 0 \\ \sin \omega t & 0 < T < C \\ 0 & t > C \end{cases}$$

Then, applying again Eqs. (4.77), (4.78), we present the field as a sum of two sinusoidal functions, having different amplitudes and 90 degrees phase shift. The radial component is

$$b_R(t) = \begin{cases} 0 & t < \tau_0 \\ \sin(\omega(t - \tau_0)) + \omega \tau_0 \cos(\omega(t - \tau_0)) & \tau_0 \leq t \leq \tau_0 + C \\ 0 & t > \tau_0 + C \end{cases}$$

Of course, if $T \gg \tau_0$ and $t \gg \tau_0$, the field becomes a quasi-stationary and changes almost synchronously with the dipole moment.

4.5 THE TRANSIENT FIELD IN A CONDUCTING MEDIUM

Next we return to study the dependence of the electric field on time in the case of a conducting medium. The expressions for the field are given by Eqs. (4.67), (4.73). As in the case of a nonconducting medium, the wave front travels with velocity $\nu = c/(\varepsilon_r)^{1/2}$ and until the moment $t = \tau_0$ the magnitude of the field is zero. The intensity $e_\phi^{(1)}$ of the signal at $t = \tau_0$ essentially depends on the parameter $q\tau_0$:

$$q\tau_0 = \frac{1}{2}\gamma\left(\frac{\mu_0}{\varepsilon}\right)^{1/2} R = b_\infty R = \frac{R}{R_0} = m \qquad (4.80)$$

where $R_0 = 1/b_\infty$ is the characteristic length, introduced in the previous section, and b_∞ coincides with the high frequency limit for the imaginary part of the wavenumber k. Taking into account (4.51), we have:

$$m = \frac{188.5}{\rho(\varepsilon_r)^{1/2}} R \qquad (4.81)$$

For the most practical cases of borehole geophysics this quantity is a very large number, and one can assume that the amplitude of the first arrival is practically zero due to the very small values of the exponential term. It is convenient to represent the function $E_\phi^{(2)}$ as

$$E_\phi^{(2)} = -\frac{M\rho}{2\pi R^4} e_\phi^{(2)} \sin\theta \qquad (4.82)$$

Here

$$e_\phi^{(2)} = m^3 \exp(-mn) \frac{I_2\left[m(n^2-1)^{1/2}\right]}{n^2 - 1} \qquad (4.83)$$

and $n = \dfrac{t}{\tau_0} \geq 1$.

First, consider the field at the moment when it arrives, $n = 1$. Applying the expansion of the function $I_2(z)$ in series and using only the leading term

$$I_2(z) \approx \frac{z^2}{8}$$

we obtain

$$e_\phi^{(2)} \approx \frac{1}{8} m^5 \exp(-m) \qquad (4.84)$$

Function $e_\phi^{(2)}$ has a maximum when $m=5$. If the distance from the dipole R does not exceed $5R_0$, the field increases with conductivity. It also increases with distance if observed at $R<5R_0$. Large values of m correspond to the extremely small fields of almost zero value. Thus, in most cases the field is equal to zero at the first arrival, $t=\tau_0$. Now, suppose that the argument of the function I_2 in Eq. (4.83) is large:

$$m(n^2-1) \gg 1$$

Replacing $I_2(z)$ by its asymptotic expression

$$I_2(z) \approx \frac{\exp(z)}{(2\pi z)^{1/2}} \quad \text{at} \quad z \gg 1$$

we obtain

$$e_\phi \approx e_\phi^{(2)} \approx \frac{1}{(2\pi)^{1/2}} \frac{m^{5/2}}{(n^2-1)^{5/4}} \exp m\left[(n^2-1)^{1/2} - n\right] \qquad (4.85)$$

This equation is applicable when the field is observed at times $t \gg \tau_0$ ($n \gg 1$).

Then we can write

$$e_\phi \approx \frac{1}{(2\pi)^{1/2}} \left(\frac{m}{n}\right)^{5/2} \exp\left(-\frac{m}{2n}\right) \qquad (4.86)$$

which describes an independent on dielectric permittivity the quasi-stationary field. Curves of the function $e_\phi(n)$ are shown in Fig. 4.5. The index of curves is parameter $m = R/R_0$. If the distance from the dipole does not exceed $5R_0$ the electric field decreases monotonically with time. However, with an increase of the parameter m (an increase of conductivity or distance, or a decrease of dielectric permittivity), the maximum of e_ϕ shifts to a later time. Therefore, observing the field in a conducting medium we can distinguish the following stages of the transient response:

1. The field is equal to zero until the moment $\tau_0 = \dfrac{R}{c}(\varepsilon_r)^{1/2}$. For example, if observations are performed at distances from the dipole around 1 m, the arrival time τ_0 is of the order of nanoseconds.

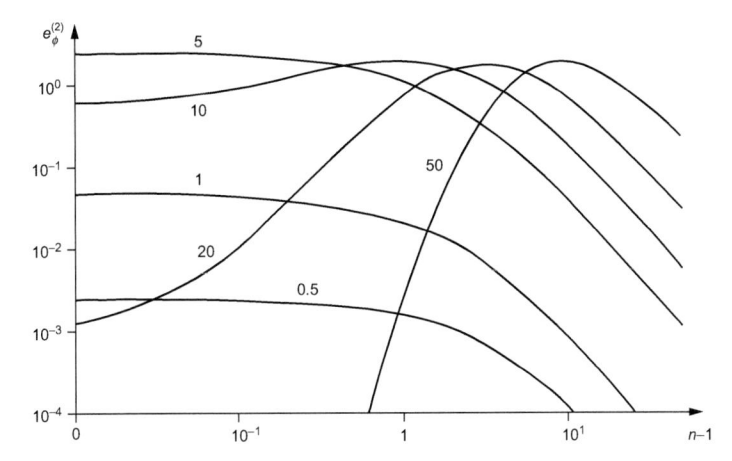

Fig. 4.5 Transient responses of the electric field $e_\phi(n)$. Index of curves is m.

2. At the instant τ_0 the field intensity is a function of the distance from the dipole and electrical parameters of the medium. After the initial wave front passes the observation point, the electric field does not disappear instantaneously. At the beginning, when the time of observation is close to τ_0, both the conduction and displacement currents generate the magnetic field. In other words, the change of the electric field with time cannot be disregarded at the early stage of the transient response when it is only several times greater than τ_0. With an increase of resistivity, the time interval, where displacement currents play an essential role, becomes wider.

3. In the last stage when the electric field varies with time relatively slowly, displacement currents can be disregarded. In this final stage, the field becomes a quasi-stationary and demonstrates features typical for diffusion phenomena. The larger the conductivity is, the earlier the time moment when transition to a quasi-stationary regime occurs.

Next, consider the electric field as a function of the distance from the dipole. Using Eq. (4.67), we have

$$E_\phi^{(2)} = -\frac{M_0 \rho}{2\pi} \frac{(qt)^3}{(vt)^4} F_E \sin\theta$$

where

$$F_E = x \exp(-qt) \frac{I_2\left[qt(1-x^2)^{1/2}\right]}{1-x^2} \quad \text{and} \quad x = \frac{R}{vt}$$

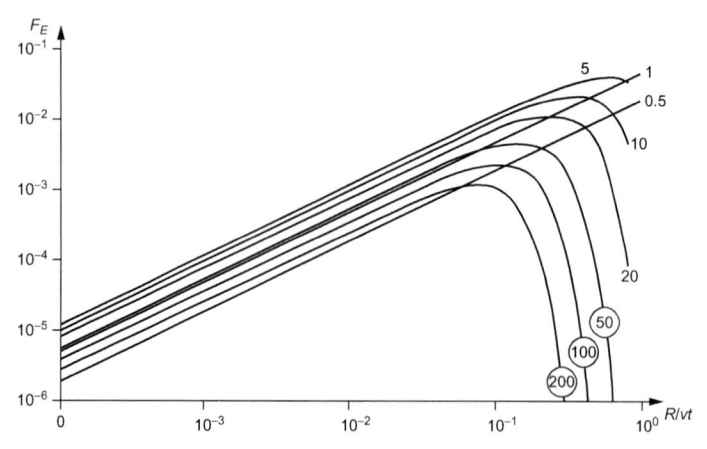

Fig. 4.6 Function F_E at different values of parameter qt.

Graphs, of the electric field as a function of distance, are shown in Fig. 4.6. The curve index is the parameter qt. For small values of qt, the maximum of the field intensity occurs near the wave front and it decreases linearly while approaching the dipole. At large values the maximum moves away from the wave front.

The diffusion equation does not allow us to study the first arrival of energy or the initial stages of the transient response. In the case of sinusoidal oscillations, the quasi-stationary approximation is described by a sinusoidal wave, whose amplitude strongly decays with the distance. An accuracy of this approximation is defined by the ratio of displacement and conduction currents. Propagation and diffusion phenomena take place not only in the case of the step-function excitation. The same phenomena are observed when the pulse is of an arbitrary shape. Of course, the shape affects behavior of the field. For example, when the pulse has a rectangular shape of a very small width comparable to τ_0, the field within the pulse is subjected to influence of both the displacement and conduction currents. In other words, this field propagates as high-frequency waves. At the same time, outside the pulse the field is relatively weak because the fields caused by the step-functions, comprising the pulse, almost cancel each other, and the quasi-stationary stage is hardly noticeable. Correspondingly, a system of such alternating step-functions, following one after another, approximately represents the high-frequency wave, when only propagation is observed. With an increase of the pulse width the quasi-stationary stage appears inside the pulse as well as outside because the cancelation effect becomes weaker.

FURTHER READING

[1] Gabbilard R. Reflexions sur le problem de le propagation d'une onde electromagnetique dans le sol. Rev Ins Fr Ret 1963;XVIII(9).
[2] Kaufman AA. Geophysical theory and methods: Part B. New York: Academic Press; 1994.
[3] Stratton JA. Electromagnetic theory. New York: McGraw-Hill; 1941.
[4] Kaufman A, Alekseev D, Oristaglio M. Principles of electromagnetic methods in surface geophysics. Amsterdam: Elsevier; 2014.

CHAPTER FIVE

Quasistationary Field of Magnetic Dipole in a Uniform Medium

Contents

5.1 Expressions for the Field	163
5.2 Low and High Frequency Asymptotic	166
5.3 Expression for Induced Currents	168
Further Reading	172

In developing the theory of induction logging, we focus our attention on quasistationary fields observed in the borehole in the presence of cylindrical and horizontal boundaries. However, to gain understanding of peculiarities of fields in complicated formations, it is useful to study fields in a uniform medium excited by a vertical magnetic dipole and obtain insight into the physical principles that form the basis for induction logging.

5.1 EXPRESSIONS FOR THE FIELD

When a magnetic dipole with a sinusoidal current is placed in a uniform conducting medium, a change of the primary magnetic field with time causes a primary vortex electric field, and the latter gives rise to the induced currents. These currents and their interaction cause an appearance of the secondary magnetic and electric fields. Due to the symmetry, the interaction does not change a current's direction, and in the spherical system of coordinates they have only a ϕ-component. Because the system is linear, the secondary field is also a sinusoidal function of the same frequency as the primary field. In Chapter 4, we derived equations for the electromagnetic field of the magnetic dipole in a uniform medium when both conduction and displacement currents are present. Taking into account Eqs. (4.42), (4.45), we have for the complex amplitudes of the quasistationary field:

Basic Principles of Induction Logging
http://dx.doi.org/10.1016/B978-0-12-802583-3.00005-8

$$E_\varphi^* = \frac{i\mu_0\omega M_0}{4\pi R^2}(1 - ikR)\exp(ikR)\sin\theta$$

$$B_R^* = \frac{\mu_0 M_0}{2\pi R^3}(1 - ikR)\exp(ikR)\cos\theta \qquad (5.1)$$

$$B_\theta^* = \frac{\mu_0 M_0}{4\pi R^3}\left(1 - ikR - k^2R^2\right)\exp(ikR)\sin\theta$$

Here the wave number is

$$k = \frac{1+i}{\delta}, \quad \delta = \left(\frac{2}{\gamma\mu_0\omega}\right)^{1/2} = \frac{10^3}{2\pi}(10\rho T)^{1/2} m \qquad (5.2)$$

where T is the period of oscillation and, as before, δ is the skin depth. The dipole moment varies as

$$M = M_0 \cos\omega t \qquad (5.3)$$

and, in accordance with the Biot-Savart law, it generates primary magnetic fields, $B_R^{(0)}$ and $B_\theta^{(0)}$:

$$B_R^{(0)} = \frac{\mu_0 M_0}{2\pi R^3}\cos\theta\cos\omega t \quad \text{and} \quad B_\theta^{(0)} = \frac{\mu_0 M_0}{4\pi R^3}\sin\theta\cos\omega t \qquad (5.4)$$

This field is confined to meridian planes and synchronously changes with the dipole current. Earlier we called this field quasistationary. Its variation with time causes the vortex electric field (Chapter 3) with complex amplitude:

$$E_\phi^{(0)*} = \frac{i\omega\mu_0 M_0}{4\pi R^2}\sin\theta \qquad (5.5)$$

and for the field $E_\phi^{(0)}$ we have:

$$E_\varphi^{(0)} = \text{Re}\left[\frac{i\omega\mu_0 M_0}{4\pi R^2}\exp(-i\omega t)\right] = \frac{\omega\mu_0 M_0}{4\pi R^2}\sin\theta\sin\omega t \qquad (5.6)$$

which is confined to horizontal planes and exists at any point in space regardless of whether the medium is conductive. The primary electric and magnetic fields are shifted in-phase with respect to each other by 90 degrees. As in the general case (Chapter 4), it is convenient to express the complex amplitudes of the field in a conducting medium in terms of the primary field, that is

$$b_R^* = (1 - ikR)\exp(ikR)$$
$$b_\theta^* = (1 - ikR - k^2R^2)\exp(ikR) \quad (5.7)$$
$$e_\varphi^* = (1 - ikR)\exp(ikR)$$

Inasmuch as the right-hand sides in Eq. (5.7) are complex numbers, we can say that there is a phase shift between the field and the dipole current. For instance, in the case of the radial component we have:

$$B_R = \frac{\mu_0 M_0}{2\pi R^3} \cos\theta \operatorname{Re}[(c_R + id_R)\exp(-i\omega t)]$$

or

$$B_R = \frac{\mu_0 M_0}{2\pi R^3} \cos\theta [c_R \cos\omega t + d_R \sin\omega t] \quad (5.8)$$

where

$$c_R + id_R = b_R^*.$$

By analogy,

$$B_\theta = \frac{\mu_0 M_0}{4\pi R^3} \sin\theta [c_\theta \cos\omega t + d_\theta \sin\omega t] \quad (5.9)$$

Here

$$c_\theta + id_\theta = b_\theta^*$$

In essence, the field is a sinusoidal wave that relatively rapidly decays with the distance from the dipole. We can also interpret fields as a sum of two harmonic functions, called the in-phase and quadrature components:

$$Inb_R = c_R \cos\omega t \quad Qb_R = d_R \sin\omega t$$
$$Inb_\theta = c_\theta \cos\omega t \quad Qb_\theta = d_\theta \sin\omega t \quad (5.10)$$

By definition, the real and imaginary parts of the complex amplitude are the amplitudes of the in-phase and quadrature components, respectively. The in-phase component changes synchronously with the primary field, whereas the quadrature component is shifted in-phase by 90 degrees. In general, these oscillations have different amplitudes. Similarly, the electric field and the current density can be represented as the sum of the quadrature and in-phase components. According to the Biot-Savart law, the quadrature component of the secondary magnetic field arises due to currents that are shifted in-phase by 90 degrees with respect to the current in the dipole, whereas the in-phase component of the field is the algebraic sum of the primary field and the in-phase component of the secondary field. The latter is contributed by induced currents in the medium that are in-phase with the dipole current. This representation is useful for understanding the physical principles of induction logging, which is based on measurements of corresponding components of the field. It is natural to distinguish two special cases when either radial or equatorial components exist: $\theta = 0$ ($b_R \neq 0$ and $b_\theta = 0$) and $\theta = \pi/2$ ($b_R = 0$ and $b_\theta \neq 0$).

5.2 LOW AND HIGH FREQUENCY ASYMPTOTIC

First, consider the low frequency spectrum (or limit) of the field. Expanding $\exp(ikR)$ in the series

$$\exp(ikR) = \sum_{n=0}^{\infty} \frac{(ikR)^n}{n!}$$

and substituting this into Eq. (5.7) after some simple algebra, we have:

$$b_R^* = 1 + \sum_{n=2}^{\infty} \frac{1-n}{n!} 2^{n/2} p^n \exp\left(i\frac{3\pi n}{4}\right) \tag{5.11}$$

Here

$$p = \left(\frac{\gamma \mu_0 \omega}{2}\right)^{1/2} \quad R = \frac{R}{\delta} \tag{5.12}$$

is the parameter characterizing the distance between the dipole and an observation point expressed in units of skin depth δ. Sometimes the parameter p is called the induction number. Taking into account Eq. (5.11), we see that the series describing the low frequency part of the spectrum contains whole and fractional powers of ω. As follows from this equation:

$$\operatorname{Im} b_R^* = d_R \approx p^2 - \frac{2}{3}p^3 \quad \text{and} \quad \operatorname{Re} b_R^* = c_R \approx 1 - \frac{2}{3}p^3 \tag{5.13}$$

or

$$\operatorname{Im} B_R^* \approx \frac{\mu_0 M_0}{2\pi R^3} \cos\theta \left[\frac{\gamma \mu_0 R^2}{2} \omega - \frac{(\gamma \mu_0 R^2)^{3/2}}{3(2)^{1/2}} \omega^{3/2} \cdots \right] \tag{5.14}$$

and

$$\operatorname{Re} B_R^* \approx \frac{\mu_0 M_0}{2\pi R^3} \cos\theta \left[1 - \frac{(\gamma\mu_0 R^2)^{3/2}}{3(2)^{1/2}} \omega^{3/2} \cdots \right] \quad (5.15)$$

Thus, within the range of small parameter p, the quadrature and in-phase components are related to the frequency, the conductivity, and the distance from the dipole in completely different manners. The first term on the right-hand side of Eq. (5.15) characterizes the primary field, which is caused only by the dipole current. The next term describes the in-phase component of the secondary magnetic field, which arises due to the currents induced in the conductive medium. At the same time, all the terms describing the quadrature component correspond to the secondary field. Comparison of the last two equations shows that the in-phase component of the secondary field is more sensitive to changes in conductivity than the first term of the quadrature component, and in this low frequency limit the in-phase component is independent of the dipole-receiver distance. In fact, this interesting feature at $p \ll 1$ indicates potentially large depth of penetration of the in-phase component compared to that of the quadrature component. In a similar manner, we obtain expressions for the azimuthal component of the field:

$$\operatorname{Im} b_\theta^* \approx -p^2 + \frac{4}{3}p^3 \quad \text{and} \quad \operatorname{Re} b_\theta^* \approx 1 + \frac{4}{3}p^3 \quad (5.16)$$

In accordance with Eq. (5.7) at the high frequency range when $p \gg 1$, the in-phase and quadrature components of the field approach zero:

$$\operatorname{Re} \mathbf{b}^* \to 0 \quad \text{or} \quad \operatorname{Re} \mathbf{b}^{s^*} = -\mathbf{b}^0 \quad \text{and} \quad \operatorname{Im} \mathbf{b}^* \to 0$$

where $\mathbf{b}^{s^*}$ is the complex amplitude of the secondary magnetic field. At such frequencies the induced currents are concentrated in the vicinity of the dipole causing strong skin effect. Correspondingly, the secondary in-phase component differs from the primary field by sign only.

Since the radial and azimuthal components behave similarly, we may focus on the radial component:

$$\begin{aligned} \operatorname{Im} b_R^* &= \exp(-p)[(1+p)\sin p - p\cos p] \\ \operatorname{Re} b_R^* &= \exp(-p)[(1+p)\cos p + p\sin p] \end{aligned} \quad (5.17)$$

The graphs, illustrating dependence of both quadrature and in-phase field components on the parameter p, are presented in Fig. 5.1A and B. With an increase in the induction number, the quadrature component ($\operatorname{Im} b_R^*$)

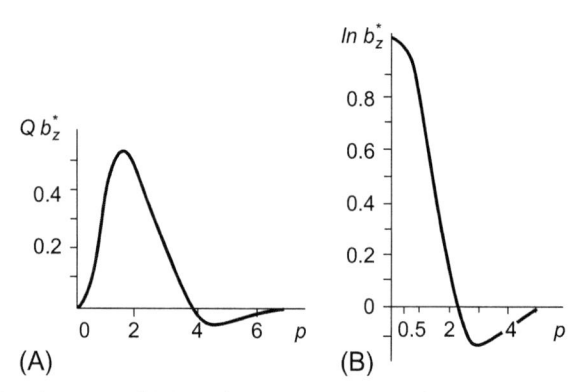

Fig. 5.1 (A) Quadrature and (B) in-phase components of the magnetic field.

increases, reaches maximum, and then tends to zero. By contrast, the in-phase component decreases and then, like the quadrature component, approaches zero in an oscillating manner. According to Eq. (5.13), at the low frequency limit, the amplitude of the quadrature component prevails over the secondary in-phase component InB_R^s, and we have:

$$QB_R = \frac{\mu_0 M_0}{4\pi R} \gamma \mu_0 \omega \cos\theta \sin\omega t, \quad p \ll 1 \qquad (5.18)$$

Hence in the range of a small parameter, the quadrature component is directly proportional to the conductivity and the frequency, and inversely proportional to the distance from the magnetic dipole. As will be shown later, some of these features of the field also remain valid in a nonuniform conducting medium. From Eq. (5.17), we also see that at $p \ll 1$ the in-phase component of the secondary field InB_z^s is much smaller than the primary field and the quadrature component of the secondary field QB_R:

$$InB_z^s \ll QB_R \ll B_R^{(0)} \qquad (5.19)$$

Because of this inequality (5.19) low frequency induction measurements require high-accuracy compensation of the primary field.

5.3 EXPRESSION FOR INDUCED CURRENTS

Let us analyze the behavior of the field in terms of the distribution of induced currents. Applying Eq. (5.7) and Ohm's law:

$$\mathbf{j} = \gamma \mathbf{E}$$

we have the following expression for the current density at any point in a uniform medium:

$$\overset{*}{j}_\phi = \frac{i\gamma\mu_0\omega M_0}{4\pi R^2} \exp(ikR)(1-ikR)\sin\theta \qquad (5.20)$$

As in the case of the magnetic field, we represent the current density as the sum of the quadrature and in-phase components, and, using Eq. (5.20), obtain:

$$\begin{aligned}\mathrm{Im}\, j_\varphi &= \frac{\gamma\mu_0\omega M_0 r}{4\pi R^3}\exp(-p)[(1+p)\cos p + p\sin p]\\ \mathrm{Re}\, j_\varphi &= -\frac{\gamma\mu_0\omega M_0 r}{4\pi R^3}\exp(-p)[(1+p)\sin p - p\cos p]\end{aligned} \qquad (5.21)$$

The distribution of currents represents a system of rings located in planes perpendicular to this axis (Fig. 5.2A) and having a common axis with that of the dipole. According to Eq. (5.5), for the density of induced currents arising due to the primary electric field, we have:

$$j_\varphi^{(0)*} = \gamma E_\varphi^{(0)*} = \frac{i\gamma\mu_0\omega M_0 r}{4\pi R^3} \qquad (5.22)$$

and their phase is shifted by 90 degrees with respect to the dipole current. If interaction between induced currents is negligible, then Eq. (5.22) describes the actual distribution. In this case, the current density at any point in the medium is a product of two terms. The first term depends on the dipole moment, frequency, and conductivity at the observation point; the second is determined by coordinates of the point of observation. Finding current distribution and magnetic field of these currents is an elementary task when interaction between induced currents is negligible and the primary electric field does not intersect any boundaries. This last condition is critical because appearance of the electric charges changes the direction of the current density; the geometry of currents becomes unknown, making it impossible to apply the Biot-Savart law.

In Chapter 6 we demonstrate that the approximation based on the use of Eq. (5.22) is the foundation of Doll's "geometrical factor theory" in "low frequency" induction logging. The behavior of amplitude of the current $j_\phi^{(0)}$ in planes perpendicular to the dipole axis is shown in Fig. 5.2B. It also illustrates that increase of the distance from the dipole along z-direction $z_1 < z_2 < z_3$ leads to the shift of the maximal density along the radial direction.

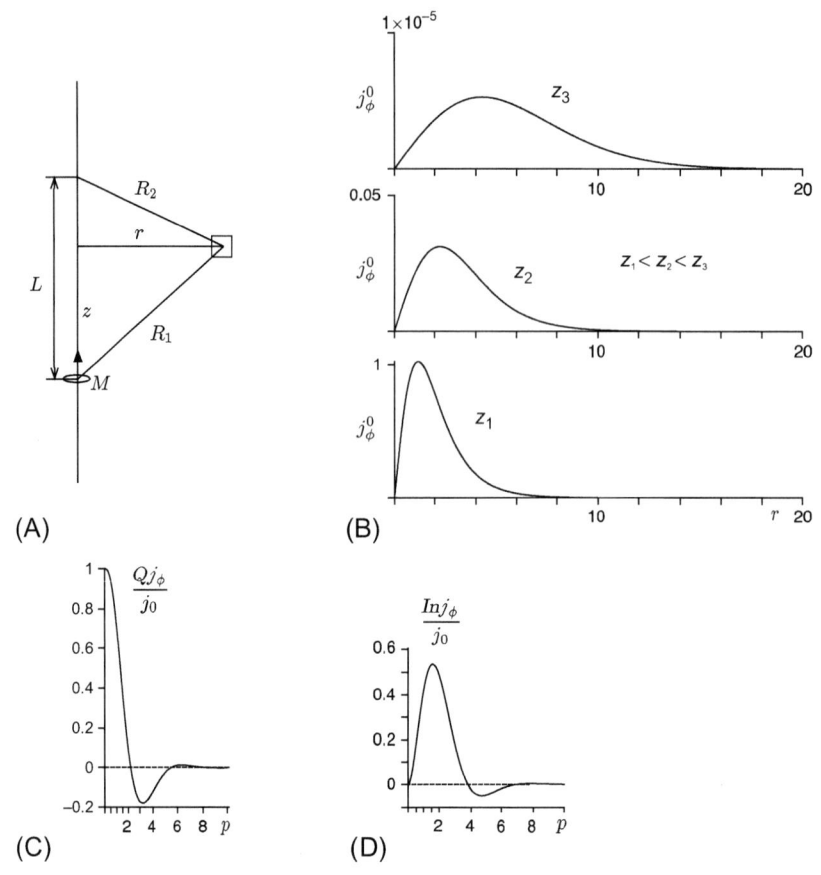

Fig. 5.2 (A) Geometry of current tubes; (B) distribution of current density, j_ϕ^0, in planes perpendicular to the dipole axes; (C), (D) quadrature and in-phase components of the current density, respectively.

Introducing notation:

$$j_0 = \frac{\gamma\mu_0\omega M_0}{4\pi}\frac{r}{R^3}$$

we may rewrite Eq. (5.21) as

$$Qj_\varphi = j_0 \exp(-p)[(1+p)\cos p + p\sin p]$$
$$Inj_\varphi = j_0 \exp(-p)[(1+p)\sin p - p\cos p] \quad (5.23)$$

Analyzing functions Eq. (5.23) we can see how the actual current density, j_ϕ, differs from j_0 for different values of p. The quadrature and in-phase components of j_ϕ normalized by j_0 are shown in Fig. 5.2C and D. For small values of the parameter $p \leq 0.7$, the quadrature component of the current density is essentially the same as j_0, indicating that interaction between induced currents is negligible. As the parameter p increases, the ratio Qj_ϕ/j_0 decreases, passes through zero, and for large p, approaches zero in an oscillating manner. The ratio $In(j_\phi/j_0)$ has a completely different character. At small p the ratio Inj_ϕ/j_0 approaches zero, then increases to a maximum value at $p \approx 1.5$ and tends to zero again in an oscillating manner. The actual distribution of currents, in contrast to j_0, is determined by both geometric factors and the interaction of currents. Although at small values of p the quadrature component of the current density is dominant (Fig. 5.2C and D), there is a range of p where the in-phase component is significantly larger.

The main features of the magnetic field can be analyzed, proceeding from the distribution of the corresponding components of the current density. If the frequency is low enough and the medium has a relatively high resistivity, the range of distances for which the actual current density Qj_ϕ is almost equal to j_0 becomes large and the magnetic field $Q\mathbf{B}$ is defined entirely by currents in this area. In this frequency limit the depth of investigation cannot be increased by lowering frequency despite increased penetration of the field into the formation. Both the current density Qj_ϕ in this area and magnetic field caused by these currents are directly proportional to the frequency, Eq. (5.22). Within some range of the parameter p, the dimensions of this volume remain much greater than the distance from the dipole to an observation point. As the parameter p increases (e.g., by an increase of the frequency), the size of this volume becomes smaller, leading to decreased growth of $Q\mathbf{B}$ with frequency. As frequency increases further, there is a rapid decrease of both ratio Qj_ϕ/j_0 and the quadrature component of magnetic field.

By analogy, the behavior of the in-phase component of the field can be explained by the in-phase component of the current. Unlike the quadrature component Qj_ϕ, which is not indicative of the diffusion in the medium, the in-phase component clearly shows a diffusion process. For instance, a maximum of Inj_ϕ moves away from the dipole when the frequency decreases, indicating an increased sensitivity of magnetic field to the distant parts of a medium. The depth of penetration of the in-phase component gradually increases with a decrease of frequency, regardless of the distance between the dipole and an observation point. This feature of the in-phase component manifests itself primarily when the separation between the dipole and receiver is comparable to or less than the thickness of the skin depth (similar behavior is observed in the transient field discussed in

Chapter 10). Inasmuch as the density of the current Inj_ϕ around the dipole is small, the field component, $In\mathbf{B}$, is defined by currents located relatively far from the probe. For this reason, a change of relatively small distance between the dipole and receiver practically has no effect on the field. However, with further increase of separation, the dipole-receiver distance has greater influence. These general features of the quadrature and in-phase components of the field remain valid for a nonuniform medium as well.

FURTHER READING
[1] Kaufman AA. About the theory of induction logging. Moscow: Geology and Prospecting; 1960.

CHAPTER SIX

Geometrical Factor Theory of Induction Logging

Contents

6.1 Two-Coil Probe	174
6.1.1 Geometrical Factor of the Elementary Ring	174
6.1.2 Solution of the Forward Problem	179
6.1.3 Apparent Conductivity	181
6.2 The Vertical Responses of the Two-Coil Probe in the Media With the Horizontal Boundaries	182
6.2.1 Geometric Factor of an Elementary Layer	182
6.2.2 Geometric Factor of a Layer With a Finite Thickness	184
6.2.3 Apparent Conductivity in the Presence of a Layer With Finite Thickness	188
6.3 Radial Characteristics of Two-Coil Induction Probe	195
6.3.1 Geometric Factor of the Borehole	195
6.3.2 Radial Characteristics of Two-Coil Probe	199
6.4 Multicoil or "Focusing" Induction Probe	204
6.4.1 Conditions for the Application of "Focusing" Probes	205
6.4.2 Three- and Multicoil Probes	206
6.5 Corrections of the Apparent Conductivity	222
6.5.1 Skin Effect Corrections	222
6.5.2 Borehole Correction	225
References	226
Further Reading	226

In 1949, Henri Doll suggested the method of induction logging for measuring the electrical resistivity of formations surrounding a borehole [1]. He also developed an approximate geometrical factor theory, allowing one to establish a relation between parameters of a geo-electric section and a signal, measured by the induction probe. The basis of this theory is the assumption that the frequency of the induction probe, located on the borehole axis, is relatively low, and the mutual interaction of currents induced in the borehole and surrounding axially symmetric medium can be neglected.

This assumption implies 90 degrees phase shift between induced currents in the medium and the current in a transmitting coil. Thus, the signal measured by a receiver coil of the induction probe consists of two parts. The first is the primary signal, caused by a transmitting coil located in free space; the second part is caused by induced currents in a medium whose amplitude depends on the conductivity of formation. Analysis of the field of a magnetic dipole in a uniform medium (Chapter 5) shows that the behavior of the field and the induced currents assumed in Doll's theory is a good approximation of the actual situation when the transmitter-receiver distance, frequency, and conductivity are relatively small. Later, we compare results of a field calculation using the Doll's theory and exact solutions and establish conditions when the theory of geometric factor is valid. Doll's theory permits a simple derivation of the quadrature (out-of-phase) component of the magnetic field in a medium with either horizontal or cylindrical interfaces, provided that the field is caused by a vertical magnetic dipole directed along the vertical axis of the borehole. In this case, there is no component of electric field perpendicular to the boundaries between regions of different conductivity and, therefore, no surface electrical charges. Of course, Doll understood very well the phenomenon of skin effect as well as the conditions under which this effect is negligible. In practice, these conditions define useful operating frequencies of induction logging, and it is not occasional that, after 60 years, most induction logging instruments still use frequencies in the range that are close to those suggested by Doll.

6.1 TWO-COIL PROBE

Describing Doll's geometrical factor theory we begin from the basic concept of the geometrical factor of a simple two-coil probe.

6.1.1 Geometrical Factor of the Elementary Ring

Let us consider the region formed by the intersection of two horizontal planes with two coaxial cylindrical surfaces having a common axis with the borehole (Fig. 6.1).

The elementary region bounded by the planes and cylindrical surfaces forms a horizontal ring, which is filled with a uniform medium. Its cross-section dS is rectangular; for convenience, we assume that the region has unit area ($dS = 1$). It is essential that dimensions of the cross-section are small compared with the ring's radius. Doll called this part of the medium an "elementary unit ring." Now we find the signal at the receiver of a two-coil

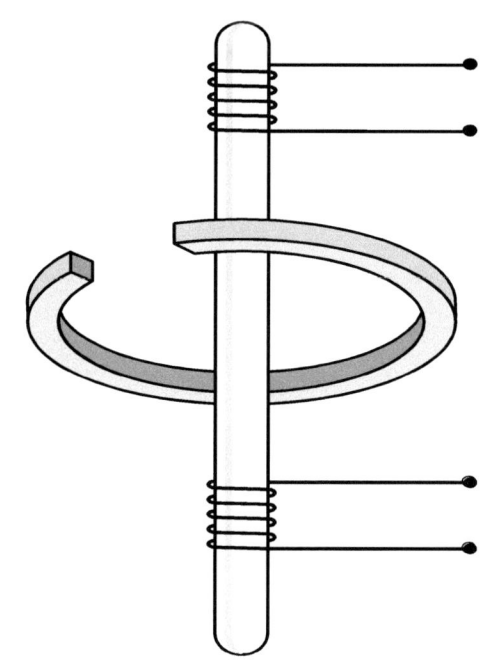

Fig. 6.1 Elementary ring with respect to the two-coil induction probe.

induction probe caused by the induced current circulating in the ring. As shown earlier (Eq. 3.42), the primary electric field of a dipole in free space is

$$E_\phi = \frac{i\mu_0 \omega r M_T}{4\pi R_1^3}$$

Here r is the radius of the ring, M_T is a magnetic moment of the transmitter coil, and R_1 is the distance from the transmitter coil to the ring. Then Ohm's law directly gives the complex amplitude of the induced current in the ring as

$$I^* = \frac{i\gamma\mu_0 \omega r M_T}{4\pi R_1^3} \quad \text{if} \quad dS = 1 \tag{6.1}$$

where γ is the conductivity of the medium occupied by the ring (Fig. 6.2).

Cross-sections of elementary rings with a maximal geometrical factor lie on the solid circle. Because the ring's cross-section is small:

$$dr \ll r \quad \text{and} \quad dz \ll z$$

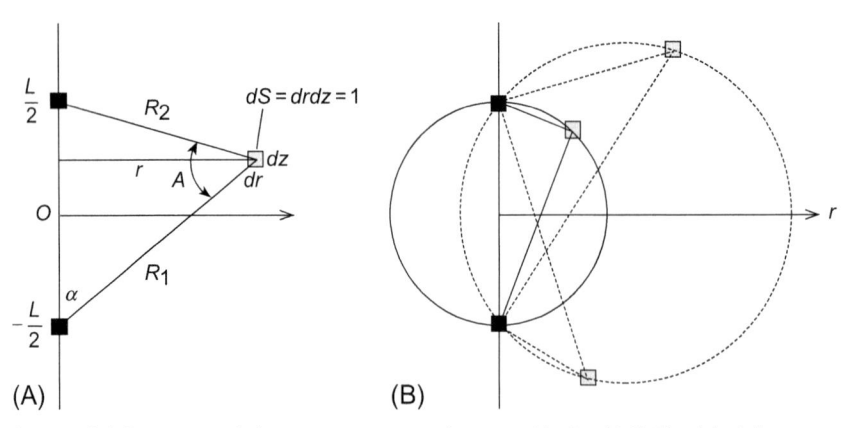

Fig. 6.2 (A) Geometry of elementary current rings used in Eq. (6.1). The black boxes on the borehole axis are the locations of the transmitter and receiver coils (dipoles). (B) Elementary rings whose cross sections lie on circles that pass through transmitter and receiver coils with centers on the r axis.

the elementary ring forms a circular loop of radius r. The current in the elementary ring with radius r generates the secondary magnetic field, which has only the vertical component at the axis:

$$B_z^* = \frac{\mu_0 I^* r^2}{2 R_2^3} \qquad (6.2)$$

Here, R_2 is the distance from points of the ring to the receiver coil. The flux of this secondary magnetic field piercing a small receiver coil along the axis is

$$\Phi^* = \frac{\mu_0}{2} \frac{I^* r^2}{R_2^3} S_2 n_2$$

where S_2 and n_2 are area and number of turns in the coil, respectively. For the complex amplitude of the electromotive force in the coil, arising due to a sinusoidal in time magnetic field, we have

$$\Xi^* = i\omega \Phi^* = \frac{i\omega \mu_0}{2} \frac{I^* r^2}{R_2^3} S_2 n_2 \qquad (6.3)$$

Substituting Eq. (6.1) into Eq. (6.3), we obtain the expression for the electromotive force in the small receiver coil, generated by the secondary magnetic field:

$$\Xi = -\frac{\pi}{2}f^2\mu_0^2\gamma I_0 S_1 S_2 n_1 n_2 \frac{r^3}{R_1^3 R_2^3} \qquad (6.4)$$

where I_0 is the current amplitude in the transmitter coil, S_1, n_1, and S_2, n_2 are areas and number of turns of the transmitter and receiver coils, respectively; f is the frequency of the field. By definition

$$M_T = I_0 S_1 n_1$$

is the magnetic moment of the transmitter coil. Also the product

$$M_R = S_2 n_2$$

is called the moment of the receiver coil, so that Eq. (6.4) can be rewritten as

$$\Xi = -\frac{\pi}{2}f^2\mu_0^2\gamma M_T M_R \frac{r^3}{R_1^3 R_2^3}$$

Note that the transmitter and receiver moments have different units (Am^2 and m^2), but this notation is still convenient, especially for discussing multicoil probes.

Let us write the last equation in the form

$$\Xi = K_0 \gamma g_0 \qquad (6.5)$$

where

$$K_0 = -\frac{\pi}{2}f^2\mu_0^2 I_0 S_1 S_2 n_1 n_2$$

is the coefficient that depends on parameters of the two-coil probe, and

$$g_0 = \frac{r^3}{R_1^3 R_2^3} \qquad (6.6)$$

is a function depending on the radius and location of the ring as well as on the probe length L. Doll called this function "the geometric factor of an elementary ring" or the "elementary geometric factor." Thus, the signal generated by the current in an elementary ring within a medium is directly proportional to the conductivity and geometric factor of the ring. Now we represent the function g_0 in a cylindrical coordinate system, r, z with its origin at the middle of the induction probe (Fig. 6.2). Because

$$R_1 = \left[r^2 + (L/2 + z)^2\right]^{1/2}, \quad R_2 = \left[r^2 + (L/2 - z)^2\right]^{1/2}$$

we have for the function g_0:

$$g_0 = \frac{r^3}{\left[r^2 + (L/2+z)^2\right]^{3/2}\left[r^2 + (L/2-z)^2\right]^{3/2}} \qquad (6.7)$$

Following Doll, we introduce a new function g:

$$g = \frac{L}{2}g_0 = \frac{L}{2}\frac{r^3}{\left[r^2 + (L/2+z)^2\right]^{3/2}\left[r^2 + (L/2-z)^2\right]^{3/2}} \qquad (6.8)$$

At the same time, the probe coefficient K_0 is multiplied by $2/L$:

$$K = \frac{2}{L}K_0$$

It will be shown later that the geometric factor of the whole space is equal to unity. In accordance with Eq. (6.8), this factor g depends on the angle under which both coils of the induction probe are seen from any point of the elementary ring, and it is equal to

$$g = \frac{\sin^3 A}{2L^2} \qquad (6.9)$$

Indeed, as follows from Fig. 6.2,

$$\frac{\sin A}{L} = \frac{\sin \alpha}{R_2}, \quad \sin \alpha = \frac{r}{R_1}, \quad \frac{\sin A}{L} = \frac{r}{R_1 R_2}$$

and, therefore,

$$g = \frac{L}{2}\frac{r^3}{R_1^3 R_2^3} = \frac{L}{2}\frac{\sin^3 A}{L^3} = \frac{\sin^3 A}{2L^2}$$

In other words, for a given probe length L the elementary geometric factor is defined by the angle under which the probe is seen from points of the elementary ring. Thus, all elementary rings have the same geometric factor, if the probe is seen under the same angle from the ring's points. Consequently, they contribute the same signal if they have the same conductivity. The cross-sections of elementary rings with the same geometrical factor lie along circles that pass through the transmitter and receiver coils and have their centers on the r axis. Fig. 6.2A illustrates this concept for such circles of different radii. Elementary rings for which $\sin A = 1$ have the maximum geometric factor, which are equal to $1/2L^2$. Cross-sections of these rings are located on the circle with radius $L/2$.

6.1.2 Solution of the Forward Problem

Now we derive the signal, caused by induced currents in a whole space. In fact, making use of the principle of superposition and neglecting interaction of induced currents, the electromotive force is equal to the sum of the signals from all elementary rings, i.e.:

$$\Xi = K \int_S \gamma g \, dS \tag{6.10}$$

where dS is the cross-section of the elementary ring. In general, the conductivity can be a continuous or discontinuous function of coordinates. In particular, if the medium is uniform, we have

$$\Xi = K\gamma \int_S g \, dS = K\gamma \int_r dr \int_z g \, dz$$

Inasmuch as radii of elementary rings change from 0 to ∞ and the coordinate z varies from $-\infty$ to $+\infty$, the expression for electromotive force in uniform medium is

$$\Xi = K\gamma \int_0^\infty dr \int_{-\infty}^\infty g \, dz \tag{6.11}$$

As follows from Eq. (6.8), this double integral gives

$$\int_0^\infty dr \int_{-\infty}^\infty g \, dz = \int_0^\infty dr \int_{-\infty}^\infty \frac{L}{2} \frac{r^3}{\left[r^2 + (L/2+z)^2\right]^{3/2} \left[r^2 + (L/2-z)^2\right]^{3/2}} dz = 1 \tag{6.12}$$

and

$$\Xi = K\gamma$$

In other words, the geometric factor of uniform medium is equal to unity. Let us consider a nonuniform medium divided into different uniform regions, as shown in Fig. 6.3.

Taking into consideration the axial symmetry, we denote the regions with letters A, B, C, D, and E. The contribution of every uniform part of the medium to the total signal is proportional to the product of the corresponding conductivity and geometric factor of this part. By definition,

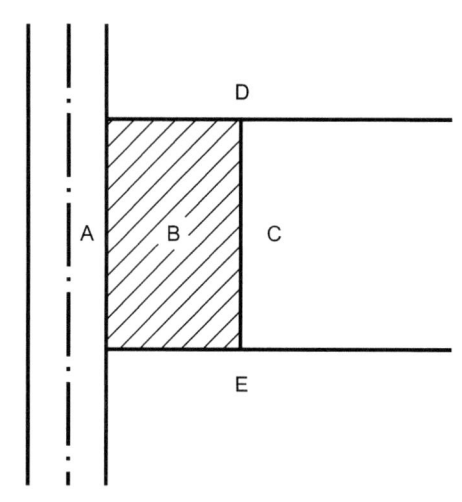

Fig. 6.3 Conductivity distribution in a nonuniform medium.

the latter is a sum of geometric factors of elementary rings over the area of the considered part of the medium. For example, if conductivities of parts A, B, C, D, and E are equal to $\gamma_A, \gamma_B, \gamma_C, \gamma_D$, and γ_E, the total electromotive force is

$$\Xi = K\left[\gamma_A\iint_A gdS + \gamma_B\iint_B gdS + \gamma_C\iint_C gdS + \gamma_D\iint_D gdS + \gamma_E\iint_E gdS\right] \quad (6.13)$$

Here

$$\iint_A gdS \cdots \iint_E gdS$$

are geometric factors of the corresponding parts of the medium. Introducing notations

$$G_A = \iint_A gdS, \quad G_B = \iint_B gdS \cdots G_E = \iint_E gdS$$

we obtain the following expression for the magnitude of the electromotive force:

$$\Xi = K(\gamma_A G_A + \gamma_B G_B + \gamma_C G_C + \gamma_D G_D + \gamma_E G_E) \quad (6.14)$$

Inasmuch as the geometric factor for a whole uniform medium is equal to unity, we always have

$$G_A + G_B + G_C + G_D + G_E = 1 \tag{6.15}$$

As follows from Eq. (6.14) the electromotive force is given by the sum of products of geometrical factors and the conductivities of different regions of the media. Thus the conductivity of a region and its geometric factor has a similar influence on the signal. For instance, a region with a high conductivity and small geometrical factor and a region with low conductivity and large geometrical factor make the same contributions to the total signal, if their conductivity-geometric factor products are the same.

6.1.3 Apparent Conductivity

By analogy with the apparent resistivity in direct-current methods, Doll introduced the apparent conductivity γ_a defined by

$$\frac{\gamma_a}{\gamma_1} = \frac{\Xi}{\Xi^{un}(\gamma_1)} \tag{6.16}$$

It characterizes how the measured electromotive force differs from that in a uniform medium with conductivity γ_1. This definition is equivalent to the more commonly used definition, utilizing K-factor:

$$\gamma_a = \Xi/K.$$

In accordance with Eqs. (6.12) and (6.14), for the apparent conductivity γ_a and apparent resistivity ρ_a we have:

$$\gamma_a = \gamma_A G_A + \gamma_B G_B + \gamma_C G_C + \gamma_D G_D + \gamma_E G_E \tag{6.17}$$

$$\rho_a = [\gamma_A G_A + \gamma_B G_B + \gamma_C G_C + \gamma_D G_D + \gamma_E G_E]^{-1} \tag{6.18}$$

The concept of an elementary geometric factor makes it easy to derive expressions for the signal caused by currents in various parts of a conducting medium. Doll also used geometrical factors of elementary horizontal and cylindrical layers to study both the vertical and radial characteristics of the two-coil probes, as well as much more complicated induction systems. The concept of geometric factor, which is often called "geometrical factor theory," can be summarized as:

1. In geometrical factor theory, induced currents are generated only by the primary electric field of the transmitter coil:

$$E_\phi^{0*} = \frac{i\omega\mu_0 M_T r}{4\pi R^3}$$

2. The theory implies that the mutual interaction between induced currents is neglected, and thus every element of a medium manifests itself independently, regardless of the resistivity of neighboring parts. This theory does not predict an in-phase component of the secondary magnetic field.
3. The Doll's approximation is more valid in regions that are close to the source where induced currents are mainly defined by the primary electric field. In areas located far from the source, currents are subject to the skin effect and geometrical factor theory is less applicable. As the length of the probe increases, the frequency must be lowered in order for Doll's approximation to remain accurate.
4. Simplicity of the theory is based on an axially-symmetric geometry of a media and absence of electrical charges at the boundaries of regions with different conductivities. In the absence of axial symmetry surface charges appear, whose density depends on the magnitude of the normal component of electric field at the boundary and on the resistivity contrast across the boundary. In such cases, it becomes impossible to retain the concept of a geometric factor, although at a sufficiently low frequency, the quadrature component of magnetic field is still directly proportional to the frequency and conductivity.

In the following chapter, we show that geometrical factor theory represents the first approximation of the integral equation, describing the response of induction tool in a medium with varying electrical conductivity. Now we use Doll's theory to study the response of induction probes in different formations and start from horizontally-layered media.

6.2 THE VERTICAL RESPONSES OF THE TWO-COIL PROBE IN THE MEDIA WITH THE HORIZONTAL BOUNDARIES

6.2.1 Geometric Factor of an Elementary Layer

By analogy with an elementary ring whose cross section is small, a layer whose thickness is much less than the probe length can be considered as an elementary layer. The geometrical factor of an elementary layer is defined by a summation of the geometrical factors of its elementary rings, which are all located at the same distance z with respect to the origin. The radius of these elementary rings ranges from zero to infinity. Thus, the geometric factor of an elementary layer G_z is

$$G_z = \int_0^\infty g\,dr \tag{6.19}$$

where g is the geometric factor of the elementary ring. Making use of Eq. (6.8), we have

$$G_z = \frac{L}{2} \int_0^\infty \frac{r^3\,dr}{\left[r^2 + (L/2+z)^2\right]^{3/2}\left[r^2 + (L/2-z)^2\right]^{3/2}}$$

Here L is the probe length. Introducing notations:

$$L/2 + z = m, \quad L/2 - z = n, \quad \text{and} \quad r^2 = x$$

we obtain

$$G_z = \frac{L}{4} \int_0^\infty \frac{x\,dx}{\left[x^2 + (m^2 + n^2)x + m^2 n^2\right]^{3/2}}$$

or

$$G_z = \frac{L}{4} \int_0^\infty \frac{x\,dx}{(x^2 + bx + c)(x^2 + bx + c)^{1/2}} \tag{6.20}$$

where

$$m^2 + n^2 = b, \quad m^2 n^2 = c$$

This integral is well known and is equal to

$$\frac{Lc}{(4c - b^2)c^{1/2}} - \frac{Lb}{2(4c - b^2)}$$

Thus, the geometric factor of an elementary layer is

$$G_z = \frac{L}{2(m+n)^2} \tag{6.21}$$

In case when the elementary layer is located between coils of the probe:

$$-L/2 < z < L/2$$

the geometric factor G_z is

$$G_z = \frac{1}{2L} \qquad (6.22)$$

and it is independent on coordinate z. When the layer is located above or below the probe:

$$z < -L/2 \quad \text{or} \quad z > L/2$$

the geometrical factor is

$$G_z = \frac{L}{8z^2} \qquad (6.23)$$

And again, the geometric factor of the whole space is equal to unity:

$$G = \frac{L}{8}\int_{L/2}^{\infty} \frac{dz}{z^2} + \frac{L}{2L} + \frac{L}{8}\int_{-\infty}^{-L/2} \frac{dz}{z^2} = \frac{L\,2}{8\,L} + \frac{1}{2} + \frac{L\,2}{8\,L} = 1$$

According to Eqs. (6.22), (6.23), the geometric factors of elementary layer located outside the probe decrease inversely proportional to z^2 while geometric factors of all elementary layers located inside the probe are equal to $G_z = 1/(2L)$. A curve, illustrating the behavior of geometric factor G_z (in units of L) as a function of z, is shown in Fig. 6.4. The middle of the two-coil probe is located at the origin of coordinates.

It is useful to notice that geometrical factors of two regions:

$$-\frac{L}{2} < z < \frac{L}{2}, \quad z < -\frac{L}{2}, \quad \text{and} \quad z > \frac{L}{2}$$

have the same value, equal to 0.5. As follows from definition of the function G_z, the geometric factor of a layer with very small thickness $dz \ll L$ is equal to $G_z dz$ and it is dimensionless. Since the function G_z gives sensitivity of the probe to induced currents in elementary layers, Doll called this function G_z the vertical response of a two-coil probe.

6.2.2 Geometric Factor of a Layer With a Finite Thickness

Using geometric factor of an elementary layer it is easy to find geometric factors of layers with a finite thickness. To proceed it is necessary to present

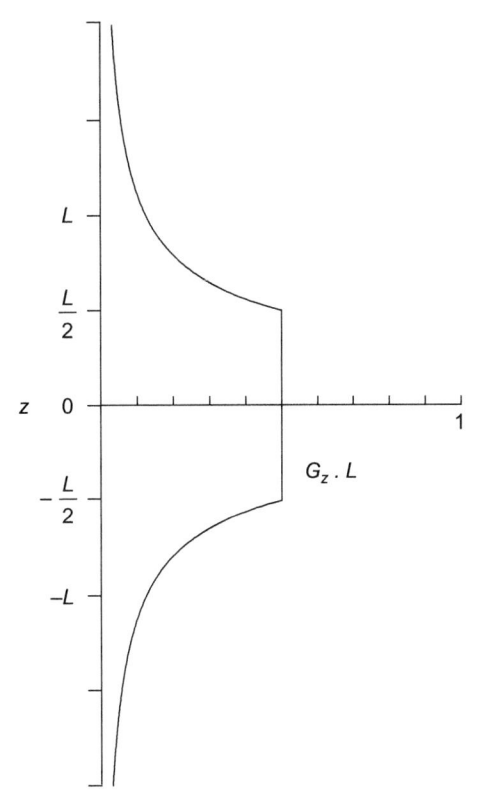

Fig. 6.4 Geometric factor of an elementary layer as a function of the layer position z.

the layer as a sum of elementary layers and perform summation of their geometric factors. Let us consider several positions of the two-coil probe with respect to the bed.

Case 1

The probe is located outside the bed of finite thickness (Fig. 6.5A). To derive the geometric factor of this bed G_b we integrate function G_z over the interval from z_1 to z_2 which characterizes the bed thickness. Then we have

$$G_b = \frac{L}{8}\int_{z_1}^{z_2}\frac{dz}{z^2} = \frac{L}{8}\left(\frac{1}{z_1} - \frac{1}{z_2}\right) \qquad (6.24)$$

Assuming that coordinate z_0 corresponds to the middle of the bed and taking into account that

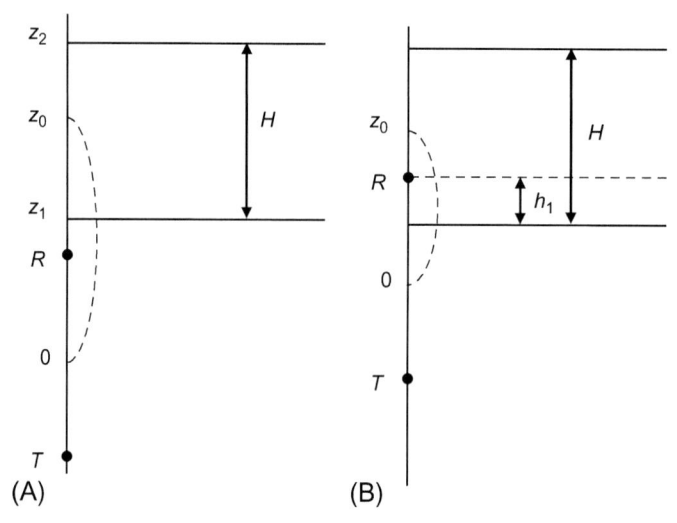

Fig. 6.5 Position of the probe with respect to the bed. (A) Probe is outside the bed. (B) One coil is inside while the other one is outside the bed.

$$z_1 = z_0 - H/2 \quad \text{and} \quad z_2 = z_0 + H/2$$

we have

$$G_b = \frac{LH}{8} \frac{1}{z_0^2 - (H/2)^2} \tag{6.25}$$

Here, H is the bed thickness, z_0 is the distance from the middle of the bed to the center of a two-coil probe. This equation is applied if the upper coil of the probe does not intersect the low boundary of the bed, i.e., it is valid if

$$z_1 \geq \frac{L}{2} \quad \text{or} \quad z_0 \geq \frac{L+H}{2}$$

Case 2

One coil is inside of the bed, while the other one is outside (Fig. 6.5B). To derive the geometric factor of a bed with thickness H, we have to sum up geometric factors of the parts of the bed located outside and inside the probe. In accordance with Eq. (6.23), the part outside the probe is

$$G_1 = \frac{L}{8}\left(\frac{1}{L/2} - \frac{1}{(z_0 + H/2)}\right) = \frac{1}{4} - \frac{L}{8(z_0 + H/2)}$$

The geometric factor of the part h_1 located inside the probe is

$$G_2 = \frac{h_1}{2L} = \frac{1}{2L}\left(\frac{L}{2} - z_0 + \frac{H}{2}\right)$$

because

$$h_1 = L/2 - (z_0 - H/2)$$

Therefore, for the geometric factor of the bed, we have

$$G_b = G_1 + G_2 = \frac{1}{2} - \frac{1}{2L}(z_0 - H/2) - \frac{L}{8(z_0 + H/2)} \qquad (6.26)$$

This formula is applicable until the upper coil of the probe is located within the bed and thickness of the bed is smaller than the probe length, $(H < L)$, i.e., when

$$(L - H)/2 \leq z_0 \leq (L + H)/2$$

When the bed thickness is greater than the probe length $(H > L)$, this formula can be used until the lower coil does not intersect the lower boundary of the bed, i.e., when $z_0 \geq (H - L)/2$.

Case 3

The probe is located against the bed (Fig. 6.6). There are two possible cases:
a. The probe length exceeds the bed thickness $(H < L)$; thus, the geometric factor G_b is

$$G_b = \frac{H}{2L} \qquad (6.27)$$

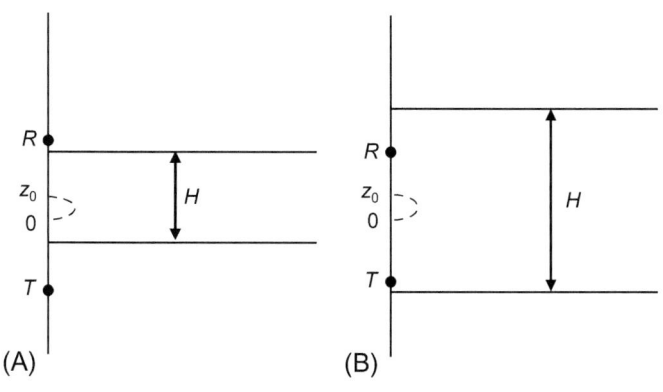

Fig. 6.6 Position of the probe with respect to the bed (Case 3). (A) Probe length exceeds the bed thickness. (B) Thickness of the bed is greater than the length of the probe.

b. The thickness of the bed is greater than the length of the probe ($H > L$); thus, the geometric factor G_b is

$$G_b = \frac{L}{2L} + \frac{L}{8}\left(\frac{2}{L} - \frac{1}{z_0 + H/2}\right) + \frac{L}{8}\left(\frac{2}{L} - \frac{1}{z_0 - H/2}\right)$$

$$= \frac{1}{2} + \frac{1}{4} - \frac{L}{8(z_0 + H/2)} + \frac{L}{8(z_0 - H/2)} + \frac{1}{4} \quad (6.28)$$

$$= 1 + \frac{LH}{8\left[z_0^2 - (H/2)^2\right]}$$

These equations can be applied, provided that

$$0 \leq z_0 < \frac{L}{2} - \frac{H}{2} \quad \text{if } H < L$$

and

$$0 \leq z_0 < \frac{H}{2} - \frac{L}{2} \quad \text{if } H > L$$

Derived formulas allow us to determine apparent conductivity for a two-coil probe located in a medium with two horizontal interfaces.

6.2.3 Apparent Conductivity in the Presence of a Layer With Finite Thickness

As follows from Eq. (6.17) in case of a layer with conductivity γ_1 surrounded by a medium of conductivity γ_2 for the apparent parameter γ_a, we have

$$\gamma_a = \gamma_1 G_1 + \gamma_2 G_2 \quad (6.29)$$

where G_1 and G_2 are geometric factors of the layer and surrounding medium, correspondingly. By definition, the sum of these factors is equal to unity, i.e.,

$$G_2 = 1 - G_1$$

Before we discuss apparent conductivity in the presence of a layer of a finite thickness, let us consider the influence of one horizontal interface. If the probe is located in a medium with conductivity γ_2 (Fig. 6.7A), then in accordance with Eq. (6.24), the geometric factors of both half-spaces are

$$G_1 = \frac{L}{8z_0}, \quad G_2 = \frac{L}{8}\left(\frac{2}{L} - \frac{1}{z_0}\right) + \frac{L}{2L} + \frac{L2}{8L} = 1 - \frac{L}{8z_0}$$

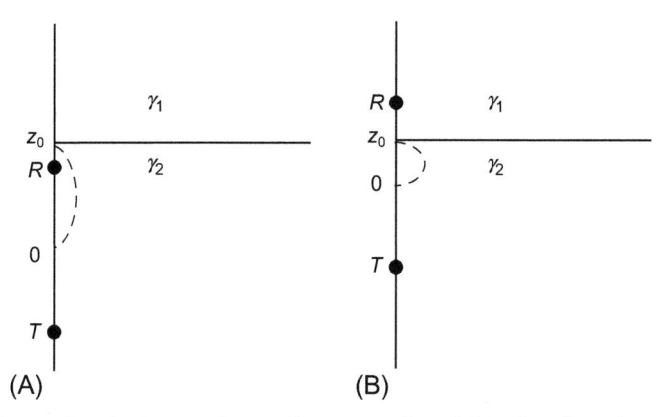

Fig. 6.7 Two-coil probe in a medium with one interface. (A) Both coils are located in the same layer. (B) Coils are located in different layers.

and

$$\gamma_a = \gamma_2 - (\gamma_2 - \gamma_1)\frac{L}{8z_0}, \quad z_0 \geq L/2$$

This formula is applicable when the interface is above the probe. In the case when the interface is below the probe ($z_0 \leq L/2$), we have

$$\gamma_a = \gamma_1 - (\gamma_1 - \gamma_2)\frac{L}{8z_0}$$

When coils are located in different layers $-L/2 \leq z_0 \leq L/2$ (Fig. 6.7B), the geometric factors are

$$G_1 = \frac{L}{8L} + \frac{1}{2L}\left(\frac{L}{2} - z_0\right) = \frac{1}{2} - \frac{z_0}{2L}, \quad G_2 = \frac{1}{2L}\left(\frac{L}{2} + z_0\right) + \frac{L}{8L} = \frac{1}{2} + \frac{z_0}{2L}$$

For the function γ_a, we have

$$\gamma_a = \frac{1}{2}(\gamma_1 + \gamma_2) + (\gamma_2 - \gamma_1)\frac{z_0}{2L}$$

Apparent conductivity curves for different positions of the probe with respect to the interface are shown in Fig. 6.8. One can notice that the value of the apparent conductivity is equal to the mean value of both conductivities when the probe center is located at the interface.

Now we turn to apparent conductivity curves in the presence of a layer. Because it is assumed that conductivity above and beneath the layer is the

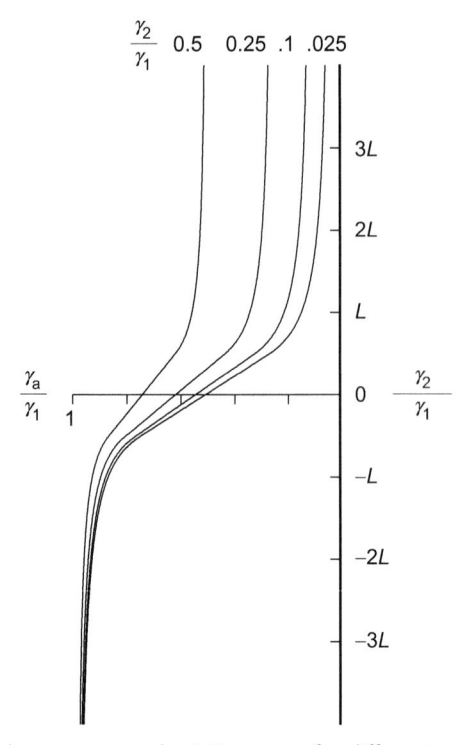

Fig. 6.8 Normalized apparent conductivity curves for different positions of a two-coil probe with respect to the interface.

same, we can restrict ourselves to cases when z_0 is positive. In deriving formulae for apparent conductivity for various positions of the probe, we use the equations of geometric factors of a layer of finite thickness.

Case 1

Probe is located outside the layer (Fig. 6.5A).

The expression for the apparent conductivity is

$$\gamma_a = \gamma_b G_b + \gamma_s G_s$$

where γ_b and G_b are the conductivity and geometric factor of the layer, and γ_s, G_s are the conductivity and geometric factor of the surrounding medium. Because

$$G_s = 1 - G_b$$

we can rewrite the expression for the apparent conductivity as

$$\gamma_a = \gamma_b G_b + \gamma_s(1 - G_b) = \gamma_s + G_b(\gamma_b - \gamma_s) \qquad (6.30)$$

According to Eq. (6.25), the geometric factor of the layer is

$$G_b = \frac{LH}{8[z_0^2 - (H/2)^2]}$$

Substituting this expression into Eq. (6.30), we obtain

$$\gamma_a = \gamma_s + (\gamma_b - \gamma_s)\frac{LH}{8[z_0^2 - (H/2)^2]} \qquad (6.31)$$

The latter applies until the upper coil does not intersect the low boundary of the layer, i.e., if $z_0 \geq H/2 + L/2$.

Case 2

One coil is located inside the layer (Fig 6.5B). In this case, according to Eq. (6.26) the geometric factor of the layer is

$$G_b = \frac{1}{2} - \frac{z_0 - H/2}{2L} - \frac{L}{8(z_0 + H/2)}$$

Substituting the expression for G_b into Eq. (6.30), we obtain

$$\gamma_a = \frac{\gamma_b + \gamma_s}{2} + \frac{\gamma_s - \gamma_b}{2L}(z_0 - H/2) + (\gamma_s - \gamma_b)\frac{L}{8(z_0 + H/2)} \qquad (6.32)$$

This formula applies when

$$\frac{L-H}{2} \leq z_0 \leq \frac{L+H}{2}$$

provided that $H \leq L$. In the case of ($H \geq L$), Eq. (6.32) is valid when

$$\frac{H-L}{2} \leq z_0 \leq \frac{L+H}{2}$$

Case 3

The layer is located either between the probe coils ($H \leq L$), or the probe is inside the layer ($H \geq L$) (Fig. 6.6A and B). For the first case (Eq. 6.27), we have

$$G_b = \frac{H}{2L}$$

Correspondingly,

$$\gamma_a = \gamma_s + (\gamma_b - \gamma_s)\frac{H}{2L} \qquad (6.33)$$

This equation is applied if $0 \leq z_0 \leq (L-H)/2$ and $H < L$. When $H \geq L$, according to Eq. (6.28), we have

$$G_b = 1 + \frac{LH}{8\left[z_0^2 - (H/2)^2\right]}$$

Thus,

$$\gamma_a = \gamma_b + (\gamma_b - \gamma_s)\frac{LH}{8\left[z_0^2 - (H/2)^2\right]} \qquad (6.34)$$

if $0 < z_0 \leq (H-L)/2$.

Note that, upon introducing new variables, one can represent these equations in the form that does not contain the length of the probe. Curves, showing dependence of γ_a/γ_b on the ratio of the layer thickness to probe length, are presented in Fig. 6.9, $H/L \geq 1$. Calculations have been made using equation

$$\frac{\gamma_a}{\gamma_b} = 1 + \left(\frac{\gamma_s}{\gamma_b} - 1\right)\frac{L}{2H}$$

With increase of conductivity of the surrounding medium and a decrease of the layer thickness, the influence of the surrounding medium

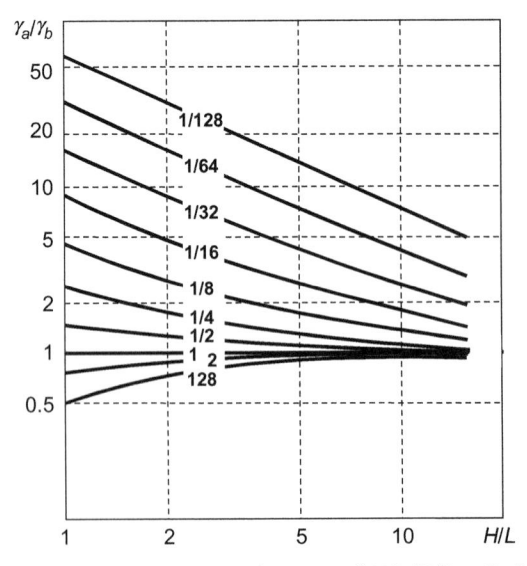

Fig. 6.9 Apparent conductivity γ_a/γ_b as a function of H/L ($H/L \geq 1$). Center of a probe coincides with the middle of the layer. Curve's index is γ_b/γ_s.

becomes greater. If the conductivity of the layer γ_b is significantly smaller than γ_s, the apparent conductivity is strongly affected by the surrounding medium and only for a large ratio (H/L) approaches the conductivity of the layer. In such cases, the vertical characteristic of the two-coil induction probe is essentially worse than the one corresponding to the response of the normal probe used in electrical logging with direct current. If the layer conductivity is greater than that of the surrounding medium, for most typical values of γ_b/γ_s, the influence of the surrounding medium becomes insignificant when $H/L > 4$. Apparent conductivity curves γ_a/γ_b for the case of the layer thinner than the probe length are shown in Fig. 6.10. They are calculated as

$$\frac{\gamma_a}{\gamma_b} = \frac{\gamma_s}{\gamma_b} + \left(1 - \frac{\gamma_s}{\gamma_b}\right)\frac{H}{2L}$$

If the layer's resistivity is higher than that of the surrounding medium and its thickness is less that $0.2L$, the layer is practically invisible to

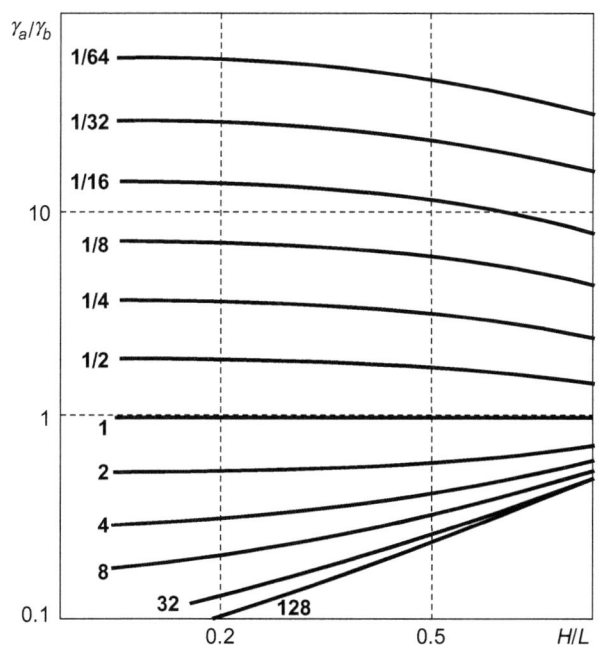

Fig. 6.10 Curves of ratio γ_a/γ_b as a function of H/L, $(H/L \leq 1)$ when the center of a probe coincides with the middle of the layer. The index is γ_b/γ_s.

induction logging. This behavior is directly opposite to the DC electrical logging, which is sensitive to the presence of the resistive layer. Thin conductive layers may have essential influence on induction tool responses; with an increase of the ratio γ_b/γ_s the apparent conductivity γ_a tends to a constant value of $S_1/2L$, where $S_1 = \gamma_b H$ is the longitudinal conductance of the layer and can become much greater than γ_s (Fig. 6.10).

Let us consider a case when the probe is located opposite to a system of very thin layers. Then, the expression for the apparent conductivity (6.33) is

$$\gamma_a = \gamma_s \left(1 - \sum_1^n \frac{h_i}{2L}\right) + \gamma_b \sum_1^n \frac{h_i}{2L} = \gamma_s \left(1 - \frac{H}{2L}\right) + \gamma_b \frac{H}{2L} \qquad (6.35)$$

where h_i is the thickness of the i-layer, n is number of layers, and $H = \sum_{i=1}^n h_i$. A set of thin layers located against the probe is equivalent to one layer having the same longitudinal conductivity and thickness equal to the sum of thicknesses of all thin layers. Eq. (6.35) can be generalized for the more general case of different conductivities and thicknesses of layers. It is noticeable that in all cases above Doll's theory have allowed us to study vertical characteristics of the probe using only elementary functions.

Finally, following Eqs. (6.4) and (6.16), we derive an expression for the quadrature component of magnetic field:

$$\Xi = K\gamma_a = -\frac{2\omega^2\mu_0^2 M_T M_R}{L \cdot 8\pi}\gamma_a = -QB_z^*\omega M_R$$

Here,

$$QB_z^* = \frac{\omega\mu_0^2 M_T}{4\pi L}\gamma_a$$

Then, for the quadrature component of magnetic field b_z, normalized by the primary field $B_z^0 = \mu_0 M_T/2\pi L^3$, we have

$$Qb_z^* = \frac{QB_z^*}{B_z^0} = \frac{\omega\mu_0 L^2}{2}\gamma_a \qquad (6.36)$$

6.3 RADIAL CHARACTERISTICS OF TWO-COIL INDUCTION PROBE

Following Doll, we begin to study the radial response of the two-coil probe located at the borehole axis, assuming that the surrounding medium is uniform. For the apparent conductivity γ_a, we have

$$\gamma_a = \gamma_1 G_1 + \gamma_2 G_2$$

where γ_1, γ_2 and G_1, G_2 are conductivity and geometric factors of the borehole and formation, respectively.

6.3.1 Geometric Factor of the Borehole

By definition, the geometric factor of the borehole G_1 is a sum of geometric factors of elementary rings, located inside the borehole and, in accordance with Eq. (6.8), we have

$$G_1 = \frac{L}{2} \int_S g dS = \frac{L}{2} \int_{-\infty}^{\infty} dz \int_0^a \frac{r^3}{R_1^3 R_2^3} dr$$

or

$$G_1 = \frac{L}{2} \int_{-\infty}^{\infty} dz \int_0^a \frac{r^3 \, dr}{\left[r^2 + (L/2+z)^2\right]^{3/2} \left[r^2 + (L/2-z)^2\right]^{3/2}} \qquad (6.37)$$

where a is the borehole radius. Unlike the geometric factor of the layer, the radial characteristics of the probe G_1 cannot be expressed through elementary functions. There were numerous attempts to simplify Eq. (6.37) and make it more convenient for calculations. For instance, Doll performed integration of geometric factors of elementary rings and obtained the following expression:

$$G_1(\alpha) = 1 - \frac{1}{(1 + 4/\alpha^2)^{1/2}} \left(E(k) + \frac{2}{\alpha^2}[E(k) - K(k)] \right) \qquad (6.38)$$

Here E and K are elliptical integrals of the first and second kind,

$$k = \frac{\alpha}{(\alpha^2 + 4)^{1/2}}$$

and $\alpha = L/a$ is the ratio of the probe length to the borehole radius. At the end of 1950s, Kudravchev (Russia) obtained another expression for the G_1:

$$G_1 = 1 - \frac{2\alpha}{\pi} \int_0^\infty A(m) \cos \alpha m \, dm \qquad (6.39)$$

where

$$A(m) = \frac{m}{2}\left[2K_0(m)K_1(m) - mK_1^2(m) + mK_0^2(m)\right] \qquad (6.40)$$

and $K_0(m), K_1(m)$ are modified Bessel functions of a second kind. The form (Eq. 6.39) is convenient for the analysis because

$$\alpha = \frac{L}{a} \qquad (6.41)$$

is a single parameter that defines the geometric factor of the borehole. Later, Eq. (6.39) will be derived rigorously, but here it is used to study radial responses of a two-coil probe. Also we need an expression for the geometric factor of the formation, which is

$$G_2 = \frac{2\alpha}{\pi} \int_0^\infty A(m) \cos \alpha m \, dm \qquad (6.42)$$

First, consider the function $A(m)$. For the sufficiently large m, we have the following asymptotes:

$$K_0(m) \approx \exp(-m)\left(\frac{\pi}{2m}\right)^{1/2}\left(1 - \frac{0.125}{m}\right),$$

$$K_1(m) \approx \exp(-m)\left(\frac{\pi}{2m}\right)^{1/2}\left(1 + \frac{0.375}{m}\right)$$

When $m \to \infty$, the integrand in Eq. (6.39) rapidly decreases, and calculation of the integral in this range of m becomes a simple matter. When $m \to 0$, the corresponding asymptotes are

$$K_0(m) \to -\left(\ln\frac{m}{2} + C\right), \quad K_1(m) \to \frac{1}{m}$$

where C is some constant. Substituting these expressions into Eq. (6.40), we obtain

$$A(m) \to K_0(m) \to -\left(\ln\frac{m}{2} + C\right), \quad \text{as } m \to 0 \qquad (6.43)$$

Thus, the integrand has a logarithmic singularity as m tends to zero. In order to remove this singularity, we make use of the following equation:

$$\frac{1}{(1+\alpha^2)^{1/2}} = \frac{2}{\pi}\int_0^\infty K_0(m)\cos(\alpha m)\,dm$$

Then, the function G_1 can be presented in the form

$$G_1(\alpha) = 1 - \frac{2\alpha}{\pi}\int_0^\infty A(m)\cos(\alpha m)\,dm = 1 - \frac{\alpha}{(1+\alpha^2)^{1/2}}$$
$$+ \frac{2\alpha}{\pi}\int_0^\infty [K_0(m) - A(m)]\cos(\alpha m)\,dm \qquad (6.44)$$

In accordance with Eq. (6.43), the integrand in Eq. (6.44) is free of singularities, and its calculation represents a relatively simple task. Next, we find the asymptotic expression for the geometric factor of the borehole, starting from the case when the probe length L is small compared with the size of the borehole. In this case, parameter α tends to zero while the function G_1 tends to unity as

$$G_1(\alpha) \approx 1 - 0.5862\alpha, \quad \text{if } \alpha \ll 1 \qquad (6.45)$$

Now we consider the case when probe length L is much greater than the borehole radius a. Let us first analyze the integrand in Eq. (6.44), which is the product of two functions:

$$\Phi(m) = K_0(m) - A(m) \quad \text{and} \quad \cos(\alpha m)$$

The first function gradually changes with m, while the second oscillates with a period,

$$m = \frac{2\pi}{\alpha}$$

and decreases with an increase of α. For this reason the integral in Eq. (6.44) is defined by the function $\Phi(m)$ and its derivatives when m approaches zero, provided that $\alpha \gg 1$. In fact, integrating by parts we receive

$$\int_0^\infty \Phi(m)\cos\alpha m\, dm = \frac{1}{\alpha}\int_0^\infty \Phi(m)d\sin\alpha m = \frac{1}{\alpha}\Phi\sin(\alpha m)\Big|_0^\infty - \frac{1}{\alpha}\int_0^\infty \Phi'(m)\sin\alpha m\, dm$$

$$= \frac{1}{\alpha}\Phi\sin\alpha m\Big|_0^\infty + \frac{1}{\alpha^2}\Phi'\cos\alpha m\Big|_0^\infty - \frac{1}{\alpha^2}\int_0^\infty \Phi''(m)\cos\alpha m\, dm$$

(6.46)

For the large values of m, the function $\Phi(m)$ and its derivatives tend to zero; therefore, instead of Eq. (6.46), we have

$$\int_0^\infty \Phi(m)\cos\alpha m\, dm = -\frac{0}{\alpha}\cdot\Phi(0) - \frac{1}{\alpha^2}\cdot\Phi'(0) - \frac{1}{\alpha^2}\int_0^\infty \Phi''(m)\cos\alpha m\, dm \quad (6.47)$$

For small values of m ($m \to 0$), we have

$$K_0(m) \approx -\ln m - \frac{m^2}{4}\ln m + \frac{m^2}{4} - C \text{ and } K_1(m) \approx \frac{1}{m} + \frac{m}{2}\ln m - \frac{m}{4}$$

Substituting these expressions into $\Phi(m)$ gives

$$\Phi(m) \approx \frac{1}{2} + \frac{1}{4}m^2\ln m, \quad \Phi'(m) \approx \frac{m}{2}\ln m, \quad \Phi''(m) \approx \frac{1}{2}\ln m, \text{ if } m \to 0$$

Thus,

$$\int_0^\infty \Phi(m)\cos\alpha m\, dm \to -\frac{1}{2\alpha^2}\int_0^\infty \ln m\cos\alpha m\, dm \approx \frac{1}{2\alpha^2}\int_0^\infty K_0(m)\cos\alpha m\, dm \to \frac{1}{2\alpha^2}\frac{\pi}{2\alpha}$$

if $\alpha \gg 1$.

Then, for the geometric factor (Eq. 6.44), we get

$$G_1(\alpha) \approx 1 - (1+\alpha^{-2})^{-1/2} + \frac{1}{2\alpha^2} = \frac{1}{\alpha^2} = \frac{a^2}{L^2}, \text{ if } \alpha \gg 1 \quad (6.48)$$

Therefore, for large values of α, the geometric factor of the borehole is inversely proportional to α^2. (This peculiarity is used to design "focusing" probes with minimal sensitivity to the borehole.) Comparison of Eq. (6.48) versus exact solution shows that Eq. (6.48) describes with sufficient accuracy the value of function $G_1(\alpha)$ if $\alpha > 4$. In other words, the asymptotic behavior is already observed when the probe length exceeds

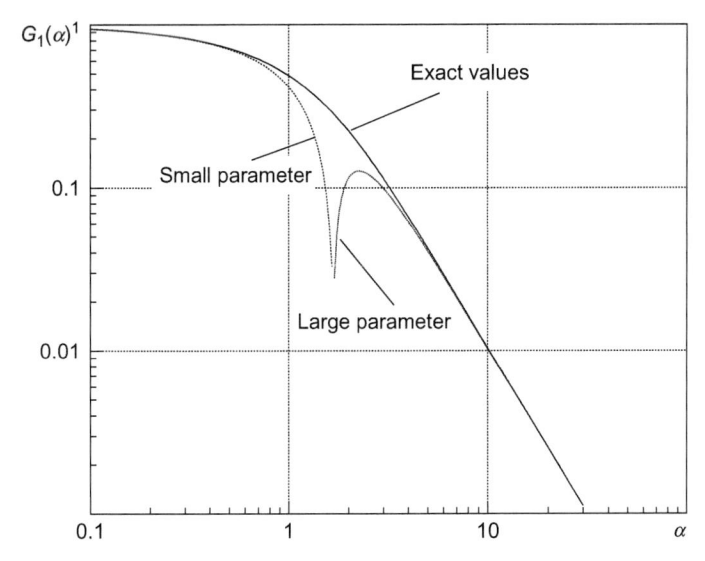

Fig. 6.11 Exact, small, and large parameter approximation of function $G_1(\alpha)$.

twice the borehole diameter. Similarly, we can obtain the following terms for the expansion of the function $G_1(\alpha)$ for the large values of α. For example, the expansion with the second term gives

$$G_1(\alpha) \approx \frac{1}{\alpha^2} + \frac{3 \ln \alpha - 4.25}{\alpha^4} \quad \text{if } \alpha \gg 1 \quad (6.49)$$

The geometric factor $G_1(\alpha)$ for the small and large values of parameter α in comparison with an exact solution, obtained through numerical calculations, is shown in Fig. 6.11. As we can see, the asymptotes describe with sufficient accuracy the value of function $G_1(\alpha)$ if $\alpha > 4$ or $\alpha < 0.6$.

6.3.2 Radial Characteristics of Two-Coil Probe

The function $G_1(\alpha)$ allows us to study the radial responses of induction probes in a medium with cylindrical interfaces. Again, we use the concept of the apparent conductivity γ_a:

$$\frac{\gamma_a}{\gamma_1} = \frac{QB_z}{QB_{z0}(\gamma_1)} \quad \text{or} \quad \frac{\gamma_a}{\gamma_1} = \frac{Q\Xi}{Q\Xi_0(\gamma_1)} \quad (6.50)$$

where γ_1 is the borehole conductivity, QB_z and $Q\Xi$ are the quadrature components of magnetic field and electromotive force, respectively, while $QB_{z0}(\gamma_1)$ and $Q\Xi_0(\gamma_1)$ are quadrature components of the field and electromotive force of the two-coil probe in medium with conductivity γ_1. According to Eq. (6.36), we have

$$Qb_{z0}(\gamma_1) = \frac{\mu_0 \omega L^2}{2}\gamma_1 \quad \text{and} \quad Qb_z = \frac{\mu_0 \omega L^2}{2}\gamma_a \qquad (6.51)$$

As follows from Eq. (6.50), the ratio γ_a/γ_1 shows how the field or the measured electromotive force differs from the same quantity in a uniform medium with conductivity γ_1. This method of introduction of the apparent conductivity is natural only within Doll's theory, where the skin effect is negligible. By definition, for the apparent conductivity, we have

$$\gamma_a = \sum_{n=1}^{N} \gamma_n G_n \qquad (6.52)$$

For a cylindrically layered medium with borehole and formation, we obtain

$$\gamma_a = \gamma_1 G_1 + \gamma_2 G_2 = \gamma_2 + G_1(\gamma_1 - \gamma_2) \qquad (6.53)$$

while in the presence of an invasion zone, we have

$$\gamma_a = \gamma_1 G_1 + \gamma_2 G_2 + \gamma_3 G_3 \qquad (6.54)$$

A very short probe is mainly sensitive to the currents in the vicinity of the probe; that is,

$$\gamma_a \to \gamma_1 \quad \text{if} \quad \alpha \to 0 \qquad (6.55)$$

On the other hand, as the probe length increases, the geometric factor of every cylindrical layer of finite thickness decreases, while the geometric factor of the formation approaches unity. Therefore, the depth of investigation in the radial direction increases; that is, for any given conductivity distribution $\gamma(r)$ there is such length L, when the probe is mainly sensitive to the conductivity of the formation γ_N:

$$\gamma_a \to \gamma_N \quad \text{if} \quad \alpha \to \infty \qquad (6.56)$$

This is illustrated in Fig. 6.12 where apparent conductivities γ_a/γ_1 are presented for the different ratios of γ_2/γ_1. It is seen that the influence of the borehole becomes greater as its conductivity or radius increases. Of course, as the probe length increases ($\alpha \to \infty$), the quadrature component of magnetic field B_z approaches that in a uniform medium having

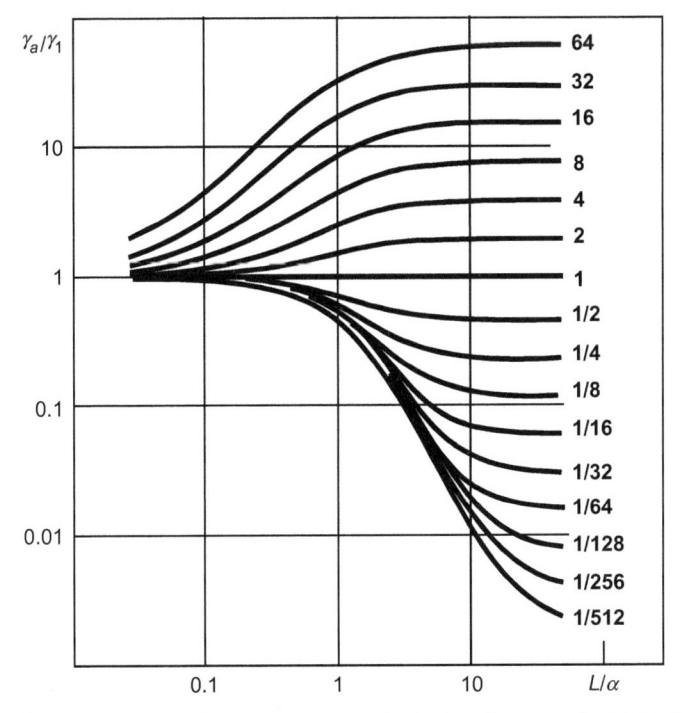

Fig. 6.12 The apparent conductivity curves γ_a/γ_1 in the absence of an invasion zone. Curve index is γ_2/γ_1.

conductivity γ_N. For illustration, let us consider one numerical example, assuming $L=1$ m, $a=0.1$ m, and $\rho_2/\rho_1=30$.

Then, as follows from Eq. (6.54), $G_1 \approx 0.01$, and, for the apparent conductivity, we have

$$\frac{\gamma_a}{\gamma_2} = 1 + \left(\frac{\gamma_1}{\gamma_2} - 1\right) G_1 \approx 1.29$$

showing a significant (29%) contribution from the borehole. In reality, it is even stronger, because the geometrical factor theory does not take into account the skin effect that reduces contribution of formation into the measurements. Also, in the presence of an invasion zone, distortion of the apparent conductivity of a two-coil probe becomes even stronger, especially in the case of the conductive invasion when $\rho_1 < \rho_3 < \rho_3$.

Although the induction logging was first introduced for the measurements in boreholes filled with water-based muds, these measurements are

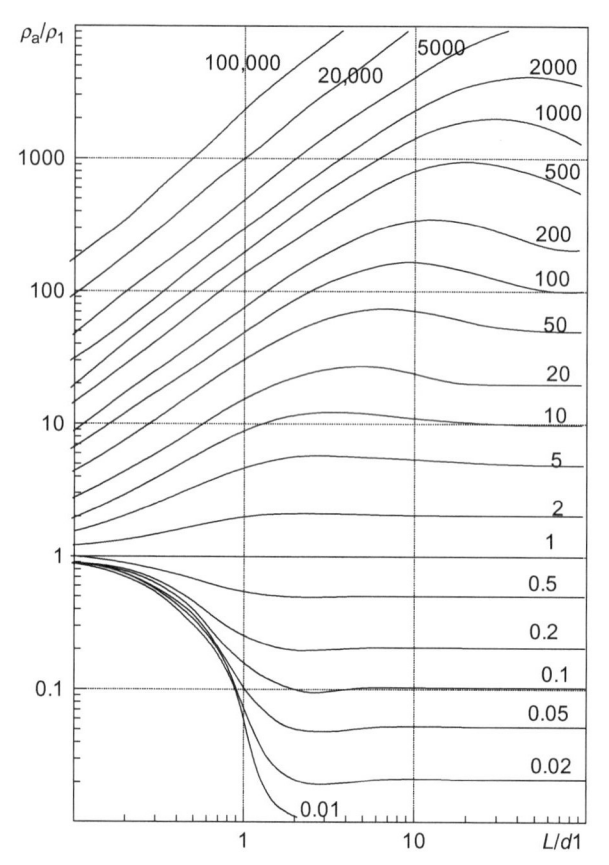

Fig. 6.13 The apparent resistivity curves for the normal probe in the absence of the invasion zone; d_1 is the borehole diameter. Curves index is ρ_2/ρ_1.

more suitable in boreholes with a nonconducting mud. Also, a comparison of radial responses of the two-coil induction probe versus responses of the standard DC two-electrode probe (Fig. 6.13) showed no advantages of induction measurements in the case of conductive borehole fluid. Naturally, the first field experiments had shown that the conductive borehole makes a strong influence on the induction measurements. To overcome the shortcomings of the two-coil induction probe, Doll suggested multicoil differential probes (1949). He also developed an approach for determining parameters of the corresponding tools, which have much better radial and sometimes vertical characteristics. This made the induction measurements one of the most successful in the logging industry and are now widely used all over the world.

We have to point out that quantitative interpretation, based on the geometrical factor concept, requires relatively low frequencies. But at such frequencies, the secondary field, containing the information about a medium conductivity, is always much smaller than the primary field, and this situation creates a serious measurement problem for detecting a small signal in the presence of a larger one. Although the primary and secondary signals are shifted in phase with respect to one another, even a small error in the detecting electronics may greatly decrease the accuracy of the measurements. To improve measurements, engineers undertook several steps. The first was compensation of the electromotive force of the primary field at the receiver coil. For this purpose, an additional coil is placed into the receiver line of the probe, as is shown in Fig. 6.14. This compensating coil has a smaller number of turns than the main receiver coil and is located relatively closer to the transmitter. Moments of receiver coils are chosen in such a way that the primary electromotive force at the compensating coil has the same amplitude but an opposite sign to the primary electromotive force in the main receiver. Thus, in place of a two-coil probe, we obtain the three-coil probe, consisting of two two-coil probes. The first probe has length L, which characterizes the distance between transmitter and the main receiver coil, while the second probe, formed by the transmitter coil and the compensating receiver coil, has a smaller length L_1. The output of the three-coil

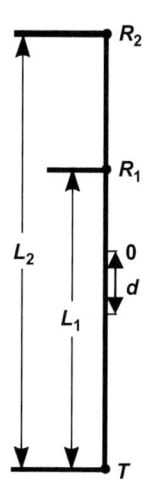

Fig. 6.14 Three-coil induction probe; d is the distance between centers of the short and long two-coil probes.

probe is the combination of the outputs of the two receiver coils. Because the number of turns in the compensating coil is considerably smaller than that of the receiver coil, the secondary signal from formation induced in this coil has only a minor effect on the total secondary signal. (The same approach of compensation of the primary field is used in multicoil probes). However, this method alone does not provide sufficient accuracy of measurements in the real borehole conditions, when changes in pressure and temperature cause fluctuation of the primary and thus affect the responses in the coils. Because electromotive forces caused by the primary and the secondary field are shifted in phase by 90 degree, the original three-coil probes designed by Doll measured only the quadrature component of the field (or in-phase component of EMF). This procedure greatly reduces an influence of the primary electromotive force instability. The next improvement was the usage of a negative feedback scheme, which permits stabilization of the measurements by reinjecting the electronic drift back into the system. Of course, with time, due to great progress in quality of materials and electronics, it became possible to measure even small in-phase components of the secondary magnetic field (or quadrature of EMF).

6.4 MULTICOIL OR "FOCUSING" INDUCTION PROBE

Analysis of the field in a media with horizontal and cylindrical interfaces shows that a two-coil induction probe has noticeable advantages over direct-current probes only in boreholes filled with nonconductive fluids. In the case of conductive fluids, the influence of induced currents in the borehole can be so strong that only very long probes can permit determination of the formation resistivity. The last circumstance motivated developers to look at the alternative approach, which permits deep depth of investigation with relatively short probes. The approach is based on the focusing of electromagnetic field into the deep part of the formation.

Proceeding from geometrical factor theory, let us analyze multicoil probes provided that currents in any part of a medium create a signal defined by the conductivity of the region in question and its geometric factor. The contribution of various parts of a medium in forming a signal, essentially depends on the probe length: with an increase of the probe length, the influence of remote parts of a medium increases; consequently, the relative contribution of induced currents near the probe becomes smaller. By applying probes of various lengths with a different number of turns and connected in series either in the same or opposite directions, one can significantly reduce the signal caused by currents in any element of the medium.

However, improvement of radial characteristics of the probe, practically, always leads to deterioration of its vertical characteristics and vice versa. Indeed, reduction of the near borehole regions and penetration of the field into the deep part of formation requires a long probe with reduced vertical characteristics. Moreover, increase in the length leads to reduction of the measured signal level. Thus, the opposite requirements have to be satisfied simultaneously to improve the radial and vertical characteristics of a multicoil probe. As will be shown later, under certain conditions it can be done, although, in general, an improvement of the radial characteristic of a probe results in deterioration of the vertical one and vice versa. A multicoil probe can be treated as a sum of two-coil probes. Early in the development of probes with improved radial response, they were called "focusing" probes by an analogy with the focusing of optical and seismic waves. Wave-field focusing uses the phenomenon of constructive and destructive interference to enhance a response: for example, by using a lens to force a set of parallel light rays converge at some point. The physics of a multicoil "focusing" probes is based on a completely different principle: the addition and subtraction of geometrical factors.

6.4.1 Conditions for the Application of "Focusing" Probes

Geometrical factor theory assumes that the interaction between currents is absent, i.e., that all currents induced in a conducting medium are shifted in phase by 90 degrees, regardless of distance from the source. For this reason, signals induced in different measuring coils are in phase with each other. In a more general case, when skin effect manifests itself, the induced currents have both in-phase and quadrature components, and the magnitude of the quadrature component depends on distribution of conductivity in a medium. Correspondingly, geometric factors become different from the ones determined by the geometric factor theory. Deviation from Doll's region leads to a serious deterioration of the "focusing" features developed under the assumption that the interaction between currents in those parts of a medium (borehole, invasion zone), the influence of which should be significantly reduced is absent.

However, the absence of such an interaction is not sufficient for application of focusing probes. Indeed, we may present a probe signal as a sum of the signal, caused by the currents, which are not affected by skin effect and a signal generated by currents in the external area, for example, in the formation. Because the skin effect increases with the distance from the primary source increases, the distribution of currents in the external area depends on the currents in the near probe area. For this reason, this first eliminated area may still indirectly affect distribution of currents in the formation. In this case, the signal from the formation is not only a function of its conductivity but also a function of the conductivity and geometry of the borehole and invasion zone. Therefore, the second condition for application of "focusing" probes is the absence of an influence of currents in the internal area on the current distribution in the formation. The skin effect in the formation has to manifest itself in the same manner, as if the borehole and invasion were absent.

It turns out that two conditions listed above are often met far beyond the range of small induction numbers, and, correspondingly, the "focusing" may perform even at higher frequencies than those dictated by Doll's original theory. In the following chapter, we discuss the approximate theory of the induction logging, accounting for the skin effect and satisfying the above two conditions. A comparison between this theory and exact solution will allow us to establish the maximal frequency when the multicoil probes are still able to reduce the influence of the borehole and invasion zone.

6.4.2 Three- and Multicoil Probes

Focusing probes consist of several two-coil probes. The main probe has maximal product of the transmitter and receiver coil moments. Also, there are some additional probes, which are located either inside or outside the main probe. In the most successful focusing probes, both types of additional two-coil probes are present. In all multicoil probes, transmitting coils as well as receiving ones, are connected in series. Because all transmitting coils have the same current, and the area of coils is the same for all transmitters and receivers, it is convenient to characterize the moments of all coils by the number of turns. The coil, wounded in the opposite direction to the main probe, is characterized by the negative number of turns.

Let us consider the geometric factor of a cylinder for such multicoil probe. For each two-coil probe the function $G_1(r, L)$ characterizes a signal caused by induced currents in a uniform cylinder with radius r. By selecting particular lengths of the two-coil probes and changing their number of turns, it is possible to achieve three goals. First is to reduce the borehole and invasion zone effect and obtain the geometric factor with minimal contribution into the signal from the cylinder with relatively large radius. Second is to maximize the signal caused by the currents outside of this cylinder. Besides, these "focusing" probes should have sufficient vertical resolution to guarantee an accurate interpretation in relatively thin layers.

The multicoil probes can be divided into symmetrical and nonsymmetrical ones, considering the location of additional coils with respect to the center of the main probe. In a symmetrical arrangement, the identical two-coil probes are located symmetrically with respect to the center of the main probe. Symmetrical probes, unlike nonsymmetrical ones, have a symmetrical response with respect to the center of a bed provided that the resistivity of a medium above and below the bed is the same. Symmetrical vertical response is also observed in a modified symmetrical multicoil probe where the number of turns in all transmitters or receivers is changed by the same coefficient. An example of such symmetrical probe will be shown considering the dual induction tool. Depending on location of additional coils, multicoil probes can also be classified as probes with internal, external, and mixed "focusing." In probes with internal "focusing," the additional coils are located between the main ones; with external "focusing," they are located outside the main probe, and probes with mixed "focusing" additional coils are placed inside and outside the main probe. An example of the simplest nonsymmetrical three-coil probe is depicted in Fig. 6.14. Examples of symmetrical probes with internal, external, and mixed "focusing" are shown in Fig. 6.15.

In 1952 Schlumberger introduced the first focusing induction probe 5FF27, which had two external "focusing" coils and one internal coil. The length of the main two-coil probe was around 0.7 m (27 in.), and it

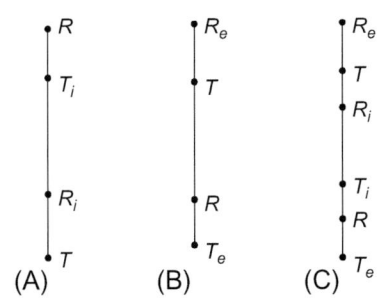

Fig. 6.15 Multicoil symmetrical "focusing" probes with (A) internal, (B) external, and (C) mixed "focusing."

provided limited depth of investigation, especially in a presence of invasion. In 1956 they introduced 5FF40 with a slightly deeper depth of investigation, but it was still insufficient in wells drilled with salty muds. Then, the Lane Well Company (Disterhoft, Hartlineand Thomsen) introduced a multicoil induction tool with a deeper depth of investigation. In 1959 Schlumberger, in turn, developed the 6FF40 probe [2], which became an industry standard and through various modifications was successfully used by well logging companies for more than 30 years all over the world. About the same time, Doll, recognizing that a multicoil probe is a superposition of two-coil probes, suggested to measure individual signals of each two-coil probe and, then, numerically form different combinations obtaining information about cylindrical layers, located at different distances from the borehole.

In the late 1950s, Russian geophysicists, following achievements by US logging companies, started developing the theory and equipment for induction logging. The first nonsymmetrical induction probe was developed in 1959 by Akselrod (Baku). Slightly later, Russian well logging operators began to use the symmetrical induction probe 6 F1, designed by Plusnin (Moscow). This probe had similar parameters to the 6FF40. At the same time, further development of induction logging theory took place in Novosibirsk [3], where it was accompanied by the design of the first high-frequency induction probe (VIK).

To illustrate the concept of "focusing" probes, we consider only two examples. Proceeding from the known expression for the electromotive force in a two-coil probe, caused by the quadrature component of the magnetic field,

$$\Xi = \frac{-\omega^2 \mu_0^2 M_T M_R}{4\pi L} \gamma_a$$

we find expressions for the measured signal in the multicoil probes. Let us start from the simplest three-coil probe.

Example One
Three-coil probes (Fig. 6.14).

This focusing probe [3] consists of one transmitting coil and two receiving coils, which have opposite direction of turns. Then, the measured electromotive force is

$$\Xi = \frac{-\omega^2 \mu_0^2 M_T}{4\pi} \left[\frac{\gamma_a(L)}{L} M_R - \frac{\gamma_a(L_1)}{L_1} M_{R1} \right] \qquad (6.57)$$

Here, L_1 and MR_1 are the length and the receiver moment of the short two-coil probe, $L_1 < L$. Bearing in mind that the primary electromotive force for the main two-coil probe is

$$\Xi_0(L) = \frac{-\omega\mu_0 M_T M_R}{2\pi L^3}$$

instead of Eq. (6.57), we have

$$\Xi = \frac{1}{2}\omega\mu_0\left[L^2|\Xi_0(L)|\gamma_a(L) - L_1^2|\Xi_0(L_1)|\gamma_a(L_1)\right] \quad (6.58)$$

Suppose that the number of turns in receiving coils is chosen to compensate the primary field:

$$|\Xi_0(L)| = |\Xi_0(L_1)| \quad (6.59)$$

Then, Eq. (6.58) is simplified to

$$\Xi = \frac{1}{2}\Xi_0(L)\omega\mu_0\left[L^2\gamma_a(L) - L_1^2\gamma_a(L_1)\right] \quad (6.60)$$

When the invasion zone is absent, we have

$$\Xi = \frac{1}{2}\Xi_0(L)\omega\mu_0\left(\gamma_1\left[L^2 G_1(L) - L_1^2 G_1(L_1)\right] + \gamma_2\left[L^2 G_2(L) - L_1^2 G_2(L_1)\right]\right) \quad (6.61)$$

Here G_1 and G_2 are geometric factors of the borehole and formation, and their sum is equal to unity. Suppose that $L_1 \gg a_1$, where a_1 is the borehole radius, then, as follows from Eq. (6.49),

$$G_1 \approx \frac{1}{\alpha^2} + \frac{3\ln\alpha - 4.25}{\alpha^4} \quad (6.62)$$

where $\alpha = L/a_1$. Substitution of the leading term of the latter into Eq. (6.61) gives an approximation for Ξ:

$$\Xi \approx \frac{1}{2}\Xi_0(L)\omega\mu_0\left(L^2 - L_1^2\right)\gamma_2 \quad (6.63)$$

which does not depend on parameters of the borehole. As soon as the radius of the invasion zone is much smaller than the probe length L_1, an influence of this zone is also negligible. Let us present Eq. (6.60) as

$$\Xi = \frac{L^2}{2}\Xi_0(L)\left(1 - p^2\right)\omega\mu_0\left(\gamma_1 G_1^* + \gamma_2 G_2^*\right) \quad (6.64)$$

Here, $p = L_1/L < 1$ and

$$G_1^* = \frac{G_1(L) - p^2 G_1(L_1)}{1 - p^2}, \quad G_2^* = \frac{G_2(L) - p^2 G_2(L_1)}{1 - p^2} \qquad (6.65)$$

are geometric factors of the three-coil probe for the borehole and formation, respectively. Assuming that $\alpha \gg 1$, we have for the function G_1^*:

$$G_1^* \approx \frac{1}{p^2 \alpha^4}\left(2.17 - \frac{3 \ln p}{1 - p^2} - 3 \ln \alpha\right) \qquad (6.66)$$

In Fig. 6.16, we show the borehole geometrical factor for a three-coil probe. It illustrates reduced sensitivity of the probe to the region close to the borehole axis. At the initial part of the radial response, G_1^* has negative values, which are much smaller than those of the geometric factors for the two-coil probes $G_1(L_1)$ and $G_1(L)$. Near the radius, where

$$3 \ln \alpha = 3 \ln \frac{L}{a_1} \approx 2.17 - \frac{3 \ln p}{1 - p^2}$$

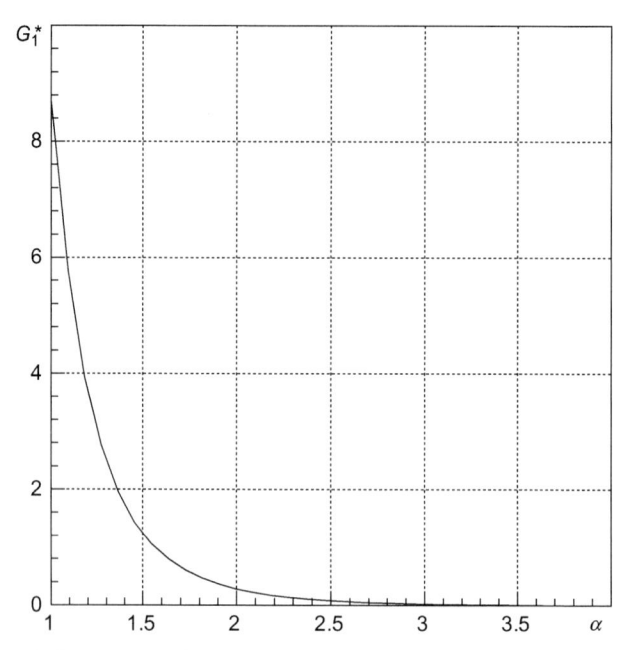

Fig. 6.16 Borehole geometric factor of the three-coil probe as a function of parameter α. Parameters of the probe are $L = 1.4$ m, $p = 0.7$.

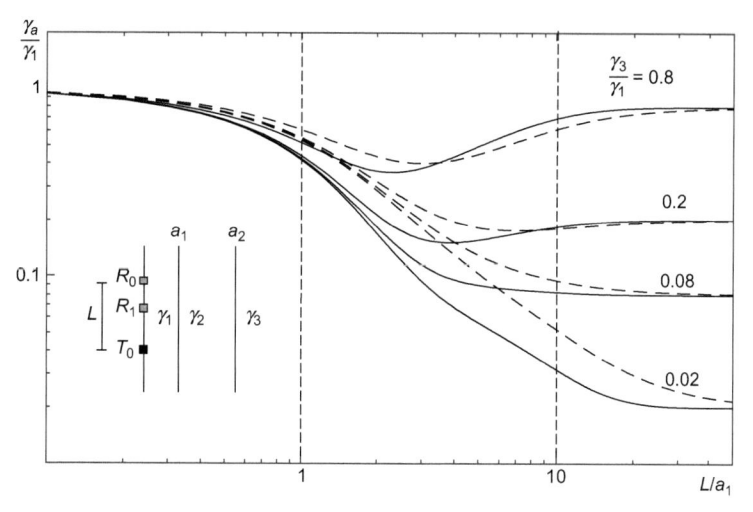

Geometrical Factor Theory of Induction Logging

Fig. 6.17 Normalized apparent conductivity curves for two-coil *(dashed)* and three-coil *(solid)* probes in a medium with an invasion zone, $L_1/L = 0.7$; $a_2/a_1 = 6$; $\gamma_2/\gamma_1 = 0.3$.

the geometric factor G_1^* is equal to zero and then rises monotonically, approaching unity. A combination of two factors—compensation of the primary field and behavior of the function $G_1(\alpha)$ as $1/\alpha^2$ provides a significant reduction of the borehole and invasion zone on the signal, if probe length L is several times greater than radius of the invasion a_2.

Of course, as the probe length increases, the effect of "focusing" manifests itself stronger. For illustration, Fig. 6.17 shows behavior of apparent conductivity curves for a medium with an invasion zone. When parameter p increases the focusing properties of the probe improve for the expense of the signal level. Let us consider the vertical response of the three-coil probe starting from the geometric factor of an elementary layer. Sometimes it is called the "vertical response function of a probe." From Eq. (6.65) we have

$$G_z^* = \frac{G_z(L, z) - p^2 G_z(L_1, z + d)}{1 - p^2} \qquad (6.67)$$

where, according to Eq. (6.21), $G_z(L, z)$ is

$$G_z(L, z) = \frac{L}{2\left(\left|\frac{L}{2} + z\right| + \left|\frac{L}{2} - z\right|\right)^2}$$

and

$$d = -L/2 + L_1/2 = -L\frac{(1-p)}{2} \quad (6.68)$$

is the distance between centers of the short and long two-coil probes (Fig. 6.14). This equation for the depth-offset d is valid for a three-coil probe with one transmitter located at the bottom of the probe (see Fig. 6.14).

Obviously, if the probe is turned upside down, the offset d changes the sign. Geometric factors of the elementary layer for three-coil and corresponding two-coil probes ($p=0.5$) are shown in Fig. 6.18. Geometric factors of an elementary layer for different three-coil probes are shown in Fig. 6.19. Corresponding offsets are $=d/L=0.7, -0.375$, and -0.125.

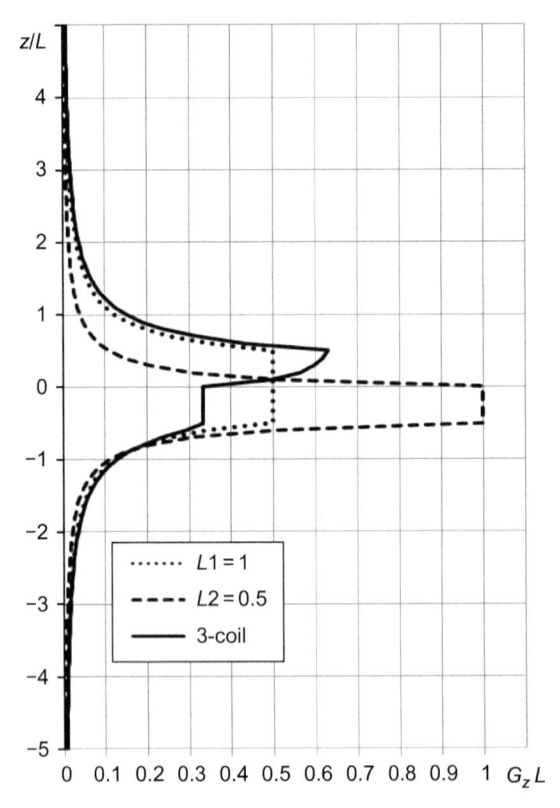

Fig. 6.18 Geometric factor of an elementary layer for a three-coil probe. The *solid line* shows the geometrical factor of a three-coil probe with $p=0.5$. Dotted and dashed curves show the geometrical factors of the corresponding two-coil probes.

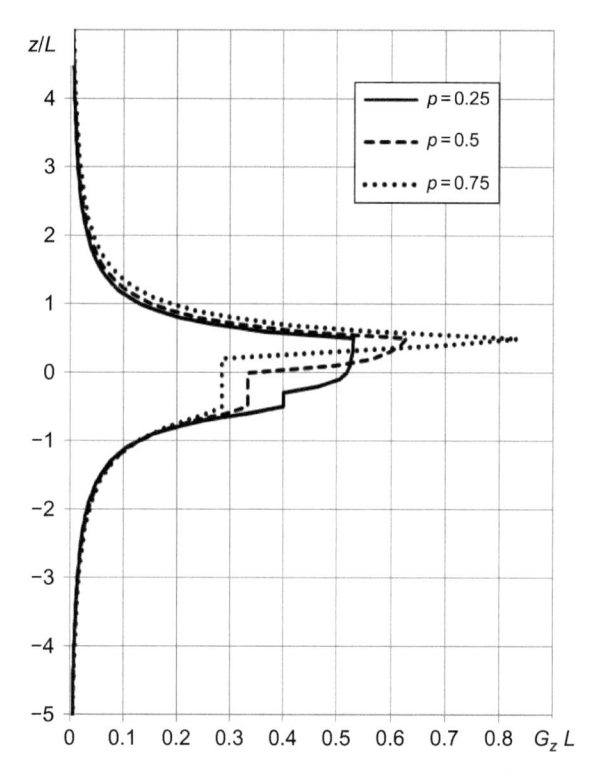

Fig. 6.19 Geometric factors of the elementary layer for three-coil probes with $p = 0.25$, 0.5, and 0.75.

As we see, the vertical characteristics of three-coil probes can be quite complicated if the layer thickness is smaller than the length of the main probe.

Suppose that the center of the three-coil probe is located against the layer and its middle coincides with the center of the main two-coil probe (Fig. 6.20A). As in the case of the two-coil probe, we consider a function $\gamma_a/\gamma_b = f(H/L)$ for different ratios of γ_b/γ_s (Fig. 6.20B). Here, L is the thickness of the layer with conductivity γ_b, L is the length of the main two-coil probe, and γ_s is the conductivity of the surrounding medium ("shoulders").

Examples of profiling curves are given in Fig. 6.21. The three-coil probe has almost the same vertical response as the two-coil probe, but displays some asymmetry with respect to the center of the layer and is slightly more influenced by surrounding medium. The latter is also true for any multicoil probe with internal "focusing" and probes with the external focusing, when the thickness of the layer is greater than the probe length. Indeed, we may to recall that the geometric factor of the two-coil probe placed in the middle

(A)

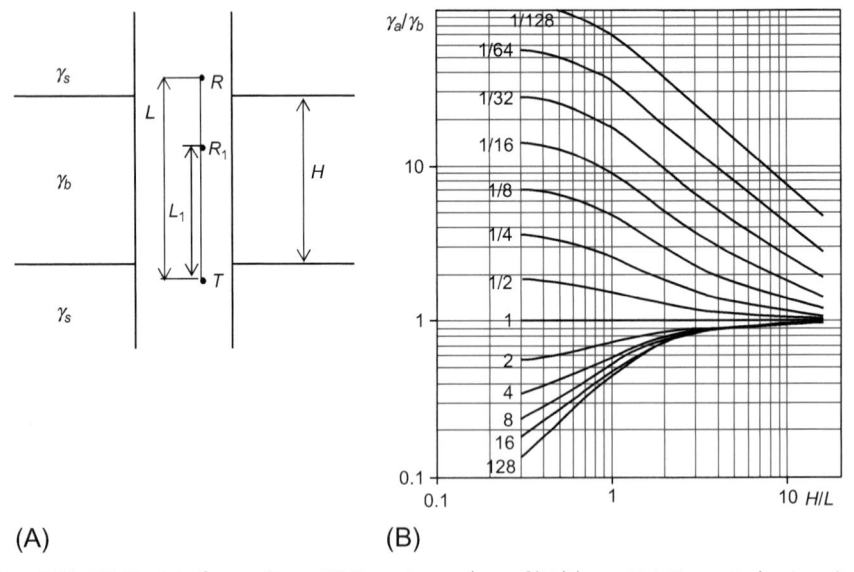

(B)

Fig. 6.20 (A) Model of a medium. (B) Function $\gamma_a/\gamma_b = f(H/L), p = 0.6$. Curve index is γ_b/γ_s.

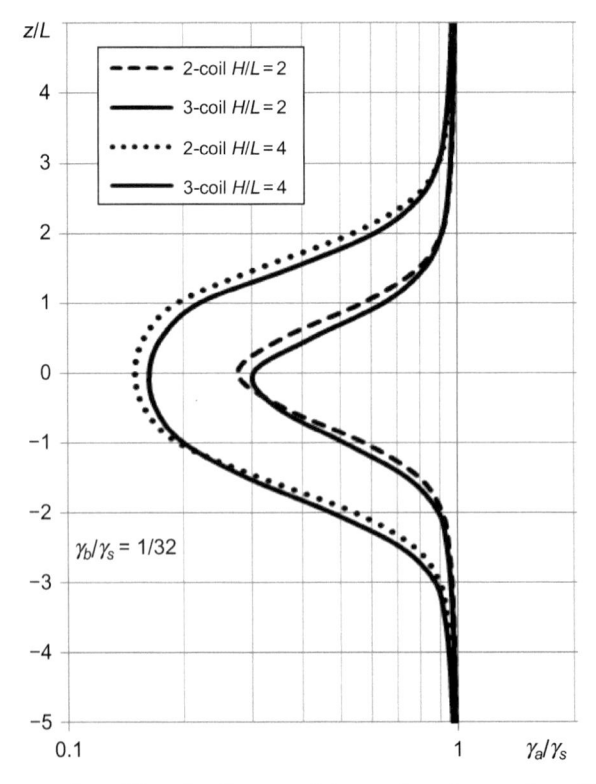

Fig. 6.21 Curves of profiling for three and two-coil probes, $\gamma_a/\gamma_s = f(z/L)$, $H/L = 2$ and 4; $\gamma_b/\gamma_s = 1/32$, $p = 0.6$.

of the layer, and having the same thickness as the probe length, is equal 0.5. Suppose that the length of the main two-coil probe is equal to the thickness of more resistive layer, and that external focusing probes are located in the surrounding medium. In this case, the vertical response of the probe is improved due to a relatively large reduction of the signal from the surrounding medium compared with the marginal reduction from the layer.

Focusing probes are more affected by the skin effect compared with the two-coil probes of the same length due to increased sensitivity to the deep part of the formation, where interaction between induced currents are the most pronounced.

Also, the vertical response of the three-coil probe is slightly worse than that of the two-coil probe.

Example Two
Multicoil probe 6FF40

The probe was introduced in 1960 and became the industry standard for 30 years. The 6FF40 array has six coils with the main transmitter-receiver pair spaced 40 in. (102 cm) apart. The main design parameters of the array are the spacing between the coils, the number of turns, and the polarity of each coil. The three transmitting and receiving coils are each connected in series to produce one signal output. The 6FF40 was designed to read deep into the formation while minimizing the signal close to the tool and maintaining reasonable vertical resolution.

This symmetrical focusing probe 6FF40 is shown in Fig. 6.22. One of the objectives of the focusing probe is to compensate primary electromotive force (EMF). In the dipole approximation with parameters presented in

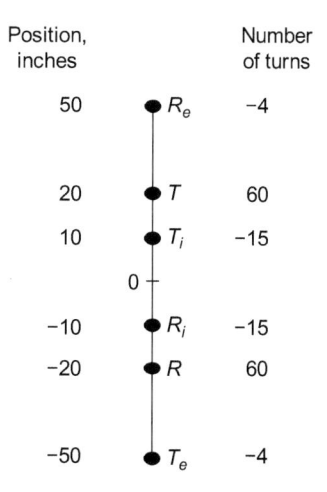

Position, inches		Number of turns
50	R_e	−4
20	T	60
10	T_i	−15
0		
−10	R_i	−15
−20	R	60
−50	T_e	−4

Fig. 6.22 Configuration of the symmetrical focusing probe 6FF40.

Fig. 6.22 (distances between coils and their moments), the uncompensated part of the primary field is more than 3% of the primary field of the main two-coil probe. Obviously, such level of compensation is insufficient for measurements of the secondary field. In reality these calculations may not be relevant because of the contribution of other factors such as the finite sizes of the coils, inaccuracy in coil positions, or uneven winding of the wire in the coils. For this reason, positions of the two external focusing coils are adjusted to account for all distorting effects to provide required compensation. At the same time, these small adjustments do not have a visible effect on the focusing properties of the probe.

The symmetrical probe 6FF40 could be defined by six parameters, which characterize the distance between coils and moments of coils. Values of these parameters were chosen graphically using the radial characteristic of two-coil probes, and they are given below in Table 6.1. As shown in Fig. 6.22, this system can be presented as a sum of nine two-coil probes with the lengths (L_i), products of number of turns in the transmitter and the receiver (M_i), and offsets of its centers from the center of the main probe (d_i), given below:

1. $L_1 = L$, $M_1 = n^2$, $d_1 = 0$;
2. $L_2 = p_i L$, $M_2 = c_i^2 n^2$, $d_2 = 0$;
3. $L_3 = p_e L$, $M_2 = c_i^2 n^2$, $d_2 = 0$;
4. $L_4 = \dfrac{L_1 + L_2}{2} = \dfrac{1 + p_i}{2}L$, $M_4 = -c_i n^2$, $d_4 = \dfrac{L_1}{2} - \dfrac{L_4}{2} = \dfrac{L_1 - L_2}{4} = \dfrac{1 - p_i}{4}L$;
5. $L_5 = \dfrac{L_2 + L_1}{2} = \dfrac{1 + p_i}{2}L$, $M_5 = -c_i n^2$, $d_5 = \dfrac{L_2}{2} - \dfrac{L_5}{2} = \dfrac{L_2 - L_1}{4} = -\dfrac{1 - p_i}{4}L$;
6. $L_6 = \dfrac{L_3 - L_1}{2} = \dfrac{p_e - 1}{2}L$, $M_6 = -c_e n^2$, $d_6 = \dfrac{L_1}{2} + \dfrac{L_6}{2} = \dfrac{L_1 + L_3}{4} = \dfrac{1 + p_e}{4}L$;
7. $L_7 = \dfrac{L_3 - L_1}{2} = \dfrac{p_e - 1}{2}L$, $M_7 = -c_e n^2$, $d_7 = -\dfrac{L_3}{2} + \dfrac{L_7}{2} = \dfrac{-L_3 - L_1}{4} = -\dfrac{1 + p_e}{4}L$;
8. $L_8 = \dfrac{L_3 - L_2}{2} = \dfrac{p_e - p_i}{2}L$, $M_8 = c_i c_e n^2$, $d_8 = \dfrac{L_2}{2} + \dfrac{L_8}{2} = \dfrac{L_2 + L_3}{4} = \dfrac{p_i + p_e}{4}L$;
9. $L_9 = \dfrac{L_3 - L_2}{2} = \dfrac{p_e - p_i}{2}L$, $M_9 = c_i c_e n^2$, $d_9 = -\dfrac{L_3}{2} + \dfrac{L_9}{2} = \dfrac{-L_3 - L_2}{4} = -\dfrac{p_i + p_e}{4}L$

The total electromotive force is equal to

$$\Xi = |\Xi(L_1)| + |\Xi(L_2)| + |\Xi(L_3)| - |\Xi(L_4)| - |\Xi(L_5)| - |\Xi(L_6)| \\ - |\Xi(L_7)| + |\Xi(L_8)| + |\Xi(L_9)| \qquad (6.69)$$

Table 6.1 Parameters of the 6FF40 Focusing Probe

L (inches)	p_i	p_e	n	c_i	c_e
40	0.5	2.5	60	0.25	0.066667

Inasmuch as $c_i < 1, c_e < 1$, terms 1, 4, 5, 6, and 7 give the main contribution into the measured signal. As shown above, the secondary signal in the main two-coil probe is

$$\Xi = \frac{-\omega^2 \mu_0^2}{4\pi L_1} M_T M_R \gamma_a(L_1)$$

where $M_T = I_0 S n$ and $M_R = S n$. Therefore, from Eq. (6.69), we have

$$\Xi = \frac{-\omega^2 \mu_0^2}{4\pi L_1} M_T M_R \left[\gamma_a(L_1) + \frac{c_i^2}{p_i} \gamma_a(L_2) + \frac{c_e^2}{p_e} \gamma_a(L_3) - \frac{2c_i}{1+p_i} \gamma_a(L_4) - \frac{2c_i}{1+p_i} \gamma_a(L_5) \right.$$
$$\left. - \frac{2c_e}{p_e - 1} \gamma_a(L_6) - \frac{2c_e}{p_e - 1} \gamma_a(L_7) + \frac{2c_i c_e}{p_e - p_i} \gamma_a(L_8) + \frac{2c_i c_e}{p_e - p_i} \gamma_a(L_9) \right]$$

(6.70)

If formation resistivity does not change in a vertical direction, the symmetrical probes 4 and 5 measure the same signals. The same is true for the pair of coils 6 and 7, 8 and 9. Then, Eq. (6.70) is transferred into

$$\Xi = \frac{-\omega^2 \mu_0^2}{4\pi L_1} M_T M_R \left[\gamma_a(L_1) + \frac{c_i^2}{p_i} \gamma_a(L_2) + \frac{c_e^2}{p_e} \gamma_a(L_3) \right.$$
$$\left. - \frac{4c_i}{1+p_i} \gamma_a(L_4) - \frac{4c_e}{p_e - 1} \gamma_a(L_6) + \frac{4c_i c_e}{p_e - p_i} \gamma_a(L_8) \right]$$

(6.71)

In particular, if a medium is an infinite cylinder with a radius r and conductivity γ_1 for the signal Ξ, we have

$$\Xi = \frac{-\omega^2 \mu_0^2}{4\pi L_1} M_T M_R \gamma_1 \left[G_1(L_1) + \frac{c_i^2}{p_i} G_1(L_2) + \frac{c_e^2}{p_e} G_1(L_3) \right.$$
$$\left. - \frac{4c_i}{1+p_i} G_1(L_4) - \frac{4c_e}{p_e - 1} G_1(L_6) + \frac{4c_i c_e}{p_e - p_i} G_1(L_8) \right].$$

(6.72)

Here, G_1 is the geometric factor of the borehole for a two-coil probe. With increase of r the function G_1 tends to unity. It is natural to introduce the geometric factor of the borehole for a multicoil probe as

$$G_1^* = \left[1 + \frac{c_i^2}{p_i} + \frac{c_e^2}{p_e} - \frac{4c_i}{1+p_i} - \frac{4c_e}{p_e - 1} + \frac{4c_i c_e}{p_e - p_i} \right]^{-1} \left[G_1(L_1) + \frac{c_i^2}{p_i} G_1(L_2) \right.$$
$$\left. + \frac{c_e^2}{p_e} G_1(L_3) - \frac{4c_i}{1+p_i} G_1(L_4) - \frac{4c_e}{p_e - 1} G_1(L_6) + \frac{4c_i c_e}{p_e - p_i} G_1(L_8) \right]$$

(6.73)

In such case, the geometric factor G_1^* changes from zero to unity; therefore, the sum of geometric factors of a borehole, invasion zone, and the formation is equal to unity. Correspondingly, the apparent conductivity in a medium with cylindrical boundaries is

$$\gamma_a = \gamma_1 G_1^* + \gamma_2 G_2^* + \gamma_3 G_3^* \tag{6.74}$$

and

$$G_2^* = G_1^*(r_2) - G_1^*(r_1), \quad G_3^* = 1 - G_1^* + G_2^*$$

Finally, for the signal Ξ, we have

$$\Xi = \frac{-\omega^2 \mu_0^2}{4\pi L_1} M_T M_R \left(1 + \frac{c_i^2}{p_i} + \frac{c_e^2}{p_e} - \frac{4c_i}{1+p_i} - \frac{4c_e}{p_e - 1} + \frac{4c_i c_e}{p_e - p_i} \right) \gamma_a \tag{6.75}$$

The geometric factor G_1^* for the probe 6FF40 is shown in Fig. 6.23.

To illustrate the radial responses of 6FF40 and two-coil probe, the curves of the apparent conductivity are shown in Fig. 6.24.

Next, we describe the vertical responses of the 6FF40 probe and first consider the geometric factor of an elementary layer for this probe. By analogy with the geometric factor of the elementary layer for a three-coil probe and making use of Eqs. (6.70) and (6.73), we obtain

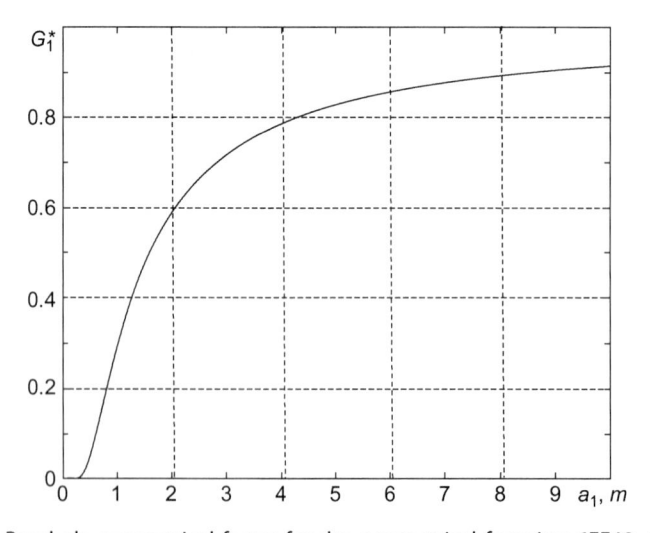

Fig. 6.23 Borehole geometrical factor for the symmetrical focusing 6FF40 probe. The function G_1^* is plotted as a function of the borehole radius for the parameters in Table 6.1.

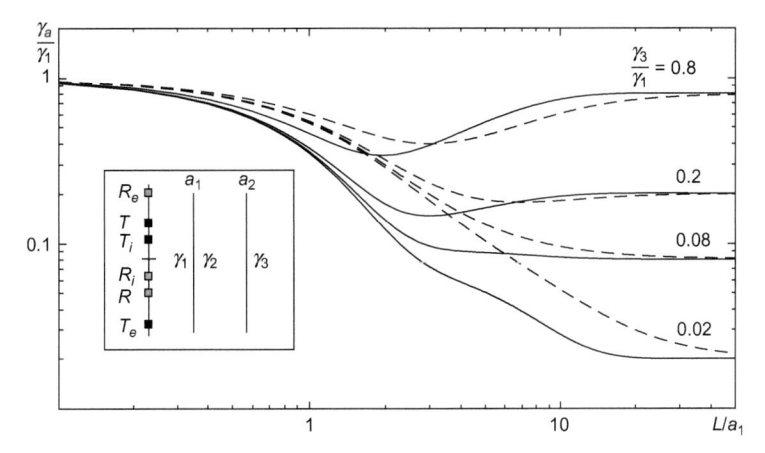

Geometrical Factor Theory of Induction Logging

Fig. 6.24 Normalized apparent conductivity curves for two-coil (*dashed*) and 6FF40 (*solid*) probes with invasion zone. The ratio $a_2/a_1 = 6$, $\gamma_2/\gamma_1 = 0.3$.

$$G_z^* = \left[1 + \frac{c_i^2}{p_i} + \frac{c_e^2}{p_e} - \frac{4c_i}{1+p_i} - \frac{4c_e}{p_e - 1} + \frac{4c_i c_e}{p_e - p_i}\right]^{-1}$$

$$\left[G_z(L_1, d_1) + \frac{c_i^2}{p_i}G_z(L_2, d_2) + \frac{c_e^2}{p_e}G_z(L_3, d_3) - \frac{2c_i}{1+p_i}G_z(L_4, d_4) - \frac{2c_i}{1+p_i}G_z(L_5, d_5)\right.$$

$$\left. - \frac{2c_e}{p_e - 1}G_z(L_6, d_6) - \frac{2c_e}{p_e - 1}G_z(L_7, d_7) + \frac{2c_i c_e}{p_e - p_i}G_z(L_8, d_8) + \frac{2c_i c_e}{p_e - p_i}G_z(L_9, d_9)\right]$$

(6.76)

Here

$$G_z(L, z) = \frac{L}{2(|L/2 + z| + |L/2 - z|)^2}$$

and offsets d_i are derived in the beginning of this subsection.

The geometric factors of the elementary layer for the 6FF40 and two-coil probe with $L = 40$ in. are shown in Fig. 6.25. A geometric factor of the 6FF40 probe has some "horn" effect caused by the positioning of the focusing coils with respect to the elementary layer boundaries (Fig. 6.25).

Profiling curves of γ_a/γ_s as function of logging depth for the 6FF40 probe, assuming no influence of the borehole and invasion are presented in Fig. 6.26. Two cases are shown: $H = 6.67$ ft and $H = 13.33$ ft, which, respectively, correspond to $H/L_1 = 2$ and $H/L_1 = 4$. For comparison, the dashed and dotted lines show responses for a two-coil probe with $L = 40$ in.

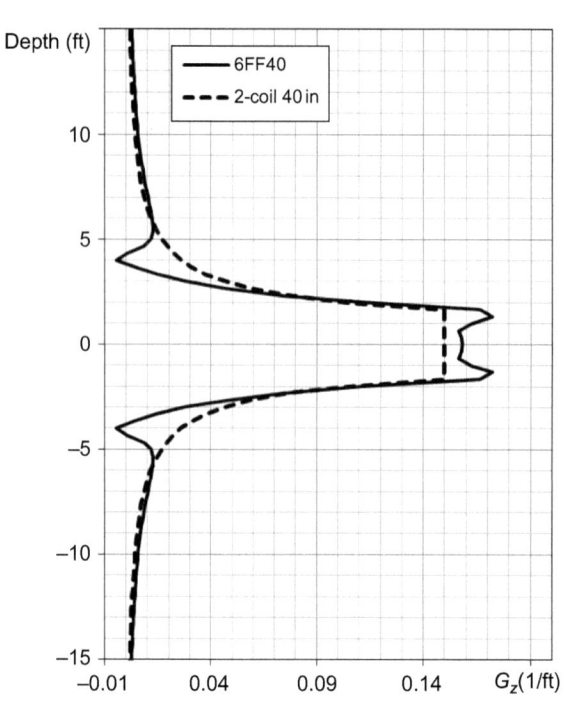

Fig. 6.25 Geometric factor of the elementary layer for 6FF40 and two-coil probe ($L=40$ in.).

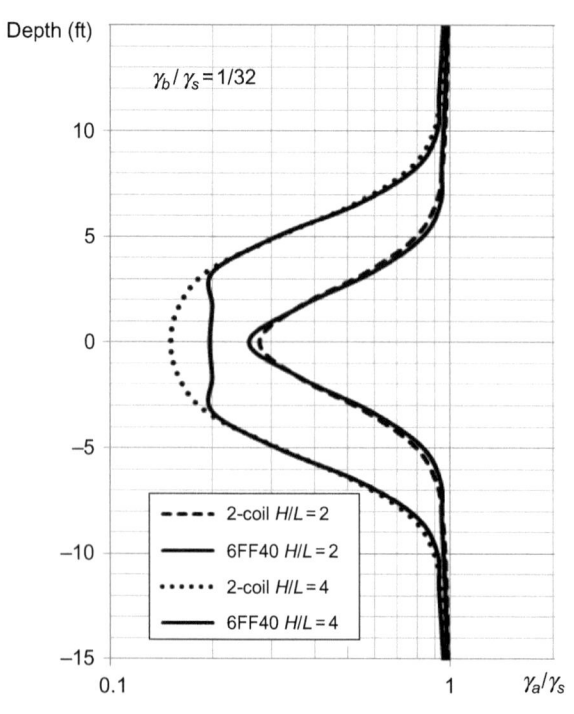

Fig. 6.26 Profiling curves for 6FF40 and two-coil probes, $\gamma_a/\gamma_s = f(z), H/L_1 = 2$ and 4; $\gamma_b/\gamma_s = 1/32$.

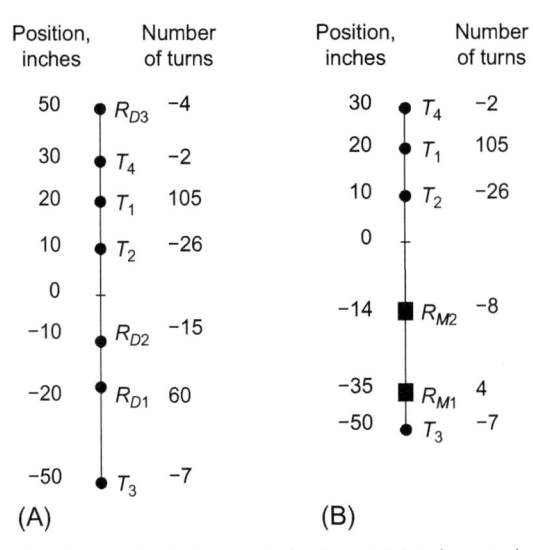

Geometrical Factor Theory of Induction Logging

Position, inches		Number of turns	Position, inches		Number of turns
50	R_{D3}	−4	30	T_4	−2
30	T_4	−2	20	T_1	105
20	T_1	105	10	T_2	−26
10	T_2	−26	0		
0			−14	R_{M2}	−8
−10	R_{D2}	−15	−35	R_{M1}	4
−20	R_{D1}	60	−50	T_3	−7
−50	T_3	−7			
(A)			(B)		

Fig. 6.27 Dual induction probe. (A) Deep induction. (B) Medium induction.

One can see that 6FF40's vertical response has a flat part over the bed with $H/L_1 = 4$.

In 1962 Schlumberger developed a dual induction probe (DIL tool), which contains 6FF40 and a smaller probe to better resolve parameters of the invasion zone. The dual induction tool (tools DIT-D and DIT-E) is shown in Fig. 6.27.

Evolution of Multicoil Focusing Probes

The dual induction system with two focusing probes was the first step towards the development of more general multicoil induction arrays. The 6FF40 and the dual induction probes had a long and continuing success, providing both deep depths of investigation and good vertical resolution. However, with time these tools were unable to satisfy all the needs of the industry, which required greater depth of investigation in the presence of large invasion. Moreover, the industry also desired tools capable of resolving thin beds with the thickness down to 0.3–0.6 m. To meet the needs of the industry, Schlumberger (AIT) and Western Atlas (HDIL) developed array induction tools, which were comprised of plurality of the tree-coil probes.

Both AIT and HDIL generate resistivity logs from measurements made at several different depths of investigation ranging from several to tens of inches. The differences between the curves enable characterization of the invaded zone and determination of the deep formation resistivity.

6.5 CORRECTIONS OF THE APPARENT CONDUCTIVITY

Focusing probes are designed to remove influence of the borehole and invasion zone and measure formation conductivity. However, some influence of these parts of a medium remains; correspondingly, the apparent conductivity differs from the conductivity of a formation. In most cases, there is an influence of medium (so-called shoulders), located above and beneath a bed, especially if their conductivity is greater than that of a bed. Also a displacement of a probe from the borehole axis causes a change of an apparent conductivity. In addition, function γ_a is equal to conductivity of a uniform medium only in the absence of skin effect, which in reality is always present. Below, we describe some approximate methods (corrections), which allow one to take into account the influence of these undesirable factors.

6.5.1 Skin Effect Corrections

As previously discussed, the skin effect increases with the probe length, formation conductivity, and frequency. Because measured signal Ξ is proportional to the frequency, we face two opposing tendencies: on one hand, it is useful to increase the frequency to generate a larger signal and improve vertical response; on the other hand, it is attractive to use a lower frequency to minimize the skin effect and benefit from the simplicity of the low-frequency approximation. To meet these opposing requirements the multicoil induction probes are equipped with the option of selecting a frequency, for example, 10, 20, and 40 kHz. Moran suggested a method to correct for the skin effect, assuming that the measured signal is caused only by currents in a formation. To make a correction, he used an apparent conductivity curve in a uniform medium (Fig. 6.28.) A horizontal axis depicts the corrected value of conductivity, while the y-axis is the apparent conductivity. By drawing a horizontal line, corresponding to a measured value of γ_a, one may find the point of intersection with the curve $\gamma_a = f(\gamma)$ and perpendicular from this point to the x axis gives a corrected conductivity. As one can see in Fig. 6.28, the skin effect is practically negligible for formations with conductivity lower than 0.1 S/m. But at conductivity of 1.0 S/m, it becomes quite visible and, for the two-coil probe, leads to more than 20%, 35%, and 56% error at 10, 20, and 40 kHz, correspondingly.

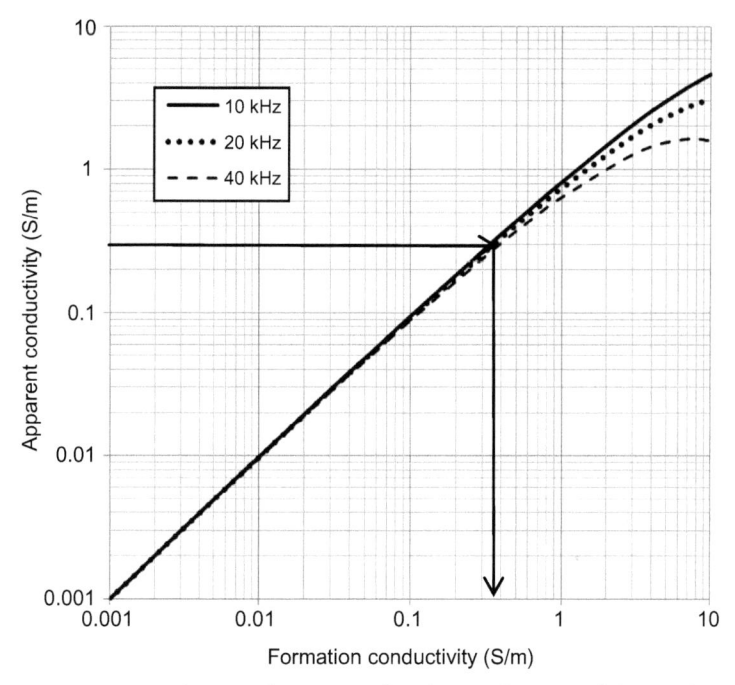

Geometrical Factor Theory of Induction Logging 223

Fig. 6.28 Apparent conductivity for a two-coil probe as a function of the conductivity of a uniform formation, $L = 40$ in.

A similar curve for the 6FF40 probe is shown in Fig. 6.29, and, as pointed out earlier, the skin effect here is larger than for the corresponding main 40-in. two-coil probe. The increased skin effect is caused by deeper penetration of the field into the formation. In a medium with high conductivity, this correction method may give two values of corrected conductivity. For example, if the apparent conductivity for the probe 6FF40 at a frequency of 20 kHz is equal to 1 S/m, it may correspond either to formation resistivity of 0.7 ohm-m or 0.125 ohm-m. In order to avoid such ambiguity, it is necessary either to use a lower frequency or combination of several frequencies.

In the case of 6FF40, instead of the graphical approach, the approximate equation

$$\log \gamma_a^{corrected} = \log(a\gamma_a) + ab\gamma_a \qquad (6.77)$$

can be used. Here, coefficients $a = 1.0899$ and $b = 0.000135$ are chosen to provide an exact conductivity reading of 6FF40 in 500 mS/m formation. The apparent conductivities in this equation are expressed in mS/m. The correction technique described above is applied only if an influence of

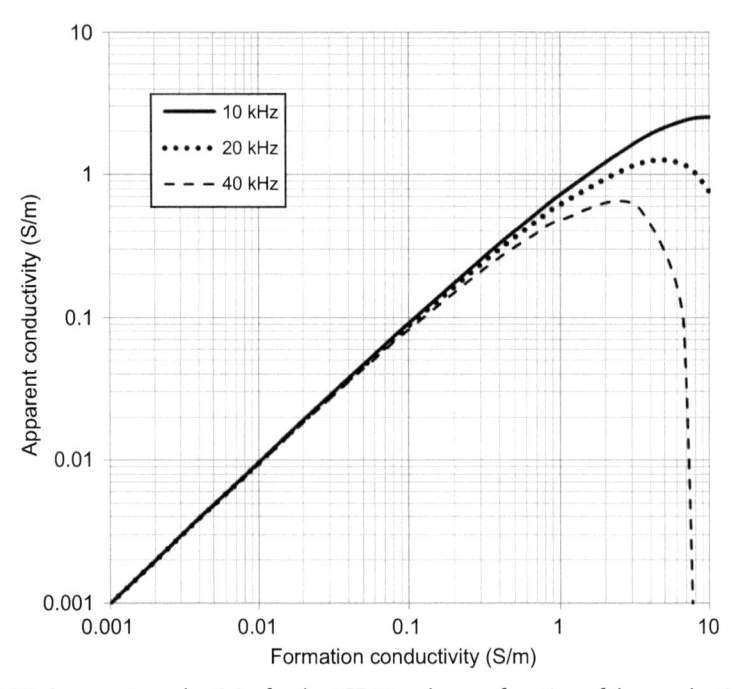

Fig. 6.29 Apparent conductivity for the 6FF40 probe as a function of the conductivity of a uniform formation.

"shoulders" is negligible. This limitation motivated development of alternative correction techniques. They are valid in a uniform and nonuniform medium, provided that the low frequency signal can be described by only two terms of the series for the both quadrature and in-phase components. The corrections rely on the fact that the second terms for both components are equal to each other. For instance, as follows from Eqs. (5.14) and (5.15) for a two-coil probe in a uniform medium, we have

$$QB_z^* \approx \frac{\mu_0 M_T}{2\pi L^3} \left\{ \frac{\gamma \mu_0 L^2}{2} \omega - \frac{(\gamma \mu_0 L^2)^{3/2}}{3(2)^{1/2}} \omega^{3/2} + \cdots \right\} \quad (6.78)$$

$$InB_z^* \approx \frac{\mu_0 M_T}{2\pi L^3} \left\{ 1 - \frac{(\gamma \mu_0 L^2)^{3/2}}{3(2)^{1/2}} \omega^{3/2} \cdots \right\} \quad (6.79)$$

The first term of the quadrature component (Eq. 6.78) corresponds to the secondary signal in geometrical factor theory, while the first term of the in-phase component (Eq. 6.79) represents the primary field, which is compensated in all multicoil probes. Because second terms are the same and a multicoil probe measures both quadrature and in-phase components, a skin effect correction can be made by subtracting the in-phase component from the quadrature component. This technique was first used in the dual induction tool, which does measure both components.

Next, consider one more approach allowing for reduction of the skin effect. Let us modify (6.78) and keep only the first two terms:

$$QB_z^* \frac{4\pi L}{\omega \mu_0^2 M_T} \approx \gamma - \frac{(\gamma \mu_0 L^2)^{1/2}}{3(2)^{-1/2}} \gamma \omega^{1/2} \qquad (6.80)$$

As follows from Eq. (6.36), the left-hand side of this equation is the measured apparent conductivity. By rewriting this equation for two frequencies, we receive

$$\gamma_a(f_1) \approx \gamma - \frac{(\gamma \mu_0 L^2)^{1/2}}{3(2)^{-1/2}} \gamma (2\pi f_1)^{1/2} \qquad (6.81)$$

$$\gamma_a(f_2) \approx \gamma - \frac{(\gamma \mu_0 L^2)^{1/2}}{3(2)^{-1/2}} \gamma (2\pi f_2)^{1/2} \qquad (6.82)$$

Multiplying Eq. (6.81) by $\sqrt{f_2}$ and Eq. (6.82) by $\sqrt{f_1}$ and subtracting one from another, we obtain the following skin effect correction formulae:

$$\gamma_a^{corrected} = \frac{\gamma_a(f_1)\sqrt{f_2} - \gamma_a(f_2)\sqrt{f_1}}{\sqrt{f_2} - \sqrt{f_1}} \qquad (6.83)$$

If measurements are performed at more than two frequencies, one can apply a least-squares technique and further improve accuracy of this approach.

6.5.2 Borehole Correction

Usually the focusing probe allows one to remove influence of the borehole, but if its radius is large and conductivity greatly exceeds that of the formation, $\gamma_m/\gamma_f > 100$, an influence of the borehole should be taken into account. Because the borehole diameter and its conductivity are known from independent measurements, it is easy to make an approximate correction. Indeed, as follows from Eq. (6.29),

$$\gamma_a = \gamma_m G_b + \gamma_f (1 - G_b) \qquad (6.84)$$

where γ_m and γ_f are the mud and formation conductivity, respectively, and G_b is the geometric factor of the borehole. Thus

$$\gamma^{corrected} = \frac{\gamma_a - \gamma_m G_b}{1 - G_b} \tag{6.85}$$

In the case of the 6FF40 probe G_b is typically smaller than 0.01, and, Eq. (6.85) is reduced to

$$\gamma^{corrected} \approx \gamma_a - \gamma_m G_b \tag{6.86}$$

REFERENCES

[1] Doll H-G. Introduction to induction logging and applications to logging of wells drilled with oil-base mud. J Petrol Technol 1949;1:148–62.
[2] Tanguy D. Induction well logging, U.S. Patent 3,067,383 A; 1962.
[3] Kaufman AA. Theory of induction logging. Novosibirsk: Nauka; 1965.

FURTHER READING

[1] Anderson B, Barber TD. Deconvolution and boosting parameters for obsolete Schlumberger induction tools. Log Analyst 1999;40:133–7.
[2] Kaufman AA, Anderson BA. Principles of electric methods in surface and borehole geophysics. Amsterdam: Elsevier; 2010.
[3] Moran JH, Kunz KS. Basic theory of induction logging and application to study of two-coil sondes. Geophysics 1962;27(6):829–58.
[4] Oristaglio M, Dorozynski A. A sixth sense: the life and science of Henri-Georges Doll. New York: Overlook; 2009.

CHAPTER SEVEN

Integral Equations and Their Approximations

Contents

7.1 Physical Principles of the Hybrid Method	227
7.2 Derivation of the Equation for the Field	228
7.2.1 Media With Cylindrical Boundaries	231
7.2.2 Media With Borehole and a Layer of Finite Thickness	233
7.2.3 Media With Horizontal Bed and Invasion	233
7.3 A Volume Integral Equation and Its Linear Approximation	235
7.4 A Surface Integral Equation for the Electric Field	238
7.4.1 Integral Equation for Cylindrically Layered Formation	238
7.4.2 Integral Equation for Horizontally Layered Formation	244
7.4.3 Integral Equation and the Born Approximation	246
References	247
Further Reading	248

Chapter 6 describes Henri Doll's theory of induction logging, which provides a good approximate representation of the field at low induction numbers when the skin effect is negligible. In this chapter, we shall consider two other approximations that in many cases can greatly simplify determination of the field, while still taking into account the skin effect. We start with the so-called hybrid method [1].

7.1 PHYSICAL PRINCIPLES OF THE HYBRID METHOD

As was shown earlier, the geometrical factor theory is based on the assumption that the secondary currents in a conducting medium are determined solely by the electric field generated by the time-varying primary magnetic field of the transmitting coil. This assumption implies that the interaction between secondary currents, which generates the skin effect, is neglected. As a result, the secondary currents have only a quadrature component. However, the analysis of the field has demonstrated that the magnitude of the actual secondary currents in a uniform medium (Chapter 5) decreases faster than predicted by the geometrical factor theory.

The same behavior is observed in a more complicated medium—the values of the quadrature component of the magnetic field are smaller than those calculated from the geometrical factor theory. There was definitely a need for approximation that would take into account the skin effect while still avoiding time-consuming numerical calculations. One so-called hybrid method was developed in Russia in 1963 [1], and is still sometimes used in solving forward and inverse problems. In this section, we describe this approximate method, which under certain conditions accurately accounts for the skin effect. The hybrid method is quite simple. Let us represent the currents in the space around the induction probe as a sum of two parts, namely: (a) currents in an "internal" region, where the induction probe is located; and (b) currents in an "external" region. For simplicity, we assume that the conductivity of the external region is constant. Suppose that two conditions are valid:

(1) The induced currents in the internal region are shifted in-phase by 90 degrees with respect to the dipole current, and their density depends only on the conductivity of the medium at a given location. In other words, mutual interaction between currents induced within this region is practically absent, and they are induced only by the primary vortex electric field of the magnetic dipole.

(2) The induced currents in the external area do not depend on the resistivity within the internal area; thus the interaction between currents located in these two different areas can be ignored. This condition emphasizes the fact that the skin effect manifests itself at relatively large distances from the source.

7.2 DERIVATION OF THE EQUATION FOR THE FIELD

Proceeding from these assumptions we derive simple expressions for the quadrature and in-phase components of the magnetic field. Let us represent the quadrature component of the magnetic field as a sum of the magnetic fields caused by currents in the internal and external areas:

$$QB_z^* = QB_z^{i*} + QB_z^{e*}$$

or

$$Qb_z^* = Qb_z^{i*} + Qb_z^{e*} \qquad (7.1)$$

where the superscripts "i" and "e" denote the components of the magnetic field caused by currents within the internal and external areas, respectively. In Eq. (7.1), all terms are normalized by the field in free space

$$B_z^{(0)} = \frac{\mu_0 M}{2\pi L^3}$$

Here, M is the magnetic moment of the source coil, and L is the distance between the source and receiving coil of a two-coil probe. Results of Chapter 6, as well as the first assumption above, enable us to represent the magnetic field Qb_z^{i*} as

$$Qb_z^{i*} = \frac{\omega\mu_0 L^2}{2}\gamma_a^i \qquad (7.2)$$

where γ_a^i is the apparent conductivity for the internal area. In accordance with Eq. (6.17), the apparent conductivity is related to the actual conductivity as:

$$\gamma_a^i = \gamma_a G_A + \gamma_B G_B + \gamma_C G_C + \cdots + \gamma_F G_F \qquad (7.3)$$

Here, $G_A, G_B, G_C,$ and G_F are geometric factors of homogeneous regions in the internal area with corresponding conductivities $\gamma_A, \gamma_B, \gamma_C,$ and γ_F. First consider the special case when the conductivities of the internal and external areas are equal to each other and we have a uniform medium. Then, the field can be presented in the form:

$$Qb_z^{un*}(\gamma_e) = Qb_z^{i*}(\gamma_e) + Qb_z^{e*}(\gamma_e) \qquad (7.4)$$

This last expression follows from the assumption that the field in the external area does not depend on the conductivity of the internal area. In Eq. (7.4), $Qb_z^{un}(\gamma_e)$ is the quadrature component of the field in a uniform medium with the conductivity of the external area, γ_e, and $Qb_z^i(\gamma_e)$ is the quadrature component of the field caused by currents of the internal area whose conductivity is also γ_e. As follows from the first assumption, this part of the field can be expressed through the geometric factor of the internal area, G_i;

$$Qb_z^{i*}(\gamma_e) = \frac{\omega\mu_0 L^2}{2}\gamma_e G_i \qquad (7.5)$$

Therefore, for the quadrature component of the field caused by currents in the external area we have:

$$Qb_z^{e*}(\gamma_e) = Qb_z^{un*}(\gamma_e) - Qb_z^{i*}(\gamma_e) = Qb_z^{un*}(\gamma_e) - \frac{\omega\mu_0 L^2}{2}\gamma_e G_i \qquad (7.6)$$

Correspondingly, for the total quadrature component of the field in a nonuniform medium, we obtain:

$$Qb_z^* = \frac{\omega\mu_0 L^2}{2}\gamma_a^i + Qb_z^{un*}(\gamma_e) - \frac{\omega\mu_0 L^2}{2}\gamma_e G_i$$
$$= Qb_z^{un*}(\gamma_e) + \frac{\omega\mu_0 L^2}{2}(\gamma_a^i - \gamma_e G_i) \qquad (7.7)$$

where γ_a^i is given by Eq. (7.3). Thus, according to the hybrid method, to determine the field it is sufficient to know the geometric factors of the corresponding parts of the internal area and the field of the magnetic dipole in a uniform medium with the conductivity of the external area, γ_e. The field in a uniform medium is well known, while calculation of geometrical factors is a simple matter, which for some typical cases was already addressed. The first term of the right-hand side in Eq. (7.7) accounts for the skin effect in a uniform medium with conductivity γ_e. It is proper to emphasize again that Eq. (7.7) corresponds to the special case of the uniform external medium, although later this limitation will be dropped and the method applied to more general cases.

Now let us show that as the induction number $p = L/\delta$ decreases, Eq. (7.7) describes the field derived from the geometrical factor theory. Here, L is again the length of the two-coil probe and δ is the skin depth in the external area. As was shown in Chapter 5, the quadrature component of the magnetic field in a uniform medium can be expressed in the form:

$$\frac{QB_z^{un*}}{B_z^0} = Qb_z^{un*}(\gamma_e) = \frac{\gamma_e \mu_0 \omega L^2}{2}, \quad \text{if } p \ll 1 \qquad (7.8)$$

Substitution of Eq. (7.8) into Eq. (7.7) gives:

$$Qb_z^* = \frac{\omega\mu_0 L^2}{2}[\gamma_e G_e + \gamma_A G_A + \gamma_B G_B + \cdots + \gamma_F G_F] \qquad (7.9)$$

where G_e is the geometrical factor of the external area. The last equation coincides with the expression derived by Doll. Now using the relation between the apparent conductivity and the field we have:

$$\gamma_a = \frac{2}{\omega\mu_0 L^2}Qb_z^* = \gamma_a^{un} + \gamma_a^i - \gamma_e G_i$$

or

$$\frac{\gamma_a}{\gamma_e} = \frac{\gamma_a^{un}}{\gamma_e} + \frac{\gamma_a^i}{\gamma_e} - G_i = \frac{\gamma_a^{un}}{\gamma_e} + \frac{\gamma_A}{\gamma_e} G_A + \frac{\gamma_B}{\gamma_e} G_B + \cdots + \frac{\gamma_F}{\gamma_e} G_F - G_i \quad (7.10)$$

Unlike geometrical factor theory, the hybrid method predicts a value for the in-phase component of the field caused by the secondary currents. In fact, because the currents in the internal area do not contribute to the in-phase component, we can write:

$$Inb_z^* = Inb_z^{un*}(\gamma_e) \quad (7.11)$$

In particular, for small values of the induction number we have:

$$Inb_z^* \approx 1 - \frac{2}{3} p_e^3 \quad (7.12)$$

Thus we obtained an in-phase component of the field that is the same as if the whole medium was uniform. Expressions (7.7), (7.11), describing in-phase and quadrature components, can be combined into the complex field:

$$b_z^* = b_z^{un*}(\gamma_e) + \frac{i\omega\mu_0 L^2}{2}\left(\gamma_a^i - \gamma_e G_i\right) \quad (7.13)$$

Next, we derive expressions for the field in some typical geo-electrical models.

7.2.1 Media With Cylindrical Boundaries

First, suppose that there is no invasion zone (Fig. 7.1A). Then, from Eq. (7.13) we obtain:

$$Qb_z^* = Qb_z^{un*}(\gamma_2) + \frac{\omega\mu_0 L^2}{2}(\gamma_1 - \gamma_2) G_1(\alpha) \quad (7.14)$$

and

$$Inb_z^* = Inb_z^{un*}(\gamma_2)$$

Correspondingly,

$$\frac{\gamma_a}{\gamma_2} = \frac{\gamma_a^{un}(\gamma_2)}{\gamma_2} + \left(\frac{\gamma_1}{\gamma_2} - 1\right) G_1(\alpha) \quad (7.15)$$

Here, γ_1, γ_2 are conductivities of the borehole and formation, respectively; G_1 is the geometric factor of the borehole; $\alpha = L/a_1$ is the ratio of

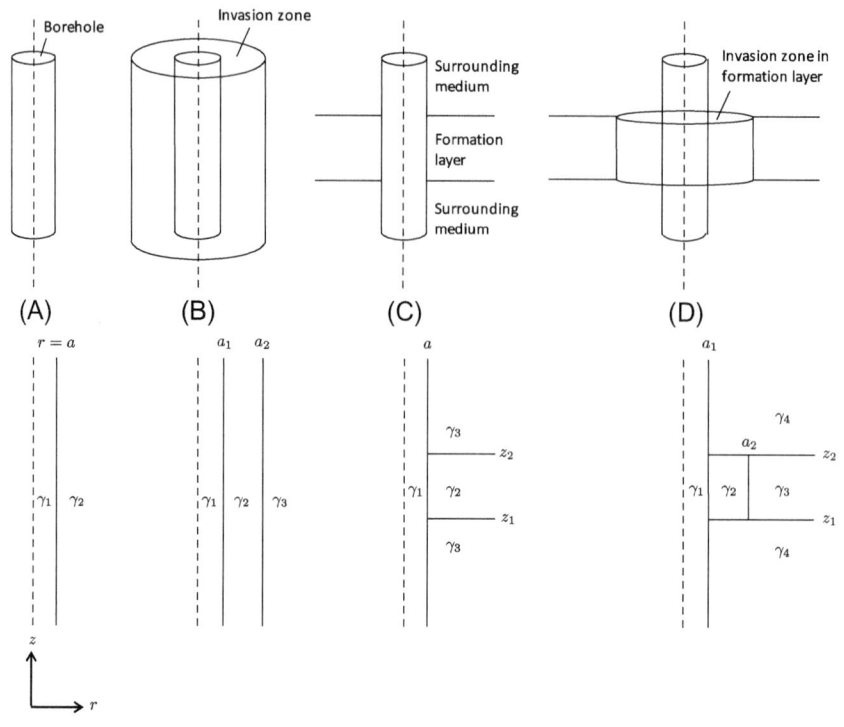

Fig. 7.1 Medium with (A) one and (B) two cylindrical boundaries; (C) medium with one cylindrical and two horizontal boundaries; (D) medium with two cylindrical and two horizontal boundaries.

the length of the two-coil probe and borehole radius; and $Qb_z^{un*}(\gamma_2)$ and $\gamma_a^{un}(\gamma_2)$ are the quadrature component of the magnetic field and the apparent conductivity in a uniform medium with conductivity γ_2. As follows from Eq. (7.14), to determine the field, we have to know the field in a uniform medium and the geometric factor of the borehole. In Chapter 8, we show that the use of Eq. (7.14) is much simpler than a rigorous numerical solution of the corresponding forward problem. Next, suppose that there is also an invasion zone, which, together with the borehole, forms the internal area (Fig. 7.1B). Then by an analogy with Eq. (7.14), we have:

$$Qb_z^* = Qb_z^{un*}(\gamma_3) + \frac{\omega\mu_0 L^2}{2}(\gamma_1 - \gamma_3)G_1 + \frac{\omega\mu_0 L^2}{2}(\gamma_2 - \gamma_3)G_2 \qquad (7.16)$$

$$Inb_z^* = Inb_z^{un*}(\gamma_3) \qquad (7.17)$$

where G_1 and G_2 are the geometrical factors of the borehole and invasion zone, respectively. Of course, in the range of the small induction number, the in-phase component is:

$$Inb_z^* \approx 1 - \frac{2}{3}p_3^3 \qquad (7.18)$$

7.2.2 Media With Borehole and a Layer of Finite Thickness

In this more general case (Fig. 7.1C), the borehole is still the internal region and a two-layered medium with horizontal boundaries is treated as the external area. Applying the same approach, we obtain:

$$Qb_z = Qb_z^*(\gamma_2, \gamma_3) + \frac{\omega\mu_0 L^2}{2}(\gamma_1 - \gamma_2)G_1^* + \frac{\omega\mu_0 L^2}{2}(\gamma_1 - \gamma_3)(G_1 - G_1^*) \qquad (7.19)$$

Here, $Qb_z^*(\gamma_2, \gamma_3)$ is the quadrature component of the magnetic field in the absence of the borehole when conductivity of the bed and surrounding medium are γ_2 and γ_3, respectively; G_1 is the geometric factor of the borehole, and G_1^* is the geometric factor of the part of borehole that is contained within the bed. For the in-phase component we have:

$$Inb_z^* = Inb_z^*(\gamma_2, \gamma_3) \qquad (7.20)$$

Later we demonstrate that in the range of small induction numbers, Eq. (7.20) becomes:

$$Inb_z^* \approx 1 - \frac{2}{3}p_3^3 \qquad (7.21)$$

In this case, the in-phase component of the secondary field is defined by conductivity γ_3 surrounding the bed. Note that it is a simple matter to generalize Eq. (7.19) for the case in which the media above and beneath the bed have different conductivities.

7.2.3 Media With Horizontal Bed and Invasion

Now suppose that the bed has an invasion zone (Fig. 7.1D). By analogy with the previous case, we have for the quadrature component:

$$Qb_z^* = Qb_z^*(\gamma_3, \gamma_4) + \frac{\omega\mu_0 L^2}{2}(\gamma_1 - \gamma_3)G_1^*$$
$$+ \frac{\omega\mu_0 L^2}{2}(\gamma_2 - \gamma_3)G_1^{**} + \frac{\omega\mu_0 L^2}{2}(\gamma_1 - \gamma_4)(G_1 - G_1^*) \quad (7.22)$$

and

$$Inb_z^* = Inb_z^*(\gamma_2, \gamma_3) \quad (7.23)$$

for the in-phase component. As shown earlier, we can expect that at the range of small parameter

$$Inb_z^* \approx 1 - \frac{2}{3}p_4^3 \quad (7.24)$$

Here, $Qb_z^*(\gamma_3, \gamma_4)$ is the quadrature component of the field in a medium with only two horizontal boundaries; G_1^* is the geometrical factor of the part of the borehole against the invasion zone, and G_1^{**} is the geometrical factor of the invasion zone, which can be presented as:

$$G_1^{**} = G_1^*(a_2) - G_1^*(a_1).$$

Some comments:
1. The method described in this section represents a natural extension of Doll's theory and is called the hybrid method because the concepts of both geometric factor and skin effect are used in derivations.
2. The hybrid method is much simpler to apply than rigorous numerical calculations, and is therefore useful for quick estimates of the field in typical geo-electrical scenarios.
3. This hybrid method is valid in the range of frequencies for which the borehole and invasion zone do not contribute to the in-phase component of the field. In particular, in the case of cylindrical boundaries, the in-phase component is defined only by the conductivity of the formation.
4. The derived equations also enable us to formulate conditions when geometrical factor theory can be applied with sufficient accuracy. As was shown in this section, the quadrature component of the field can be written in the form:

$$Qb_z^* \approx p^2(\gamma_1) f\left(\frac{\gamma_i}{\gamma_1}, G_i\right) - \frac{2}{3}p^3(\gamma_s) \quad (7.25)$$

Here, γ_1 is conductivity of the borehole, γ_i characterize the conductivity of an invasion zone, formation bed, and a surrounding medium; γ_s is the conductivity of the whole space surrounding a bed. When conductivities above and below the bed are different from each other, γ_s corresponds to the largest value of conductivity. Functions G_i describe geometric factors of the borehole, invasion zone, and formation. Thus, for each model of a medium, the geometrical factor theory can be applied if

$$p^2(\gamma_1) f\left(\frac{\gamma_i}{\gamma_1}, G_i\right) \gg \frac{2}{3} p^3(\gamma_s) \tag{7.26}$$

For instance, in a uniform medium we obtain the known conditions

$$p \ll 1$$

since $f\left(\dfrac{\gamma_i}{\gamma_1}, G_i\right) = 1$ and $\gamma_1 = \gamma_s = 1$.

5. In accordance with Eqs. (7.24), (7.25), the second term of the quadrature component coincides with the in-phase component of the secondary field. Correspondingly, by measuring the in-phase component, we can correct for the skin effect and determine the first term of the quadrature component (7.25) (Chapter 6).
6. Limits of the hybrid method were established by conducting rigorous numerical calculations and comparing results versus approximate solutions. The comparison was carried out for the layered formations with cylindrical boundaries and it was shown that for some typical cases, a satisfactory accuracy can be reached if

$$f < 0.2 \times \rho_{\min}/\left(\mu_0 L^2\right)$$

where $\rho_{\min}$ is the minimal resistivity comprising the medium and $L \sim 1$ m. The method might be useful in studying focusing systems because the last inequality coincides with conditions favorable for application of focusing probes.

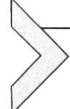

7.3 A VOLUME INTEGRAL EQUATION AND ITS LINEAR APPROXIMATION

The geometrical factor theory and the hybrid method are derived from specific assumptions about the distribution of induced currents in a medium. In this section we demonstrate that both approaches follow from the first (linear) approximation of the integral equation for the electrical field. To derive an integral equation for the electric field, we assume that a vertical magnetic dipole is located on the borehole axis and the medium is axially-symmetric. The time-variable moment of magnetic dipole creates a magnetic field, which, in accordance with Faradey's law, produces primary vortex electric field $\mathbf{E}_0^*$. Because of the axial symmetry, the electric field does not intersect boundaries between media of different conductivities and, therefore surface charges are absent. As a result, the electromagnetic field is generated solely by the primary source and induced currents in the medium.

The density of the induced current is determined by the Ohm's law:

$$\mathbf{j}^* = \gamma\left(\mathbf{E}_0^{*0} + \mathbf{E}_s^*\right) \tag{7.27}$$

Here, $\mathbf{E}_0^*$ and $\mathbf{E}_s^*$ are the complex amplitudes of the primary and secondary electric fields and γ is conductivity at a given point. Induced currents and the primary electric field have only the azimuthal component j_φ, which drastically simplifies the derivation of the integral equation. Visually, we can imagine the whole space filled with an infinite set of elementary current tubes of circular shape, whose centers are located on the borehole axis. Each tube creates the electric field at the point of observation p equal to:

$$dE_{s\varphi}^*(p) = i\omega\mu_0 G(p, q) j_\varphi^*(q) dS \tag{7.28}$$

Here dS is the cross-sectional area of the elementary tube, $G(p,q)$ is a Green's function that depends on geometrical parameters and can be presented in explicit form; while $j_\varphi^*(q)$ is the complex amplitude of the total current density at the point q. Performing integration of Eq. (7.28) over the entire cross section occupied by current tubes, we receive:

$$E_{s\varphi}^*(p) = i\omega\mu_0 \int_S G(p, q) j_\varphi^*(q) dS$$

or

$$E_{s\varphi}^*(p) = i\mu_0\omega \int_S \gamma(q) E_{0\varphi}^*(q) G(p, q) dS + i\omega\mu_0 \int_S \gamma(q) E_{s\varphi}^*(q) G(p, q) dS \tag{7.29}$$

where integration is performed over the region ($r > 0$ and $-\infty < z < \infty$). This is a Fredholm integral equation of the second kind for the secondary field $E^*_{s\varphi}(q)$. The equation connects the secondary electric field at any point in the medium with the electric field of the primary source and induced currents. Let us rewrite Eq. (7.29) in the form:

$$E^*_{s\varphi}(p) = F(p) + i\omega\mu_0 \int_{-\infty}^{\infty} dz \int_{0}^{\infty} \gamma(q) G(q,p) E^*_{s\varphi}(q) dr \qquad (7.30)$$

where

$$F(p) = i\omega\mu_0 \int_{-\infty}^{\infty} dz \int_{0}^{\infty} \gamma(q) E^*_{0\varphi}(q) G(p,q) dS \qquad (7.31)$$

is the secondary electric field created by the primary field $E^*_{0\varphi}$. By discarding the second term in Eq. (7.30), we arrive at the first approximation of the volume integral equation

$$E^*_{s\varphi}(p) \approx F(p) = i\omega\mu_0 \int_{-\infty}^{\infty} dz \int_{0}^{\infty} \gamma(q) E^*_{0\varphi}(q) G(p,q) dS \qquad (7.32)$$

It is essential that the integrand on the right-hand side of this equation is known, so that Eq. (7.32) represents a formula for calculating an approximation to the secondary electric field. By disregarding the second term in Eq. (7.30), we assume that the induced currents arise only due to the primary field in free space, which exactly coincides with the main assumption of geometrical factor theory. Thus Doll's theory represents a first order approximation in solving the integral equation (7.29).

Although the electric field on the axis of the borehole is equal to zero, the electromotive force Ξ^* arising in a horizontal loop of a finite size r can be estimated as

$$\Xi^* = 2\pi m E^*_{\varphi}$$

where n is the number of turns in the receiver coil.

The approximation described by Eq. (7.32) implies that induced currents are caused only by the primary electric field, and that the skin effect is absent, because no interaction between induced currents is taken into account. Inasmuch as these assumptions are made regardless of the distance from the primary source, it is natural to expect that the function $F(p)$ correctly describes the field at sufficiently low frequencies only when the probe is insensitive to remote parts of a medium where the skin effect is always present.

7.4 A SURFACE INTEGRAL EQUATION FOR THE ELECTRIC FIELD

Eq. (7.30) is not very convenient to use because of numerical complications caused by the infinite limits in r and z directions. Furthermore, it does not allow derivation of simple asymptotic expressions that take into account the skin effect. To facilitate calculation of the field and obtain more accurate expression valid at any frequencies, we derive an integral equation along the surfaces located at the fixed distances of r.

7.4.1 Integral Equation for Cylindrically Layered Formation

We start by assuming that there is no invasion zone and that the medium surrounding the borehole is uniform. Then, proceeding from Green's formula we obtain an integral equation for the component E_ϕ^* in which the integration is performed over the cross section of the borehole only. The vector electric field $\mathbf{E}^*$ at any regular point of a homogeneous medium satisfies the vector Helmholtz equation:

$$\nabla^2 \mathbf{E}^* + k^2 \mathbf{E}^* = 0 \qquad (7.33)$$

Let us represent the electric field as a sum of two components:

$$\mathbf{E}^* = \mathbf{E}_0^* + \mathbf{E}_1^* \qquad (7.34)$$

where $\mathbf{E}_0^*$ is a function that obeys the following equation outside and inside the borehole:

$$\nabla^2 \mathbf{E}_0^* + k_2^2 \mathbf{E}_0^* = 0 \qquad (7.35)$$

except at location of the dipole, and describes the electric field of a magnetic dipole in a uniform medium with the conductivity of the formation γ_2. The field $\mathbf{E}_0^*$ consists of the field of the dipole source in free space and the field of the currents induced in the uniform medium. The second term $\mathbf{E}_1^*$ in Eq. (7.34) appears because of the presence of the borehole with conductivity γ_1 and radius a. Substituting Eq. (7.34) in Eq. (7.35), we find that the field $\mathbf{E}_1^*$ satisfies the equation

$$\nabla^2 \mathbf{E}_1^* = -k^2 \mathbf{E}_1^* - k^2 \mathbf{E}_0^* - \nabla^2 \mathbf{E}_0^* \qquad (7.36)$$

Integral Equations and Their Approximations

Taking into account Eq. (7.35) for the region of the formation and the borehole we have:

$$\nabla^2 \mathbf{E}_1^* = -k_2^2 \mathbf{E}_1^*, \quad \text{if } r \geq a \tag{7.37}$$

and

$$\nabla^2 \mathbf{E}_1^* = -k_1^2 \mathbf{E}_1^* + \left(k_2^2 - k_1^2\right)\mathbf{E}_0^*, \quad \text{if } r \leq a \tag{7.38}$$

Unlike the total field $\mathbf{E}^*$, the functions $\mathbf{E}_0^*$ and $\mathbf{E}_1^*$ do not characterize the actual electric field in the medium, and application of the Biot-Savart law directly to the terms $\gamma_1 \mathbf{E}_0^*, \gamma_2 \mathbf{E}_0^*$ and $\gamma_1 \mathbf{E}_1^*, \gamma_2 \mathbf{E}_1^*$ is not straightforward. At the same time, by applying the Biot-Savart law for the total current densities

$$\gamma_1 \left(\mathbf{E}_0^* + \mathbf{E}_1^*\right) \quad \text{and} \quad \gamma_2 \left(\mathbf{E}_0^* + \mathbf{E}_1^*\right)$$

we may calculate the magnetic field. By definition, the complex amplitudes of the electric field are

$$\mathbf{E}_0^* = E_0^* \mathbf{i}_\varphi \quad \text{and} \quad \mathbf{E}_1^* = E_1^* \mathbf{i}_\varphi \tag{7.39}$$

where $\mathbf{i}_\phi$ is a unit vector directed along the φ-coordinate line. Next, we introduce a vector function $\mathbf{P}^* = P^* \mathbf{i}_\phi$, which along with its derivative, is a continuous function and satisfies the equation

$$\nabla^2 \mathbf{P}^* + k_2^2 \mathbf{P}^* = 0 \tag{7.40}$$

except at the point p, at which the field is determined. Also at this point the function $\mathbf{P}^* = P^* \mathbf{i}_\phi$ has a singularity of logarithmic type. Consider the expression

$$\mathbf{P}^* \nabla^2 \mathbf{E}_1^* - \mathbf{E}_1^* \nabla^2 \mathbf{P}^*$$

It is obvious that

$$\mathbf{P}^* \nabla^2 \mathbf{E}_1^* = P^* \mathbf{i}_\phi \left(\mathbf{i}_\phi \nabla^2 E_1^* + E_1^* \nabla^2 \mathbf{i}_\phi\right)$$

By analogy,

$$\mathbf{E}_1^* \nabla^* \mathbf{P}^* = E_1^* \mathbf{i}_\varphi \left(\mathbf{i}_\varphi \nabla^2 P^* - P^* \nabla^2 \mathbf{i}_\varphi\right)$$

Thus, we have proved that

$$\mathbf{P}^* \nabla^2 \mathbf{E}_1^* - \mathbf{E}_1^* \nabla^2 \mathbf{P}^* = P^* \nabla^2 E_1^* - E_1^* \nabla^2 P^* \tag{7.41}$$

The next step is to use the two-dimensional Green's formula and derive an integral equation in the form we are interested in. We may notice that the anomalous field $\mathbf{E}_1^*$ is a continuous function of a point of integration q, but the function $P(p,q)$ depends on both the point q and an observation point p where the electric field is determined.

The Green's formula is given as:

$$\int_S (\varphi \nabla^2 \phi - \phi \nabla^2 \varphi) dS = \oint_l \left(\varphi \frac{\partial \phi}{\partial n} - \phi \frac{\partial \varphi}{\partial n} \right) dl$$

Here, functions ϕ and φ are continuous at any point of the surface S, and l is a contour surrounding the surface. The normal n is directed outward at the area of integration. To apply this formula to functions E_1 and P, we have to surround an observation point p by a small circle l_p, because the latter has a singularity at the point p. Then, for the borehole and formation we have:

$$\int_{S_1} (P^* \nabla^2 E_1^* - E_1^* \nabla^2 P^*) dS = \oint_{l_0} \left(P^* \frac{\partial E_1^*}{\partial n} - E_1^* \frac{\partial P^*}{\partial n} \right) dl$$
$$+ \oint_l \left(P^* \frac{\partial E_1^*}{\partial r} - E_1^* \frac{\partial P^*}{\partial r} \right) dl \quad \text{if } r < a \quad (7.42)$$

and

$$\int_{S_2} (P^* \nabla^2 E_1^* - E_1^* \nabla^2 P^*) dS = -\oint_l \left(P^* \frac{\partial E_1^*}{\partial n} - E_1^* \frac{\partial P^*}{\partial n} \right) dl \quad \text{if } r > a \quad (7.43)$$

Here, l is the line parallel to the z-axis at $r = a$ that corresponds to the radius of the borehole, and l_p is a small circular contour around the point p (Fig. 7.2); $\partial/\partial n$ is the normal derivative. The negative sign at the right-hand side of Eq. (7.43) is selected because of the opposite direction along the line l in integrals of Eqs. (7.42), (7.43). As follows from Eqs. (7.37), (7.38):

$$P^* \nabla^2 E_1^* - E_1^* \nabla^2 P^* = 0, \quad \text{if } r > a$$

and

$$P^* \nabla^2 E_1^* - E_1^* \nabla^2 P^* = (k_2^2 - k_1^2) E_1^* P^* + (k_2^2 - k_1^2) E_0^* P^*, \quad \text{if } r < a$$

Integral Equations and Their Approximations

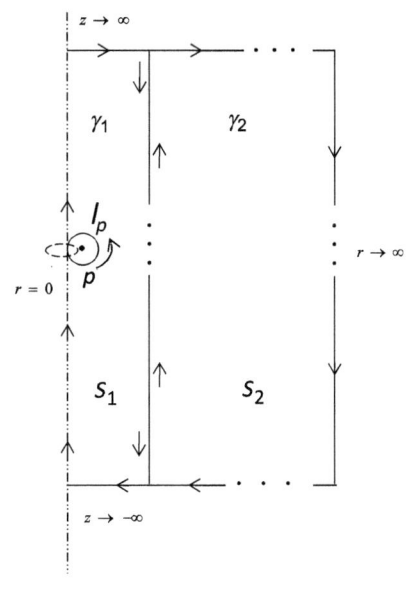

Fig. 7.2 Integration contours in Eq. (7.42).

From continuity of functions E and P and their derivatives at the borehole boundary, from the last two equations we have:

$$\left(k_2^2 - k_1^2\right) \int_{S_1} E_1^* P^* \, dS + \left(k_2^2 - k_1^2\right) \int_{S_1} E_0^* P \, dS = \oint_{l_p} \left(P^* \frac{\partial E_1^*}{\partial n} - E_1^* \frac{\partial P^*}{\partial n}\right) dl \quad (7.44)$$

To proceed in deriving an equation for the anomalous field E_1 for any point p in the borehole, let us define the function

$$G = i\omega\mu_0 P$$

as the electric field caused by the circular unit current with radius r_p. Later we demonstrate that it can be presented as:

$$\frac{i\omega\mu_0}{\pi} Ir_k \int_0^\infty I_1(r_k v) K_1(rv) \cos mz \, dm \quad \text{if } r \geq r_p$$

$$\frac{i\omega\mu_0}{\pi} Ir_k \int_0^\infty I_1(rv) K_1(r_k v) \cos mz \, dm \quad \text{if } r \leq r_p$$

Correspondingly, in place of Eq. (7.44), we have:

$$(\gamma_2 - \gamma_1) \int_{S_1} E_1^*(q) G^*(q,p) dS + (\gamma_2 - \gamma_1) \int_{S_1} E_0^*(q) G^*(q,p) dS$$
$$= \frac{1}{i\omega\mu_0} \oint_{l_p} \left(G^* \frac{\partial E_1^*}{\partial n} - E_1^* \frac{\partial G^*}{\partial n} \right) dl \quad (7.45)$$

Next consider the integral on the right-hand side of the last equation when point q of the contour l_p approaches an observation point p and radius r of the circle tends to zero. Taking into account that radius vector r and normal n have opposite directions we obtain:

$$\oint_{l_p} \left(G^* \frac{\partial E_1^*}{\partial n} - E_1^* \frac{\partial G^*}{\partial n} \right) dl = \oint_{l_p} \left(E_1^* \frac{\partial G^*}{\partial r} - G^* \frac{\partial E_1^*}{\partial r} \right) dl. \quad (7.46)$$

In approaching the current circle, the electric field G^* is defined only by the current element located in the vicinity of the point q. Inasmuch as the distance between the point q and this element tends to zero, it can be treated as infinitely long current line. It is well known [2] that the electric field of such source placed in a uniform medium is

$$G^* = \frac{i\omega\mu_0}{2\pi} K_0(k_2 r) \quad (7.47)$$

where $K_0(k_2 r)$ is a modified Bessel function of the second kind, and

$$K_0(k_2 r) \to -\ln r \quad \text{if } r \to 0$$

Bearing in mind that the field E_1^* and $\partial E_1^* / \partial r$ have finite values and

$$\frac{\partial G^*}{\partial r} \to -\frac{i\omega\mu_0}{2\pi r} \quad \text{if } r \to 0$$

the contour integral in Eq. (7.45) can be replaced with $-E_1^*(p)$. Thus, this equation becomes

$$E_1^*(p) = (\gamma_1 - \gamma_2) \int_{S_1} E_1^*(q) G^*(q, p, k_2) dS + (\gamma_1 - \gamma_2) \int_{S_1} E_0^*(q) G^*(q, p, k_2) dS$$

$$(7.48)$$

The latter is the Fredholm integral equation of the second kind, where integration is performed over the region of the borehole only. Next suppose

Integral Equations and Their Approximations

that there is an invasion zone with conductivity γ_2, surrounded by the formation with conductivity γ_3. The function G^* has the same meaning as before and satisfies the equation

$$\nabla^2 G^* + k_2^2 G^* = 0 \tag{7.49}$$

In accordance with Eqs. (7.37), (7.38), for the anomalous field $\mathbf{E}_1^*$ we have:

$$\nabla^2 \mathbf{E}_1^* = -k_1^2 \mathbf{E}_1^* + \left(k_3^2 - k_1^2\right)\mathbf{E}_0^* \quad \text{if } 0 < r < a_1$$
$$\nabla^2 \mathbf{E}_1^* = -k_2^2 \mathbf{E}_1^* + \left(k_3^2 - k_2^2\right)\mathbf{E}_0^* \quad \text{if } a_1 < r < a_2 \tag{7.50}$$
$$\nabla^2 \mathbf{E}_1^* = -k_3^2 \mathbf{E}_1^*, \quad \text{if } r > a_2$$

where a_1 is the radius of the borehole and a_2 is the outer radius of the invasion zone. Applying Green's formula to the regions of the borehole, the invasion zone, and the formation, respectively, we have the following equations:

$$\int_{S_1} \left(G^* \nabla^2 E_1^* - E_1^* \nabla^2 G^*\right) dS = -2\pi E_1^*(p) + \int_{l_1} \left(G^* \frac{\partial E_1^*}{\partial r} - E_1^* \frac{\partial G^*}{\partial r}\right) dl \tag{7.51}$$

$$\int_{S_2} \left(G^* \nabla^2 E_1^* - E_1^* \nabla^2 G^*\right) dS = \int_{l_1} \left(-G^* \frac{\partial E_1^*}{\partial r} + E_1^* \frac{\partial G^*}{\partial r}\right) dl$$
$$+ \int_{l_2} \left(G^* \frac{\partial E_1^*}{\partial r} - E_1^* \frac{\partial G^*}{\partial r}\right) dl \tag{7.52}$$

$$\int_{S_3} \left(G^* \nabla^2 E_1^* - E_1^* \nabla^2 G^*\right) dS = \int_{l_2} \left(-G^* \frac{\partial E_1^*}{\partial r} + E_1^* \frac{\partial G^*}{\partial r}\right) dl \tag{7.53}$$

Here, l_1 and l_2 are straight lines located at the boundaries between the borehole region S_1 and the invasion zone S_2 and between the invasion zone and formation region S_3, respectively. Now taking into account Eq. (7.46) and performing a summation of Eqs. (7.51)–(7.53), we obtain an integral equation that includes two surface integrals over half cross sections of the borehole and invasion zone:

$$E_1^*(p) = (\gamma_1 - \gamma_3) \int_{S_1} E_0^*(q) G^*(k_3, p, q) dS + (\gamma_2 - \gamma_3) \int_{S_2} E_0^*(q) G^*(k_3, p, q) dS$$
$$+ (\gamma_1 - \gamma_3) \int_{S_1} E_1^*(q) G^*(k_3, p, q) dS + (\gamma_2 - \gamma_3) \int_{S_2} E_1^*(q) G^*(k_3, p, q) dS$$
$$\tag{7.54}$$

It is clear that the integral equations (7.44), (7.54) coincide with each other if $k_1 = k_2$ or $k_2 = k_3$. Thus, we derived integral equations for two cases when the solution of the boundary value problem can be obtained in explicit form. In both cases the Green's function corresponds to a uniform medium with the conductivity of the formation.

7.4.2 Integral Equation for Horizontally Layered Formation

Next we derive the integral equation for the case in which the formation with conductivity of γ_2 has a finite thickness. Let us introduce a new Green's function that, outside the source region, is a solution of the equations

$$\nabla^2 \mathbf{G}^* + k_2^2 \mathbf{G}^* = 0 \quad \text{if } z_1 < z < z_2$$
$$\nabla^2 \mathbf{G}^* + k_3^2 \mathbf{G}^* = 0 \quad \text{if } z < z_1 \text{ or } z > z_2 \tag{7.55}$$

where z_1 and z_2 are the lower and upper boundaries of the layer; k_2 and k_3 are the wave numbers of the layer and the surrounding medium, respectively. Also assume that the function $\mathbf{G}^* = G^* \mathbf{i}_\phi$ and its first derivative with respect to z are continuous at the interfaces between the formation and the adjacent medium. From the physical point of view, the function $\mathbf{G}^*$ represents the electric field of a circular filament in a horizontally layered medium, and it can be expressed in an explicit form as an integral. As before, we represent the total electric field as a sum:

$$\mathbf{E}^* = \mathbf{E}_1^* + \mathbf{E}_0^* \tag{7.56}$$

where $\mathbf{E}_0^* = E_0^* \mathbf{i}_\phi$ the electric field of the magnetic dipole in the layered medium, and $\mathbf{E}_1^* = E_1^* \mathbf{i}_\phi$ is the secondary electric field caused by the presence of the borehole. Therefore, in the formation layer and in the adjacent media, respectively, we have:

$$\nabla^2 \mathbf{E}_0^* = -k_2^2 \mathbf{E}_0 \quad \text{and} \quad \nabla^2 \mathbf{E}_0^* = -k_3^2 \mathbf{E}_0 \tag{7.57}$$

Taking into account Eqs. (7.36), (7.57), we have the following equation in the formation:

$$\nabla^2 \mathbf{E}_1^* = -k_2^2 \mathbf{E}_1^* \tag{7.58}$$

and in the adjacent medium:

$$\nabla^2 \mathbf{E}_1^* = -k_3^2 \mathbf{E}_1^* \tag{7.59}$$

Integral Equations and Their Approximations

In the part of the borehole that is located against the formation layer we have:

$$\nabla^2 \mathbf{E}_1^* = (k_2^2 - k_1^2)\mathbf{E}_1^* \tag{7.60}$$

and in the part of the borehole that is located against the adjacent medium:

$$\nabla^2 \mathbf{E}_1^* = (k_3^2 - k_1^2)\mathbf{E}_1^* \tag{7.61}$$

Correspondingly, the function

$$G^* \nabla^2 E_1^* - E_1^* \nabla^2 G^*$$

is equal to zero within the formation layer and the surrounding medium.
At the same time, it is equal to

$$(k_2^2 - k_1^2)E_0^* G^* + (k_2^2 - k_1^2)E_1^* G^*$$

in the part of the borehole located against the formation layer and to

$$(k_3^2 - k_1^2)E_0^* G^* + (k_3^2 - k_1^2)E_1^* G^*$$

in the adjacent medium. Then, applying Green's formula we obtain an integral equation for the secondary electric field

$$E_1^*(p) = F_1^*(p) + (\gamma_3 - \gamma_1)\int_{S_2} E_1(q)G^*(p,q)dS \\ + (\gamma_2 - \gamma_1)\int_{S_1} E_1(q)G^*(p,q)dS \tag{7.62}$$

Here

$$F_1^*(p) = (\gamma_3 - \gamma_1)\int_{S_2} E_0^*(q)G^*(p,q)dS + (\gamma_2 - \gamma_1)\int_{S_1} E_0^*(q)G^*(p,q)dS, \tag{7.63}$$

where S_1 and S_2 are the regions of the borehole located against and outside the formation layer correspondingly.

The solution of the integral equation (7.62) enables us to determine the electric field, and therefore the total electric field in the receiver. In the presence of an invasion zone, the integral equation has the form:

$$E_1^*(p) = F_2^* + (\gamma_3 - \gamma_1)\int_{S_2} E_1^*(q)G^*(p,q)dS + (\gamma_2 - \gamma_1)\int_{S_1} E_1^*(q)G^*(p,q)dS$$
$$+ (\gamma_4 - \gamma_2)\int_{S_3} E_1^*(q)G^*(p,q)dS$$

Here, S_3 is the region representing the invasion zone in the formation with the wave number k_2. Also

$$F_2^* = (\gamma_3 - \gamma_1)\int_{S_2} E_0^*(q)G^*(p,q)dS + (\gamma_2 - \gamma_1)\int_{S_1} E_0^*(q)G^*(p,q)dS$$
$$+ (\gamma_4 - \gamma_2)\int_{S_3} E_0^*(q)G^*(p,q)dS \qquad (7.65)$$

It is obvious that all previous cases follow from Eqs. (7.64), (7.65).

7.4.3 Integral Equation and the Born Approximation

At the end of the 19th century the mathematician Carl Neumann developed the theory of integral equations for potential fields and, in particular, constructed formal solutions to these integral equations as an infinite series of terms, which is now called the Neumann series. The first term on the right-hand side of this series is called the first or linear approximation of the solution, similar to linear term in a Taylor series expansion of a function. Before Neumann's work, the physicist Lord Rayleigh had used the linear term as a first approximation to the integral equation that describes the scattering of light by small objects. In 1926, Max Born applied this approach in the approximate solution of integral equations that describe the scattering of quantum mechanical wave functions, and now it is commonly called the Born approximation in the physics literature. When applied to our case, the Born approximation for the field E_1^* is

$$E_1^*(p) \approx F_2^*(p) \qquad (7.66)$$

with F_2^* given by Eq. (7.65). For simplicity, consider the simplest model with one cylindrical boundary. Then Eqs. (7.64), (7.65) give:

$$E_1^* = (\gamma_2 - \gamma_1)\int_{S_1} E_0^*(k_2, q)G^*(k_2, p, q)dS \qquad (7.67)$$

In this approximation the sources of the secondary field arise due to the field $E_0^*(k_2, q)$ created by a magnetic dipole located in a uniform medium characterized by the wave number k_2. The function $G^*(p, q, k_2)$ describes the electric field of a current ring, whose cross section passes through the point q in the (r, z) plane. The function $E_0^*(q, k_2)$ has a very simple expression, and $G^*(p, q, k_2)$ can also be expressed with elementary functions. Thus, calculation of the field E_1^* by integration over the borehole region is a relatively simple task. The situation is not much more complicated in the presence of an invasion zone. When the formation has a finite thickness, solving the forward problem in the presence of the borehole requires complicated numerical techniques. At the same time the use of Born approximation is much simpler. Note that when frequencies are relatively low E_0^* can be replaced by the primary field in a free space with G^* describing the field of the current ring in a free space. For instance, for the case of one cylindrical interface, instead of Eq. (7.65), we obtain:

$$E_1^* = (\gamma_2 - \gamma_1) \int_{S_1} E_0^*(q) G^*(p, q) dS \qquad (7.68)$$

This is the same expression that was derived earlier using the hybrid method. Finally, replacing the field $E_0^*(p)$ in Eq. (7.56) by the first term of its expansion in series of a small parameter (the induction number), we arrive at the expression, which corresponds to the geometrical factor theory. Thus, the geometrical factor theory and the hybrid method are particular cases of the first approximation of the integral equation (Born approximation). In conclusion, we note that in a medium with only cylindrical boundaries, the Born and hybrid approximations require practically the same computation effort, but in the presence of the horizontal interfaces, the hybrid method is simpler to apply.

Numerical calculations show that for typical in induction logging frequencies, the hybrid method describes the field with error less than 5% unless the ratio $\gamma_1/\gamma_2 \leq 200$. At the same time, the Born approximation is even better and permits accurate calculations in a wider range of frequencies and conductivity contrasts.

REFERENCES

[1] Kaufman A. Method for approximate calculation of the field on the borehole axis. USSR, Geology and Geophysics; 1964.
[2] Kaufman A, Keller G. The magnetotelluric method. Amsterdam: Elsevier; 1982.

FURTHER READING
[1] Jackson JD. Classical electrodynamics. New York: Wiley; 1975.
[2] Born M. Quantenmechanik der Stossvorgängem. Zeitschrift fur Physik 1926;38:803.

CHAPTER EIGHT

Electromagnetic Field of a Vertical Magnetic Dipole in Cylindrically Layered Formation

Contents

8.1 The Boundary Value Problem for the Vector Potential	249
8.2 Expressions for the Field Components	254
8.3 The Magnetic Field in the Range of Small Induction Number	257
8.3.1 The First Approach for Deriving the Leading Term of the Quadrature Component (Transition to Doll's Formula)	258
8.3.2 The Second Approach for Deriving the Leading Term of the In-Phase Component	260
8.3.3 The Third Approach to Deriving Asymptotic Expressions of the Field	265
8.4 Far Zone of Magnetic Field on the Axis of Borehole	266
8.4.1 Cauchy's Formula and Deformation of Integration Contour	267
8.4.2 Validity of the Approximate Solution	274
8.4.3 Sensitivity to the Formation of Amplitudes Ratio and Phase Difference (Three-Coil Probe)	277
8.4.4 The Main Features of the Field of the Two-Coil Probe	279
8.5 Displacement of the Probe from the Borehole Axis	284
References	288
Further Reading	288

We derive an expression for the vertical component of a magnetic field on the axis of a borehole when the source of the primary field is a vertical magnetic dipole and the formation has an infinite thickness and several radial zones. Special attention is paid to the asymptotic behavior of both quadrature and in-phase components.

8.1 THE BOUNDARY VALUE PROBLEM FOR THE VECTOR POTENTIAL

In the formulation of the boundary value problem it is assumed that:

1. The borehole surrounding the induction probe is uniform and isotropic.
2. The electrical properties of the medium do not change in the direction parallel to the borehole axis. This means that the top and bottom of the bed are significantly distant from the probe.
3. The borehole shape is an infinitely long circular cylinder.
4. A medium located between the borehole and the bed represents a system of coaxial cylindrical layers with the axis coinciding with the borehole axis.
5. The transmitter and receiver coils of the probe are located on the borehole axis, and they can be considered dipoles because they are small compared to both the probe length and the borehole radius.

Thus, the boundary problem is formulated as follows. The medium comprises a set of $(n-1)$ coaxial cylindrical surfaces with radii $a_1, a_2, a_3, \ldots, a_{n-1}$, separating n isotropic cylindrical layers having conductivity γ_i $(i=1,\ldots,n)$. Magnetic permeability and dielectric constant are usually assumed to be equal to those in a free space, μ_0, ε_0. The vertical magnetic dipole is located at the borehole axis and its moment is a sinusoidal function of time, causing a primary electrical field to have only an azimuthal component $E_\phi^{(0)}$. The currents, induced in the horizontal planes of the medium, also have only an azimuthal component. Therefore, the vector lines of the currents are circles with a center on the borehole axis, and the corresponding boundary value problem can be solved by using only one component of the vector potential. As shown in Chapter 2, for the complex amplitude of the vector potential $\mathbf{A}_z^*$, we have:

$$\nabla^2 \mathbf{A}_z^* + k^2 \mathbf{A}_z^* = 0 \qquad (8.1)$$

and

$$\mathbf{E}^* = \operatorname{curl} \mathbf{A}^*, \quad i\omega \mathbf{B}^* = k^2 \mathbf{A}^* + \operatorname{grad} \operatorname{div} \mathbf{A}^* \qquad (8.2)$$

Here k is a wave number

$$k^2 = i\gamma\mu_0\omega \quad \text{and} \quad k = \left(\frac{i\gamma\mu_0\omega}{2}\right)^{1/2}(1+i) = \frac{(1+i)}{\delta} \qquad (8.3)$$

where δ is the skin depth.

Let us choose a cylindrical system of coordinates (r, ϕ, z) with a magnetic dipole, placed at the origin of this system (Fig. 8.1). The moment of the magnetic dipole is oriented along the z-axis. As mentioned in the previous

Electromagnetic Field of a Vertical Magnetic Dipole 251

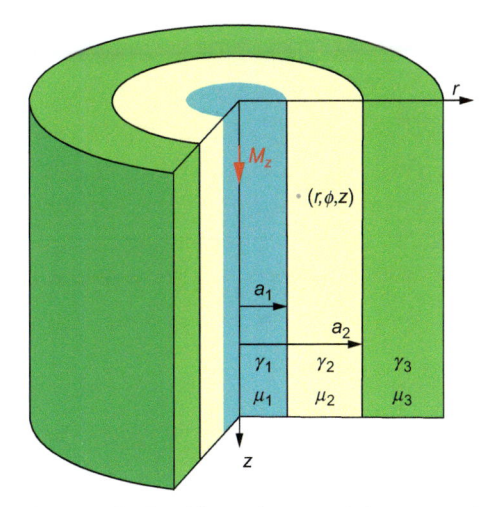

Fig. 8.1 Medium with two cylindrical boundaries and the magnetic dipole on the axis.

section, we look for a solution using only the z-component of the vector potential A_z^*.

According to Maxwell's equations, the vector potential must satisfy several conditions:

1. Function A_z^* is a solution of Helmholtz's equation in every part of the medium:

$$\nabla^2 A_z^* + k^2 A_z^* = 0 \text{ if } R \neq 0$$

This equation can be written in the form:

$$\frac{1}{r}\frac{\partial}{\partial r}\left(r\frac{\partial A_z^*}{\partial r}\right) + \frac{1}{r^2}\frac{\partial^2 A_z^*}{\partial \phi^2} + \frac{\partial^2 A_z^*}{\partial z^2} + k^2 A_z^* = 0 \quad (8.4)$$

2. Near the origin of coordinates system the function A_z^* tends to the vector potential of magnetic dipole in a uniform medium, that is:

$$A_z^* = \frac{i\omega\mu_0 M_0^0}{4\pi R} \exp(ikR)$$

3. At the interface $r = a_m$ tangential components of both the electric field E and function $\mathbf{B}/\mu$ are continuous functions. The electrical field has only E_ϕ component, but the magnetic field is characterized by two components B_r and B_z, and they are expressed through the vector potential as:

$$E_\phi^* = -\frac{\partial A_z^*}{\partial r}, \quad i\omega B_r^* = \frac{\partial^2 A_z^*}{\partial r \partial z}, \quad i\omega B_z^* = k^2 A_z^* + \frac{\partial^2 A_z^*}{\partial z^2}, \quad E_\phi^* = -\frac{\partial A_z^*}{\partial r} \quad (8.5)$$

Here A_z^*, E_ϕ^*, B_r^*, and B_z^* are complex amplitudes of the corresponding vectors. Therefore, boundary conditions for the vector potential at the interface of a medium of a different conductivity and magnetic permeability can be written in the form:

$$\frac{\partial A_{z,m}^*}{\partial r} = \frac{\partial A_{z,m+1}^*}{\partial r}$$

and

$$\frac{1}{\mu_i}\left(k_m^2 A_{z,m}^* + \frac{\partial^2 A_{z,m}^*}{\partial z^2}\right) = \frac{1}{\mu_{m+1}}\left(k_{m+1}^2 A_{z,m+1}^* + \frac{\partial^2 A_{z,m+1}^*}{\partial z^2}\right) \quad (8.6)$$

4. With an increase of the distance from the magnetic dipole the function A_z^* tends to zero. Moreover, the function A_z^* has to obey the following conditions, related to the medium and the source. First, due to the axial symmetry the vector potential and all the field components do not depend on the ϕ coordinate, that is $A_z^* = A_z^*(r, z)$. Also, the vector potential does not depend on the sign of the z-coordinate because of a symmetry of a primary source with respect to the plane $z = 0$:

$$A_z^*(r, z) = A_z^*(r, -z)$$

To find the field we have to solve Helmholtz's equation, which is a differential equation of the second order with partial derivatives with respect to coordinates r and z. To solve this equation we represent the solution as the product of two functions depending on one argument only. Consequently, we have:

$$A_z^* = T(r)\Phi(z)$$

Substituting the latter into Eq. (8.4) and taking into account that A_z^* is independent of ϕ we obtain

$$\frac{\Phi}{r}\frac{\partial}{\partial r}\left(r\frac{\partial T}{\partial r}\right) + T(r)\frac{\partial^2 \Phi}{\partial z^2} + k^2 T(r)\Phi(z) = 0 \quad (8.7)$$

Dividing both sides by $T(r)\Phi(z)$ we have

$$\frac{1}{rT}\frac{\partial}{\partial r}\left(r\frac{\partial T}{\partial r}\right) + \frac{1}{\Phi}\frac{\partial^2 \Phi}{\partial z^2} + k^2 = 0$$

Electromagnetic Field of a Vertical Magnetic Dipole

Let us represent the left hand side of this equation as a sum of two terms:

$$\text{Term} 1 = \frac{1}{rT}\frac{\partial}{\partial r}\left(r\frac{\partial T}{\partial r}\right) + k^2, \quad \text{Term} 2 = \frac{1}{\Phi}\frac{\partial^2 \Phi}{\partial z^2}$$

At first glance each term depends on the argument r or z, and Eq. (8.7) can be written as

$$\text{Term} 1(r) + \text{Term} 2(z) = 0$$

Obviously, the last equation might hold if each term does not depend on the coordinate and represents a constant value. For convenience we designate this constant in the form $\pm m^2$, where m is a constant of separation. Thus, instead of Helmholtz's equation, we obtain two ordinary differential equations of the second order:

$$\frac{1}{Tr}\frac{d}{dr}\left(r\frac{dT}{dr}\right) + k^2 = \pm m^2$$

and (8.8)

$$\frac{1}{\Phi}\frac{d^2\Phi}{dz^2} = \mp m^2$$

Reduction of partial differential equation down to two ordinary differential equations represents the essence of the method of separation of variables.

The symmetry of the field with respect to coordinate z suggests the negative sign in the equation for the function $\Phi(z)$:

$$\frac{d^2\Phi}{dz^2} + m^2\Phi = 0 \qquad (8.9)$$

The solutions to Eq. (8.9) are the trigonometric functions $\sin mz$ and $\cos mz$. In particular, the function $\cos mz$ provides a symmetry of the potential with respect to the plane $z=0$. Correspondingly, the equation for the function $T(r)$ becomes

$$\frac{1}{r}\frac{d}{dr}\left(r\frac{dT}{dr}\right) - \left(m^2 - k^2\right)T = 0$$

Introducing a new variable y

$$y = \left(m^2 - k^2\right)^{1/2} r$$

and performing differentiation we obtain

$$\frac{d^2 T(y)}{dr^2} + \frac{1}{y}\frac{dT(y)}{dy} - T(y) = 0$$

The solutions to the last equation are modified Bessel functions of zero order $I_0(y)$ and $K_0(y)$. Bearing in mind that $A_z^*(r, z) = A_z^*(r, -z)$, we should use only the function $\cos mz$ for the solution of Eq. (8.9) and, correspondingly, for each value of the separation constant we have:

$$\begin{aligned}A_z^*(r, z, m, k) &= T(r, k)\Phi(z)\\ &= [C_m I_0(r, k, m) + D_m K_0(r, k, m)] \cos mz\end{aligned} \quad (8.10)$$

By definition the function $A_z^*(r, z, m)$ satisfies the Helmholtz equation, and we may think that the first step in solving the boundary value problem is accomplished. However, this assumption is incorrect because the function $A_z^*(r, z, m)$ depends on m, which appears as a result of transformation of Helmholtz's equation into two ordinary differential equations. At the same time, the vector potential A_z^*, describing an electromagnetic field in the medium, is independent of m. Inasmuch the function $A_z^*(r, z, m)$, given by Eq. (8.10), obeys the Helmholtz equation for any m, we present the solution in the form of an integral as a superposition of partial solutions $A_z^*(r, z, m)$ corresponding to different values of m $(0 \leq m < \infty)$:

$$A_z^* = \int_0^\infty [C_m I_0(m, k, r) + D_m K_0(m, k, r)] \cos mz\, dm \quad (8.11)$$

which becomes independent of m after integration.

8.2 EXPRESSIONS FOR THE FIELD COMPONENTS

Taking into account the symmetry of the field with respect to the plane $z = 0$, the expression for the vector potential within the borehole can be written as:

$$A_{z1}^* = \frac{i\omega\mu_1 M_0}{4\pi}\left[\frac{\exp(ik_1 R)}{R} + \frac{2}{\pi}\int_0^\infty CI_0(m_1 r)\cos mz\, dm\right] \quad (8.12)$$

Electromagnetic Field of a Vertical Magnetic Dipole

because the function $K_0(m_1 r)$ tends to infinity as $r \to 0$. Here $m_1 = (m^2 - k_1^2)^{1/2}$, and C is some function, which does not depend on coordinates. From the theory of Bessel's function it follows that

$$\frac{\exp(ik_1 R)}{R} = \frac{2}{\pi} \int_0^\infty K_0(m_1 r) \cos mz \, dm$$

Thus

$$A_{z1}^* = \frac{i\omega\mu_1 M_0}{2\pi^2} \int_0^\infty [K_0(m_1 r) + CI_0(m_1 r)] \cos mz \, dm \qquad (8.13)$$

The right-hand side of this equation represents the Fourier's integral. First, consider a solution when the invasion zone is absent. Inasmuch as the function $I_0(mr)$ increases to infinity when $r \to \infty$, the vector potential within the formation is

$$A_{z2}^* = \frac{i\omega\mu_2 M_0}{2\pi^2} \int_0^\infty DK_0(m_2 r) \cos mz \, dm \qquad (8.14)$$

Let us recall one remarkable property of Fourier's integrals. Considering the equality:

$$\int_0^\infty \Psi_1(m) \cos mz \, dm = \int_0^\infty \Psi_2(m) \cos mz \, dm$$

we derive that

$$\Psi_1(m) = \Psi_2(m)$$

Then substitution of Eqs. (8.13), (8.14) into Eq. (8.6) gives

$$m_1[-K_1(m_1 a_1) + CI_1(m_1 a)] = -m_2 K_1(m_2 a_1) D$$
$$\mu_2 m_1^2 [K_0(m_1 a_1) + CI_0(m_1 a_1)] = \mu_1 m_2^2 K_0(m_2 a_1) D$$

since

$$I_0'(x) = \frac{dI_0(x)}{dx} = I_1(x), \quad K_0'(x) = \frac{dK_0(x)}{dx} = -K_1(x)$$

Here $I_1(x), K_1(x)$ are Bessel functions of the first order. Solving the system we obtain

$$C = \frac{\mu_1 m_2 K_0(m_2 a_1) K_1(m_1 a_1) - \mu_2 m_1 K_0(m_1 a_1) K_1(m_2 a_1)}{\mu_1 m_2 K_0(m_2 a_1) I_1(m_1 a_1) + \mu_2 m_1 I_0(m_1 a_1) K_1(m_2 a_1)} \quad (8.15)$$

$$D = \frac{\mu_2 m_1}{m_2 a_1 [\mu_1 m_2 K_0(m_2 a_1) I_1(m_1 a_1) + \mu_2 m_1 I_0(m_1 a_1) K_1(m_2 a_1)]} \quad (8.16)$$

The function A_z^* along with coefficients in Eqs. (8.15), (8.16) satisfies all conditions of the boundary value problem and thus describes the vector potential and components of the electromagnetic field. As follows from Eq. (8.5) the complex amplitudes of the field within the borehole are

$$E_\phi^* = E_{0\phi}^* - \frac{i\omega\mu_1 M_0}{2\pi^2} \int_0^\infty m_1 C I_1(m_1 r) \cos mz \, dm$$

$$B_z^* = B_{0z}^* - \frac{\mu_1 M_0}{2\pi^2} \int_0^\infty m_1^2 C I_0(m_1 r) \cos mz \, dm \quad (8.17)$$

$$B_r^* = B_{0r}^* - \frac{\mu_1 M_0}{2\pi^2} \int_0^\infty m m_1 C I_1(mr) \sin mz \, dm$$

Here $E_{0\phi}^*, B_{0z}^*$, and E_{0r}^* are complex amplitudes of the field in a uniform medium with parameters γ_1, μ_1. In particular, at the borehole axis we have

$$E_\phi^* = 0 \quad \text{and} \quad B_r^* = 0$$

$$B_z^* = B_{0z}^* - \frac{\mu_1 M_0}{2\pi^2} \int_0^\infty m_1^2 C \cos mz \, dm \quad (8.18)$$

The primary magnetic field in a nonconducting medium along the z-axis caused by the magnetic dipole is

$$B_z^{(0)} = \frac{\mu_1 M_0}{2\pi L^3}$$

and, correspondingly, the vertical component of the normalized magnetic field is

$$b_z^* = \frac{B_z^*}{B_z^{(0)}} = b_{0z}^* - \frac{L^3}{\pi} \int_0^\infty m_1^2 C \cos mL \, dm \quad (8.19)$$

Here L is the length of the two-coil probe, while the function b_{0z}^* was described in detail in Chapter 5. It is obvious that in the presence of several cylindrical interfaces we arrive at the expression similar to Eq. (8.19), but with the modified function C. For instance, in the case of two cylindrical interfaces and a nonmagnetic medium the function C is

$$C = \frac{\Delta_1}{\Delta} \tag{8.20}$$

Here

$$\begin{aligned}
\Delta_1 &= [-m_2 I_0(m_2 a_1) K_1(m_1 a_1) - m_1 K_0(m_1 a_1) I_1(m_2 a_1)] \\
&\quad \times [m_3 K_1(m_2 a_2) K_0(m_3 a_2) - m_2 K_0(m_2 a_2) K_1(m_3 a_2)] \\
&\quad + [m_2 K_0(m_2 a_1) K_1(m_1 a_1) - m_1 K_0(m_1 a_1) K_1(m_2 a_1)] \\
&\quad \times [-m_3 I_1(m_2 a_2) K_0(m_3 a_2) - m_2 I_0(m_2 a_2) K_1(m_3 a_2)] \\
\Delta &= [-m_2 I_0(m_2 a_1) I_1(m_1 a_1) + m_1 I_0(m_1 a_1) I_1(m_2 a_1)] \\
&\quad \times [m_3 K_1(m_2 a_2) K_0(m_3 a_2) - m_2 K_0(m_2 a_2) K_1(m_3 a_2)] \\
&\quad + [m_2 K_0(m_2 a_1) I_1(m_1 a_1) + m_1 I_0(m_1 a_1) K_1(m_2 a_1)] \\
&\quad \times [-m_3 I_1(m_2 a_2) K_0(m_3 a_2) - m_2 I_0(m_2 a_2) K_1(m_3 a_2)]
\end{aligned} \tag{8.21}$$

and

$$m_1 = \left(m^2 - k_1^2\right)^{1/2}, \quad m_2 = \left(m^2 - k_2^2\right)^{1/2}, \quad k_3^2 = \left(m^2 - k_3^2\right)^{1/2}$$

Also a_1 and a_2 are the radii of the borehole and invasion zone, respectively. Thus, the complex amplitude of the magnetic field on the borehole axis is expressed in terms of an improper integral, and its integrand represents the product of complex function $m_1^2 C$ and the oscillating multiplier $\cos mL$. Let us study the frequency response of the B_z field and start from the case when the induction number is either too small or too large.

8.3 THE MAGNETIC FIELD IN THE RANGE OF SMALL INDUCTION NUMBER

A small induction number corresponds to the near zone, when transmitter to receiver spacing, L, is much smaller than the wave length λ or the wave number k tends to zero:

$$p \ll 1 \quad \text{or} \quad |kL| \ll 1 \tag{8.22}$$

To analyze the asymptotic behavior, we apply three different approaches.

8.3.1 The First Approach for Deriving the Leading Term of the Quadrature Component (Transition to Doll's Formula)

At the beginning consider the case when an invasion zone is absent and $\mu_1 = \mu_2 = \mu_0$. Then as follows from Eq. (8.15) we have

$$m_1^2 C = m_1^2 \frac{m_2 K_0(m_2 a_1) K_1(m_1 a_1) - m_1 K_1(m_2 a_1) K_0(m_1 a_1)}{m_2 K_0(m_2 a_1) I_1(m_1 a_1) + m_1 K_1(m_2 a_1) I_0(m_1 a_1)} \quad (8.23)$$

If $|k_1| \ll m$ and $|k_2| \ll m$ then keeping the first two terms in Taylor's expansion for the functions m_1 and m_2 we obtain:

$$m_1 = (m^2 - k_1^2)^{1/2} = m\left(1 - \frac{k_1^2}{m^2}\right)^{1/2} \approx m - \frac{1}{2}\frac{k_1^2}{m}$$

$$m_2 = (m^2 - k_2^2)^{1/2} = m\left(1 - \frac{k_2^2}{m^2}\right)^{1/2} \approx m - \frac{1}{2}\frac{k_2^2}{m} \quad (8.24)$$

By analogy we have:

$$I_0(m_1 a_1) \approx I_0(m a_1) - \frac{1}{2}\frac{k_1^2 a_1}{m} I_0'(m a_1)$$

$$I_1(m_1 a_1) \approx I_1(m a_1) - \frac{1}{2}\frac{k_1^2 a_1}{m} I_1'(m a_1)$$

$$K_0(m_1 a_1) \approx K_0(m a_1) - \frac{1}{2}\frac{k_1^2 a_1}{m} K_0'(m a_1) \quad (8.25)$$

$$K_1(m_1 a_1) \approx K_1(m a_1) - \frac{1}{2}\frac{k_1^2 a_1}{m} K_1'(m a_1)$$

Substituting Eqs. (8.24), (8.25) into Eq. (8.23) and making use of recurrence relations of Bessel functions:

$$I_0'(x) = I_1(x), \quad K_0'(x) = -K_1(x),$$

$$I_{\nu-1}(x) - I_{\nu+1}(x) = \frac{2\nu}{x} I_\nu(x), \quad I_{\nu-1}(x) + I_{\nu+1}(x) = 2I_\nu'(x),$$

$$K_{\nu-1}(x) - K_{\nu+1}(x) = -\frac{2\nu}{x} K_\nu(x), \quad K_{\nu-1}(x) + K_{\nu+1}(x) = -2K_\nu'(x)$$

after simple algebra we obtain

$$m_1^2 C = (k_2^2 - k_1^2)\frac{m a_1}{2} \times \left\{2 K_0(m a_1) K_1(m a_1) - m a_1 \left[K_1^2(m a_1) - K_0^2(m a_1)\right]\right\} \quad (8.26)$$

Thus, the quadrature component of the magnetic field expressed in terms of the primary field (8.19) is

$$Qb_z = \frac{\gamma_1 \mu_0 \omega L^2}{2} + \frac{L^3}{2\pi}(s-1)\gamma_1\mu_0\omega \int_0^\infty ma_1 \qquad (8.27)$$
$$\times \left[2K_0 K_1 - ma_1\left(K_1^2 - K_0^2\right)\right] \cos mL\,dm$$

Here $s = \gamma_2/\gamma_1$. Let us introduce notations

$$x = ma_1, \quad \alpha = L/a_1$$

Then Eq. (8.27) can be rewritten as

$$Qb_z = \frac{\omega\mu_0 L^2}{2}\left\{\gamma_1 + (\gamma_2 - \gamma_1)\frac{2\alpha}{\pi}\int_0^\infty \frac{x}{2}\left[2K_0 K_1 - x\left(K_1^2 - K_0^2\right)\right]\cos\alpha x\,dx\right\}$$

or

$$Qb_z = \frac{\omega\mu_0 L^2}{2}(\gamma_1 G_1 + \gamma_2 G_2) \qquad (8.28)$$

where

$$G_2 = \frac{2\alpha}{\pi}\int_0^\infty \frac{x}{2}\left[2K_0(x)K_1(x) - x\left(K_1^2 - K_0^2\right)\right]\cos\alpha x\,dx \qquad (8.29)$$

and

$$G_1 = 1 - G_2$$

In the same manner we can derive Doll's theory or low frequency asymptotic for the medium with several cylindrical interfaces. Making expansion of radicals m_i with respect to a small parameter, it was assumed that k_i^2/m^2 is less than unity. Since integration is performed from 0, there are always some small values of m when ratio k_i^2/m^2 exceeds unity and our assumption is not valid. But it turns out that for very small values of k, contribution of this part of integration can be neglected, provided that only the leading term of the field is calculated. At the same time, if we are interested in the following terms of the low-frequency spectrum, it is advisable to use a different approach.

8.3.2 The Second Approach for Deriving the Leading Term of the In-Phase Component

By applying the first approach we were able to derive the leading term of the series describing the quadrature component of the magnetic field. To obtain the leading term for the in-phase component of the secondary field, we have to recall that magnetic field B_{0z}^* on the axis of the magnetic dipole in a uniform medium can be presented as

$$B_{0z}^* = \frac{\mu_0 M_0}{2\pi L^3}\left[1 + \sum_{n=2}^{\infty} \frac{1-n}{n!}(ikL)^n\right]$$

Neglecting all terms except the first three, we have

$$B_{0z}^* \approx \frac{\mu_0 M_0}{2\pi L^3}\left[1 + \frac{k^2 L^2}{2} + \frac{1}{3}i(kL)^3 + \cdots\right] \tag{8.30}$$

The second term of this series is the leading term for the quadrature component, $k^2 = i\gamma\mu_0\omega$, while the last term

$$\frac{\mu_0 M_0}{2\pi L^3}\frac{i}{3}(kL)^3 = \frac{\mu_0 M_0}{6\pi}ik^3 \tag{8.31}$$

defines the leading term of the in-phase component, as well as the second term of a quadrature component. Now we demonstrate that in a more general case, when there are cylindrical boundaries, the leading term of the in-phase component of the secondary field B_z^* is also defined by Eq. (8.31), provided that k corresponds to an external medium of the formation. To proceed let us represent the integral on the right-hand side of Eq. (8.17) as a sum of two integrals

$$\int_0^\infty m_1^2 C \cos mz\, dm = \int_0^{m_0} m_1^2 C \cos mz\, dm + \int_{m_0}^\infty m_1^2 C \cos mz\, dm \tag{8.32}$$

where m_0 is a very small number. In the case of the second integral when the value of m is greater than the magnitude of wave numbers: $m > k$, the radicals can be expanded in series by powers k^2/m^2. Correspondingly, the integrand C can be presented as:

$$C = \sum_{n=1}^{\infty} a_n \left(\frac{k_1}{m}\right)^{2n} \quad \text{if } m > m_0 \tag{8.33}$$

where a_n are coefficients that depend on the parameters of the medium. Because the external integral

$$\int_{m_0}^{\infty} m_1^2 C \cos mz \, dm$$

does not contain point $m=0$, we can replace it with the series

$$\int_{m_0}^{\infty} m_1^2 C \cos mz = \sum_{n=1}^{\infty} b_n k_1^{2n} \tag{8.34}$$

Therefore, the series describing the external integral has only terms of even powers of wave number, k, and the integer powers of ω. This suggests that the terms of the series with odd powers of k, in particular k^3, can be derived from an expansion of the internal integral only, provided that

$$k \to 0, \quad m \to 0 \tag{8.35}$$

Taking into account the behavior of modified Bessel functions for a small argument:

$$I_0(x) \approx 1, \quad I_1(x) \approx \frac{x}{2}, \quad K_0(x) \approx -\ln x, \quad K_1(x) \approx \frac{1}{x}$$

the function C can be presented as

$$C \approx \frac{m_2 K_0(m_2 a_1) K_1(m_1 a_1) - m_1 K_0(m_1 a_1) K_1(m_2 a_1)}{m_1 K_1(m_2 a_1)}$$

or

$$C \approx \frac{m_2}{m_1} \frac{K_1(m_1 a_1)}{K_1(m_2 a_1)} K_0(m_2 a_1) - K_0(m_1 a_1) \tag{8.36}$$

Replacing the ratio $K_1(m_1 a_1)/K_1(m_2 a_1)$ with its asymptotic value we finally have:

$$C \approx \frac{m_2^2}{m_1^2} K_0(m_2 a_1) - K_0(m_1 a_1) \quad \text{and} \quad m_1^2 C \approx m_2^2 K_0(m_2 a_1) - m_1^2 K_0(m_1 a_1)$$

Thus, the internal integral can be presented as

$$\int_0^{m_0} m_1^2 C \cos mz \, dm \approx \int_0^{m_0} m_2^2 K_0(m_2 a_1) \cos mz \, dm - \int_0^{m_0} m_1^2 K_0(m_1 a_1) \cos mz \, dm \tag{8.37}$$

Taking into account that

$$\frac{\exp(ik_i R)}{R} = \frac{2}{\pi}\int_0^\infty K_0(m_i r)\cos mz\,dm$$

and keeping in mind that we are interested in odd powers of k, the following equality can be written:

$$\frac{\mu_0 M_0}{2\pi^2}\int_0^\infty m_1^2 C\cos mz\,dm = -B_{0z}^*(k_2 R) + B_{0z}^*(k_1 R) \qquad (8.38)$$

where $B_{0z}^*(k_2 R)$ and $B_{0z}^*(k_1 R)$ are magnetic fields on the surface of the borehole in a uniform medium with resistivity of a formation and borehole, respectively, $R = (a^2 + z^2)^{1/2}$. Substituting Eq. (8.38) into Eq. (8.17) we have:

$$B_z^* \approx B_{0z}^*(k_1 z) + B_{0z}^*(k_2 R) - B_{0z}^*(k_1 R) \qquad (8.39)$$

Again, the latter is valid in the range of small induction numbers only when the terms of a series proportional to odd powers of the wave number are considered. As follows from Eq. (8.31), the second term of a series, describing the magnetic field on the borehole axis, is:

$$i\frac{\mu_0 M_0}{6\pi}k_2^3$$

Thus, we see that the leading term of the in-phase component of the secondary field B_z coincides with that in a uniform medium with conductivity of a formation:

$$InB_z \rightarrow InB_{0z}(k_2 L) \qquad (8.40)$$

This result does not depend on the ratio of conductivities as well as the probe length. In other words, at the range of small parameter the borehole becomes "transparent" and does not contribute into the in-phase component. Next, we demonstrate that the same result is valid for a three-layered medium. Let us proceed from Eqs. (8.20), (8.21), again assuming that $m \to 0$, $k \to 0$. Introducing notations

Electromagnetic Field of a Vertical Magnetic Dipole

$$b_1 = -m_2 I_0(m_2 a_1) K_1(m_1 a_1) - m_1 K_0(m_1 a_1) I_1(m_2 a_1)$$

$$c_1 = m_3 K_1(m_2 a_2) K_0(m_3 a_2) - m_2 K_0(m_2 a_2) K_1(m_3 a_2)$$

$$b_2 = m_2 K_0(m_2 a_1) K_1(m_1 a_1) - m_1 K_0(m_1 a_1) K_1(m_2 a_1)$$

$$c_2 = -m_3 I_1(m_2 a_2) K_0(m_3 a_2) - m_2 I_0(m_2 a_2) K_1(m_3 a_2)$$

$$b_3 = -m_2 I_0(m_2 a_1) I_1(m_1 a_1) + m_1 I_0(m_1 a_1) I_1(m_2 a_1)$$

$$c_3 = m_2 I_1(m_1 a_1) K_0(m_2 a_1) + m_1 I_0(m_1 a_1) K_1(m_2 a_1)$$

and taking into account the behavior of modified Bessel functions for the small argument we have

$$b_1 \approx -m_2 K_1(m_1 a_1) = \frac{m_2}{m_1 a_1}, \quad c_1 \approx \frac{m_3}{m_2 a_2} K_0(m_3 a_2) - \frac{m_2}{m_3 a_2} K_0(m_2 a_2)$$

$$b_2 \approx \frac{m_2}{m_1 a_1} K_0(m_2 a_1) - \frac{m_1}{m_2 a_1} K_0(m_1 a_1), \quad c_2 \approx -\frac{m_2}{m_3 a_2}$$

$$b_3 \approx -\frac{m_1 m_2 a_1}{2} + \frac{m_1 m_2}{2} \to 0, \quad c_3 \to \frac{m_1}{m_2 a_1}$$

Whence, for small values of m and k we obtain

$$C \approx \frac{b_1 c_1 + b_2 c_2}{c_2 c_3} = \frac{b_1 c_1}{c_2 c_3} + \frac{b_2}{c_3}$$

Inasmuch as

$$c_2 c_3 = -\frac{m_1}{m_3 a_1 a_2}, \quad b_1 c_1 = -\frac{m_3}{m_1 a_1 a_2} K_0(m_3 a_2) + \frac{m_2^2}{m_1 m_3 a_1 a_2} K_0(m_2 a_2),$$

$$\frac{b_1 c_1}{c_2 c_3} = \frac{m_3^2}{m_1^2} K_0(m_3 a_2) - \frac{m_2^2}{m_1^2} K_0(m_2 a_2), \quad \frac{b_2}{c_3} = \frac{m_2^2}{m_1^2} K_0(m_2 a_1) - K_0(m_1 a_1),$$

we have the following expression for the function $m_1^2 C$:

$$m_1^2 C \approx m_3^2 K_0(m_3 a_2) - m_2^2 K_0(m_2 a_2) + m_2^2 K_0(m_2 a_1) - m_1^2 K_0(m_1 a_1)$$

Thus, the internal integral has the form

$$\int_0^{m_0} m_1^2 C \cos mz \, dm \approx \int_0^{m_0} m_3^2 K_0(m_3 a_2) \cos mz \, dm - \int_0^{m_0} m_2^2 K_0(m_2 a_2) \cos mz \, dm +$$

$$\int_0^{m_0} m_2^2 K_0(m_2 a_1) \cos mz \, dm - \int_0^{m_0} m_1^2 K_0(m_1 a_1) \cos mz \, dm$$

(8.41)

which leads to the following expression for the in-phase component of the field:

$$B_z^* = B_{0z}^*(k_1 z) + B_{0z}^*(k_3 R_2) - B_{0z}^*(k_2 R_2) + B_{0z}^*(k_2 R_1) - B_{0z}^*(k_1 R_1) \quad (8.42)$$

Here

$$R_1 = (z^2 + a_1^2)^{1/2} \text{ and } R_2 = (z^2 + a_2^2)^{1/2}$$

Similarly to the case of a two-layered medium, we derive the k^3 term

$$i\frac{\mu_0 M}{6\pi} k_3^3$$

which corresponds to the in-phase component of the secondary field in a uniform formation with conductivity γ_3. Bearing in mind the expression for the quadrature component derived earlier, we have the following expression for the secondary magnetic field ($k \to 0$):

$$B_z^* \approx \frac{\mu_0 M_0}{4\pi} \left[\frac{1}{L} \sum_{n=1}^{3} k_i^2 G_i + \frac{2}{3} i k_3^3 \right] \quad (8.43)$$

Here $k_i^2 = i\gamma_i \mu_0 \omega$, L is the probe length. For the quadrature and in-phase component we have:

$$QB_z \approx \frac{\mu_0 M_0}{4\pi} \left[\frac{\omega\mu_0}{L} \sum_{n=1}^{3} \gamma_i G_i - \frac{2^{1/2}}{3} (\gamma_3 \mu_0 \omega)^{3/2} \right]$$

and
$$(8.44)$$

$$InB_z^s \approx -\frac{\mu_0 M_0}{4\pi} \frac{2^{1/2}}{3} (\gamma_3 \mu_0 \omega)^{3/2}$$

where γ_i, G_i are conductivity and geometric factor of the corresponding part of a medium such as borehole, invasion, and formation. The last result can be generalized and applied to the case of invasion zone with resistivity varying in a radial direction. Also, we have to notice that Eq. (8.44) can be derived using the hybrid method, which gives

$$B_z^* \approx \frac{\mu_0 M_0}{2\pi L^3} b_z^*(\gamma_3) + \frac{i\omega\mu_0 M_0}{4\pi L} \sum_{n=1}^{2} (\gamma_i - \gamma_3) G_i \quad (8.45)$$

By expanding the right-hand side of the latter into a series and keeping the leading terms only, we arrive at Eq. (8.44).

1. The first and second approaches give the leading terms for the quadrature and in-phase components of the series, describing the field on the borehole axis.
2. Application of the Doll theory almost always requires correction for the skin effect. At the same time, as follows from Eq. (8.44), at the low frequency limit the second term for the quadrature component has the same magnitude as the in-phase component. Therefore, measurements of InB_z^s enable us to correct the quadrature component for the skin effect.
3. We demonstrated that in a medium with cylindrical interfaces the leading term of the series for the in-phase component is defined by the conductivity of the external part of the formation. In other words, the borehole and invasion zone become transparent and do not affect the measurements. As we see later, such behavior is also observed at the low frequency limit in a medium with horizontal boundaries, as well as in more complicated cases.
4. This discussion underlines again that the quadrature and in-phase components depend quite differently on parameters of the formation, and thus they have a different depth of investigation.
5. The series, Eq. (8.43), is valid regardless of the probe length.
6. As follows from Eq. (8.44), the second term of the quadrature component and the leading term of the in-phase component do not depend on either probe length or the parameters of the borehole and invasion zone. Therefore, by measuring these quantities, one can essentially measure properties of the deepest part of the formation.

8.3.3 The Third Approach to Deriving Asymptotic Expressions of the Field

We have derived only two terms of the series describing the quadrature component of the field and the leading term of the in-phase component of the secondary field. To obtain subsequent terms of both series, it is necessary to perform more cumbersome transformations on expansion of the internal and external integrals in Eq. (8.32). In the internal integral, Bessel functions can be expanded in the series because their argument is small. This reduces the integral to a sum of simple integrals of elementary functions. The integration of the external integral is based on Eq. (8.34) and calculation of coefficients b_n. Finally, we have the following series describing the field at the low frequencies:

$$B_z^* = B_z^{(0)} \left(\sum_{n=1}^{\infty} a_{1n} k^{2n} + \sum_{n=1}^{\infty} a_{2n} k^{2n+1} + \ln k \sum_{n=1}^{\infty} a_{3n} k^{2n} \right), \qquad (8.46)$$

where $B_z^{(0)}$ is the field, caused by the primary source in a free space. Later it will be shown that only the second and third sum in Eq. (8.46), corresponding to the internal integral, contribute to the late stage of the transient field. Holding only first terms of the last two sums in Eq. (8.46), we arrive at the following asymptotic expressions for $b_z^* - 1$:

(1) two-layered medium

$$b_z^* - 1 \approx f_3 k_1^3 + f_5 k_1^5 + f_7 k_1^7 + l_7 k_1^7 \ln k_1, \qquad (8.47)$$

where

$$f_3 = \frac{\alpha^3 s^{3/2}}{3}, \quad f_5 = f_3 \left(\frac{\alpha^2 s}{10} - \frac{1-s}{2} \right)$$

$$f_7 = f_3 \left[\frac{\alpha^4 s^2}{280} - \frac{\alpha^2 s(1-s)}{20} + \frac{5}{32}(1-s)^2 - \frac{s(1-s)}{10} \left(C - \frac{77}{60} + \frac{\ln s}{2} \right) \right]$$

$$l_7 = -f_3 \frac{s}{10}(1-s) \qquad (8.48)$$

C is Euler's constant,

$$s = \gamma_2 / \gamma_1 \text{ and } \alpha = L/a_1$$

(2) three-layered medium

$$b_z^* - 1 \approx d_3 k_1^3 + d_5 k_1^5$$

Here

$$d_3 = \frac{1}{3} \alpha^3 s_1^{3/2}, \quad d_5 = d_3 \left(\frac{\alpha^2 s_1}{10} - \frac{s_{12}}{2} \right), \quad s_1 = \gamma_3/\gamma_1, \quad s_2 = \gamma_2/\gamma_1$$

$$s_{12} = 1 - s_2 + (s_2 - s_1)\beta^2, \quad \beta = a_2/a_1$$

8.4 FAR ZONE OF MAGNETIC FIELD ON THE AXIS OF BOREHOLE

Now we focus our attention on the case of a large parameter L/a_1 when the probe length exceeds several times the borehole radius. The purpose is to find from an asymptotic representation some specific features of the field that can be further utilized for increasing depth of investigation. Derivation of asymptotic expression of the field is based on a proper treatment of singularities of the integrand $m_1^2 C$ on the complex plane of m. In accordance with Eq. (8.19) the variable of integration m has only real values $0 \leq m < \infty$, while the probe length L is the multiplayer in the argument of the oscillating term $\cos Lm$. For small values of $|kL|$ the function $m_1^2 C$ rapidly decreases with an increase of m. In addition, the presence of the oscillating factor $\cos Lm$ also reduces contribution from the integrand at large values of m. For this reason the integral

$$\int_0^\infty m_1^2 C \cos Lm\, dm \qquad (8.49)$$

is mainly defined by the integrand $m_1^2 C$ near small values of m, allowing a derivation of the geometric factors of the borehole, invasion zone, and formation. With an increase of the wave number $|k|$ the integrand $m_1^2 C$ decreases slowly and for $m < |k|$ it does not practically change. Correspondingly, despite an increased number of oscillations the integral is not defined anymore by the integrand at the initial part of integration and additional transformations of the integral (8.49) are needed to treat the case of $|kL| > 1$.

8.4.1 Cauchy's Formula and Deformation of Integration Contour

To obtain asymptotical expression for the field at $|kL| > 1$ we use an approach based on the Cauchy formula. Since

$$\cos mL = \frac{1}{2}[\exp(imL) + \exp(-imL)]$$

we have

$$\int_0^\infty m_1^2 C \cos mL\, dm = \frac{1}{2}\left[\int_0^\infty m_1^2 C \exp(imL)\, dm + \int_0^\infty m_1^2 C \exp(-imL)\, dm\right]$$

The latter describes the secondary field, Eq. (8.18). Taking into account that $m_1^2 C$ is even a function of m, the last equality can be represented as

$$\int_0^\infty m_1^2 C \cos mL\, dm = \frac{1}{2}\int_{-\infty}^\infty m_1^2 C \exp(imL)\, dm \qquad (8.50)$$

In accordance with Cauchy's theorem an integral from an analytical function $f(z)$ around a closed path l is equal to zero:

$$\oint_l f(z)\,dz = 0 \qquad (8.51)$$

which corresponds to a single-valued function $f(z)$ with no singularities inside l. In other words, the deformation of the contour does not change the integral if the integration path doesn't intersect singularities on the complex plane of the variable m. Note that with deformation of the contour of integration in the upper half plane ($\operatorname{Im} m > 0$), the exponent $\exp(imL)$ tends to zero with increase of $\operatorname{Im}(m)$. In general, the integrand in Eq. (8.50) has two types of singularities, namely, branch points and poles. From Eq. (8.18) we have:

$$b_z^* = b_{0z}^* - \frac{L^3}{\pi} \int_0^\infty m_1^2 C \cos mL\,dm \qquad (8.52)$$

and the function $m_1^2 C$ has two branch points at the upper half plane of m:

$$m = k_1 \quad \text{and} \quad m = k_2$$

where this function is not an analytical one. Now consider a closed path D, shown in Fig. 8.2, consisting of several paths, namely: (1) the original path from $-\infty$ to ∞, (2) the path D_1, which includes two lines in the vicinity of branch cut $m_1 = 0$, (3) the path D_2 which also has two lines near branch cut $m_2 = 0$, and finally (4) the semicircle of an infinitely large radius.

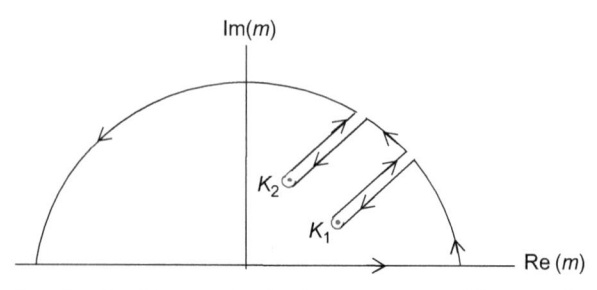

Fig. 8.2 The closed path of integration in the upper part of the complex plane of m.

Since the integrand in Eq. (8.50) has the term $\exp(imL)$, the integral along this last part is equal to zero. Inside the closed path D the integrand is an analytical function and we can write

$$\oint_D m_1^2 C \exp(imL) dm = 0 \quad \text{or}$$

$$\int_{-\infty}^{\infty} m_1^2 C \exp(imL) dm = \int_{D_1} m_1^2 C \exp(imL) dm + \int_{D_2} m_2^2 C \exp(im_2 L) dm \tag{8.53}$$

Integrating along the path, where $\operatorname{Re} m_1 = 0$, we introduce a new variable of integration $m_1 = it$. Here t is the parameter of the branch line, which varies from 0 to ∞ on the right side of the branch line and from $-\infty$ to 0 on its left, since the radical changes sign bypassing around the branch point. The variable of integration m along the contour D_1 can be presented as

$$m = \left(-t^2 + k_1^2\right)^{1/2} = i\left(t^2 - in_1^2\right)^{1/2}$$

and correspondingly

$$dm = \frac{itdt}{\left(t^2 - in_1^2\right)^{1/2}} \quad \text{and} \quad m_2 = \left(-t^2 + in_1^2 - in_2^2\right)^{1/2}$$

where

$$n_1^2 = i\gamma_1\mu_0\omega \quad \text{and} \quad n_2^2 = i\gamma_2\mu_0\omega$$

Thus, for the integral along both sides around the branch cut $m_1 = 0$ we have the following expression:

$$\int_0^\infty (-t^2) \left[\frac{m_2 K_0(m_2 a_1) K_1(ita_1) - it K_0(ita_1) K_1(m_2 a_1)}{m_2 K_0(m_2 a_1) I_1(ita_1) + it K_1(m_2 a_1) I_0(ita_1)} - \right.$$

$$\left. \frac{m_2 K_0(m_2 a_1) K_1(-ita_1) + it K_0(-ita_1) K_1(m_2 a_1)}{m_2 K_0(m_2 a_1) I_1(-ita_1) - it K_1(m_2 a_1) I_0(-ita_1)} \right] \frac{it \exp\left[L(t^2 - in_1^2)^{1/2}\right]}{(t^2 - in_1^2)^{1/2}} dt \tag{8.54}$$

Making use of relations

$$I_0(-ita_1) = I_0(ita_1), \quad K_0(-ita_1) = K_0(ita_1) + i\pi I_0(ita_1)$$
$$I_1(-ita_1) = -I_1(ita_1), \quad K_1(-ita_1) = -K_1(ita_1) + i\pi I_1(ita_1) \quad (8.55)$$

we can present the second term in parentheses of Eq. (8.54) in the following form:

$$\frac{m_2 K_0(m_2 a_1)[-K_1(ita_1) + i\pi I_1(ita_1)] + it K_1(m_2 a_1)[K_0(ita_1) + i\pi I_0(ita_1)]}{-m_2 K_0(m_2 a_1) I_1(ita_1) - it K_1(m_2 a_1) I_0(ita_1)}$$

$$= \frac{m_2 K_0(m_2) K_1(it) - it K_1(m_2) K_0(it)}{m_2 K_0(m_2) I_1(it) + it K_1(m_2) I_0(it)} - i\pi$$

(8.56)

Inasmuch as the first terms in Eqs. (8.56), (8.54) are the same, the integral along the path D_1 is greatly simplified and we have:

$$\pi \int_{-\infty}^{\infty} t^3 \frac{\exp\left[-L(t^2 - in_1^2)\right]}{(t^2 - in_1^2)^{1/2}} dt$$

This integral, which is being multiplied by $1/\pi$, represents the field of magnetic dipole b_{0z}^* in a uniform medium with conductivity γ_1. Thus, as follows from Eqs. (8.52), (8.53) the field on the borehole axis is expressed in terms of the integral along the branch cut D_2, $\mathrm{Re}\, m_2 = 0$ only. Replacing the variable $m_2 = it$ we have:

$$m = i(t^2 - in_2^2)^{1/2}, \quad dm = \frac{it\, dt}{(t^2 - in_2^2)^{1/2}}, \quad m_1 = \left[-t^2 + i(n_2^2 - n_1^2)\right]^{1/2}$$

Respectively, the integral along the path D_2 can be rewritten as

$$\int_0^\infty m_1^2 \left[\frac{it K_0(ita_1) K_1(m_1 a_1) - m_1 K_0(m_1 a_1) K_1(ita_1)}{it K_0(ita_1) I_1(m_1 a_1) + m_1 I_0(m_1 a_1) K_1(ita_1)} \right.$$

$$\left. - \frac{-it K_0(-ita_1) K_1(m_1 a_1) - m_1 K_0(m_1 a_1) K_1(-ita_1)}{-it K_0(-ita_1) I_1(m_1 a_1) + m_1 I_0(m_1 a_1) K_1(ita_1)} \right] \frac{it \exp\left[-L(t^2 - in_2^2)^{1/2}\right]}{(t^2 - in_2^2)^{1/2}} dt$$

(8.57)

Electromagnetic Field of a Vertical Magnetic Dipole

Using Eq. (8.55) we obtain for the numerator inside large square brackets of Eq. (8.57), the following expression:

$$m_1 it [I_0(m_1 a_1) K_1(m_1 a_1) + I_1(m_1 a_1) K_0(m_1 a_1)]$$
$$\times [K_0(-ita_1) K_1(ita_1) + K_0(ita_1) K_1(-ita_1)]$$

Inasmuch as

$$I_0(x) K_1(x) + I_1(x) K_0(x) = \frac{1}{x}$$

the numerator is further reduced to $i\pi/a_1^2$, and the field b_z^* on the borehole axis is expressed through the integral along the right-hand side of the branch cut D_2

$$b_z^* = \frac{L^3}{2} \int_0^\infty \frac{m_1^2 t \exp\left[-L(t^2 - n_2^2)^{1/2}\right]}{(t^2 - in_2^2)^{1/2}[it K_0(ita_1) I_1(m_1 a_1) + m_1 K_1(ita_1) I_0(m_1 a_1)]} \quad (8.58)$$

$$\times \frac{dt}{[-it K_0(-ita_1) I_1(m_1 a_1) + m_1 K_1(-ita_1) I_0(m_1 a_1)]}$$

The integrand can be presented as a product of two functions: $F(m_1, t)$ and

$$\frac{t^3 \exp\left[-L(t^2 - n_2^2)\right]}{(t^2 - n_2^2)^{1/2}}$$

The last function is the integrand of Somerfield integral describing field in a uniform medium with conductivity γ_2. For the sufficiently long probes this integral is mainly defined by an initial part of the integration path when m is small and function $F(m_1, t)$ varies gradually. By taking this slow-varying function out of the integral and assuming $m = 0$, we receive

$$b_z^* \approx \frac{1}{I_0^2 \left[(k_2^2 - k_1^2)^{1/2} a_1\right]} b_z^*(k_2 L)$$

or

$$b_z^* \approx \frac{1}{I_0^2 \left[(k_2^2 - k_1^2)^{1/2} a_1\right]} \exp(ik_2 L)(1 - ik_2 L) \quad (8.59)$$

For the conductivity of the borehole being much greater than conductivity of a surrounding medium we have:

$$b_z^* \approx \frac{1}{I_0^2(ik_1a_1)} \exp(ik_2L)(1 - ik_2L) \qquad (8.60)$$

The expression (8.59) was first derived by V. Sokolov [1]. He also obtained the asymptotic formula for the medium with an invasion zone. In this case there are three contours around the branch cuts: along $m_1 = 0$, $m_2 = 0$, and $m_3 = 0$. Integration along $m_1 = 0$ gives the field $b_{0z}^*(\gamma_1)$, the integral along branch cut $m_2 = 0$ is equal to zero, and an expression for the field contains the integral along branch cut $m_3 = 0$ only. In more general cases of n cylindrical or plane interfaces the integration is also reduced to that along the branch cut $m_{n+1} = 0$ only. For the two cylindrical interfaces we have

$$b_z^* = \frac{1}{I_0^2\left[(k_1^2 - k_2^2)^{1/2}a_1\right]I_0^2\left[(k_2^2 - k_3^2)^{1/2}a_2\right]} b_z^*(k_3L) \qquad (8.61)$$

and it is reduced to Eq. (8.59) either at $k_2 = k_3$ or $k_1 = k_2$. Here $b_z^*(k_3L)$ is the complex amplitude of the field in a uniform medium with a resistivity of a formation.

As was pointed out in Chapter 1, the behavior of the quasistationary field often reflects some features of a propagation of the field. Suppose that there is one cylindrical interface. Then one may imagine that the electromagnetic field travels from the dipole to an observation point by two passes. One is the wave moving through the borehole, while the second wave moves from the dipole to the boundary, then along the borehole surface inside the formation, and the last interval is located between the borehole surface and an observation point, Fig. 8.3A and B.

The last path suggests that the field can be described by the equation

$$b_z^* = f^2(k_1, k_2) b_z^{*un}(k_2L)$$

Comparison with Eq. (8.59) shows that

$$f = I_0^{-1}\left[(k_1^2 - k_2^2)^{1/2}a_1\right]$$

Assuming that the resistivity of the formation is larger than the resistivity of the borehole, it is natural to expect that with increase of the probe length the second wave plays the dominant role, while the influence of the wave, propagating through the borehole, is negligible. Similar interpretation can be given to Eq. (8.61). Note that wave paths for the three-coil probe have

Electromagnetic Field of a Vertical Magnetic Dipole

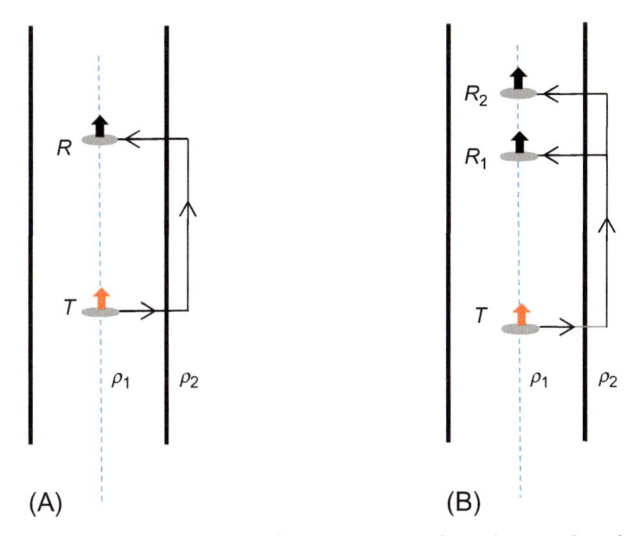

Fig. 8.3 (A) Wave path in two-coil probe. (B) Wave path in three-coil probe.

common elements, located in the borehole, and this leads to a very important practical application. Let us present the complex amplitude of the field, given by Eq. (8.61), as

$$b_z^* = A_* e^{i\phi_*} \cdot A(k_3 L) \exp\left[i\phi_0(k_3 L)\right] \tag{8.62}$$

where functions A^* and ϕ^* depend on the conductivity and radii of the borehole and invasion zone and have no dependency on the probe length L. The rest of Eq. (8.62) coincides with the complex amplitude of the field in a uniform medium with resistivity of the formation. Suppose that the field is measured at two distances L_1 and L_2 from the dipole source, corresponding to the far zone. By definition the electromotive force in the receiver is equal to

$$\Xi(L) = \Xi_0(L) b_z^*(L)$$

and their ratio is

$$\frac{\Xi(L_2)}{\Xi(L_1)} = \frac{\Xi_0(L_2)}{\Xi_0(L_1)} \frac{b_z^*(L_2)}{b_z^*(L_1)}$$

Then Eq. (8.62) gives

$$\frac{\Xi(L_2)}{\Xi(L_1)} = \frac{\Xi_0(L_2)}{\Xi_0(L_1)} \frac{A(k_3 L_2)}{A(k_3 L_1)} \exp(i\Delta\phi)$$

In particular, if moments of receiver coils are chosen in such a way that primary electromotive forces Ξ_0 are the same, Eq. (8.62) gives

$$\left|\frac{\Xi(L_2)}{\Xi(L_1)}\right| = \frac{|b_z^*(L_2)|}{|b_z^*(L_1)|} = \frac{A(k_3 L_2)}{A(k_3 L_1)}$$

and

$$\Delta\phi = \phi_0(k_3 L_2) - \phi_0(k_3 L_1) \qquad (8.63)$$

As is seen from Eq. (8.63), these two quantities at the far zone, are insensitive to parameters of the borehole and invasion zone, and this remarkable fact is the main reason why these measurements are used in some modifications of the induction logging (for example, VIKIZ system). Since amplitudes are measured in the presence of the primary field, the operating frequencies in such logging systems should be high enough to increase the secondary field and provide sufficient sensitivity to properties of the formation.

As soon as an observation point is located at the far zone the further increase of the probe length doesn't influence the depth of investigation. From Eq. (8.63) we see that at this zone the ratio of amplitudes and difference of phases are independent of parameters of the borehole and invasion zone. To utilize these measurements the three-coil probe, described earlier as the simplest "focusing" probe, is used. (Historically the measurements of amplitudes ratio and phase differences were first introduced in dielectric and later were also applied in induction logging).

8.4.2 Validity of the Approximate Solution

Now we evaluate a range of medium parameters, frequency, and probe length where approximation (8.59) is valid. First, consider the low frequency part of spectrum when

$$\left|k_2^2 - k_1^2\right| a_1^2 \ll 1$$

Then, bearing in mind that

$$I_0(x) \approx 1 + \frac{x^2}{4}, \quad \text{if } x < 1$$

we have:

$$\frac{1}{I_0^2\left[(k_2^2 - k_1^2)^{1/2} a_1\right]} \approx 1 + \frac{1}{2}(k_1^2 - k_2^2) a_1^2$$

Thus

$$b_z^* \approx \frac{1}{2}i\omega\mu_0(\gamma_1 - \gamma_2)a_1^2 b_z^{un}(k_2 L) + b_z^{un}(k_2 L)$$

We are interested in the low-frequency spectrum, thus the first term can be simplified to

$$b_z^* \approx i\frac{\omega\mu_0}{2}(\gamma_1 - \gamma_2)a_1^2 + b_z^{un}(k_2 L)$$

or

$$b_z^* = i\frac{\omega\mu_0 L^2}{2}(\gamma_1 - \gamma_2)\frac{1}{\alpha^2} + b_z^{un}(k_2 L)$$

The latter coincides with an equation derived by hybrid method for the case when the probe length exceeds several times the size of the borehole (Chapter 7). In particular, considering the quadrature component only the second term $b_z^{un}(k_2 L)$ can be replaced with

$$\frac{i\omega\mu_0 \gamma_2 L^2}{2}$$

yielding to the expression for

$$b_z^* = \frac{i\omega\mu_o L^2}{2}\left[\gamma_1 \frac{1}{\alpha^2} + \gamma_2\left(1 - \frac{1}{\alpha^2}\right)\right]$$

corresponding to the Doll's approximation. Therefore, Eq. (8.59) certainly gives the correct result at the range of small parameters when the probe length is sufficiently large. Also, this equation describes the field when the argument of Bessel function is very small and the distance L is large

$$\left|(k_2^2 - k_1^2)^{1/2} a_1\right| \ll 1 \text{ and } \alpha \gg 1$$

Considering propagation of waves along the paths, shown in Fig. 8.3A, one may assume that Eq. (8.59) is also valid when

$$|k_1 L| > 1 \text{ and } \gamma_1 > \gamma_2$$

In fact, in such case the wave propagating inside of the borehole decays more rapidly. To confirm these assumptions we compare exact solution versus an approximate one using Eq. (8.59). The comparison is conducted for the three-coil probe using two functions $T_1(\alpha)$ and $T_2(\alpha)$:

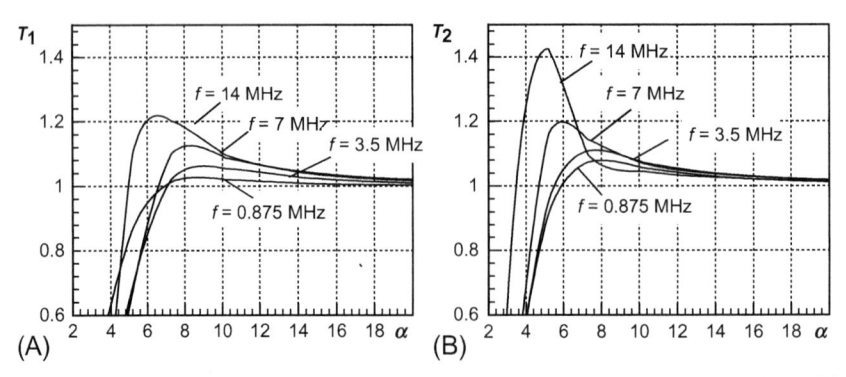

Fig. 8.4 Comparison of approximate and exact solution. Attenuation (A) and phase difference (B). $\rho_1 = 0.1$ ohmm, $\gamma_2/\gamma_1 = 1/100$.

$$T_1(\alpha) = \left|\frac{A(L_2)}{A(L_1)}\right| \bigg/ \left|\frac{A^d(L_2)}{A^d(L_1)}\right|, \quad \text{and} \quad T_2(\alpha) = \frac{\Delta\phi}{\Delta\phi^d}$$

shown in Figs. 8.4 and 8.5. $A^d(L)$ and $A(L)$ are field amplitudes, corresponding to the approximate (8.59) and exact solution, respectively. In practice, instead of ratio of amplitudes we use the attenuation:

$$20\log\left|\frac{A_2}{A_1}\right|$$

assuming that receiving moments $\dfrac{M_1}{M_2} = \left(\dfrac{L_2}{L_1}\right)^3$ selected to provide zero attenuation of the field in the air. Index of curves is frequencies, used in the VIKIZ. In the calculations: $a_1 = 0.1$ m, $L_2/L_1 = 0.7$.

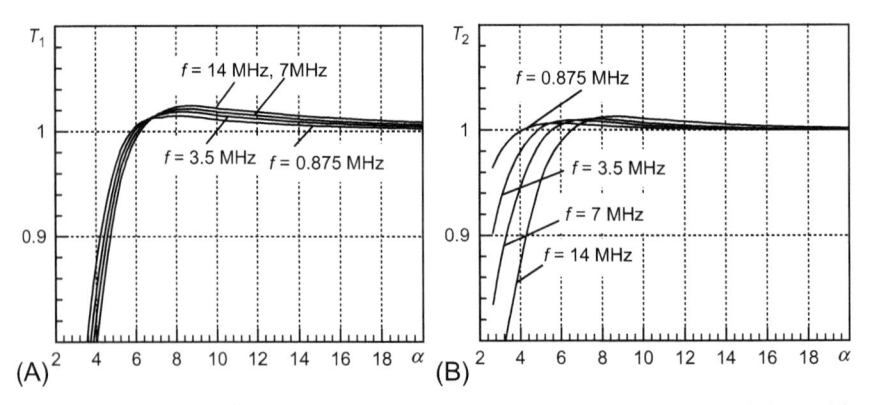

Fig. 8.5 Comparison of approximate and exact solution. Attenuation (A) and phase difference (B). $\rho_1 = 0.5$ ohmm, $\gamma_2/\gamma_1 = 1/20$.

As we see even in a quite conducive borehole ($\gamma_2/\gamma_1 = 1/100$) the approximate solution (8.59) provides an accurate estimate of both attenuation (Fig. 8.4A) and phase difference (Fig. 8.4) with an error less than 10% if $\alpha > 10$: the lower the frequency, lesser the error. In the case of less conductive borehole ($\gamma_2/\gamma_1 = 1/20$) the approximate solution describes responses with an error not exceeding 5% (Fig. 8.5), when

$$\alpha > 5 \tag{8.64}$$

Similar conditions can be derived when there is an invasion zone: the validity of approximation is shifted toward greater distances from the dipole and it deteriorates with increase of conductivity and radius of the invasion.

8.4.3 Sensitivity to the Formation of Amplitudes Ratio and Phase Difference (Three-Coil Probe)

As was pointed out earlier Eqs. (8.59), (8.61) show that attenuation and phase difference at the far zone allows one essentially to reduce an influence of the borehole and invasion. Let us consider sensitivity of these two quantities to a change of a formation conductivity and, as example, choose frequencies used in the VIKIZ system, Fig. 8.6, assuming that $L_1 = 1$ m, $L_2 = 0.7$ m, and $\rho_1 = 0.5$ ohmm. Along the vertical axis we plot either attenuation Fig. 8.6A or phase difference Fig. 8.6B, while a ratio of conductivities is plotted along the horizontal axis. Index of curves is frequency. At relatively small resistivity contrast $\gamma_2/\gamma_1 \geq 0.1$ attenuation and phase difference have practically the same sensitivity (ramp of the curve) to the conductivity of formation, but attenuation is more sensitive when the contrast increases, $\gamma_2/\gamma_1 \leq 0.1$.

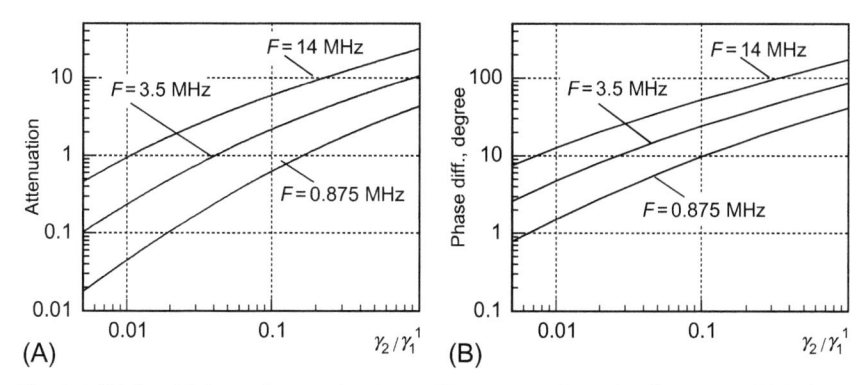

Fig. 8.6 (A) Sensitivity to the conductivity of formation of amplitude ratio and (B) phase difference.

It is also useful to compare sensitivity of attenuation $At(B_z^*)$ and phase difference $\Delta\phi(B_z^*)$ of probes with different length to the conductivity of formation. With this purpose in mind we calculate normalized attenuation and phase difference by the corresponding values measured in uniform medium with resistivity of formation:

$$P_A(\alpha) = \left|\frac{At(B_z^*)}{At(B_z^{*un})}\right| \quad \text{and} \quad P_{\Delta\phi}(\alpha) = \frac{\Delta\phi(B_z^*)}{\Delta\phi(B_z^{*un})}$$

shown in Figs. 8.7 and 8.8. Here $At(B_z^{*un})$ and $\Delta\phi(B_z^{*un})$ are attenuation and phase difference measured in uniform medium with resistivity of a formation. Calculations are performed for the case when $f = 3.5$ MHz, $a_1 = 0.1$ m, $\rho_1 = 0.5$ ohmm, $L_2/L_1 = 0.7$. Index of curves γ_2/γ_1.

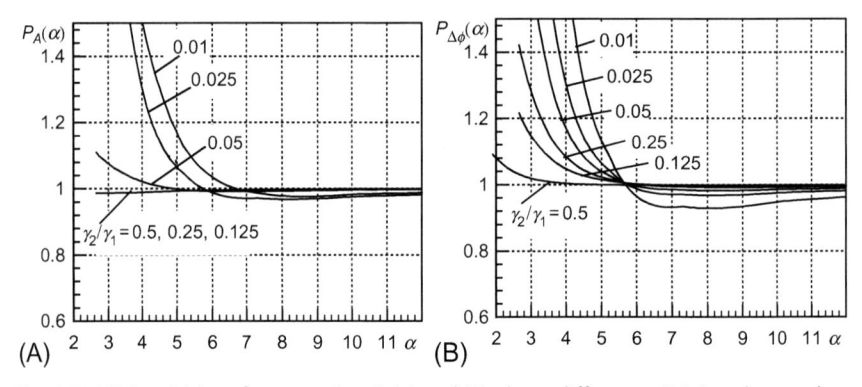

Fig. 8.7 (A) Sensitivity of attenuation $P_1(\alpha)$ and (B) phase difference $P_2(\alpha)$ to the conductivity of formation.

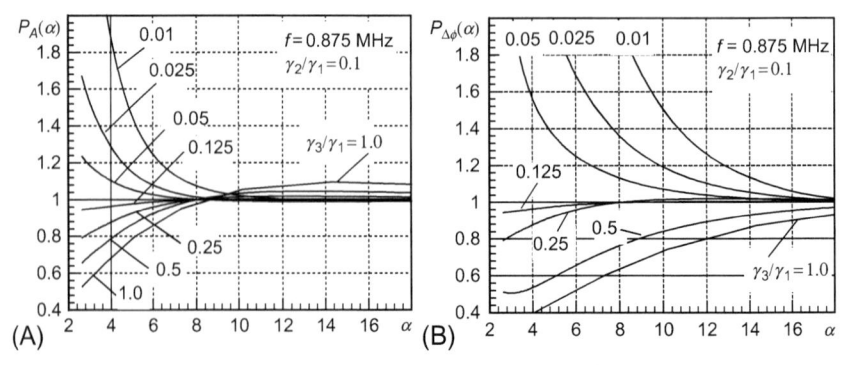

Fig. 8.8 (A) Sensitivity of attenuation, $P_A(\alpha)$, and (B) phase difference $P_{\Delta\phi}(\alpha)$ to the conductivity of formation in the presence of invasion zone.

Comparison shows that when $\alpha \geq 5$, attenuation and phase difference have similar sensitivity to the formation and almost the same probe length is needed to eliminate an influence of the borehole. Also, Fig. 8.8 illustrates behavior of these functions in the presence of invasion zone, when γ_3/γ_1 is varying and $a_2/a_1 = 4, \gamma_2/\gamma_1 = 0.1$.

Clearly, in the presence of invasion the attenuation is less than phase difference affected by the borehole, especially when parameter $\alpha \geq 7$.

8.4.4 The Main Features of the Field of the Two-Coil Probe

Now consider the main features of frequency responses of this field on the borehole axis. Results of numerical modeling presented in this section are based on calculations of the field b_z, given by Eq. (8.19) for the models of a medium with one and two cylindrical interfaces. As we already know:

1. Vertical magnetic dipole induces eddy currents located in horizontal planes that have shape of circles with the common center on the borehole axis. An electrical field has only azimuthal component E_ϕ, but the magnetic field has two components, B_r and B_z. On the borehole axis the magnetic field is oriented vertically, while both the electrical field, E_ϕ, and the radial component of magnetic field, B_r are equal to zero.
2. Induced currents density, j_ϕ, at any point of medium is characterized by the in-phase and quadrature component. Unlike the in-phase component the quadrature components is shifted in phase by 90 degrees with respect to the dipole current. Distributions of these components, Inj_ϕ, and, Qj_ϕ, are essentially different. The quadrature component is dominant near the source and rapidly decreases with an increase of the distance from the dipole, frequency, and conductivity of formation. In the range where Qj_ϕ dominates, the skin effect manifests itself similarly to a uniform medium with resistivity of the formation.
3. Near the source the quadrature component Qj_ϕ of the current density is directly proportional to the frequency, but with an increase of the distance from the source it becomes stronger subjected to an influence of the skin effect.
4. Near the dipole the in-phase component Inj_ϕ is significantly less than the quadrature one; with an increase of the distance it reaches a maximum and then rapidly approaches zero.
5. In accordance with the Biot-Savart law both the quadrature and the in-phase component of the field are determined by the distribution of the quadrature and in-phase component of current density,

respectively. Examples of the vertical component of magnetic field, expressed in units of the primary field, are presented in Figs. 8.9 and 8.10 where the quadrature Qb_z^* and in-phase Inb_z^* components are expressed in units of the primary field. The ratio a_1/λ_1 is plotted along the abscissa; $\lambda_1 = 2\pi\delta_1$ is the wave length in the borehole, δ_1 is the skin depth. The index of curves is ρ_2/ρ_1 (Fig. 8.9) and ρ_3/ρ_1 (Fig. 8.10). For a three-layered medium, calculations are performed for $\rho_2/\rho_1 = 4, a_2/a_1 = 4, L/a_1 = 10$.

The quadrature component increases linearly with frequency, reaches a maximum and then tends to zero. (The oscillating behavior of Qb_z at the right part of the response $(a_1/\lambda_1 \to \infty)$ is not shown because of a logarithmic scale.)

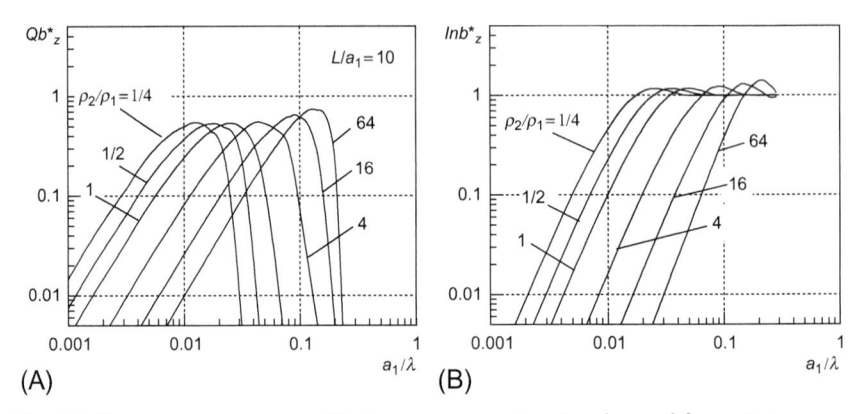

Fig. 8.9 Frequency responses of field components in a two-layered formation.

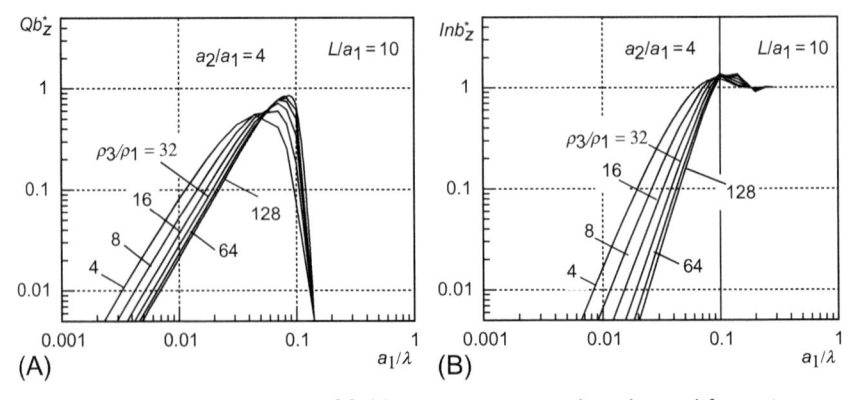

Fig. 8.10 Frequency responses of field components in a three-layered formation.

6. The left-hand asymptote of the frequency response of the quadrature component is a straight line with a slope of 63°30′ with respect to the horizontal axis. This part of the response corresponds to the case when intensity of induced currents is defined only by the primary magnetic flux and resistivity of a medium. As was mentioned above, the area where induced currents, shifted in phase by 90 degrees, increases with a decrease of frequency and an increase of formation resistivity. At the same time, with an increase of the probe length the volume of the formation contributing to the measured signal increases and, correspondingly, the influence of the medium near the probe becomes smaller. For this reason the longer the probe the earlier the deviation of the quadrature component from its left-hand asymptote begins.

7. The part of the frequency response, Qb_z which practically coincides with its left-hand asymptote is called Doll's range. Within this range the quadrature component is significantly larger than the secondary in-phase component.

8. In a two-layered medium when resistivity of the borehole exceeds that of the formation ($\rho_1 > \rho_2$) the departure from Doll's range takes place at the same values of parameter L/δ_2 as in a uniform medium with conductivity γ_2.

9. If conductivity of the borehole exceeds that of the formation, $\gamma_1 > \gamma_2$, and the skin depth in the borehole is significantly larger than its radius, the Doll's range is shifted toward larger values of parameter L/δ_2.

    In this case a relative contribution of induced currents, subjected to the skin effect in the formation, is smaller than in the case of a uniform medium with conductivity γ_2. Similar features are observed for a three-layered medium: with an increase of the conductivity of the borehole and invasion, as well as its radius a_2, the Doll's range is shifted towards larger values of L/δ_3 compared to a uniform medium with conductivity γ_3.

10. With an increase of parameter a_1/λ_1 the frequency response Qb_z departs from the left-hand asymptote and within a certain range of parameter a_1/λ_1 there is practically no skin effect neither in the borehole nor in the invasion zone. But in the formation the skin effect manifests itself in the same manner as in a uniform medium with the resistivity of formation. This low-frequency range is the most favorable one for the "focusing" probes. The main features of the field within this range have been described in detail earlier. This range of a_1/λ_1 is favorable for application of the hybrid method.

11. Frequency responses of the quadrature component, Qb_z, for a two-layered medium has one maximum, which to some extent increases with an increase of resistivity of the borehole. The position of the maximum is mainly defined by resistivity of the formation. For example, an increase of the borehole conductivity by a factor of 100 only slightly shifts the maximum to a range of lower frequencies. In some cases, when the invasion zones are relatively large, we can observe two maxima.
12. With increase of the frequency the skin effect leads to increased influence of the borehole and reduced sensitivity to the formation; in the presence of thick conductive invasion frequency response in a three-layered medium almost coincides with response in a two-layered medium with resistivity of invasion ρ_2. Within Doll's range, the influence of the borehole is defined by geometric factors and distribution of resistivity in the medium. Within a broad range of frequencies, far beyond the Doll's range, the influence of the borehole and invasion zone depend on their geometric factors and resistivity, but the influence of the formation is determined by the skin depth in a medium with resistivity of formation ρ_3.
13. In a wide range of frequencies, when the skin depth δ_1 is several times larger than the borehole radius, the influence of resistive borehole is not significant and the frequency response of the field, Qb_z, practically coincides with that corresponding to a uniform medium with the resistivity of the formation.
14. Since the response Qb_z has a maximum the same value of the quadrature component can be observed at two different values of a_1/λ_1. The ambiguity can be removed by using either an additional measurement or prior information.
15. Selection of frequencies for induction logging cannot be based only on the study of the field in a medium with cylindrical interfaces. However, these calculations allow us to study radial characteristics of two-coil probes, as well as probes consisting of several coils. In particular, the calculations permit to establish a range of frequencies and resistivities favorable for application of "focusing" multicoil probes.
16. Although both in-phase and quadrature component depend on the same geo-electric properties of formation, their frequency responses are quite different. At the range of small parameter a_1/λ_1 (low frequencies, high resistivity) the function Inb_z^s tends to zero as $\omega^{3/2}$, and with a decrease of frequency the ratio of the in-phase and quadrature

components rapidly decreases. In this range the in-phase component Inb_z^s depends on the conductivity of formation as $\gamma^{3/2}$. With an increase of the ratio a_1/λ_1 the in-phase component of the secondary field increases and then becomes greater than the quadrature component. In particular, when skin depth is smaller than the borehole radius, the in-phase component prevails and function Inb_z^s approaches -1 indicating the concentration of the induced currents in the borehole.

17. With a decrease of frequency maximum of the induced currents is shifted deeper from the borehole providing increased sensitivity of the in-phase component to the remote part of the formation. But complexity in the cancelation of the primary field, coinciding with the in-phase secondary component, makes application of this component quite difficult. For this reason it is more practical to use a quadrature component. Because of the strong skin effect at the high frequency, the induced currents are concentrated near the borehole, and quadrature component of the field mainly provides information about an invasion zone and the borehole. At the same time decrease of the frequency leads to increase of the depth of investigation only up to a certain limit. Below this limit contribution of the borehole and invasion into the measurements remains practically the same and quite significant, especially for the short-spaced probes. In other words, the depth of investigation is limited, regardless of frequency. To overcome the limit one should increase a length of the probe. This outlines a current trend in the development of advanced logging tools toward multiarray systems with frequencies in the range of tens to thousands of kilohertz. These systems mainly rely either upon measurements of the quadrature component or attenuation and phase difference. Also, as follows from Eq. (8.43), combination of quadrature components at two different frequencies

$$\frac{QB_z(\omega_2)}{\omega_2} - \frac{QB_z(\omega_1)}{\omega_1}$$

allows us to remove leading linear term and achieve a depth of investigation, similar to that of the in-phase component. As was mentioned above, the presence of this linear term doesn't permit an increased depth of investigation by a simple decrease of the frequency.

18. When invasion zone is absent and measurements are performed in the far zone, the sensitivities of the attenuation and phase difference to the

formation are practically the same. But in the presence of invasion attenuation is significantly more sensitive to the properties of formation.

8.5 DISPLACEMENT OF THE PROBE FROM THE BOREHOLE AXIS

In real conditions the induction probe may be shifted from the borehole axis. Below we study the effect of this displacement on the readings of the two- and three-coil probes. Shift of the probe leads to change in the geometry of the induced currents and appearance of the surface charges at the boundary between borehole and formation. Since the surface electric charges give rise to the electric field, the induced currents are generated by both an inductive electric field and charges. In general, these vortex and galvanic parts are related to each other, but at the range of small induction number they are independent.

Inasmuch as the density of charges depends on the conductivity of the borehole and formation the concept of the geometric factor is not applicable anymore. However, there is one exception, namely when conductivity of the formation γ_2 is small and the coefficient

$$K_{12} = \frac{\gamma_1 - \gamma_2}{\gamma_1 + \gamma_2}$$

is close to unity. At the range of small induction number the charges are mainly created by the primary electric field, and correspondingly we can expect that the in-phase component of the secondary field tends to that in a uniform medium with resistivity of the formation:

$$Inb_z^* \to Inb_z^{*un}(\gamma_2), \quad \text{if } p \to 0$$

In other words, it is less sensitive to the probe displacement than the quadrature component. The same tendency is observed in dual-frequency transformation:

$$\frac{QB_z^*(\omega_1)}{\omega_1} - \frac{QB_z^*(\omega_2)}{\omega_2}$$

which has reduced sensitivity to the near borehole zone. Of course, when the probe is shifted from the borehole axis the current lines aren't circles anymore and have some vertical component. In general, determination of the magnetic field with no symmetry is rather a complicated problem, which is usually solved numerically either by a finite element or finite difference techniques. But if the medium is symmetrical one, the semianalytical approach allows us to reduce the original non 1D problem to the series of 1D problems [2]. Our analysis of an influence of the probe displacement is based on the use of the last approach. First we consider the range of small induction number when interaction of currents inside the borehole can be neglected. Then, by analogy with the response of the two-coil probe we have:

$$Qb_z^* = \frac{\omega\mu_0\gamma_1 a_1^2 \alpha^2}{2}\left[G_1^d(\alpha, s, d) + sG_2^d(\alpha, s, d)\right] \quad (8.64)$$

Here

$$\alpha = L/a_1, \quad s = \gamma_2/\gamma_1, \quad d = r/a_1$$

The parameter d characterizes a displacement of the probe, normalized by the borehole radius. Inasmuch as analytical expression for the field is absent, we cannot derive formulas for the function $G^d(\alpha, 0, d)$, and particularly the asymptote for $\alpha \gg 1$. Nevertheless we can try to determine the function G_1^d approximately for the limiting case of a nonconductive formation, $s = 0$. In such case the influence of galvanic part of the field is maximal and there is no normal component of the current on the borehole surface. From Eq. (8.64) we obtain

$$F = \alpha^2 G_1^d(\alpha, 0, d) = \frac{2}{\omega\mu_0 a_1^2 \gamma_1} Qb_z^* \quad (8.65)$$

Behavior of the function $F(\alpha)$ is shown in Fig. 8.11, where index of curves is the parameter d and $\rho_1 = 1$ ohmm, $a_1 = 0.1$ m.

The displacement of the probe makes strong influence on the function $F(\alpha)$, which approaches the constant value with increase of the probe length. In other words, the function G_1^d, characterizing an influence of the borehole, decreases as $1/\alpha^2$:

$$G_1^d \to 1/\alpha^2, \quad \text{if } \alpha \gg 1$$

regardless of the displacement. Correspondingly, with an increase of α the normalized response of the two-coil probe Qb_z^* also tends to the constant, but the field decays as the primary field. With an increase of d transition to the asymptote is observed at larger values of the probe length. Bearing in mind the main features of the three-coil probe, we can expect reduction of the effect of displacement on this probe.

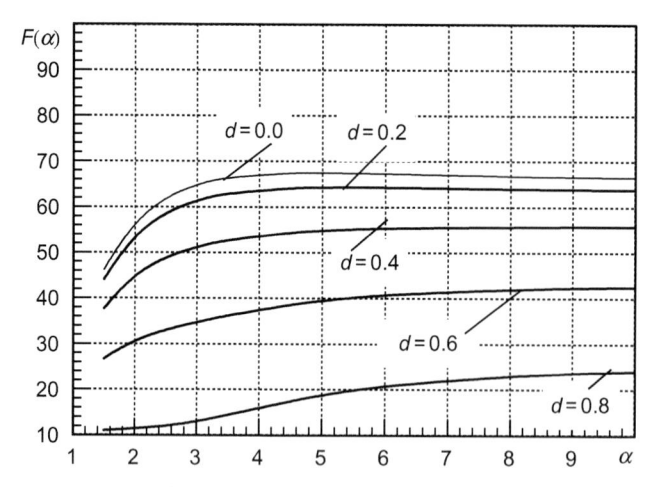

Fig. 8.11 Function $F(\alpha)$. Index is a displacement of the probe.

When parameter $s = \gamma_2/\gamma_1$ is small and charge density reaches a maximal value, the function G_1^d has almost the same asymptotic behavior as in the case when $s = 0$. To confirm this behavior we use a ratio

$$r(Q_Z) = Qb_z^*(\alpha, d, s)/Qb_z^{*un}(\gamma_2)$$

indicating closeness of the response Qb_z^* to that in the homogeneous medium with conductivity γ_2. Results of calculations for two-coil, Fig. 8.12A, and three-coil, Fig. 8.12B, probes at small induction number are presented below. Parameter α varies from 2 to 20, while other

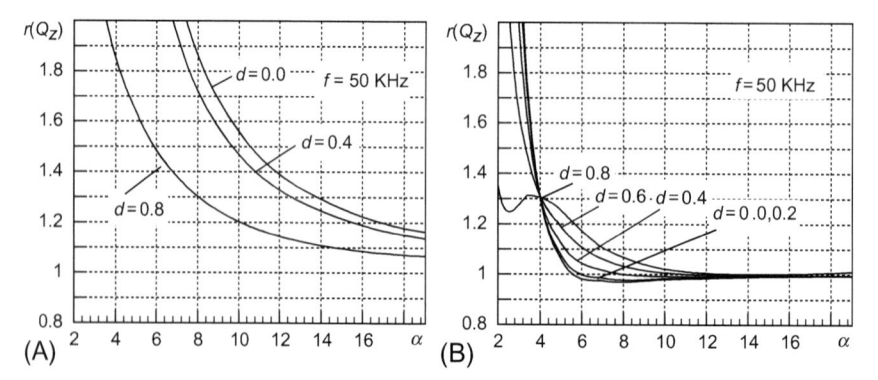

Fig. 8.12 (A) Effect of probe displacement on two-coil and (B) three-coil probe. Function $r(Q_Z) = Qb_z^*(\alpha, d, s)/Qb_z^{*un}(\gamma_2)$. Index is a displacement of the probe.

parameters are: $a_1 = 0.1$ m, $\rho_1 = 0.2$ ohmm and $s = 0.02$. The index of curves is the parameter d.

With an increase of the probe length all curves approach unity, but in the case of three-coil probe Fig. 8.12B it takes place at much smaller probe length. Bearing in mind the main feature of the latter, we can expect such asymptotic behavior of the function G_1^d regardless of the contrast s. In other words, the three-coil probe demonstrates "focusing" properties.

It is noticeable that an increase of the displacement for the two-coil probe leads to the reduced influence of the borehole, while for the three-coil probe the opposite tendency takes place. This behavior is observed in a wide range of frequencies typical for induction logging.

Earlier we pointed out that at the low frequency limit the in-phase component of the secondary field is less sensitive to the borehole than the quadrature component. This is confirmed by Fig. 8.13, where effect of displacement on the two-coil probe at frequency $f = 50$ kHz is presented in the form of a ratio

$$r(In_z) = (Inb_z^* - 1) / (Inb_z^{*un}(\gamma_2) - 1)$$

indicating closeness of the response Inb_z^* to that in the homogeneous medium with conductivity γ_2. Here $\rho_1 = 0.2$ ohmm, $s = 0.02$, $a_1 = 0.1$ m.

The data clearly show reduced impact of the displacement on the in-phase component compared to the two-coil quadrature component

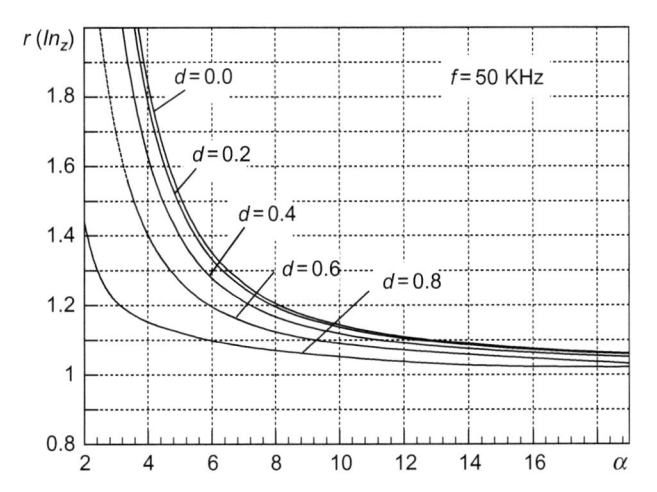

Fig. 8.13 Effect of the displacement on the in-phase component of the two-coil probe $r(In_z) = (Inb_z^* - 1) / (Inb_z^{*un}(\gamma_2) - 1)$. Index is a displacement of the probe.

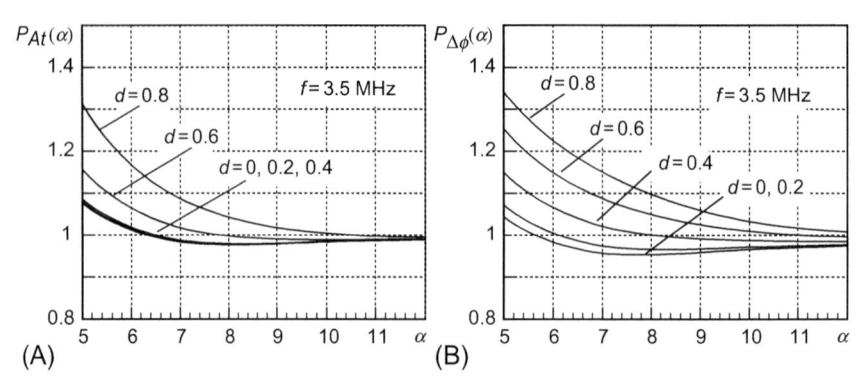

Fig. 8.14 Effects of displacement on (A) attenuation and (B) phase difference. Index is a displacement of the probe.

(Fig. 8.12A), although it is more pronounced compared to the three-coil quadrature component (Fig. 8.12B). Finally, we compare the effect of displacement on the attenuation and phase difference for the VIKIZ at the frequency of 3.5 MHz. The formation model is the same as in Fig. 8.13. The results of calculations are in Fig. 8.14.

Along y-axis we show normalized functions $P_{At}(\alpha)$ and $P_{\Delta\phi}(\alpha)$ representing attenuation and phase difference normalized by that in a uniform medium with conductivity γ_2. Fig. 8.14A and B shows that the attenuation is less sensitive to the probe displacement compared to the phase difference. The graphs above can be used to perform corrections for the displacement when the parameters of the borehole and value of the displacement are known.

REFERENCES

[1] Kaufman A, Sokolov V. Theory of induction logging based on the use of transient field. Novosibirsk: Nauka; 1972.
[2] Nikitenko M, Itskovich G, Seryakov A. Fast electromagnetic modeling in cylindrically layered media excited by eccentred magnetic dipole. Radio Sci 2016;51(6):573–88.

FURTHER READING

[1] Kaufman A, Keller G. Induction logging. Amsterdam: Elsevier; 1989.

CHAPTER NINE

Quasistationary Field of the Vertical Magnetic Dipole in a Bed of a Finite Thickness

Contents

9.1 Vertical Component of the Field of a Magnetic Dipole	289
9.2 The Field of the Vertical Magnetic Dipole in the Presence of a Thin Conducting Plane	295
9.3 The Two-Coil Induction Probe in Beds With a Finite thickness	297
9.3.1 Dependence of the Field on the Parameter $p = L/\delta_1$	297
9.3.2 Some Features of the Apparent Conductivity Curves	301
9.4 Profiling Curves for a Two-Coil Probe in a Bed of Finite Thickness	306
Reference	310

In this chapter we consider vertical responses of two–coil induction probes located at different locations with respect to the horizontal interfaces between a bed and a surrounding medium. Special attention is paid to the effects that the frequency has on the vertical responses of the probe as well as the ratio of conductivities, and geometric parameters such as formation thickness and probe length and position.

9.1 VERTICAL COMPONENT OF THE FIELD OF A MAGNETIC DIPOLE

Suppose that there are two parallel interfaces that divide a space into three parts as shown in Fig. 9.1. The vertical magnetic dipole is placed at the origin of the cylindrical system of coordinates, and its moment is oriented along the z-axis.

The magnetic permeability of the medium is equal to $4\pi \times 10^{-7} H/m$. As in the case of a medium with cylindrical interfaces, we introduce the vector

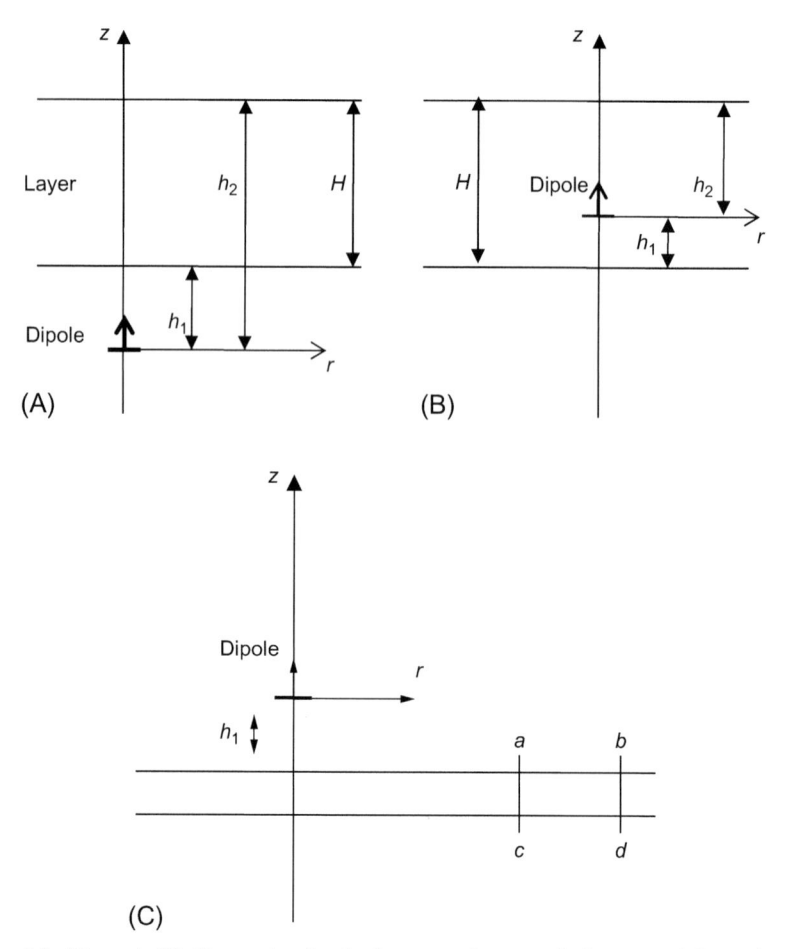

Fig. 9.1 (A) and (B) Magnetic dipole in a medium with horizontal boundaries. (C) Magnetic dipole in a medium with thin layer.

potential of the electrical type with complex amplitude obeying the Helmholtz equation:

$$\nabla^2 \mathbf{A}^* + k^2 \mathbf{A}^* = 0 \tag{9.1}$$

As was shown before

$$\mathbf{E}^* = curl\,\mathbf{A}^*, \quad \mathbf{B}^* = k^2 \mathbf{A}^* + grad\,div\,\mathbf{A}^* \tag{9.2}$$

Due to the axial symmetry the boundary value problem can be solved using only the z-component of the vector potential:

$$\mathbf{A}^* = \left(0, 0, A_z^*\right) \tag{9.3}$$

which depends on two coordinates: r and z. Then Eqs. (9.2), (9.3) give

$$B_r^* = \frac{\partial^2 A_z^*}{\partial r \partial z}, \quad B_z^* = k^2 A_z^* + \frac{\partial^2 A_z^*}{\partial z^2}, \quad B_\phi^* = 0$$

and

$$E_r^* = E_\phi^* = 0, \quad E_\phi^* = -i\omega\mu_0 \frac{\partial A_z^*}{\partial r} \tag{9.4}$$

From the continuity of tangential components of the field, boundary conditions for the vector potential at interfaces are:

$$A_{iz}^* = A_{kz}^* \quad \text{and} \quad \frac{\partial A_{iz}^*}{\partial z} = \frac{\partial A_{kz}^*}{\partial z} \quad \text{if} \quad z = h_i \tag{9.5}$$

Near the origin of the coordinate system, where the dipole is located, the field tends to that of a magnetic dipole in a uniform medium. Therefore, for the vector potential we have:

$$A_z^* \to \frac{\mu_0 M_0}{4\pi} \frac{e^{ikR}}{R}, \quad \text{as } R \to 0 \tag{9.6}$$

where $R^2 = r^2 + z^2$. At infinity ($R \to \infty$), the field and correspondingly, the vector potential, vanish. Thus, to find the field, it is necessary to solve the equation:

$$\nabla^2 A_{iz}^* + k_i^2 A_{iz}^* = 0 \tag{9.7}$$

and satisfy conditions (9.5) as well as a corresponding behavior of the field near the dipole and at infinity. The equation above is the Helmholtz equation and, in a cylindrical system of coordinates, it can be presented in the form:

$$\frac{\partial^2 A_z^*}{\partial r^2} + \frac{1}{r}\frac{\partial A_z^*}{\partial r} + \frac{\partial^2 A_z^*}{\partial z^2} + k_i^2 A_z^* = 0$$

because $\frac{\partial A_z^*}{\partial \phi} = 0$. Letting $A_z^* = U(r)V(z)$ and applying the method of separation of variables, we obtain two ordinary differential equations:

$$\frac{d^2 U(r)}{dr^2} + \frac{1}{r}\frac{dU(r)}{dr} + m^2 U = 0 \quad \text{and} \quad \frac{d^2 V}{dz^2} - (m^2 - k_i^2) V = 0 \tag{9.8}$$

where m is the separation constant. The first equation of the set (9.8) is a Bessel equation, and its solutions are Bessel functions of the first and second kind:

$$U(r) = A J_0(mr) + B Y_0(mr)$$

Function $Y_0(mr)$ tends to infinity as $r \to 0$, and therefore it cannot describe a field. The solution of the second equation is:

$$V(z) = Ce^{-(m^2-k^2)^{1/2}z} + De^{(m^2-k^2)^{1/2}z}$$

Thus, the general solution of Eq. (9.7) can be presented in the form:

$$A_{zi}^*(r, z) = \int_0^\infty \left[N_1 e^{(m^2-k_i^2)^{1/2}z} + N_2 e^{-(m^2-k_i^2)^{1/2}z} \right] J_0(mr) dm \quad (9.9)$$

The sign of radical $(m^2 - k_i^2)^{1/2}$ is chosen in such way that its real part is positive:

$$\mathrm{Re}(m^2 - k_i^2)^{1/2} > 0 \quad (9.10)$$

We present the field in a medium where the dipole is located as a sum:

$$A_{1z}^* = \frac{\mu_0 M_0}{4\pi} \frac{\exp(ikR)}{R} + A_z^{*s} \quad (9.11)$$

where A_z^{*s} describes the secondary field. It is known that the vector potential of the magnetic dipole is expressed as a Sommerfeld integral:

$$\frac{\exp(ikR)}{R} = \int_0^\infty \frac{m}{(m^2 - k^2)^{1/2}} \exp\left[(m^2 - k^2)^{1/2} z\right] J_0(mr) dm$$

Now we derive formulas for the vector potential for various positions of the dipole with respect to the interfaces:

Case 1: The magnetic dipole is located outside the bed, as shown in Fig. 9.1A

In accordance with Eqs. (9.9), (9.11) and taking into account the condition at infinity, expressions for the vector potential in each part of a medium can be written in the form:

$$A_{1z}^* = \frac{\mu_0 M_0}{4\pi} \int_0^\infty \left[\frac{m}{m_1} e^{-m_1|z|} + D_1 e^{m_1 z} \right] J_0(mr) dm \quad z \leq h_1$$

$$A_{2z}^* = \frac{\mu_0 M_0}{4\pi} \int_0^\infty \left[D_2 e^{m_2 z} + D_3 e^{-m_2 z} \right] J_0(mr) dm \quad h_1 \leq z \leq h_2 \quad (9.12)$$

$$A_{3z}^* = \frac{\mu_0 M_0}{4\pi} \int_0^\infty D_4 e^{-m_1 z} J_0(mr) dm \quad z \geq h_2$$

Magnetic Dipole in a Layer of a Finite Thickness

Here h_1 is the distance from the dipole to the nearest interface, $h_2 = h_1 + H$, and H is the bed thickness. From boundary conditions, Eq. (9.5), we obtain a system of linear equations with respect to D_1, D_2, D_3, and D_4:

$$\frac{m}{m_1}e^{-m_1 h_1} + D_1 e^{m_1 h_1} = D_2 e^{m_2 h_1} + D_3 e^{-m_2 h_1}$$

$$-me^{m_1 h_1} + m_1 e^{m_1 h_1} = m_2 D_2 e^{m_2 h_2} - m_2 D_3 e^{m_2 h_2} \quad (9.13)$$

$$D_2 e^{m_2 h_2} + D_3 e^{-m_2 h_2} = D_4 e^{-m_1 h_2}$$

$$m_2 D_2 e^{m_2 h_2} - m_2 D_3 e^{-m_1 h_2} = -m_1 D_4 e^{-m_1 h_2}$$

Solving this system we have:

$$D_1 = -\frac{mK_{12}e^{-2m_1 h_1}(1-e^{-2m_2 H})}{m_1(1-K_{12}^2 e^{-2m_1 H})}, \quad D_2 = \frac{2mK_{12}e^{-(m_1+m_2)h_1}e^{-2m_2 H}}{(m_1+m_2)(1-K_{12}^2 e^{-2m_1 H})},$$

$$D_3 = \frac{2me^{-(m_1-m_2)h_1}}{(m_1+m_2)(1-K_{12}^2 e^{-2m_1 H})}, \quad D_4 = \frac{4m_1 m_2 e^{-(m_1-m_2)H}}{(m_1+m_2)^2(1-K_{12}^2 e^{-2m_1 H})}$$

(9.14)

Substituting these expressions for the coefficients in Eq. (9.12), we have:

$$A_{1z}^* = \frac{\mu_0 M_0}{4\pi}\int_0^\infty \frac{m}{m_1}\left[e^{m_1|z|} - \frac{K_{12}(1-e^{-2m_2 H})e^{-m_1(2h_1-z)}}{1-K_{12}^2 e^{-2m_2 H}}\right] J_0(mr)\,dm$$

$$A_{2z}^* = \frac{\mu_0 M_0}{4\pi}\int_0^\infty \frac{2me^{-m_1 h_1}e^{-m_2(z-h_1)}\left[1+K_{12}e^{2m_2(z-h_1-H)}\right]}{(m_1+m_2)(1-K_{12}^2 e^{2m_2 H})} J_0(mr)\,dm \quad (9.15)$$

$$A_{3z}^* = \frac{\mu_0 M_0}{4\pi}\int_0^\infty \frac{4mm_2 e^{-(m_2-m_1)H}e^{-m_1 z}}{(m_1+m_2)^2(1-K_{12}^2 e^{-2m_2 H})} J_0(mr)\,dm$$

Here z is the distance between the dipole and an observation point. As follows from these equations, the vertical component of the magnetic field, measured with the two-coil probe at the z-axis when $r=0$ is:

$$b_z^{(1)*} = b_{0z}^*(\gamma_1) - \frac{1}{2}\int_0^\infty \frac{m^3}{m_1}\frac{K_{12}(1-e^{-2m_2\alpha})e^{-m_1(2\beta-1)}}{1-K_{12}^2 e^{-2m_2\alpha}}\,dm \quad \text{if } \beta \geq 1$$

$$b_z^{(2)*} = \int_0^\infty \frac{m^3 e^{-m_1\beta}e^{-m_2(1-\beta)}\left[1+K_{12}e^{2m_2(1-\beta-\alpha)}\right]}{(m_1+m_2)(1-K_{12}^2 e^{-2m_2\alpha})}\,dm \quad \text{if } 1 \geq \beta \geq 1-\alpha$$

$$b_z^{(3)*} = \int_0^\infty \frac{2m^3 m_2 e^{-(m_2-m_1)\alpha}e^{-m_1}}{(m_1+m_2)^2(1-K_{12}^2 e^{-2m_2\alpha})}\,dm \quad \text{if } \beta \leq 1$$

(9.16)

Here, b_z^* is the complex amplitude of the field, expressed in units of the primary field equal to $b_{z0}^0 = 2\mu_0 M_0/4\pi z^3$. Also:

$$K_{12} = \frac{m_2 - m_1}{m_2 + m_1}, \quad m_i = (m^2 - k_i^2 z^2)^{1/2}, \quad \alpha = \frac{H}{z}, \quad \beta = \frac{h_1}{z}$$

H is the bed thickness, and b_{0z}^* is the field in a uniform medium with conductivity γ_1. The last equation of the set (9.16) corresponds to the case of the layer located between the dipole and the observation point, and as it follows from this formula, the field does not depend on the position of the layer with respect to the probe coils.

Case 2: The magnetic dipole is located within the bed, as shown in Fig. 9.1B

For the dipole located within the bed, expressions for the vector potential can be written in the form:

$$A_{1z}^* = \frac{\mu_0 M_0}{4\pi} \int_0^\infty D_1 e^{m_1 z} J_0(mr) dm \qquad \text{if } z \leq h_2$$

$$A_{2z}^* = \frac{\mu_0 M_0}{4\pi} \int_0^\infty \left[\frac{m}{m_2} e^{-m_2|z|} + D_2 e^{m_2 z} + D_3 e^{-m_2 z}\right] J_0(mr) dm \qquad \text{if } h_2 \leq z \leq h_1$$

$$A_{3z}^* = \frac{\mu_0 M_0}{4\pi} \int_0^\infty D_4 e^{-m_1 z} J_0(mr) dm, \qquad \text{if } z \geq h_1$$

(9.17)

where h_1 is the distance from the dipole to the upper interface of the bed, $h_2 = H - h_1$, and H is the bed thickness. To determine the unknown coefficients, we use the boundary conditions, which lead to the following system:

$$D_1 e^{-m_1 h_2} = \frac{m}{m_2} e^{-m_2 h_2} + D_2 e^{-m_2 h_2} + D_3 e^{m_2 h_2}$$

$$m_1 D_1 e^{-m_1 h_2} = m e^{-m_2 h_2} + m_2 D_2 e^{-m_2 h_2} - m_2 D_3 e^{m_2 h_2}$$

$$D_4 e^{-m_1 h_1} = \frac{m}{m_2} e^{-m_2 h_1} + D_2 e^{m_2 h_1} + D_3 e^{-m_2 h_1}$$

$$-m_1 D_4 e^{-m_1 h_1} = -m e^{-m_2 h_1} + m_2 D_2 e^{m_2 h_1} - m_2 D_3 e^{-m_2 h_1}$$

(9.18)

In this case, the field is considered only inside the bed, inasmuch as expressions for the field outside the bed can be derived from the set of Eq. (9.16). Solving the system (9.18) we find:

Magnetic Dipole in a Layer of a Finite Thickness

$$D_2 = \frac{mK_{12}e^{-2m_2h_1}(1+K_{12}e^{-2m_2h_2})}{m_2(1-K_{12}^2e^{-2m_2H})} \quad \text{and}$$

$$D_3 = \frac{mK_{12}e^{-2m_2h_2}(1+K_{12}e^{-2m_2h_1})}{m_2(1-K_{12}^2e^{-2m_2H})}$$

(9.19)

Substituting these expressions into the second equation of (9.17), we obtain:

$$A_{2z}^* = \frac{\mu_0 M_0}{4\pi} \int_0^\infty \frac{m}{m_2} e^{-m_2|z|}$$

$$+ \frac{mK_{12}\left[e^{-m_2(2h_1-z)} + e^{-m_2(2h_2+z)} + 2K_{12}e^{-2m_2H}\cosh m_2 z\right]}{m_2(1-K_{12}^2e^{-2m_2H})} J_0(mr)\,dm$$

(9.20)

In accordance with Eqs. (9.4), (9.20), the expression for the vertical component of the magnetic field on the dipole axis related to the primary field is:

$$b_z^* = b_{0z}^*(\gamma_2)$$

$$+ \frac{1}{2}\int_0^\infty \frac{m^3 K_{12}\left[e^{-(1+2\alpha)m_2} + e^{-(2\alpha-2\beta-1)m_2} + 2K_{12}e^{-2\alpha m_2}\cosh m_2\right]}{m_2(1-K_{12}^2e^{-2\alpha m_2})}\,dm,$$

(9.21)

where $\alpha = H/z$, $\beta = h_2/z$. If coils of the probe are located symmetrically with respect to the interfaces, $2\beta = \alpha - 1$, the latter equation can be presented as:

$$b_z^* = b_{0z}^* + \int_0^\infty \frac{m^3 K_{12}e^{-2\alpha m_2}}{m_2}\frac{e^{\alpha m_2} + K_{12}\cosh m_2}{1-K_{12}^2 e^{-2\alpha m_2}}\,dm \quad (9.22)$$

Next we derive equations for the field in one special case.

9.2 THE FIELD OF THE VERTICAL MAGNETIC DIPOLE IN THE PRESENCE OF A THIN CONDUCTING PLANE

Let us assume that the length of the probe is significantly greater than the thickness of the bed (Fig. 9.1C). Then, if its conductivity is much larger than that of the surrounding medium and the skin depth inside the bed is much greater than its thickness, the bed can be replaced by a thin conducting plane with conductance S, equal to the product of the conductivity and thickness of this layer.

Such replacement enables one to use approximate conditions—which do not rely on the field inside the bed—instead of the exact boundary conditions. From a continuity of the tangential component of the electrical field we have:

$$E_{1\phi}^* = E_{2\phi}^* \qquad (9.23)$$

Circulation of the magnetic field along the contour *abcd* is equal to the current piercing this contour (Fig. 9.1C), hence we can write:

$$\oint B dl = B_{1r} dr - B_{2r} dr = \mu_0 dr dh E_\phi, \quad h \to 0$$

or

$$B_{1r}^* - B_{2r}^* = \mu_0 S E_\phi^* \qquad (9.24)$$

where S is the conductance of the thin layer. Correspondingly, the boundary conditions for the vector potential have the form:

$$A_{1z}^* = A_{2z}^*, \quad \frac{\partial A_{1z}^*}{\partial z} - \frac{\partial A_{2z}^*}{\partial z} = -i\omega\mu_0 S A_{2z}^* \qquad (9.25)$$

For the function A_z^* outside the conducting plane we have:

$$A_{1z}^* = \frac{\mu_0 M_0}{4\pi} \int_0^\infty \left[\frac{m}{m_1} e^{-m_1|z|} + D_1 e^{m_1 z}\right] J_0(mr) \quad \text{if } z \le h_1$$

$$A_{2z}^* = \frac{\mu_0 M_0}{4\pi} \int_0^\infty D_2 e^{-m_2 z} J_0(mr) dm \quad \text{if } z \ge h_1 \qquad (9.26)$$

Substituting these expressions into Eq. (9.25), we obtain the system of equations for determination of D_1 and D_2:

$$-D_1 e^{m_1 h_1} + D_2 e^{-m_1 h_1} = \frac{m}{m_1} e^{-m_1 h_1}, \quad m_1 D_1 e^{m_1 h_1}$$
$$+ (m_1 - i\omega\mu_0 S) D_2 e^{-m_1 h_1} = m e^{-m_1 h_1}$$

Solving this system we have:

$$D_1 = \frac{m K_s^2 e^{-2m_1 h_1}}{m_1 (2m_1 - K_s^2)} \quad \text{and} \quad D_2 = \frac{m}{(2m_1 - K_s^2)} \qquad (9.27)$$

here $K_s^2 = i\omega\mu_0 S$. Therefore, for the vector potential we obtain:

$$A_{1z}^* = \frac{\mu_0 M_0}{4\pi} \int_0^\infty \left[\frac{m}{m_1} e^{-m_1|z|} + \frac{mK_s^2 e^{-m_1 h_1}}{m_1(2m_1 - K_s^2)} e^{m_1 z} \right] J_0(mr) dm$$

and

$$A_{2z}^* = \frac{\mu_0 M_0}{4\pi} \int_0^\infty \frac{2m}{2m_1 - K_s^2} e^{m_1 z} J_0(mr) dm$$

Correspondingly, for the vertical component of the magnetic dipole along its axis we have:

$$b_{1z}^* = b_{0z}^*(\gamma_1) + \frac{n_s}{2} \int_0^\infty \frac{m^3 e^{m_1(1+2\alpha)}}{2m_1 - n_s} dm, \quad b_{2z}^* = \int_0^\infty \frac{m^3 e^{m_1}}{2m_1 - n_s} dm \quad (9.28)$$

where $m_1 = (m^2 - k_1^2 z^2)^{1/2}$, $\alpha = h_1/Z$, and $n_s = i\mu_0 \omega S z$.

The derived formulas enable us to study the vertical characteristics of the two-coil probes in the presence of a thin conducting layer surrounded by the uniform medium.

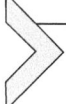

9.3 THE TWO-COIL INDUCTION PROBE IN BEDS WITH A FINITE THICKNESS

9.3.1 Dependence of the Field on the Parameter $p = L/\delta_1$

First we assume that the two-coil probe is located symmetrically with respect to the interfaces of the bed. The vertical component of the magnetic field on the z-axis, expressed in units of the primary field, is defined by three parameters: the ratio of the probe length, L, to the skin depth, δ_1, in the bed; the ratio of conductivity of the bed to that of the surrounding medium; and the ratio of the bed thickness, H, to the probe length.

Examples of the quadrature and in-phase components of the field $b_z^*(L/\delta_1)$, are presented in Figs. 9.2 and 9.3. Similar to the cases of a uniform medium and a medium with cylindrical interfaces, the in-phase component of the secondary field $|Inb_z^*|$ gradually increases with an increase of the argument, then reaches a value slightly exceeding the primary field, and finally, approaches unity in an oscillating manner. The quadrature component also increases initially, then reaches a maximum before decreasing and approaching zero. Such behavior of both components of the complex amplitude of the quasistationary field is typical for any conducting media.

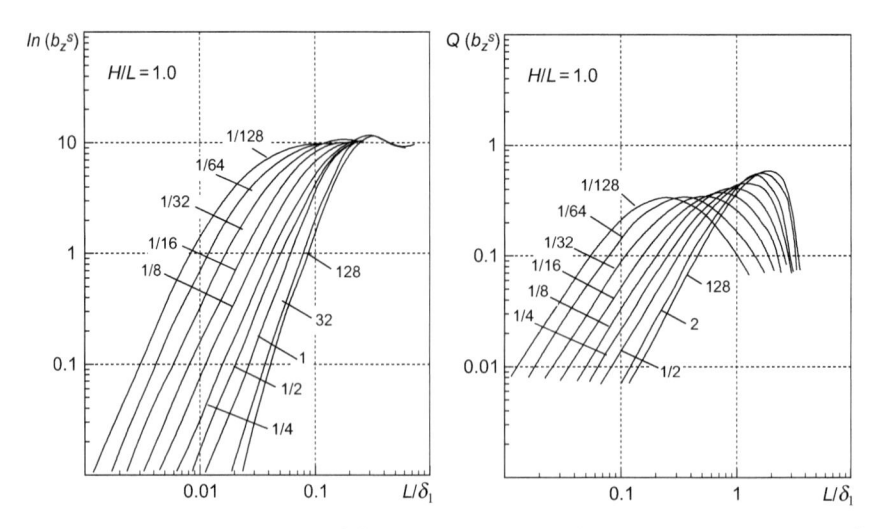

Fig. 9.2 Frequency responses of the in-phase and quadrature component. Index of curves is γ_1/γ_2.

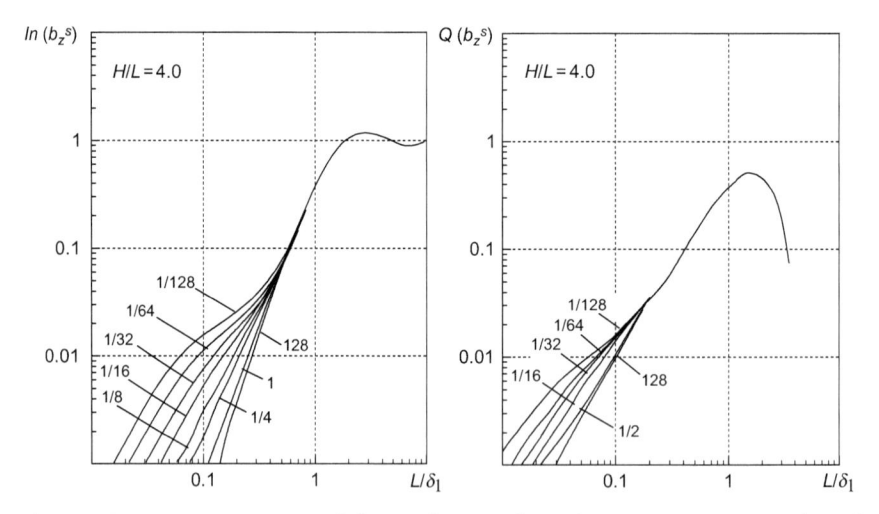

Fig. 9.3 Frequency responses of the in-phase and quadrature component. Index of curves is γ_1/γ_2.

An analysis of the results of calculations allows us to outline the following features of the field:

1. For small values of parameter L/δ_1 (low frequency, high resistivity) the in-phase component of the secondary field is much smaller than the quadrature component: $In b_z^s \ll Q b_z$. Comparison of quadrature and

in-phase components leads to the conclusion that in the range of small parameter, induced currents in the medium surrounding the bed have a much stronger influence on the in-phase component than on the quadrature component. In the limit when parameter L/δ_1 tends to zero, the in-phase component of the magnetic field approaches that of a uniform medium with the conductivity of the surrounding medium (Chapter 7):

$$Inb_z^{s*} \to -\frac{2}{3}p_2^3 \qquad (9.29)$$

Here

$$p_2^3 = \left(\frac{\gamma_2 \mu_0 \omega}{2}\right)^{3/2} L^3$$

It is essential that this result does not depend on the ratio of the bed thickness and the probe length (H/L), as well as the ratio of conductivities. In other words, with a decrease of L/δ_1, the bed becomes transparent to the in-phase component regardless of the probe length. Within this range of L/δ_1 the in-phase component is much less sensitive to the bed than the quadrature component. Its asymptotic behavior according to Eq. (9.29), is mainly defined by the conductivity of a surrounding medium.

2. In the range of small parameter, the quadrature component of the field is directly proportional to the frequency and conductivity following Doll's theory. The left-hand asymptote of the frequency response is described by the function $Qb_z^*(L/\delta_1)$:

$$Qb_z^* = \frac{\omega \mu_0 L^2}{2}(\gamma_1 G_1 + \gamma_2 G_2) = p_1^2 \left(G_1 + \frac{\gamma_2}{\gamma_1} G_2\right) \qquad (9.30)$$

Earlier, we used this equation to study in detail the vertical responses of the two-coil probe and, in particular, demonstrated that the influence of a surrounding medium is rather strong when a bed is more resistive and has a relatively small thickness. By analogy with the case of a medium with cylindrical boundaries, we will derive Eq. (9.30) from an exact solution. In fact, according to Eq. (9.22), the vertical component of the magnetic field along the dipole axis is:

$$b_z^* = b_{0z}^*(\gamma_1) + \int_0^\infty \frac{m^3 K_{12} e^{-2am_2} e^{am_2} + K_{12} \cosh m_2}{m_2} \cdot \frac{dm}{1 - K_{12}^2 e^{-2am_2}}, \quad \alpha \geq 1$$

where $b_{0z}^*(\gamma_1)$ is the field in a uniform medium with a bed conductivity expressed in units of the primary field and

$$m_1 = (m^2 - in_1)^{1/2}, \quad m_2 = (m^2 - in_2)^{1/2}, \quad K_{12} = (m_2 - m_1)/(m_2 + m_1),$$
$$n_1 = \gamma_1 \mu_0 \omega L^2, \quad n_2 = \gamma_2 \mu_0 \omega L^2, \quad s = \gamma_1/\gamma_2, \quad \alpha = H/L$$

Expanding the radicals in a series by small parameters n_i/m^2:

$$m_1 \approx m - \frac{in_1}{2m}, \quad m_2 \approx m - \frac{in_2}{2m},$$

and considering only the first term, we obtain for the integrand:

$$-\frac{1}{4}(s-1)n_1 e^{-\alpha m_1}$$

Correspondingly, taking into account the expression for $b_{0z}^*(\gamma_1)$, the integral becomes equal to

$$-\frac{i}{4\alpha}(s-1)\gamma_2 \mu_0 \omega L^2$$

Thus, the field in the range of small parameter n_1 is:

$$b_z^* = \frac{i\gamma_1 \mu_0 \omega L^2}{2} - \frac{i}{4\alpha}(s-1)\gamma_2 \mu_0 \omega L^2$$

and

$$Qb_z^* = \frac{\omega \mu_0 L^2}{2}(\gamma_1 G_1 + \gamma_2 G_2) = p_1^2 \left(G_1 + \frac{\gamma_2}{\gamma_1} G_2\right)$$

where

$$G_1 = 1 - 1/2\alpha \quad \text{and} \quad G_2 = 1/2\alpha$$

Of course, the latter was already derived in Chapter 6, proceeding from the concept of a geometric factor. Therefore, Doll's theory is in fact the theory of a very small parameter, which characterizes the dimensions of a model, expressed in units of the skin depth. For example, with a decrease of the probe length, parameter L/δ also decreases. From the physical point of view, this means that the effect of induced currents near the dipole, which are shifted in-phase by 90 degrees and do not interact with each other, increases.

3. Until now we have considered the range of parameter p when the function Qb_z^* increases in direct proportion to p^2. As was shown in Chapter 7, there is a range in which components of the field behave as:

$$Qb_z^* = \frac{\omega\mu_0 L^2}{2}(\gamma_1 - \gamma_2)G_1 + Qb_{0z}^*(\gamma_2) \text{ and } Inb_z^* = Inb_{0z}^*(\gamma_2) \quad (9.31)$$

Here G_1 is the geometric factor of the bed, and b_{0z}^* is the field in a uniform medium with a conductivity of the surrounding medium γ_2. With further increase of the parameter p, induced currents in the bed become more subject to the skin effect, and the quadrature component Qb_z^* grows at a decreasing rate. Finally, it reaches a maximum and then decreases in value, approaching zero in an oscillating manner. An increase of the parameter p can be caused by either an increase of the probe length or a change in frequency. The former leads to increased sensitivity of the probe to remote parts of the medium; the latter causes an increase of the skin effect near the probe.

With an increase of the parameter L/δ_1, the in-phase component of the secondary field also increases, and at the upper limit, it approaches that of the primary field

$$|b_z^*| \rightarrow \frac{\mu_0 M_0}{2\pi L^3}$$

Correspondingly, induced currents are concentrated in the vicinity of the source, and their direction is opposite to the direction of the current in the source dipole. In other words, the primary and secondary fields cancel each other, and the resulting field in a conducting medium is equal to zero.

9.3.2 Some Features of the Apparent Conductivity Curves

Next we consider dependence of apparent conductivity on the frequency. In general, apparent conductivity can be introduced in different ways, but we follow Doll's approach:

$$\gamma_a = \frac{2}{\omega\mu_0 L^2} Qb_z^* \quad (9.32)$$

Such apparent conductivity reads true conductivity in a uniform medium only within Doll's domain, but outside of this range the apparent conductivity differs from the true conductivity.

In most practical cases, the field behavior either corresponds to the Doll approximation or relatively close to it. At the same time, when frequency is sufficiently high and the resistivity of a bed is known, it is appropriate to introduce apparent conductivity as

$$\gamma_a = \gamma_1 \left| \frac{Qb_z^*}{Qb_{0z}^*(\gamma_1)} \right| \qquad (9.33)$$

Of course, within the range of small parameter, Eqs. (9.32), (9.33) lead to the same results for γ_a.

Let us consider the case when the bed is thicker than the length of the two-coil probe located in the middle of the bed Fig. 9.4. Normalized apparent conductivity (9.32) is plotted as a function of L/δ_1. The index of curves is γ_1/γ_2. This figure shows that all curves of the apparent conductivity at the left-hand part are in the Doll's range and parallel to the horizontal axis. In this range, an increase of the conductivity of the surrounding medium γ_2

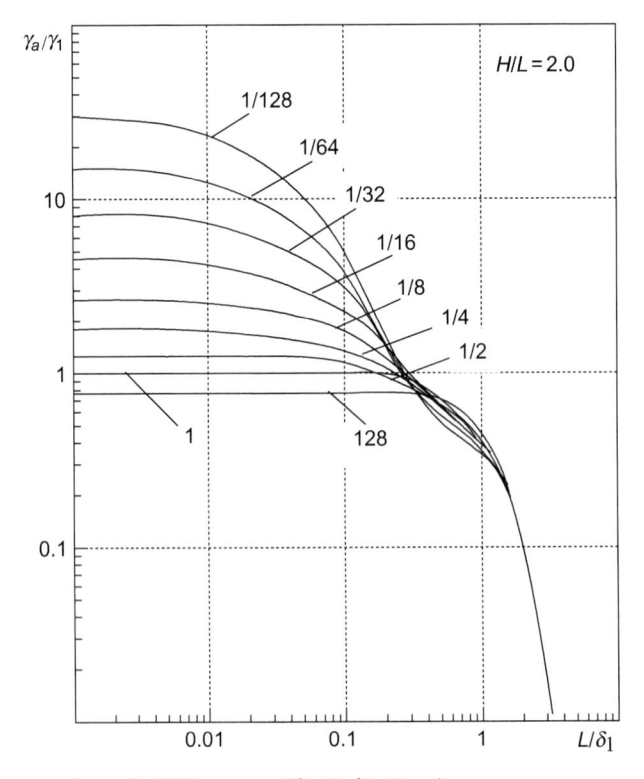

Fig. 9.4 Apparent conductivity curves. The index is γ_1/γ_2.

leads to increased readings of apparent conductivity. But with increase of the frequency, the skin effect in the surrounding medium becomes stronger, causing a decrease of γ_a. This means that a relative contribution of currents in the bed increases and the influence of the conductive surrounding medium on the response becomes less than what follows from the Doll's theory. This observation was made many years ago [1] and motivated the usage of frequencies higher than those in the original system. Of course, the frequencies should not be increased at the expense of reduced depth of investigation in the radial direction. This is one reason why in the latest systems of the lateral induction soundings, the highest frequencies are used for the shortest probes having shallow depth of investigation, and lowest frequencies in the long-spaced probes permitting deep lateral sounding.

Comprehensive numerical calculations show that decrease of the bed thickness leads to a deviation from the left-hand asymptote at the smaller frequencies if $\gamma_2 > \gamma_1$. It is appropriate to relate the maximal values of the parameter L/δ_1 to the resistivity and frequency, when the Doll's approximation is valid. As a result, for the frequency of 20 kHz and $\gamma_1 > \gamma_2$ corrections due to the internal skin effect are small, and for the relatively thick bed ($\alpha > 8$) and low resistivity ($\rho_1 \approx 1$ ohmm), they range between 10% and 20%. But if conductivity of the bed is smaller than that of the surrounding medium, then the influence of the skin effect can be significant. For instance, for $f = 20$ kHz, $\rho_1 = 20$ ohmm, and $\rho_2 = 2.5$ ohmm, $L = 1$ m and $H = 2$ m, the value of apparent conductivity, γ_2/γ_1, is equal to 2.0 versus 2.8, corresponding to the Doll's theory:

$$\gamma_a = \gamma_1 + \frac{1}{2\alpha}(\gamma_2 - \gamma_1) \qquad (9.34)$$

The influence of the internal skin effect manifests itself to an even greater extent if higher frequencies, for example 60 kHz and above, are used. Also, apparent conductivity curves show that with an increase of the probe length the influence of the skin effect becomes stronger. As mentioned previously, this is related to the increased sensitivity of the field to the remote parts of the medium. To minimize interpretation error, it is highly advisable to use exact numerical solutions, which accurately take into account the skin effect in the surrounding medium. Analysis of the field in a medium with cylindrical interfaces (borehole, invasion zone, formation) shows that the skin effect in a radial direction also has to be taken into account. However, the impact of the skin effect on the radial responses is usually less pronounced than that on the vertical responses. In theory, we can use such a high frequency that the influence of induced currents in the surrounding medium is practically negligible and thus detection of resistive beds ($\gamma_1 \ll \gamma_2$), having relatively small thickness, ($L \approx H$) will be improved. However, the intent to eliminate the influence of the surrounding medium may require frequencies of dozens of megahertz deteriorating a radial response of the probe, especially when the invasion zone presents and has intermediate resistivity ρ_2 between the borehole and formation ($\rho_1 < \rho_2 < \rho_3$). Also, high frequencies increase the influence of the borehole and the dielectric properties of formation. These issues essentially reduce attractiveness of very high frequencies for reduction of the influence of the surrounding medium on γ_a. But an increase of frequency within certain limits, when the radial response practically does not change, can significantly improve characteristics of the vertical response of the probe. It is not a coincidence that all modern systems of induction logging use frequencies that are much higher than 20 kHz.

Now let us consider the main features of the apparent conductivity γ_a/γ_1 when the bed is thinner than the probe located in the middle of the bed (Fig. 9.5). As was shown earlier, in this case the field does not depend on the position of the bed inside the probe, and it can be presented in the form:

$$b_z^* = 2 \int_0^\infty \frac{m^3 m_2 e^{-(m_2 - m_1)\alpha} e^{-m_1}}{(m_1 + m_2)^2 (1 - K_{12}^2 e^{-2\alpha m_1})} dm, \quad \alpha \leq 1$$

Comparing the curves of apparent conductivity for both thick ($\alpha > 1$) and thin ($\alpha < 1$) beds, we conclude that in the latter case, the low-frequency asymptote takes place for larger values of the parameter L/δ_1 ($\gamma_2 < \gamma_1$). Asymptotic representation for the function γ_a/γ_1, as $L/\delta_1 \to 0$, can be derived in the same manner as Eq. (9.34). Omitting intermediate manipulations we receive:

$$\frac{\gamma_a}{\gamma_1} = \frac{\gamma_2}{\gamma_1} - \left(\frac{\gamma_2}{\gamma_1} - 1\right)\frac{\alpha}{2} \qquad (9.35)$$

Of course, this equation coincides with the one derived using the geometric factor (Chapter 6). By analogy, it is a simple matter to obtain an expression for the apparent conductivity, which is valid for larger values of the parameter L/δ_1 (Chapter 7).

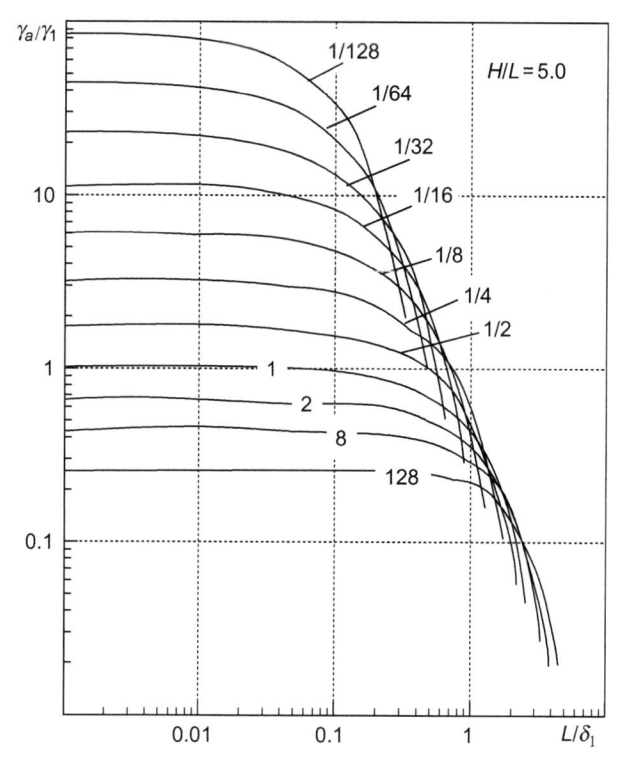

Fig. 9.5 Apparent conductivity curves. Index of curves is γ_1/γ_2.

Assuming a 90 degrees phase shift in the induced currents inside the bed, and no interaction between them and currents in the surrounding medium, we have:

$$\frac{\gamma_a}{\gamma_1} = \frac{\gamma_a^{un}}{\gamma_1} - \left(\frac{\gamma_2}{\gamma_1} - 1\right)\frac{\alpha}{2} \qquad (9.36)$$

where γ_a^{un} is an apparent conductivity in a uniform medium with conductivity γ_2. Obviously, within Doll's domain this value coincides with γ_2. Eq. (9.36) is valid for larger values of parameter L/δ_1 than Eq. (9.35), and this fact becomes more noticeable for relatively resistive beds. Analysis of apparent conductivities shows that thin beds with resistivity greater than that of the surrounding medium are hardly noticeable when the parameter L/δ_1 is small. For example, if $\alpha \leq 0.3$ and $\gamma_1/\gamma_2 \leq 1/8$, the influence of the bed does not exceed 5%–10%. On the contrary, the presence of thin conductive layers is more pronounced. For example, for small values of L/δ_1, when $\alpha \approx 0.3$ and $\gamma_1/\gamma_2 = 8$, the influence of the bed reaches 50%.

9.4 PROFILING CURVES FOR A TWO-COIL PROBE IN A BED OF FINITE THICKNESS

As was shown in the first section, the signal and the apparent conductivity depend on
- the ratio between the probe length, L, and the skin depth in a surrounding medium L/δ_2.
- the ratio between the bed thickness and the probe length: H/L.
- the ratio between conductivity of the bed and the surrounding medium γ_1/γ_2.
- the position of the probe with respect to the bed, which can be characterized by the distance between the middle of the bed and the center of the probe. (Needless to say, the measured electromotive force depends on the moments of the transmitter and receiver, as well as on the probe length and frequency). The formulas for calculation of the fields were derived earlier. We again use the following definition for the apparent conductivity:

$$\frac{\gamma_a}{\gamma_1} = \frac{2}{\gamma_1 \mu_0 \omega L^2} Qb_z^*$$

Let us consider the influence of the main factors, mentioned above, on the shape of the profiling curves corresponding to certain values of γ_1/γ_2 and H/L. In the analysis it is advisable to distinguish four typical cases.

Case I: Conductive bed

Profiling curves for the layers with thickness $H/L = 4, 2, 1$ are presented in Fig. 9.6A–C. The index of curves is the parameter $n_2 = \gamma_2 \mu_0 \omega L^2$. Along the x-axis we depict the apparent conductivity. The y-axis depicts the distance between the middle of the bed and center of the probe, expressed in units of the bed thickness. The curves are symmetrical with respect to the middle of the bed. In the case of the thick layer $H/L = 4$ and frequency, corresponding to Doll's region limit ($n_2 = 0.01$), the apparent conductivity readings are only 20% below the true value. The deviation from the true value increases as the layer becomes thinner (Fig. 9.6B) and it reaches 50% in the case of $H/L = 1$. In accordance with Doll's theory, the ratio of γ_a/γ_1 corresponding to the bed interface and its middle point is:

$$\eta = 0.5 \cdot \frac{1 - 1/4\alpha}{1 - 1/2\alpha}, \quad \alpha = H/L \geq 1$$

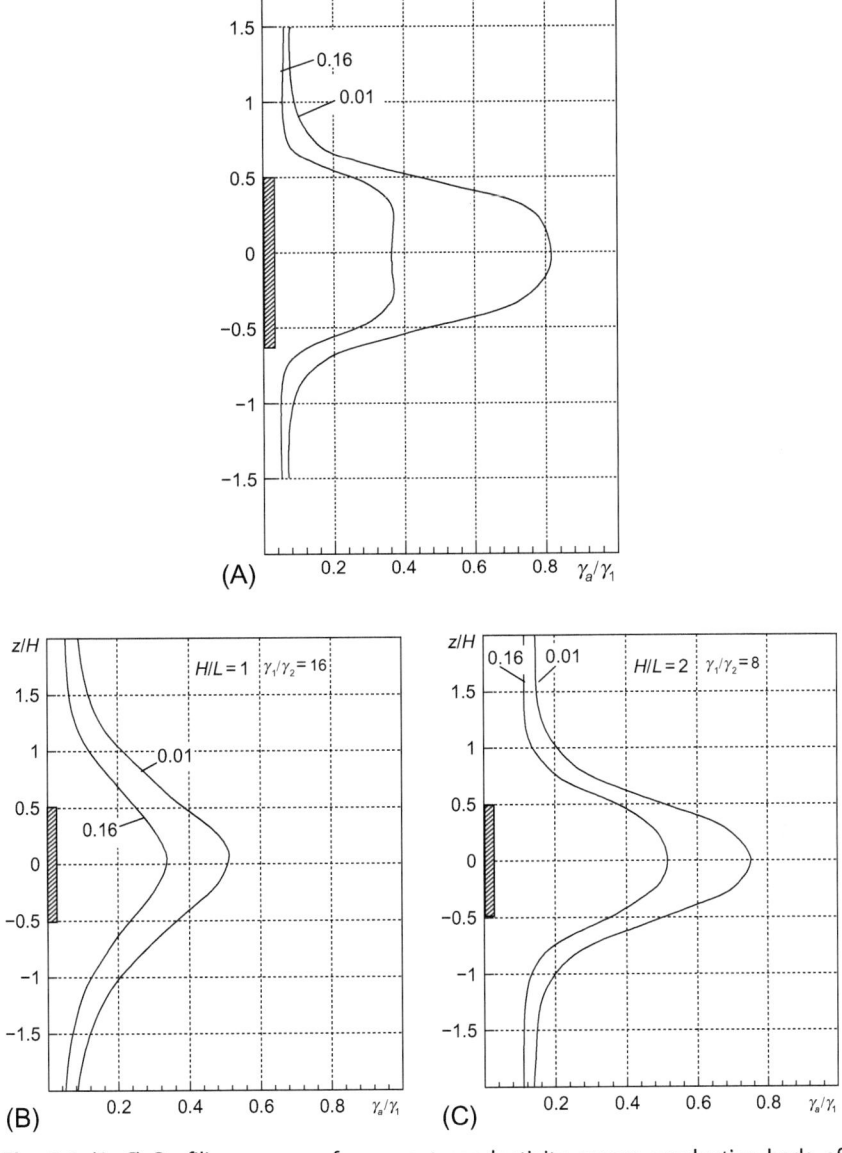

Fig. 9.6 (A–C) Profiling curves of apparent conductivity across conductive beds of different thicknesses. Index of curves is $\gamma_2\mu\omega L^2$.

According to this formula, only for the relatively thick layers the parameter η approaches the value of 0.5. In such cases the bed thickness can be determined by using points of the profiling curves, corresponding to half of the maximal value. For example, an error in H does not exceed 3% if $\alpha = 4.0$ (Fig. 9.6A), but it increases to 10% for $\alpha = 2.0$ (Fig. 9.6B) and 80% for $\alpha = 1.0$ (Fig. 9.6C). With an increase of frequency, the apparent conductivity readings γ_a/γ_1 at the middle of the bed experience strong skin effect and significantly deviate from the true conductivity of the bed. Furthermore, the width of an intermediate zone, where apparent conductivity γ_a differs from a uniform medium with conductivity γ_2, becomes narrower and the readings are closer to the true conductivity γ_2.

Case II: Thick resistive bed

Typical profiling curves for the thick $H/L = 4$ bed are presented in Fig. 9.7. With an increase of parameter n_2, the width of the intermediate zone decreases. Unlike the previous case, an increase of the frequency leads to a better detection of the resistivity of the bed. In the example from Fig. 9.7, apparent conductivities γ_a/γ_1 against the bed practically approach the true conductivity, while outside of the bed the readings become lower due to the strong skin effect. In particular, when $(n_2 = 0.64)$, $\gamma_1/\gamma_2 = 1/16$ and $H/L = 4$, the readings outside of the bed γ_a/γ_1 are 1.6 times lower compared to the true value, Fig. 9.7.

Case III: Thin conductive bed

Examples of profiling curves, corresponding to this case, are presented in Fig. 9.8A and B. Similarly to the first case, when parameter $n_2 = 0.01$ is in the Doll's range, the readings outside of the bed correspond to the true conductivity of the surrounding formation. In the middle of the profiling curves, readings are largely off the true conductivity values (Fig. 9.8A). With increase of the frequency ($n_2 = 0.32$), the skin effect becomes pronounced, reducing apparent conductivity along the entire profiling curves (Fig. 9.8A) and making determination of the thin bed impossible.

The situation improves with a gradual increase of the layer thickness to $H/L \geq 0.5$.

Case IV: Thin resistive bed

The profiling curves, shown in Fig. 9.9, behave similarly to the curves corresponding to the case of a thick resistive layer (second case). It is clear that skin effect is pronounced even at the relatively low frequency ($n_2 = 0.01$), leading to the lower apparent conductivity inside and outside the bed. An increase in the frequency ($n_2 = 0.64$) shifts the ratio γ_a/γ_1 closer

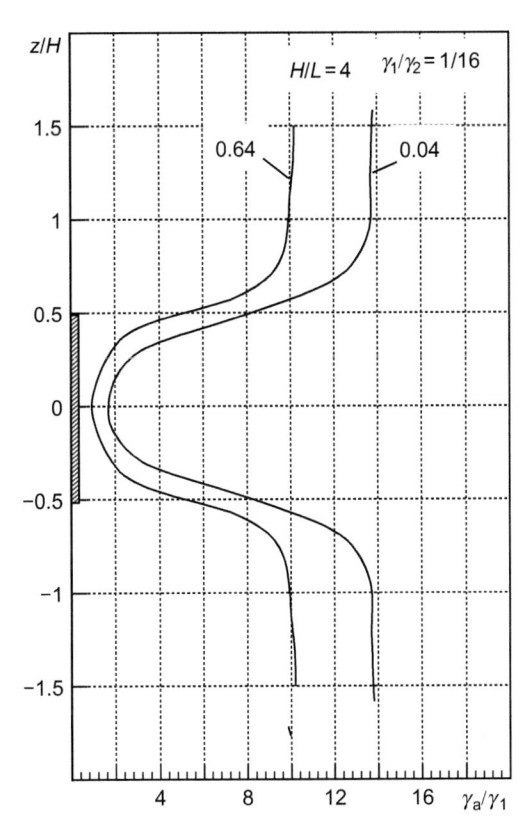

Fig. 9.7 Profiling curves across thick resistive bed. Index of curves is $\gamma_2\mu\omega L^2$.

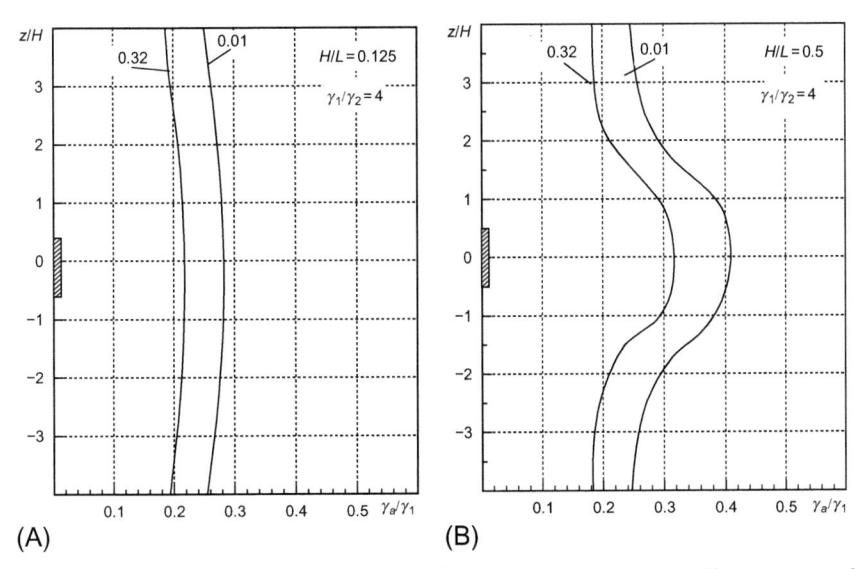

Fig. 9.8 (A) and (B) Profiling curves across thin conductive layer for different ratios of H/L. Index of curves is $\gamma_2\mu\omega L^2$.

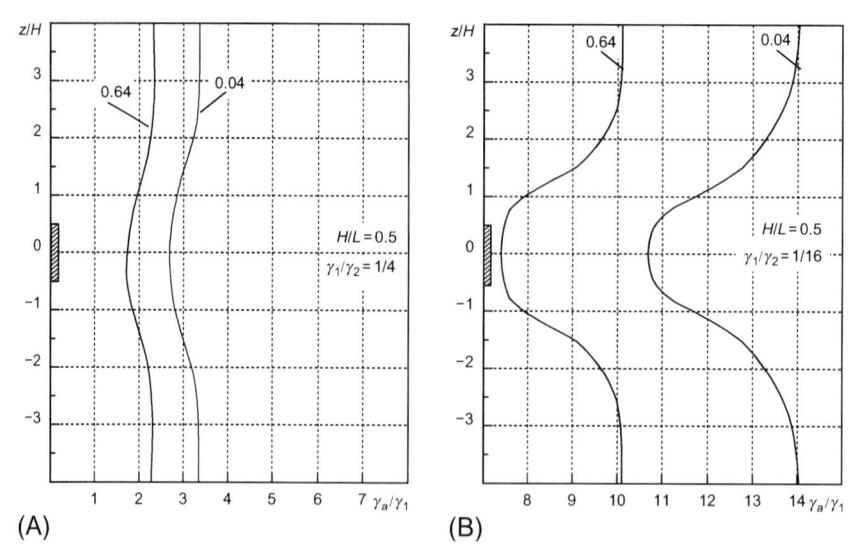

Fig. 9.9 (A) and (B) Profiling curves across thin resistive layer for different ratios of γ_1/γ_2. Index of curves is $\gamma_2\mu\omega L^2$.

to unity, although the value of γ_a still significantly deviates from γ_1. For example, when $\gamma_1/\gamma_2 = 16$ and $H/L = 0.5$, the apparent conductivity in the middle of the bed is more than seven times larger than the true value.

Determination of the thickness of such thin resistive beds using points, corresponding to half of the maximal value of γ_a/γ_1, is practically impossible.

REFERENCE

[1] Kaufman AA. Theory of induction logging. Moscow: Nedra; 1965.

CHAPTER TEN

Induction Logging Based on Transient EM Measurements

Contents

10.1 Transient Field of the Magnetic Dipole in a Uniform Medium	312
10.1.1 Expressions for the Field	313
10.1.2 Main Features of the Transient Field	314
10.2 Transient Field of the Magnetic Dipole in a Medium With Cylindrical Interfaces	321
10.2.1 Fourier Integral and Calculation of the Transient Field	321
10.2.2 The Early and Late Stage of Magnetic Field on the Borehole Axis	325
10.2.3 Apparent Resistivity Curves of the Transient Signals in a Medium With Cylindrical Interfaces	333
10.3 Transient Field of the Vertical Magnetic Dipole in a Medium With Horizontal Boundaries	337
10.3.1 Transient Field in a Medium With One Horizontal Boundary	337
10.3.2 Transient Field of the Vertical Magnetic Dipole Inside a Layer of Finite Thickness	340
10.4 Transient Field in Application to Deep-Reading Measurements While Drilling	343
10.4.1 Normal Field of the Current Ring in a Uniform Conducting Medium	344
10.4.2 Boundary Value Problem in the Presence of an Ideally Conductive Cylinder	347
10.4.3 Influence on the Finite Conductivity of the Cylinder	352
10.4.4 Effect of Spacing on the Pipe Signal	356
10.4.5 Effect of the Increased Pipe Conductivity on the Transient Response	359
10.4.6 Reduction of the Pipe Signal Using Finite Size Copper Shield and Bucking	361
10.4.7 Improving Formation/Pipe Signals Ratio Using Magnetic Shielding	364
10.5 Inversion of Transient Data in the Task of Geo-Steering	365
10.5.1 Well- and Ill-Posed Problems	366
10.5.2 Main Elements of the Inversion Algorithm	367

10.5.3 Table-Based Inversion — 368
10.5.4 Stability of the Inverse Problem Solution — 370
10.5.5 Multiparametric Inversion — 375
10.5.6 Estimation of Parameter Uncertainties — 380
References — 383
Further Reading — 383

In 1963, mathematician P.P. Frolov (Moscow), investigating the transient electromagnetic field on the earth's surface, showed that measuring the transient field with relatively small separation between the transmitter and receiver reveals information about the distribution of resistivity beneath the earth's surface. This was not an expected result because in surface and borehole geophysics, experts strongly believed that only an increased separation between transmitter and receiver permits increased depth of investigation. Soon after his publication, a new method of surface geophysics called transient soundings in the near zone was developed and found broad application in Russia and globally. The possibility of studying resistivity of the formation around the borehole with a short two-coil probe triggered an interest in transient electromagnetic measurements within the logging industry, resulting in several publications and development of the transient induction logging theory in the 1970s. Like the induction phenomena in a frequency domain, the main features of the transient field can be studied by analyzing the field in a simple model of a uniform medium.

10.1 TRANSIENT FIELD OF THE MAGNETIC DIPOLE IN A UNIFORM MEDIUM

Suppose that the constant current in the small loop (magnetic dipole) vanishes instantly following a shape of a step function:

$$I = \begin{cases} I_0 & \text{if } t \leq 0 \\ 0 & \text{if } t > 0 \end{cases} \quad (10.1)$$

In contrast to a general case discussed in Chapter 4, we focus our attention here on the quasi-stationary field when displacement currents are disregarded and the propagation speed is infinitely large. Under these assumptions, the field instantly appears at any point of a medium regardless of the distance from the source. In reality, this corresponds to the time of observation greatly exceeding the time needed for the field to arrive at the observation point. We proceed from the expression for the vector potential A_z^* for the quasi-stationary field assuming $k^2 \approx i\omega\gamma\mu_0$:

$$A_z^* = \frac{i\omega\mu_0 M_0}{4\pi} \frac{\exp(ikR)}{R}$$

Fourier's transform of the last expression gives the following for the time domain:

$$A_z(t) = \frac{\mu_0 M_0}{4\pi(2\pi)^{1/2} R} \frac{u}{t} \exp(-u^2/2) \tag{10.2}$$

and

$$u = \frac{2\pi R}{\tau}, \quad \tau = (2\pi\rho t \times 10^7)^{1/2} \tag{10.3}$$

The same result can be derived from a general expression for the vector potential (Eq. 4.61), letting ε, and therefore the time of arrival of the wave, τ_0, to be zero.

10.1.1 Expressions for the Field

Taking into account Eq. (10.2) and the known relations between the vector potential and field components:

$$\mathbf{E}^* = \operatorname{curl} \mathbf{A}^*$$

we obtain the following expressions for the components of the electromagnetic field in the time domain:

$$B_R = \frac{2\mu_0 M_0}{4\pi R^3} b_R \cos\theta = \frac{2\mu_0 M_0}{4\pi R^3}\left[\Phi(u) - \left(\frac{2}{\pi}\right)^{1/2} u \exp(-u^2/2)\right]\cos\theta$$

$$B_\theta = \frac{\mu_0 M_0}{4\pi R^3} b_\theta \sin\theta = \frac{\mu_0 M_0}{4\pi R^3}\left[\Phi(u) - \left(\frac{2}{\pi}\right)^{1/2} u(1+u^2)\exp(-u^2/2)\right]\sin\theta$$

$$E_\phi = \frac{M_0 \rho}{4\pi R^4} e_\phi \sin\theta = \left(\frac{2}{\pi}\right)^{1/2} \frac{M_0 \rho}{4\pi R^4} u^5 \exp(-u^2/2)\sin\theta$$

$$\tag{10.4}$$

where

$$\Phi(u) = \left(\frac{2}{\pi}\right)^{1/2} \int_0^u \exp(-x^2/2)\,dx$$

is the probability integral.

10.1.2 Main Features of the Transient Field

To illustrate the main features of the field components (Eq. 10.4) we present numerical data in Table 10.1, containing values of b_R, b_θ, and e_ϕ as functions of the parameter u.

The corresponding curves are shown in Fig. 10.1.

First, let us study the transient response at the early transient stage ($t \to 0$) after the current in the source is switched off. In this case, $u \to \infty$ and the function $\Phi(u)$ tends to unity, so that we have

$$B_R = \frac{2\mu_0 M_0}{4\pi R^3}\cos\theta, \quad B_\theta = \frac{\mu_0 M_0}{4\pi R^3}\sin\theta, \quad E_\phi = 0 \qquad (10.5)$$

As predicted by Faraday's law, at the early stage, the induced currents arise initially near the source and attempt to maintain magnetic field unchanged. From Eq. (4.61), which takes into account displacement currents, it follows that at the initial moment the field is absent in all parts of a medium. But in accordance with equations for the quasi-stationary field, the field propagates instantaneously and has a finite value, even when $t < \tau_0$, while in fact it equals to zero. To derive the late stage of the transient field, we should expand the probability integral in a series of a small parameter u. Performing an expansion, we have

$$\Phi(u) \approx \left(\frac{2}{\pi}\right)^{1/2}\left(u - \frac{u^3}{6} + \frac{u^5}{40} - \cdots\right) \qquad (10.6)$$

Substituting this series into Eq. (10.4), we obtain approximate formulas for components of the secondary field:

Table 10.1 Field Components as Functions of the Parameter u

u	R/τ	$1-b_R$	$1-b_\theta$	e_ϕ	b_R	b_θ
0.0500	0.796E−02	0.3300E−04	−0.6661E−04	0.249E−06	1.0000	1.000
0.0595	0.946E−02	0.5561E−04	−0.1118E−03	0.592E−06	0.9999	1.000
0.0707	0.113E−01	0.9364E−04	−0.1878E−03	0.140E−05	0.9999	1.000
0.0841	0.0134	0.1575E−03	−0.3152E−03	0.334E−05	0.9998	1.000
0.100	0.0159	0.2649E−03	−0.5290E−03	0.793E−05	0.9997	1.001
0.119	0.0189	0.4452E−03	−0.8873E−03	0.188E−04	0.9996	1.001
0.141	0.0225	0.7476E−03	−0.1487E−02	0.446E−04	0.9993	1.001
0.168	0.0268	0.1254E−02	−0.2488E−02	0.1058E−03	0.9987	1.002
0.238	0.0379	0.3518E−02	−0.6917E−02	0.590E−03	0.9965	1.007
0.283	0.0450	0.8760E−02	−0.1147E−01	0.138E−02	0.9941	1.011
0.336	0.0535	0.9780E−02	−0.1891E−01	0.324E−02	0.9902	1.019
0.400	0.0630	0.1623E−01	−0.3091E−01	0.754E−02	0.9838	1.031
0.476	0.0770	0.2676E−01	−0.4993E−01	0.173E−01	0.9732	1.050
0.556	0.0900	0.4378E−01	−0.7930E−01	0.393E−01	0.9562	1.080
0.673	0.1070	0.7081E−01	−0.1229	0.877E−01	0.9292	1.123
0.80	0.1270	0.1128	−0.1839	0.1899	0.8872	1.184
0.951	0.1514	0.1758	−0.2612	0.3955	0.8242	1.261
1.13	0.1801	0.2661	−0.3432	0.7799	0.7339	1.343
1.35	0.2141	0.3873	−0.3988	1.423	0.6127	1.399
1.60	0.2560	0.5355	−0.3732	2.326	0.4645	1.373
1.90	0.3028	0.6945	−0.2048	3.256	0.3055	1.205

Continued

Table 10.1 Field Components as Functions of the Parameter u—cont'd

u	R/τ	$1-b_R$	$1-b_\theta$	e_ϕ	b_R	b_θ
2.263	0.3601	0.8368	0.1222	3.659	0.1632	0.878
2.691	0.4283	0.9354	0.5192	3.014	0.0646	0.481
3.20	0.5093	0.9834	0.8271	1.600	0.0166	0.173
3.805	0.6057	0.9977	0.9662	0.457	0.232E−02	0.338E−01
4.525	0.7203	0.9999	0.9972	0.541E−01	0.135E−03	0.278E−02
5.382	0.8650	0.9999	0.9999	0.185E−02	0.229E−05	0.662E−04
6.400	1.019	1.0000	1.0000	0.109E−04	0.673E−08	0.273E−06
7.611	1.211	1.0000	1.0000	0.537E−08	0.160E−11	0.944E−10
9.051	1.441	1.0000	1.0000	0.788E−13	0.117E−16	0.974E−15

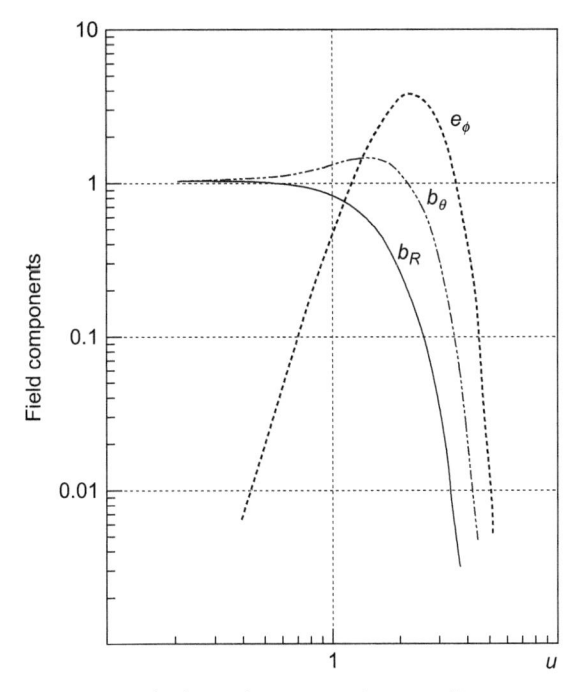

Induction Logging Based on Transient EM Measurements

Fig. 10.1 Field components $b_R, b_\theta,$ and e_ϕ in a uniform medium.

$$B_R \approx -\frac{\mu_0 M_0}{6\pi R^3} \left(\frac{2}{\pi}\right)^{1/2} u^3 \left(1 - \frac{3}{10}u^2\right) \cos\theta$$

$$B_\theta \approx \frac{\mu_0 M_0}{6\pi R^3} \left(\frac{2}{\pi}\right)^{1/2} u^3 \left(1 - \frac{3}{5}u^2\right) \sin\theta \quad (10.7)$$

$$E_\phi \approx -\frac{M_0 \rho}{4\pi R^4} \left(\frac{2}{\pi}\right)^{1/2} u^5 \left(1 - \frac{u^2}{2}\right) \sin\theta$$

By keeping only the first term in Eq. (10.7), we receive

$$B_R \approx -\frac{\mu_0 M_0}{6\pi R^3} \left(\frac{2}{\pi}\right)^{1/2} u^3 \cos\theta = -\frac{\mu_0 M_0}{12\pi(\pi)^{1/2}} \frac{\mu_0^{3/2} \gamma^{3/2}}{t^{3/2}} \cos\theta$$

$$B_\theta \approx \frac{\mu_0 M_0}{6\pi R^3} \left(\frac{2}{\pi}\right)^{1/2} u^3 \sin\theta = \frac{\mu_0 M_0}{12\pi(\pi)^{1/2}} \frac{\mu_0^{3/2} \gamma^{3/2}}{t^{3/2}} \sin\theta \quad (10.8)$$

$$E_\phi \approx -\left(\frac{2}{\pi}\right)^{1/2} \frac{M_0 \rho}{4\pi R^4} u^5 \sin\theta = -\frac{M_0}{16\pi(\pi)^{1/2}} \frac{\mu_0^{5/2} \gamma^{3/2}}{t^{5/2}}$$

These expressions describe the field at a late stage with acceptable accuracy when $u < 0.2$. During this time interval, the transient field does not depend on the separation between the source and observation point, exhibiting a stronger dependency on conductivity $\propto \gamma^{3/2}$ than the quadrature component, which is typically measured in a frequency regime. For illustration, some values of parameter u as a function of resistivity ρ and time t, if $R = 1$ m, are given in Table 10.2.

Table 10.2 shows that the value of $u = 0.2$, corresponding to the late stage, is reached quite soon at $t = 16\,\mu s$, even in the conductive medium with resistivity of 1 ohm m. The independence of the late stage field from the distance between transmitter and receiver suggests that the sources of the secondary field are located at distances from an observation point significantly larger than the probe length R. Now, consider the behavior of the current density. As follows from Eq. (10.4), the current density in a whole space is

$$j_\phi = -\left(\frac{2}{\pi}\right)^{1/2} \frac{M_0 \sin\theta}{4\pi\, R^4} u^5 \exp\left(-u^2/2\right) \tag{10.9}$$

Graphs of function

$$F = \frac{1}{R^4} u^5 \exp\left(-u^2/2\right)$$

are shown in Fig. 10.2. As time increases, the maximum of the current density shifts toward the deeper part of the medium. For this reason, electromagnetic fields on the axis of the transmitting dipole, become more sensitive over time to the remote parts of the medium.

Let us confirm this assumption through the following consideration. We mentally represent whole uniform space as a system of concentric spherical shells. At any moment, a measured magnetic field is defined by the distribution of currents in the shells. By applying Biot-Savart's law and omitting intermediate transformations related to the calculation of the magnetic field, we may find the ratio between the electromotive force, caused by the currents in shells with the radius larger than R_2, and the electromotive force in the coil, located at the distance R_1:

$$G(u_1, \alpha) = \left(1 - \frac{1}{3}u_1^2\right) \exp\left[-u_2^2(\alpha^2 - 1)/2\right]$$

where $\alpha = R_2/R_1$ and $u_1 = 2\pi R_1/\tau$. Curves $G(u_1, \alpha)$ are shown in Fig. 10.3.

Table 10.2 Parameter u for Different Values of Resistivity and Time

ρ (ohm m)	$t=1$ μs	$t=4$ μs	$t=9$ μs	$t=16$ μs	$t=25$ μs	$t=36$ μs	$t=49$ μs	$t=64$ μs	$t=81$ μs	$t=100$ μs
0.1	2.50	1.25	0.84	0.63	0.50	0.42	0.36	0.31	0.28	0.25
0.5	1.11	0.56	0.7	0.28	0.22	0.19	0.16	0.14	0.12	0.11
1.0	0.80	0.40	0.27	0.20	0.16	0.13	0.11	0.10	0.09	0.08
5.0	0.35	0.18	0.12	0.088	0.071	0.059	0.051	0.044	0.039	0.035
10.0	0.25	0.125	0.084	0.063	0.050	0.042	0.036	0.01	0.028	0.025

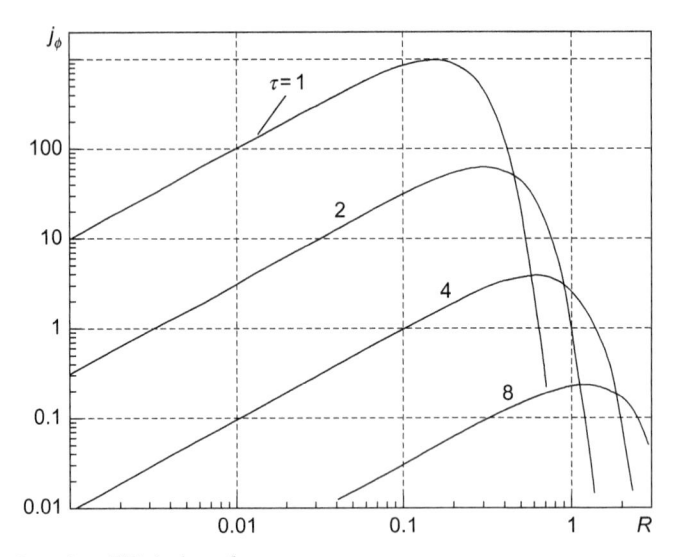

Fig. 10.2 Function F(R). Index of curves τ.

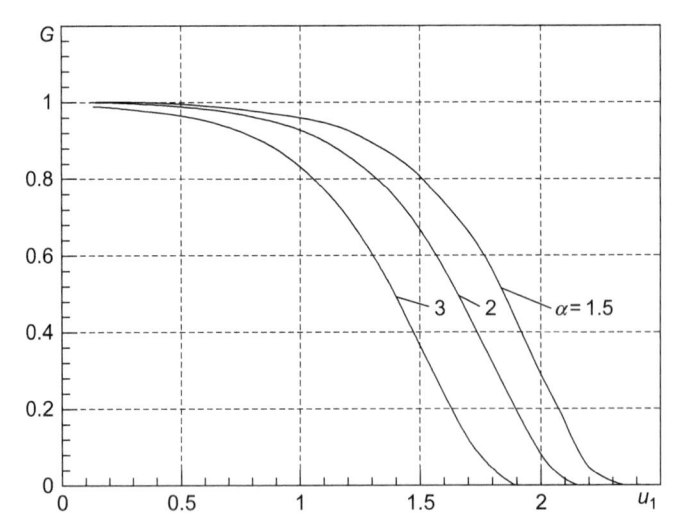

Fig. 10.3 Function $G(u_1, \alpha)$ illustrates concentration of currents in an external area at the late stage. Index of curves α.

At the earlier times, currents are concentrated mainly near the dipole, and the field, measured at point R_1, does not practically depend on induced currents located in remote parts of a medium ($u_1 \to \infty$, $G \to 0$). By contrast, for the late stage ($u_1 \to 0$), the field is mainly defined by currents induced in an external area ($R > R_2$) and $G(u_1, \alpha) \to 1$. Thus the measurements

Table 10.3 Electromotive Force (μV)

ρ (ohm m)	$t=1$ μs	$t=4$ μs	$t=9$ μs	$t=16$ μs	$t=25$ μs
0.1	0.543E+04	0.179E+04	0.365E+03	0.101E+03	0.355E+02
0.5	0.600E+04	0.300E+03	0.431E+02	0.106E+02	0.351E+01
1.0	0.290E+04	0.115E+03	0.158E+02	0.381E+01	0.126E+01
5.0	0.334E+03	0.109E+02	0.145E+01	0.346E+00	0.113E+00
10.0	0.122E+03	0.390E+01	0.515E+00	0.122E+00	0.402E−01

performed at the late stage are defined by currents of the remote parts of a medium, providing increased depth of investigation. Such behavior can also be expected in a nonuniform medium. As the depth of investigation increases radially, the sensitivity of the probe to the parts of formation located above and below the probe also increases, at, of course, the expense of the reduced signal level.

Let us estimate the signal level at different moments of time in the 1 m long two-coil probe. We assume an effective transmitter-receiver moment of $M_T M_R = 0.1$ Am^4. Calculated electromotive force Ξ values are presented in Table 10.3.

The values of Ξ are calculated using the following equations:

$$\Xi = \frac{M_T M_R}{2\pi L^3} \rho e_\phi \quad \text{and} \quad e_\phi = \left(\frac{2}{\pi}\right)^{1/2} u^5 \exp\left(-u^2/2\right)$$

The data in Table 10.3 demonstrate asymptotic behavior $\Xi \propto 1/t^{5/2}$ of the signal at the late stage.

10.2 TRANSIENT FIELD OF THE MAGNETIC DIPOLE IN A MEDIUM WITH CYLINDRICAL INTERFACES

10.2.1 Fourier Integral and Calculation of the Transient Field

In Chapter 8, we studied frequency responses of a magnetic dipole in the presence of cylindrical interfaces. Here, we use those results to study the transient field, proceeding from the Fourier integral:

$$F(t) = \frac{1}{2\pi} \int_{-\infty}^{\infty} F^*(\omega) \exp(-i\omega t) d\omega$$

and

$$F^*(\omega) = \int_0^\infty F(t)\exp(i\omega t)dt \qquad (10.10)$$

where $F(t)$ some component of the transient is field, and $F^*(\omega)$ is the product of the complex amplitudes of the field and the spectrum of excitation. When the primary magnetic field of the dipole is varying as the step function:

$$B_0(t) = B_0 \text{ if } t < 0 \quad \text{and} \quad B_0(t) = 0 \text{ if } t > 0$$

the excitation spectrum is defined by Eq. (10.10):

$$F_0(\omega) = \frac{1}{i\omega} \qquad (10.11)$$

Harmonic amplitudes (Eq. 10.11) have the same phase and decrease inversely with the frequency. Because the low-frequency harmonics dominate in the spectrum of the step function, this type of excitation is an efficient way of delivering energy to the remote parts of a medium. As follows from Eq. (10.10), the primary magnetic field can be written as:

$$B(t) = \frac{B_0}{2\pi} \int_{-\infty}^{\infty} \frac{1}{i\omega} \exp(-i\omega t) d\omega \qquad (10.12)$$

where the integration path does not include the point $\omega = 0$. Let us present the integral (Eq. 10.12) as a sum:

$$\frac{1}{2\pi i}\int_{-\infty}^{\infty} \frac{\exp(-i\omega t)}{\omega} d\omega = \frac{1}{2\pi i}\int_{-\infty}^{-\varepsilon} \frac{\exp(-i\omega t)}{\omega} d\omega$$

$$+ \frac{1}{2\pi i}\int_{-\varepsilon}^{+\varepsilon} \frac{\exp(-i\omega t)}{\omega} d\omega + \frac{1}{2\pi i}\int_{+\varepsilon}^{\infty} \frac{\exp(-i\omega t)}{\omega} d\omega$$

We choose a semicircular path of integration surrounding the origin $\omega = 0$, and let the radius of the semicircle tend to zero. It is convenient to introduce a new variable ϕ:

$$\omega = \rho \exp(i\phi)$$

Induction Logging Based on Transient EM Measurements

that gives

$$d\omega = i\rho \exp(i\phi) d\phi$$

and for the second integral we have

$$\frac{1}{2\pi i} \int_{-\varepsilon}^{+\varepsilon} \frac{\exp(i\omega t)}{\omega} d\omega = \frac{1}{2\pi i} \int_{\pi}^{2\pi} \frac{i\rho \exp(i\phi)}{\rho \exp(i\phi)} d\phi = \frac{1}{2}$$

Thus the expression for the primary field in which the variable of integration ω takes only real values is

$$B(t) = \frac{B_{\omega=0}}{2} + \frac{B_0}{2\pi i} \int_{-\infty}^{\infty} \frac{\exp(-i\omega t)}{\omega} d\omega \tag{10.13}$$

Correspondingly, for the secondary transient field caused by the step function we have

$$B(t) = \frac{B^*_{\omega=0}}{2} + \frac{1}{2\pi i} \int_{-\infty}^{\infty} \frac{B^*(\omega)}{\omega} \exp(-i\omega t) d\omega \tag{10.14}$$

Here,

$$B^*(\omega) = \operatorname{Re} B^*(\omega) + i \operatorname{Im} B^*(\omega)$$

is the complex amplitude of the field. The expression (10.14) is convenient for numerical calculations because it is carried out along the real values of frequencies. Let us write Eq. (10.14) in the form:

$$B(t) = \frac{B^*_{\omega=0}}{2} + \frac{1}{2\pi} \int_{-\infty}^{\infty} \frac{\operatorname{Im} B^*(\omega) \cos \omega t - \operatorname{Re} B^*(\omega) \sin \omega t}{\omega} d\omega -$$

$$\frac{i}{2\pi} \int_{-\infty}^{\infty} \frac{\operatorname{Im} B^*(\omega) \sin \omega t + \operatorname{Re} B^*(\omega) \cos \omega t}{\omega} d\omega \tag{10.15}$$

As follows from Eq. (10.10)

$$\operatorname{Re} B^*(\omega) = \operatorname{Re} B^*(-\omega) \quad \text{and} \quad \operatorname{Im} B^*(\omega) = -\operatorname{Im} B^*(-\omega)$$

Correspondingly, the second integral in Eq. (10.15) is equal to zero, and because the integrand is an even function of the frequency, we further receive

$$B(t) = \frac{B^*_{\omega=0}}{2} + \frac{1}{\pi}\int_0^\infty \frac{\operatorname{Im} B^*(\omega)\cos\omega t - \operatorname{Re} B^*(\omega)\sin\omega t}{\omega}d\omega \qquad (10.16)$$

Next, we make one more simplification. Because the current source changes as a step function, and there is no secondary field at $t<0$ ($B(t) = B_0$ at $t<0$), Eq. (10.16) yields

$$0 = -\frac{B^*_{\omega=0}}{2} + \frac{1}{\pi}\int_0^\infty \frac{\operatorname{Im} B^*(\omega)\cos\omega t + \operatorname{Im} B^*(\omega)\sin\omega t}{\omega}d\omega \qquad (10.17)$$

From the last two expressions, we obtain

$$B(t) = \frac{2}{\pi}\int_0^\infty \frac{\operatorname{Im} B^*(\omega)}{\omega}\cos\omega t\, d\omega \quad \text{and}$$

$$B(t) = B_{(\omega=0)} - \frac{2}{\pi}\int_0^\infty \frac{\operatorname{Re} B^*(\omega)}{\omega}\sin\omega t\, d\omega \qquad (10.18)$$

For the time derivatives we have

$$\frac{\partial B}{\partial t} = -\frac{2}{\pi}\int_0^\infty \operatorname{Im} B^*(\omega)\sin\omega t\, d\omega \quad \text{and}$$

$$\frac{\partial B}{\partial t} = -\frac{2}{\pi}\int_0^\infty \operatorname{Re} B^*(\omega)\cos\omega t\, d\omega \qquad (10.19)$$

Because the primary electric field is zero, the corresponding transformations for the electric field take the form:

$$E_\phi(t) = \frac{2}{\pi}\int_0^\infty \frac{\operatorname{Im} E^*_\phi}{\omega}\cos\omega t\, d\omega \quad \text{and} \quad E_\phi(t) = -\frac{2}{\pi}\int_0^\infty \frac{\operatorname{Re} E^*_\phi}{\omega}\sin\omega t\, d\omega \qquad (10.20)$$

The last set of equations (Eqs. 10.18–10.20) is used to calculate the transient field from the frequency responses. Next, we analyze some important asymptotic features of the transient field at the early and late transient stage.

10.2.2 The Early and Late Stage of Magnetic Field on the Borehole Axis

As was mentioned, at the beginning of the transient process the internal skin effect leads to appearance of induced currents mainly in the vicinity of the source, and thus the measured magnetic field contains information about the conductivity of the borehole only. With time the diffusion manifests itself and induced currents appear in the surrounding medium. To investigate the asymptotic behavior of the magnetic field caused by these currents, we proceed from the first equation of (10.18) assuming that the parameter t is large. First, we introduce a new notation:

$$\phi_1(\omega) = \frac{\operatorname{Im} B^*(\omega)}{\omega}$$

and it gives

$$B(t) = \frac{2}{\pi} \int_0^\infty \phi_1(\omega) \cos \omega t \, d\omega \qquad (10.21)$$

Assuming that the value of t is large, and performing integration by parts, we obtain

$$\begin{aligned} B(t) &= \frac{2}{\pi} \left[\frac{\phi_1 \sin \omega t}{\omega} \bigg|_0^\infty - \frac{1}{t} \int_0^\infty \phi_1'(\omega) \sin \omega t \, d\omega \right] \\ &= \frac{\phi_1 \sin \omega t}{\omega} \bigg|_0^\infty + \frac{1}{t^2} \phi_1' \cos \omega t \bigg|_0^\infty - \frac{1}{t^2} \int_0^\infty \phi_1''(\omega) \cos \omega t \, d\omega \end{aligned} \qquad (10.22)$$

By continuing integration by parts, we can obtain the following terms of this expansion. At first glance, calculation of these terms requires knowledge of spectrum at the high frequencies. However, because the integrands in Eq. (10.21) contain rapidly oscillating functions at $t \to \infty$, the value of the integral is mainly defined by the initial part of the integration corresponding to the low-frequency range of the spectrum. Thus such an approach does not require function ϕ_1 at the high frequencies, and the obtained series (Eq. 10.22) is suitable for derivation of an asymptote at the late stage. This asymptote is controlled by the low frequency of the spectrum and its derivatives with respect to the frequency; the intermediate and high-frequency parts of the spectrum have practically no control over the late stage of the transient field. If the derivative is a dominant factor, then the field behaves quite differently from that at the low frequency; this case will be discussed later. In the same manner, we can obtain a series expansion using a real component of the spectrum:

$$\phi_2 = \frac{\mathrm{Re}\, B^*}{\omega}$$

As was shown in Chapter 8, the low-frequency spectrum for any component can be presented as a sum:

$$\sum_{n=1}^{\infty} c_1^{(n)} k^{2n} + \sum_{n=1}^{\infty} c_2^{(n)} k^{2n+1} + \sum_{n=1}^{\infty} c_3^{(n)} k^{2n} \ln k \qquad (10.23)$$

Here, $k = (i\gamma\mu_0\omega)^{1/2}$ and c_1, c_2, c_3 are coefficients depending on the geoelectric parameters, distance, and the moment of the dipole source. Note that the first sum of Eq. (10.23)

$$\sum_{n=1}^{\infty} c_1^{(n)} k^{2n}$$

has no effect on the late stage of the transient field. In fact let us rewrite this term as a sum of the real and imaginary parts:

$$\sum_{n=1}^{\infty} c_1^{(n)} k^{2n} = \sum_{n=1}^{\infty} a_1^{(n)} \omega^{2n} + i \sum_{n=1}^{\infty} b_1^{(n)} \omega^{2n-1} \qquad (10.24)$$

For the Fourier transform of Eq. (10.24), we obtain two types of integrals, namely

$$L_n = \int_0^{\infty} \omega^{2n-1} \sin \omega t\, d\omega \quad \text{and} \quad M_n = \int_0^{\infty} \omega^{2n-2} \cos \omega t\, d\omega \qquad (10.25)$$

which are the limiting cases ($\beta \to 0$ and $t \to \infty$) of more general integrals:

$$L_n = \lim \int_0^{\infty} \omega^{2n-1} \exp(-\beta\omega) \sin \omega t\, d\omega$$

$$M_n = \int_0^{\infty} \omega^{2n-2} \exp(-\beta\omega) \cos \omega t\, d\omega \qquad (10.26)$$

Presentation (Eq. 10.26) is valid because introduction of the exponential term $\exp(-\beta\omega)$ does not change the initial part of the integration, which defines the integrals (Eq. 10.26) at t tending to infinity. These integrals are expressed through elementary functions, and they approach zero when β tends to zero. Thus we have shown that the first sum in Eq. (10.23), which contains only the integer power of ω, makes no contribution to the late stage of the transient field. At this stage, only fractional powers of ω and logarithmic terms determine the transient response. This fact plays a fundamental role in understanding the relationship between the frequency and time domain responses of the field. For example, the quadrature component at the low frequency is controlled by the leading linear term of ω, while the following terms, containing fractional powers of ω and $\ln\omega$, have a negligible effect. However, these less significant terms affect the behavior of the transient field at the late stage, making it difficult to establish an intuitive one-to-one relationship between the time and frequency responses. Specifically, the first linear term in the series for the quadrature component essentially differs from the rest of the terms by not contributing at all into the late stage of the transient signal. By contrast, the leading term in the series expansion for the in-phase component of the secondary field contains either a fractional power of ω or $\ln\omega$. For this reason, we may expect that the behavior of this component of the secondary magnetic field at the low frequency is practically the same as that of the transient field during the late stage. Indeed, such similarity is observed in a uniform whole space and in more complex media as well.

Next, using the second and third sum on the right-hand side of Eq. (10.23), we may determine the series that describes the late stage. The second sum can be written in the form:

$$\sum_{n=1}^{\infty} c_2^{(n)} k^{2n+1} = \sum_{n=1}^{\infty} a_2^{(n)} \omega^{(2n+1)/2} + i \sum_{n=1}^{\infty} b_2^{(n)} \omega^{(2n+1)/2}$$

For calculation of the series in the time domain, we can use either the first or the second sum of the last expression. For example, let us use the in-phase component of the field:

$$\sum_{n=1}^{\infty} a_2^{(n)} \omega^{(2n+1)/2} = a_2^{(1)} \omega^{3/2} + a_2^{(2)} \omega^{5/2} + \cdots$$

Substituting this sum into the Fourier integral (Eq. 10.18) we obtain

$$-\frac{2}{\pi}\sum_{n=1}^{\infty} a_2^{(n)} \int_0^{\infty} \omega^{n-1/2} \sin \omega t \, d\omega \qquad (10.27)$$

Since we are mainly concerned with the behavior of the integral at $t \to \infty$, we need to consider only the initial part of the integration path. Letting $n=1$ we have

$$I_1 = \int_0^{\infty} \omega^{1/2} \sin \omega t \, d\omega$$

Integrating I_1 by parts and discarding the high-frequency portion of the spectrum, we obtain

$$I_1 = -\frac{1}{t}\int_0^{\infty} \omega^{1/2} d\cos \omega t = -\frac{1}{t}\left[\omega^{1/2} \cos \omega t \Big|_0^{\infty} - \frac{1}{2}\int_0^{\infty} \frac{\cos \omega t}{\omega^{1/2}} d\omega\right]$$

$$= \frac{1}{2t}\int_0^{\infty} \frac{\cos \omega t}{\omega^{1/2}} d\omega = \frac{1}{2t^{3/2}}\int_0^{\infty} \frac{\cos x}{x^{1/2}} dx$$

The last integral is well defined:

$$\int_0^{\infty} \frac{\cos x}{x^{1/2}} dx = \left(\frac{\pi}{2}\right)^{1/2}$$

Thus

$$I_1 = \frac{1}{2}\left(\frac{\pi}{2}\right)^{1/2} \frac{1}{t^{3/2}} \qquad (10.28)$$

For $n=2$ we have

$$I_2 = \int_0^{\infty} \omega^{3/2} \sin \omega t \, d\omega$$

Integrating twice by parts, we obtain

$$I_2 = -\frac{1}{t}\left[\left(\omega^{3/2}\cos\omega t\right)\Big|_0^\infty - \frac{3}{2}\int_0^\infty \omega^{1/2}\cos\omega t\, d\omega\right] = \frac{3}{2t}\int_0^\infty \omega^{1/2}\cos\omega t\, d\omega$$

$$= \frac{3}{2t^2}\int_0^\infty \omega^{1/2}\, d\sin\omega t\, d\omega = \frac{3}{2t^2}\omega^{1/2}\sin\omega t\Big|_0^\infty - \frac{1}{2}\int_0^\infty \frac{\sin\omega t}{\omega^{1/2}}d\omega = -\frac{3}{4t^{5/2}}\int_0^\infty \frac{\sin x}{x^{1/2}}dx$$

In as much as

$$\int_0^\infty \frac{\sin x}{x^{1/2}}dx = \left(\frac{\pi}{2}\right)^{1/2}$$

we have

$$I_2 = -\frac{3}{4}\left(\frac{\pi}{2}\right)^{1/2}\frac{1}{t^{5/2}} \tag{10.29}$$

Using the same approach, any term in the sum (Eq. 10.27) can be calculated. We can see that a term proportional to $\omega^{3/2}$ generates a term in the time domain proportional to $t^{-3/2}$. Therefore, the portion of the spectrum described by the sum:

$$\sum a_2^{(n)}\omega^{n+1/2}$$

is responsible for the appearance of a sum

$$\sum \tilde{a}_2^{(n)}\frac{1}{t^{n+1/2}} \tag{10.30}$$

in the expression for the late stage of a transient field.

The third term in Eq. (10.23) can be written as:

$$\sum_{n=1}^\infty c_3^{(n)}k^{2n}\ln k = \ln k \sum_{n=1}^\infty c_3^{(n)}k^{2n} = \ln\left[(\gamma\mu_0\omega)^{1/2}\exp\left(i\frac{\pi}{4}\right)\right]$$

$$\sum_{n=1}^\infty c_3^{(n)}(\gamma\mu_0\omega)^n \exp\left(i\frac{\pi}{2}\right)n = \left[\ln(\gamma\mu_0\omega)^{1/2} + i\frac{\pi}{4}\right]$$

$$\left[\sum_{n=1}^\infty c_3^{(n)}(\gamma\mu_0\omega)^n \cos\frac{\pi}{2}n + i\sum_{n=1}^\infty c_3^{(n)}(\gamma\mu_0\omega)^n \sin\frac{\pi}{2}n\right]$$

Letting $n=2p$ and $n=2p-1$ in the first and second sums, respectively, and taking into account that

$$\cos \pi p = (-1)^p \text{ and } \sin \frac{2p-1}{2}\pi = (-1)^{p-1}$$

we receive the following expressions for the real and imaginary parts of the third sum in Eq. (10.23):

$$\left[\frac{1}{2}\ln(\gamma\mu_0\omega)\sum_{p=1}^{\infty}(-1)^p c_p(\gamma\mu_0\omega)^{2p} - \frac{\pi}{4}\sum_{p=1}^{\infty}(-1)^{p-1}c_p(\gamma\mu_0\omega)^{2p-1}\right] + \\ i\left[\frac{\pi}{4}\sum_{p=1}^{\infty}(-1)^p c_p(\gamma\mu_0\omega)^{2p} + \frac{1}{2}\ln(\gamma\mu_0\omega)\sum_{p=1}^{\infty}(-1)^{p-1}c_p(\gamma\mu_0\omega)^{2p-1}\right] \quad (10.31)$$

Substituting the real part of the last equation into the Fourier transform, we obtain two types of integrals:

$$A_p = \int_0^{\infty} \omega^{2p-2}\sin\omega t\, d\omega \quad \text{and} \quad B_p = \int_0^{\infty}(\omega^{2p-1}\ln\omega)\sin\omega t\, d\omega \quad (10.32)$$

For example, when $p=1$ we have

$$A_1 = \int_0^{\infty}\sin\omega t\, d\omega = \lim\int_0^{\infty}\exp(-\beta\omega)\sin\omega t = \frac{1}{t} \text{ if } \beta\to 0 \text{ and } t\to\infty$$

and

$$B_1 = \int_0^{\infty}(\omega\ln\omega)\sin\omega t\, d\omega = \int_0^{\infty}F(\omega)\sin\omega t\, d\omega$$

where $F(\omega) = \omega\ln\omega$.

Integrating by parts, we obtain

$$B_1 = -\frac{1}{t}\int_0^{\infty}F(\omega)d\cos\omega t = -\frac{1}{t}\left[F(\omega)\cos\omega t\Big|_0^{\infty} - \int_0^{\infty}F'(\omega)\cos\omega t\, d\omega\right]$$

$$= -\frac{1}{t}\left[F(\omega)\cos\omega t\Big|_0^{\infty} - \frac{1}{t}\int_0^{\infty}F'(\omega)d\sin\omega t\right]$$

$$= -\frac{1}{t}F(\omega)\cos\omega t\Big|_0^{\infty} + \frac{1}{t^2}F'(\omega)\sin\omega t\Big|_0^{\infty} - \frac{1}{t^2}\int_0^{\infty}F''(\omega)\sin\omega t\, d\omega$$

Because

$$F'(\omega) = 1 + \ln\omega \quad \text{and} \quad F''(\omega) = \frac{1}{\omega}$$

we have

$$B_1 = -\frac{1}{t^2}\int_0^\infty \frac{\sin\omega t}{\omega}d\omega = -\frac{\pi}{2}\frac{1}{t^2}$$

Similarly, we can derive integrals A_p and B_p for any values of p. It is readily seen that the portion of the low-frequency spectrum described by the last sum in Eq. (10.23) gives a rise to the sum of terms proportional to $1/t^n$ in the representation for the late stage of the transient field:

$$\sum_{n=1}^\infty \tilde{a}_3^{(n)} \frac{1}{t^n} \tag{10.33}$$

Therefore, as follows from Eqs. (10.30), (10.33), the late stage of the transient electric and magnetic fields of the magnetic dipole in a conducting medium can be presented in the form [1]:

$$\sum_{n=1}^\infty \tilde{a}_2^{(n)} \frac{1}{t^{n+1/2}} + \sum_{n=1}^\infty \tilde{a}_3^{(n)} \frac{1}{t^n} \tag{10.34}$$

In Chapter 8, we showed that the part of the low-frequency spectrum that does not contain even powers of k is

$$f_3 k_1^3 + f_5 k_1^5 + f_7 k_1^7 + l_7 k_1^7 \ln k_1 + \cdots \tag{10.35}$$

When the invasion zone is absent, we have

$$f_3 = \frac{\alpha^3 s^{3/2}}{3}, \quad f_5 = f_3\left(\frac{\alpha^2 s}{10} - \frac{1-s}{2}\right)$$

$$f_7 = f_3\left[\frac{\alpha^4 s^2}{280} - \frac{\alpha^2 s(1-s)}{20} + \frac{5}{32}(1-s)^2 - \frac{s(1-s)}{10}\left(C - \frac{77}{66} - \frac{\ln s}{2}\right)\right],$$

$$l_7 = -f_3\frac{s}{10}(1-s)$$

(10.36)

Here $\alpha = L/a_1$, $s = \gamma_2/\gamma_1$.

In the case of two cylindrical interfaces for the first two terms ϕ_3 and ϕ_5, corresponding to the low-frequency spectrum, we have

$$\phi_3 = \frac{1}{3}\alpha^3 s_1^{3/2}, \quad \phi_5 = \phi_3 \left(\frac{\alpha^2 s_1}{10} - \frac{s_{12}}{2} \right) \tag{10.37}$$

Here,

$$s_1 = \frac{\gamma_3}{\gamma_1}, \quad s_2 = \frac{\gamma_2}{\gamma_1}, \quad s_{12} = 1 - s_2 + (s_2 - s_1)\beta^2, \quad \beta = \frac{a_2}{a_1} \tag{10.38}$$

Now, using the procedure described earlier, we can present the field and its derivative at the late stage in a form similar to Eq. (10.34). First, consider the leading term of this sum:

$$B_z \approx \frac{\mu_0 M_0}{12\pi(\pi)^{1/2}} \frac{\mu_0^{3/2} \gamma_3^{3/2}}{t^{3/2}} \quad \text{and} \quad \frac{\partial B_z}{\partial t} \approx -\frac{\mu_0 M_0}{8\pi(\pi)^{1/2}} \frac{\mu_0^{3/2} \gamma_3^{3/2}}{t^{5/2}} \tag{10.39}$$

The latter coincides exactly with expressions for the field in a uniform medium with resistivity of the deepest part of the formation. In other words, the field does not depend on the resistivity and radius of either the borehole or the invasion zone; and such behavior occurs regardless of the probe length. In principle, the transient induction probe may consist of one coil only. Although the possibility of using a single-coil probe sounds very attractive, there are some serious technical challenges to overcome in implementing this approach, such as large signal dynamic range, ultra-fast current switch, etc.

The approximate expression for $\partial B_z/\partial t$, which takes into account the first two terms of the series, has the form:

$$\frac{\partial B_z}{\partial t} \approx -\left(\frac{2}{\pi}\right)^{1/2} \frac{M_0 \rho_1}{2\pi L^3 a_1^3} \left(\frac{2\pi}{\tau_1/a_1}\right)^5 \left[3\phi_3 - \frac{15}{2}\phi_5 \frac{8\pi^2}{(\tau_1/a_1)^2} \right] \tag{10.40}$$

Here ρ_1 and a_1 are resistivity and radius of a borehole, respectively; L is the probe length; and functions ϕ_3 and ϕ_5 are given by Eqs. (10.37), (10.38). Comparison with the exact solution for the three-layered models shows that the asymptotic Eq. (10.40) describes the field with accuracy sufficient for practical needs, if $\tau_1/a_1 > 20$ or

$$t > \frac{a_1^2}{\rho_1} 10^{-5} \; [\text{s}]$$

For example, for $\rho_1 = 3$ ohmm and $a_1 = 0.1$ m, the late stage occurs quite early, at $0.03\,\mu s$. In other words, the time range that contains information about the borehole is very limited.

10.2.3 Apparent Resistivity Curves of the Transient Signals in a Medium With Cylindrical Interfaces

We present results of calculations of the field, $\partial B_z/\partial t$, in the form of apparent resistivity defined as:

$$\frac{\rho_\tau}{\rho_1} = \left(\frac{\dot{B}_z^{ls}(t)}{\dot{B}_z(t)}\right)^{2/3} \tag{10.41}$$

where ρ_τ and ρ_1 are apparent resistivity and borehole resistivity, respectively; $\dot{B}_z^{ls}(\rho_1, t)$ is the time derivative of function $B_z(\rho_1, t)$ at the late stage in a uniform medium with resistivity of the borehole; and $\dot{B}_z = \partial B_z/\partial t$ is the signal observed on the borehole axis. As was shown earlier:

$$\dot{B}_z^{ls}(\rho_1, t) = \frac{1}{a_1^5}\left(\frac{2}{\pi}\right)^{1/2}\frac{M_0\rho_1}{2\pi\,\alpha^5}u_1^5 \tag{10.42}$$

Because

$$u_1 = 2\pi\alpha\frac{a_1}{\tau_1}, \quad \alpha = L/a_1, \quad \tau_1 = (2\pi\rho_1 t \times 10^7)^{1/2},$$

we have

$$\frac{\rho_\tau}{\rho_1} = \frac{8\pi^3}{\tau_1^3}\left(\frac{\pi}{\tau_1}\right)^{1/3}\left(\frac{M_0\rho_1}{t\dot{B}_z(t)}\right)^{2/3} \tag{10.43}$$

or

$$\rho_\tau = \frac{\mu_o}{4\pi t}\left(\frac{\mu_0 M_0}{t\dot{B}_z(t)}\right)^{2/3} = K\left(\dot{B}_z(t)\right)^{-2/3} \tag{10.44}$$

An advantage of introducing apparent resistivity according to Eq. (10.44) is independence of the probe coefficient K from the resistivity of a medium. Examples of apparent resistivity curves for the two-layered media, when $\rho_2/\rho_1 = 64$ and the invasion zone is absent, are given in Fig. 10.4. The code is parameter $\alpha = L/a_1$.

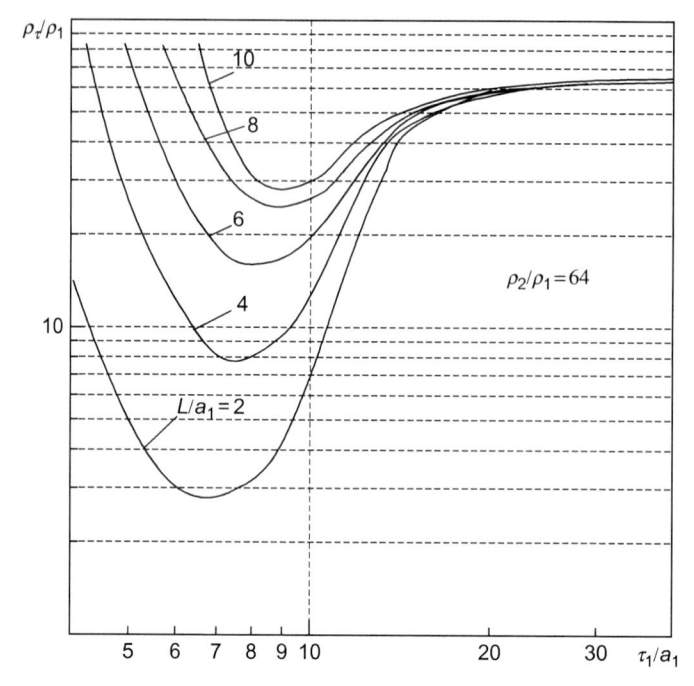

Fig. 10.4 Apparent resistivity on the borehole axis (invasion zone is absent).

All calculations are performed for the relatively long probes, exceeding the diameter of the borehole, $\alpha > 2$. For this reason, even at the early stage, the transient field does not tend to that in a uniform medium with resistivity of the borehole. With a decrease of time, a value of ρ_τ increases infinitely, due to the field at the early stage being much smaller than that, calculated using the formula for the late stage. The shape of the curves essentially depends on the probe length and conductivity of the medium. With an increase of time, these curves display a minimum, which becomes deeper with a decrease of the probe length and an increase of the formation resistivity. Then, with an increase of time, ρ_τ rapidly increases and approaches the right-hand asymptote equal to the formation resistivity. Within this range of time, the smaller the probe length and larger resistivity of the formation, the earlier the time when the main contribution into the measurements comes from the currents induced in the far-located region away from the borehole and the probe. At the same time, the density of these currents still depends on the borehole resistivity. The larger the resistivity of the external area, the more rapidly a transient field and induced currents decay near the probe. Correspondingly, the influence of the probe length on the field reduces at earlier times. The second term in Eq. (10.35) has a form:

$$f_5 = f_3 \left(\frac{\alpha^2 s}{10} - \frac{1-s}{2} \right)$$

For sufficiently large formation resistivity and relatively small probe length, when conditions

$$\alpha^2 s < 1 \quad \text{and} \quad s < 1$$

are met, the second term of the asymptotic (Eq. 10.35) is independent of formation resistivity and the probe length, and it is mainly defined by the resistivity of the borehole. Thus if parameters of the borehole are known, it is possible to correct ρ_τ for the effect of the borehole at the time range when the field differs from that in a uniform medium with resistivity ρ_2. As follows from Eqs. (10.40), (10.41) the corrected apparent resistivity is

$$\frac{\rho_\tau}{\rho_1} \approx \frac{\rho_2}{\rho_1} \left(1 - \frac{5}{3} u_1^2 \right) \qquad (10.45)$$

Here,

$$u_1 = \frac{2\pi a_1}{\tau_1}$$

The second term in Eq. (10.45), which defines a correction of ρ_τ at the late stage, is directly proportional to the conductance of the borehole $\pi a^2 \gamma_1$. Behavior of the function in Eq. (10.45) is shown in Fig. 10.5. From comparison with the exact solution, it follows that Eq. (10.45) provides sufficient accuracy in determination of ρ_2 for a relatively resistive formation, $\rho_2/\rho_1 > 10$, if $\alpha = L/a_1 < 4$ and $\tau_1/a_1 > 15$.

A comparison of the exact solution and asymptotic formula shows that the field in a two-layer medium becomes practically the same as in a uniform whole space with resistivity of formation ρ_2 if

$$\tau_1/a_1 > 30 \quad \text{or} \quad t_{\mu\text{sec}} > 90 a_1^2 / 2\pi \rho_1 \qquad (10.46)$$

Apparent Resistivity Curves in the Presence of an Invasion Zone

Examples of apparent resistivity curves in this case are shown in Fig. 10.6. The data presented are for two sets of the models $\frac{\rho_2}{\rho_1} - \frac{a_2}{a_1} - \frac{\rho_3}{\rho_1}$.

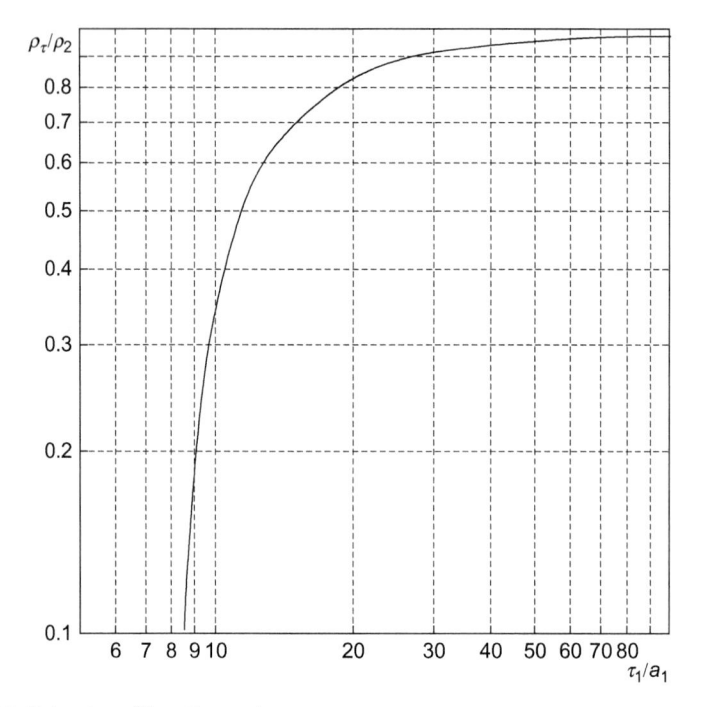

Fig. 10.5 Behavior of function ρ_τ/ρ_2.

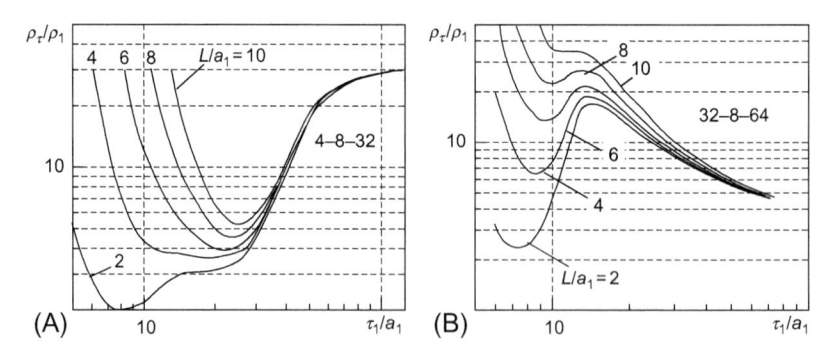

Fig. 10.6 Apparent resistivity curves in the presence of invasion zone for two sets of $\dfrac{\rho_2}{\rho_1} - \dfrac{a_2}{a_1} - \dfrac{\rho_3}{\rho_1}$. The code is $\alpha = L/a_1$.

If the resistivity of the invasion is less than that of the formation, the shape of the ρ_τ curves is the same as that of the case of two layers, but the curves are approaching the right asymptote at later times. In the case of $\rho_2/\rho_3 > 1$, the change in the shape of ρ_τ is noticeable for short probes

Induction Logging Based on Transient EM Measurements

and large radii of the invasion zone: at the early stage ρ_τ increases with time, reaches some maximum defined by the resistivity of the invasion zone, ρ_2, and then, when currents are mainly located in the conductive formation, asymptotically decreases approaching its right-hand asymptote, ρ_3.

10.3 TRANSIENT FIELD OF THE VERTICAL MAGNETIC DIPOLE IN A MEDIUM WITH HORIZONTAL BOUNDARIES

The study of the transient field in a medium with cylindrical boundaries enabled us to obtain information about radial responses of two-coil probes. As time increases, the influence on the field of a surrounding medium (shoulders), located above and beneath of the layer, becomes greater and this influence increases with increase of shoulders conductivity. It is important to establish the maximum time when measurements with a two-coil probe, located inside the layer of finite thickness, are practically independent of conductivity of the surrounding medium. With this purpose in mind, we consider the behavior of the transient field in a medium with one and two horizontal interfaces.

10.3.1 Transient Field in a Medium With One Horizontal Boundary

As we showed in Chapter 9, an expression for the harmonic field on the axis of the magnetic dipole is

$$B_z^{(1)*} = B_z^{un*}(k_1 z) + \frac{\mu_0 M_0}{4\pi} \int_0^\infty \frac{m^3}{m_1} m_{12} \exp\left[-(2\alpha - 1)m_1 z\right]dm \quad \text{if } \alpha \geq 1$$

$$B_z^{(2)*} = \frac{\mu_0 M_0}{4\pi} \int_0^\infty \frac{2m^3}{m_1 + m_2} \exp\left(-\alpha m_1 z\right) \exp\left[-(1 - \alpha) m_2 z\right]dm \quad \text{if } 0 \leq \alpha \leq 1$$

(10.47)

Here, k_1 and k_2 are wave numbers of the first and second medium, and the dipole is located in the first medium:

$$m_1 = \left(m^2 - k_1^2\right)^{1/2}, \quad m_2 = \left(m^2 - k_2^2\right)^{1/2}, \quad \text{and} \quad m_{12} = \frac{m_1 - m_2}{m_1 + m_2}$$

Also, z is the probe length, L is the vertical distance from the dipole to the boundary between two layers and

$$\alpha = \frac{L}{z}$$

Assuming that the current in the dipole changes as a step function:

$$I = I_0 \text{ if } t < 0 \text{ and } I = 0 \text{ if } t > 0$$

for the transient field $\dot{B}_z(t)$ we have

$$\dot{B}_z(t) = -\frac{2}{\pi} \int_0^\infty \operatorname{Im} B_z^*(\omega) \sin \omega \, d\omega$$

Applying this formula and omitting rather simple algebra, we derive the following asymptotic expression for the late stage:

$$\frac{\partial B_z}{\partial t} \approx -\frac{M_0 \rho_1}{\pi z^5} \left(\frac{2}{\pi}\right)^{1/2} \frac{u_1^5}{s-1} \left[\frac{s^{1/2}-1}{5} - \left(\frac{\pi}{2}\right)^{1/2} \frac{(2\alpha-1)(s+1)(s-1)^2}{4} \right]$$

(10.48)

Here, $s = \frac{\gamma_2}{\gamma_1}$, $u_1 = \frac{2\pi z}{\tau_1}$, $\tau_1 = (2\pi \rho_1 t \times 10^7)^{1/2}$.

Examples of the apparent resistivity curves, calculated using Eq. (10.47), are shown in Fig. 10.7. Apparent resistivity is related with the field as

$$\frac{\rho_a}{\rho_1} = \left(\frac{\dot{B}_z^{un}}{\dot{B}_z}\right)^{2/3}$$

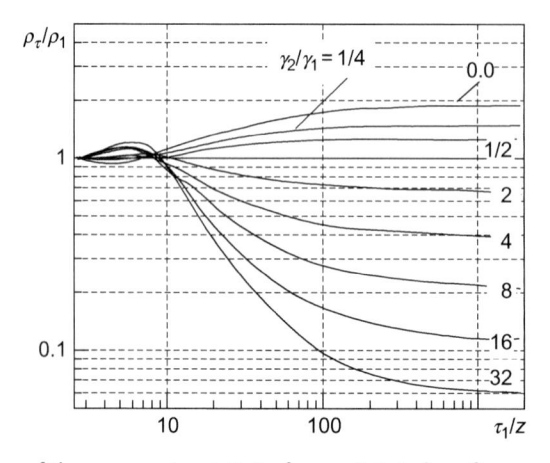

Fig. 10.7 Curves of the apparent resistivity for $\alpha = 1.2$. Index of curves is $s = \gamma_2/\gamma_1$.

where

$$\dot{B}_z^{un} = \frac{M_0 \rho_1}{2\pi z^5} \left(\frac{2}{\pi}\right)^{1/2} u_1^5 \exp\left(-\frac{u_1^2}{2}\right)$$

is the field in a uniform medium with resistivity ρ_1.

For the small values of parameter τ_1/z, when both the source and observation point are located in the medium with resistivity ρ_1, $(z < L)$, induced currents are concentrated near the source and curves approach the same asymptote $\rho_a = \rho_1$ (Fig. 10.7). As time increases, the influence on the second medium becomes stronger. Moreover, the higher its conductivity, the earlier it manifests itself. The right asymptotes correspond to the late stage, which depends on the resistivity of both media. As we can expect, with an increase of time, the induced currents are located at distances greatly exceeding L and the field is practically independent of the distance between the probe and the boundary. Within the early stage, we observe an extremum (maximum if $\rho_2/\rho_1 < 1$ and minimum when $\rho_2/\rho_1 > 1$). Appearance of the extremum can be explained as follows: at the earlier times, the near-borehole currents produce their own magnetic field, which affects the currents at some distance from the dipole and, eventually, the field in the receiver. The effect is the most pronounced in the case of high contrast between conductivities of the medium. Now, consider the case when the dipole and observation point are located in the different media (Fig. 10.8).

The left asymptote tends to zero if $\rho_2/\rho_1 > 1$ and to infinity when the second medium is more conductive, $\rho_2/\rho_1 < 1$. At the early stage, the field,

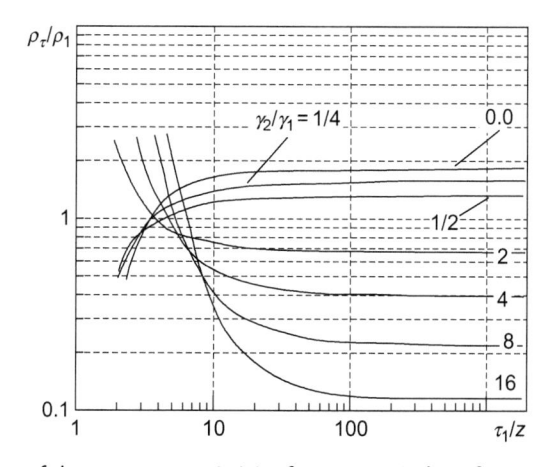

Fig. 10.8 Curves of the apparent resistivity for $\alpha = 0.6$. Index of curves is s.

measured in the second medium, depends on both resistivity near the source and resistivity in the vicinity of the observation point. By contrast, at the late stage, all curves, regardless of the position of an observation point, approach the horizontal asymptote, which depends on resistivity of both media. As follows from Eq. (10.48) for the asymptotic value of apparent resistivity, we have

$$\frac{\rho_\tau}{\rho_1} = \left(\frac{5}{2}\frac{s-1}{s^{5/2}-1}\right)^{2/3}$$

10.3.2 Transient Field of the Vertical Magnetic Dipole Inside a Layer of Finite Thickness

Suppose that the center of the two-coil probe is located in the middle of a layer and that the resistivity of a medium above and beneath the layer is the same. In this case an expression for the vertical component of the field, normalized by the primary field, $B_z^0 = \mu_0 M/2\pi L^3$, is equal to

$$b_z^* = b_z^{un*}(k_1 z) + \int_0^\infty \frac{m^3 m_{12}}{m_1} \exp(-2am_1) \frac{\exp(am_1) + m_{12} chm_1}{1 - m_{12}^2 \exp(-2am_1)} dm \quad (10.49)$$

Here, L and H are probe length and the layer thickness, respectively

$$m_1 = (m^2 - k_1^2)^{1/2}, \quad m_2 = (m^2 - k_2^2)^{1/2}, \quad m_{12} = \frac{m_1 - m_2}{m_1 + m_2}$$

k_1 and k_2 are the wave numbers of the layer and surrounding medium. Presentation of the field as a sum of two terms, (Eq. 10.49), is sufficient for the calculation of the frequency regime and becomes problematic for the calculation of the transient response at the late stage. The numerical problem is especially severed when resistivity of the surrounding medium becomes much larger than the resistivity of the layer: the two terms in Eq. (10.49) having opposite signs practically cancel each other and an accurate estimation of the spectrum becomes very time consuming. To overcome this problem it is advisable to modify Eq. (10.49) to the following form:

Induction Logging Based on Transient EM Measurements

$$b_z^* = b_z^{un*}(k_1 z) + \int_0^\infty \left[F - \frac{m^3 \exp(-\alpha m_1)}{2m_2} + \frac{m^3 \exp(-\alpha m_1)}{2m_1} \right] dm$$

$$+ \int_0^\infty \left[\frac{m^3 \exp(-\alpha m_1)}{2m_2} - \frac{m^3 \exp(-\alpha m_1)}{2m_1} \right] dm$$

To calculate a transient field, the Fourier transform is further applied to the spectrum. The major challenge here is the calculation of the signal at the late stage, which requires hardly achievable accuracy of calculations at low frequencies if Eq. (10.50) is applied literally. For this reason, we derive an asymptotic expression for the low frequency spectrum and use it further for the estimation of the late stage of the transient field. Applying Taylor's expansions in series by powers k^2 for all functions in the integrand Eq. (10.49) and performing analytical integration of the first several terms, for the case of the conducting surrounding medium we arrive to following the low-frequency asymptotic:

$$b_z^* \approx c_1 k_1^2 + c_2 k_1^3 + c_3 k_1^4 \ln 2k_2 + c_4 k_1^4 + c_5 k_1^5$$

where

$$c_2 = \frac{1}{3} s^{3/2}, \quad c_3 = \frac{\alpha s(s-1)}{4},$$

$$c_5 = -\frac{1}{15} \left[5\alpha^2 \left(s^{5/2} - \frac{7}{4} s^{3/2} + \frac{3}{4} s^{1/2} \right) - \left(\frac{5}{4} s^{3/2} - \frac{3}{4} s^{5/2} \right) \right]$$

and $s = \dfrac{\rho_1}{\rho_2}$, $\alpha = \dfrac{H}{z}$.

Coefficients of the first and fourth terms are not given because the latter contain even powers of frequency and, therefore, do not contribute into the late stage. Integration by parts of the Fourier integral gives for the late stage:

$$\frac{\partial B_z}{\partial t} = -\frac{M_0 \rho_1}{2\pi z^5} \left(\frac{2}{\pi} \right)^{1/2} u_1^5 \left\{ s^{3/2} - \left(\frac{\pi}{2} \right)^{1/2} 2\alpha u_1 s(s-1) \right.$$

$$\left. + u_1^2 \left[5\alpha^2 \left(s^{5/2} - \frac{7}{4} s^{3/2} + \frac{3}{4} s^{1/2} \right) - \left(\frac{5}{4} s^{3/2} - \frac{3}{4} s^{5/2} \right) \right] \right\}$$

(10.50)

Here

$$u_1 = \frac{2\pi z}{\tau_1}, \quad \tau_1 = (2\pi \rho_1 t \times 10^7)^{1/2}$$

The field at the late stage becomes practically equal to that in a uniform medium with resistivity of a surrounding medium independently on the probe length. Similarly, we may derive an asymptotic expression describing the late stage of the transient field in the presence of the non-conducting surrounding medium ($s=0$):

$$\frac{\partial B_z}{\partial t} = -\frac{3M_0\rho_1}{\pi z^5}\alpha^3 u_1^8 \left(1 - 8\alpha^2 u_1^2 - 4\alpha u_1^2\right) \tag{10.51}$$

In that case, the currents are uniformly distributed along the z-axis, and the field is directly proportional to the cube of the longitudinal conductance, $S = \gamma_1 H$. Results of calculation of the apparent resistivity when the two-coil probe is located symmetrically inside the layer are shown in Fig. 10.9A and B.

Each family of curves is characterized by the same parameter α. As we can see

1. At the early stage, when currents are concentrated near the source, the field only depends on the resistivity of the layer. Correspondingly, the left asymptote of curves ρ_τ/ρ_1 is equal to unity.
2. In the late stage, when $\tau_1/z \gg 1$, the currents are practically absent in the layer, and the curves approach the right asymptote equal ρ_2/ρ_1, ($\rho_2 \neq \infty$).
3. If $\rho_2/\rho_1 > 20$, then the curves of the apparent resistivity also have an intermediate asymptote, which corresponds to the case of a nonconductive surrounding medium. This asymptote occurs at a time interval

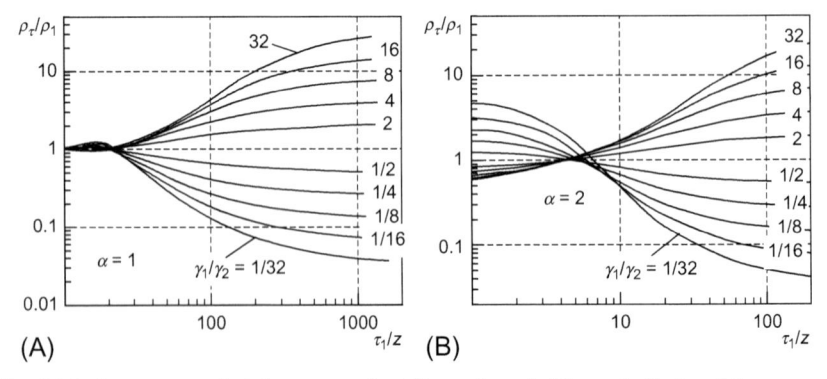

Fig. 10.9 Apparent resistivity curves for (A) $\alpha = 1$ and (B) $\alpha = 2$. Code of the curves is γ_1/γ_2.

in which currents have not yet penetrated the highly resistive surrounding medium, but inside the layer, they are distributed almost uniformly along the z-axis.
4. In the case of relatively high conductivity of the surrounding medium, the curves have a maximum even at relatively small times; such phenomenon was explained earlier.
5. Over time, the induced currents move away from an observation point and the influence on the probe length becomes very small. The calculation shows that for the given value of ρ_2/ρ_1, the field is practically defined by the parameter τ_1/H.
6. Apparent resistivity only slightly differs from the resistivity of the layer if the following conditions are met:

$$\frac{\tau_1}{H} < 6 \quad \text{or} \quad t_{ms}^{max} < 0.6 \frac{H^2}{\rho_1}$$

provided that $\alpha \geq 2$ and $16 \geq \dfrac{\rho_2}{\rho_1} \geq \dfrac{1}{16}$.

10.4 TRANSIENT FIELD IN APPLICATION TO DEEP-READING MEASUREMENTS WHILE DRILLING

During the last decade, the petroleum industry made significant progress toward developing deep-reading resistivity measurements while drilling (MWD). All major service companies rely on induction tools, which use a sinusoidal excitation source, to provide information about directional resistivity on a scale several times greater than conventional logging tools. Specifically, deep-reading tools developed by service companies such as Schlumberger, Ltd and Baker Hughes Incorporated identify resistivity contrasts at tens of meters away from the wellbore. The primary application of these tools is detection of up to 20 m away from the borehole of the oil-water contact and the reservoir faulting. Measurements are performed in the presence of conductive drill pipe, which creates a large induction signal in the receivers by diminishing sensitivity to the properties of the formation. To reduce contribution of the signal from the drill pipe while still providing greater depth of investigation, long three-coil systems (20–30 m) at frequencies between several to hundreds of kilohertz are used. The measurements are inverted to obtain distances to boundaries, resistivity of the reservoir, and the resistivity of the beds above and below the penetrated layer.

In many geo-steering scenarios, it is also desirable to detect the presence of a formation anomaly ahead of the bottom hole assembly. Traditional frequency-based measurements have limited potential to accomplish these tasks. The limitation is dictated by a controversy between requirements of having deep-reading capabilities and the necessity of increasing tool length. Indeed, deep-reading induction tools require long transmitter-receiver spacing, which immediately reduces capabilities of detecting anomalies ahead of the bottom hole assembly.

In the following paragraphs we explore an alternative approach aimed to resolve this controversy by using relatively short systems (approximately 7 m) based on transient electromagnetic measurements. Specifically, our focus is on a deep-reading transient system that is capable of looking ahead of the drill bit. We show how the effect of the drill pipe in a short system might be reduced to preserve sensitivity of the measured signals to the properties of the formation ahead.

The asymptotic formulas, describing both frequency and transient responses of the field, are very useful for understanding how to suppress signal from the drill pipe. Unfortunately, deriving them for logging-while-drilling (LWD) measurements is extremely difficult because the corresponding forward problem becomes two-dimensional and can only be solved by applying advanced numerical techniques. At the same time, solutions for some idealized models are still available, making it possible to study field characteristics at frequency and time limits that are deemed important. To simplify the study, we can assume that the drill pipe is a cylinder with a constant radius a, whereas transmitter and receiver coils have the same radius r_k, which slightly exceeds the radius of the cylinder. Such an approximation enables us to apply the method of separation of variables and derive formulas for the field at distances exceeding the radius of the cylinder. First, we study an electric field of the current ring placed in a uniform formation and then consider the field of this ring symmetrically placed around a conductive cylinder.

10.4.1 Normal Field of the Current Ring in a Uniform Conducting Medium

As in the case of the magnetic dipole, the vector potential of the electric type $\mathbf{A}$, caused by the current element Idl, is equal to:

$$d\mathbf{A}^* = \frac{\mu_0 Idl}{4\pi} \frac{\exp(ikR)}{R} \mathbf{i} \qquad (10.52)$$

Induction Logging Based on Transient EM Measurements

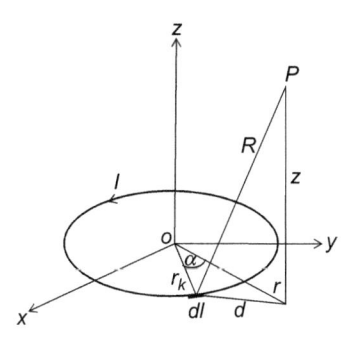

Fig. 10.10 Position of the current element and an observation point.

Here, I is the current of the element length dl, and R is the distance between an observation point p and the current element:

$$R = \left[r_k^2 + r^2 - 2ar\cos\phi + z^2\right]^{1/2}$$

while **i** is a unit vector indicating the direction of the element dl (Fig. 10.10). In the cylindrical system of coordinates, the point p and current element have coordinates $(r, 0, z)$ and $(r_k, \phi, 0)$, respectively.

The vector potential of the current ring located in a horizontal plane has the ϕ component only (Chapter 1), and for the current element dl we have

$$dA_\phi^*(p) = \frac{\exp(ikR) I\mu_0 \cos\alpha \, dl}{R \quad 4\pi} = \frac{I\mu_0 r_k}{4\pi} \frac{\exp(ikR)}{R} \cos\alpha \, d\alpha \quad (10.53)$$

The angle α is shown in Fig. 10.10. Similar to the case of the magnetic dipole, we express the field through cylindrical functions and use the following representation for the term $\exp(ikR)/R$:

$$\frac{\exp(ikR)}{R} = \frac{2}{\pi}\int_0^\infty K_0\left[\left(m^2 - k^2\right)^{1/2} d\right] \cos mz \, dm \quad (10.54)$$

Here, $k^2 = i\gamma\mu_0\omega$ and $R = (z^2 + d^2)^{1/2}$. Substituting Eq. (10.54) into Eq. (10.53) and integrating along the ring we obtain

$$A_\phi^* = \frac{I\mu_0 r_k}{4\pi}\frac{2}{\pi}\int_0^\infty \cos mz\,dm \int_0^{2\pi} K_0\left[\left(m^2 - k^2\right)^{1/2} d\right] \cos\alpha \, d\alpha \quad (10.55)$$

In accordance with the addition theorem of modified Bessel functions of the second kind, we have

$$K_0(dv) = K_0(rv)I_0(r_k v) + 2\sum_{n=1}^{\infty} K_n(rv)I_n(r_k v) \cos n\alpha \quad \text{if } r \geq r_k$$

and (10.56)

$$K_0(dv) = K_0(r_k v)I_0(rv) + 2\sum_{n=1}^{\infty} K_n(r_k v)I_0(rv) \cos n\alpha \quad \text{if } r \leq r_k$$

Replacing in Eq. (10.55) the function $K_0(dv)$ by the right-hand side of Eq. (10.56), and applying the condition of orthogonality of trigonometric functions

$$\int_0^{2\pi} \cos n\alpha \cdot \cos m\alpha \, d\alpha = 0 \quad \text{if } m \neq n$$

$$\int_0^{2\pi} \cos n\alpha \cdot \cos m\alpha \, d\alpha = \pi, \quad \text{if } m = n$$

we obtain the integral representation for the vector potential of the current ring in a uniform conductive medium:

$$A_\phi^* = \begin{cases} \dfrac{4I\mu_0 r_k}{4\pi} \displaystyle\int_0^\infty I_1(r_k v) K_1(rv) \cos mz \, dm & r \geq r_k \\[2ex] \dfrac{4I\mu_0 r_k}{4\pi} \displaystyle\int_0^\infty I_1(rv) K_1(r_k v) \cos mz \, dm & r \leq r_k \end{cases} \quad (10.57)$$

Here, $v = (m^2 - k^2)^{1/2}$. Considering that $\text{div} \mathbf{A} = 0$, for the electric field E_ϕ^* we have

$$E_\phi^* = -\frac{\partial A_\phi^*}{\partial t} = i\omega A_\phi^*$$

or

$$E_\phi^* = \frac{i\omega\mu_0}{\pi} Ir_k \int_0^\infty I_1(r_k v) K_1(rv) \cos mz \, dm \quad \text{if } r \geq r_k \quad (10.58)$$

$$E_\phi^* = \frac{i\omega\mu_0}{\pi} Ir_k \int_0^\infty I_1(rv)K_1(r_k v)\cos mz\, dm \quad \text{if} \quad r \le r_k \qquad (10.59)$$

With decrease of the radius r_k, the field tends to the field of the magnetic dipole. In such cases, the ratio z/r_k becomes very large and the integral is defined by small values of m. Then, replacing the Bessel function $I_1(x)$ with the asymptotic value of $x/2$, we have

$$E_\phi^* \approx \frac{i\omega\mu_0}{2\pi} Ir_k^2 \int_0^\infty v K_1(rv)\cos mz\, dm = -\frac{i\omega\mu_0}{2\pi^2} I\pi r_k^2 \frac{\pi}{2} \frac{\partial}{\partial r}\frac{\exp(ikR)}{R}$$

$$= \frac{i\omega\mu_0}{4\pi R^2} I\pi r_k^2 \exp(1 - ikR)\sin\theta \quad \text{if} \quad r \ge r_k$$

Thus we arrived at the expression for the electric field of the magnetic dipole. Assuming that the radius of the receiver coil is small and it is located sufficiently far from the transmitter, we have

$$B_z^* \approx \frac{\mu_0 M_0}{2\pi z^3} \exp(ikz)(1 - ikz) \qquad (10.60)$$

The asymptotic behavior of this field was studied in detail in Chapter 4 and, in particular, we found that at a low frequency and, correspondingly, at the late transient, the expressions for the field are

$$QB_z \propto \frac{\gamma\omega}{z}, \quad InB_z \propto (\gamma\omega)^{3/2}, \quad B_z(t) \propto \frac{\gamma^{3/2}}{t^{3/2}} \qquad (10.61)$$

10.4.2 Boundary Value Problem in the Presence of an Ideally Conductive Cylinder

Now we begin to study the influence on the field of a conductive cylinder and, accounting for its high conductivity. First, consider the limiting case when the cylinder is an ideal conductor. Thus we have to solve the boundary value problem for a cylinder with the radius a, surrounded by a medium with conductivity γ and excited by the current ring with radius $r_k > a$, located in the plane perpendicular to the axis of the cylinder. In Chapter 8, we showed that in that case, the solution to the Helmholtz equation can be presented as a combination of modified Bessel and trigonometric functions. Taking into account (Eq. 10.59), the total electric field at $r \leq r_k$ can presented in the form:

$$E_\phi^* = \frac{i\omega\mu}{\pi} Ir_k \left[\int_0^\infty I_1(rv) K_1(r_k v) \cos mz\, dm + \int_0^\infty C(v) K_1(rv) \cos mz\, dm \right]$$

(10.62)

Here, the second integral describes the electric field caused by currents induced in the cylinder, but the first integral represents the normal field of the current ring surrounded by a uniform medium. The function E_ϕ^* satisfies the Helmholtz equation and radiation boundary conditions at infinity. At the surface of the ideal conductor, the electric field is equal to zero, thus we have the following boundary condition to determine the unknown function $C(v)$:

$$I_1(av) K_1(r_k v) + C(v) K_1(av) = 0$$

Whence

$$C(v) = -\frac{I_1(av) K_1(r_k v)}{K_1(av)} \quad (10.63)$$

and for the total electric field we have

$$E_\phi^* = E_\phi^{*n} - \frac{Ir_k}{\pi} i\omega\mu_0 \int_0^\infty \frac{I_1(av)}{K_1(av)} K_1(r_k v) K_1(rv) \cos mz\, dm \quad (10.64)$$

The last equation enables us to estimate an electric field as a function of the z-coordinate. In particular, for the secondary field in the receiver of the radius $r = r_k$, we obtain

$$E_\phi^{*s} = -\frac{Ir_k}{\pi} i\omega\mu_0 \int_0^\infty \frac{I_1(av)}{K_1(av)} K_1^2(r_k v) \cos mz\, dm \quad (10.65)$$

In the case of a nonconductive surrounding medium, the equation earlier leads to:

$$E_\phi^{*s} = -\frac{Ir_k}{\pi} i\omega\mu_0 \int_0^\infty \frac{I_1(am)}{K_1(am)} K_1^2(r_k m) \cos mz\, dm$$

while for the normal field we have

$$E_\phi^{*n} = \frac{Ir_k}{\pi} i\omega\mu_0 Ir_k \int_0^\infty I_1(r_k m) K_1(r_k m) \cos mz \, dm$$

As follows from the last two equations the secondary electric field and the normal electric field at the surface of the cylinder differ only by a sign. In other words, the electric field of induced currents on the surface of the ideally conductive cylinder $r = a$ completely compensates the primary electric field caused by the current ring. In the case of the primary source varying as the step function, the surface currents almost instantaneously arise and then remain constant with time because there is no diffusion of the currents into the ideal conductor and correspondingly, there is no conversion of electromagnetic energy into heat.

Now we again assume that a surrounding medium is conductive and it is experiencing harmonic excitation. The current ring r_k induces primary volume currents in the medium and surface currents on the cylinder. By definition, the normal field is caused by the current ring and the induced primary volume currents in the medium, while the anomalous field is due to the surface currents and their interaction with the volume currents. Because the current ring is located near the cylinder, and the total field at the surface of the ideal conductor is equal to zero, the surface currents are mainly concentrated near the source.

The direction of the surface currents is opposite that of the current source, and they decay as $\propto 1/R^3$ with the distance from the ring. The smaller the difference between the radius of the ring and the cylinder $(r_k - a)$, the larger is the concentration of the surface currents near the ring. The phase difference between the surface currents and that of the primary source is 180 degrees.

In the case of the transient excitation, when the source current I_0 is abruptly turned off, the induced currents of the same direction instantly arise in the vicinity of the source and begin diffusion inside the formation. The induced currents in the formation cause the normal field, which gives rise to the currents on the surface of the cylinder. These surface currents create a secondary field of the opposite direction to the normal field, maintaining a zero electric field along the entire surface of the cylinder. Similar to the frequency domain, an influence on the ideally conductive cylinder, even at the relatively small distance from the source, is practically described by the transient field of the equivalent magnetic dipole, located at the center of the current ring. Thus, at distances several times greater than the size of the ring, the secondary and normal fields have the same dependence on the conductivity of the formation, observation time, and the distance from the source. In particular, at the late stage, both the normal and secondary fields decay as $\propto t^{5/2}$. To confirm this qualitative analysis, let us assume that $r = r_k$, $z \gg r$. In such case, the integral in Eq. (10.65) is mainly defined by small values of m and its integrand can be approximated as:

$$\frac{I_1(a\nu)}{K_1(a\nu)} K_1^2(r_k\nu) \cos mz \approx \frac{a^2}{2}\nu^2 K_1^2(r_k\nu) \cos mz \approx \frac{a^2}{2r_k}\nu K_1(r_k\nu) \cos mz$$

Correspondingly, for the electric field E_ϕ^{*s} we have

$$E_\phi^{*s} = \frac{Ia^2}{2\pi} i\omega\mu_0 \frac{\partial}{\partial r_k} \int_0^\infty K_0(\nu r_k) \cos mz \, dm = \frac{Ia^2}{2\pi} i\omega\mu_0 \sin\theta \frac{\partial}{\partial R} \frac{1}{R} \exp(ikR)\frac{\pi}{2}$$

or

$$E_\phi^{*s} = -\frac{Ia^2\pi}{4\pi R^2} i\omega\mu_0 \exp(ikR)(1 - ikR) \sin\theta \qquad (10.66)$$

Therefore, we arrived at the expression for the electric field of the magnetic dipole in a uniform medium. As was mentioned earlier (Eq. 10.4), the expression for the transient field is

$$E_\phi(t) = -\left(\frac{2}{\pi}\right)^{1/2} \frac{M_0\rho}{4\pi R^4} u^5 \exp\left(-\frac{u^2}{2}\right) \sin\theta$$

Here,

$$u = \frac{2\pi R}{\tau}, \quad \tau = (2\pi\rho t 10^7)^{1/2} \qquad (10.67)$$

Now we compare the dipole transient response $E_\phi(t)$ in a uniform medium with that from the ring of a finite radius that can be easily obtained using the frequency response for $E_\phi(\omega)$ (Eq. 10.59) and Fourier transform. In the case of a step function, the spectrum is $1/(-i\omega)$, and for the transient response, we have

$$E_\phi(t) = \frac{1}{2\pi} \int_0^\infty E_\phi(\omega) \frac{\exp(i\omega t)}{-i\omega} d\omega$$

In the simulation, for the case of dipole excitation we use a small receiving loop with a radius of $r_0 = 0.01$ m. For the ring excitation, transmitting and receiving coils of the same radius (e.g., $r_k = 0.1$ m) are selected. To perform a comparison of the electromotive forces excited by the dipole and current ring, they are both normalized by the product of transmitter M_t and receiver M_r moment:

$$e_\phi(t) = E_\phi(t) \cdot 2\pi r_0 / M_t M_r$$

Normalized transient responses at a very early stage ($t \leq 1$ μs), when the size of the ring is the most pronounced, are shown in Fig. 10.11A and B for formations of 1 and 100 ohm m, correspondingly. In both cases, the signals are calculated at two transmitter-receiver spacing of 0.5 and 2 m. The deviation of the ring response from the dipole response is most noticeable in the conductive formation at the relatively short spacing of 0.5 m. But even at 0.5 m spacing, this difference becomes negligible when the observation time is greater than 0.1 μs. In the case of the resistive formation, the effect of the finite size of the ring is practically negligible even at $t \geq 0.03$ μs (Fig. 10.11B). Overall, we see that if the distance between the transmitter

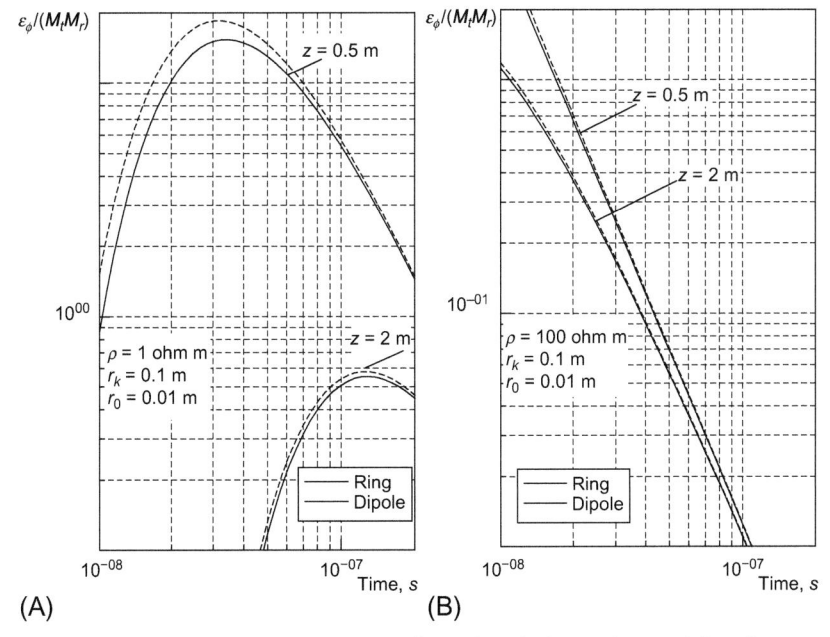

Fig. 10.11 Early stage transient responses from the whole conductive (A) and resistive (B) medium excited by the dipole and ring. Index is spacing.

and receiver is more than five times greater than the radius of the ring, the dipole and ring responses practically coincide over the entire time range.

Bearing in mind the similarity between the normal field and the field caused by the ideally conductive cylinder, we may also conclude that for all practically important transmitter/receiver spacing, the response from the cylinder surrounded by a conductive medium is described by the field of magnetic dipole. Later we will see how this fact can be used for the practical design of the transient logging tool.

10.4.3 Influence on the Finite Conductivity of the Cylinder

Boundary Condition

To find the field in the case of a cylinder of finite conductivity, solving the boundary value problem requires determination of the magnetic field inside and outside the cylinder. But we take advantage of two facts and simplify the problem. First, our goal is the field outside the cylinder only. Secondly, we are dealing with the cylinder whose conductivity in orders of magnitude is greater than that of the surrounding medium. Those two factors enable us to apply the approximate impedance boundary condition that is known as the Leontovich boundary condition. To proceed let us assume that the plane wave in a medium with the wave number k_1 approaches at some arbitrary angle the boundary that separates the medium with wave number k_2. Then in accordance with Snell's law, the direction of the refraction wave is practically normal to the boundary, provided that

$$|k_2| \gg |k_1|$$

Correspondingly, the electric and magnetic fields of the refraction wave are parallel to the boundary, and by definition their ratio is equal to the impedance of the plane wave in the second medium:

$$\frac{E_{2\phi}^*}{B_{2z}} = \frac{Z_2^*}{\mu_0}$$

Inasmuch as tangential components of both fields are continuous at the boundary, we have

$$\frac{E_{1\phi}^*}{B_{1z}^*} = \frac{E_{2\phi}^*}{B_{2z}} = \frac{Z_2^*}{\mu_0} \qquad (10.68)$$

where $E_{1\phi}^*$ and B_{1z}^* are complex amplitudes of the tangential mutually orthogonal components of the field in the first medium outside the cylinder. Leontovich had shown that Eq. (10.68) is valid for the more general case of an arbitrary boundary and nonplane incident wave, provided that the radius of the surface curvature greatly exceeds the skin depth. Eq. (10.68) implies that in the vicinity of the boundary from the side of the second medium, the field behaves as the plane wave and moves in the direction perpendicular to this boundary. The practical importance of Eq. (10.68) is the possibility of avoiding determination of the field in the second medium. Thus, for application to our problem of axial-symmetrical excitation of the cylinder, we have the following expression for the complex amplitude of the impedance:

$$Z_2^* = \mu_0 \frac{E_{1\phi}^*}{B_{1z}^*}$$

which relates the field in the surrounding medium with the impedance of the cylinder. On the other hand, for the intrinsic wave impedance Z_2^* of the plane wave, we have

$$Z_2^* = \frac{\omega\mu_0}{k_2} = (\omega\mu_0\rho_2)^{1/2} \exp(-i\pi/4) = (\omega\mu_0\rho_2/2)^{1/2} - i(\omega\mu_0\rho_2/2)^{1/2}$$

Bearing in mind that the Leontovich boundary condition is an approximate one, it is useful to estimate a minimal frequency f_{min} when the skin depth $\delta_2 = (2/\gamma_2\mu_0\omega)^{1/2}$ in the highly conductive cylinder becomes high enough to satisfy the condition. Suppose that $\rho_2 = 1 \times 10^{-6}$ ohm m and $\delta_2 = 2 \times 10^{-3}$ m. Then, for the frequency f_{min}, we have an estimate:

$$f_{min} \geq \frac{1}{(4 \times 10^{-6})} \left(\frac{4}{8\pi^2}\right) \approx \frac{1}{8\pi^2} 10^6 \approx 10^4 \, \text{Hz}$$

At such frequency, the corresponding amplitude of the complex impedance $|Z_2^*|$ is quite a small value:

$$|Z_2^*| = \frac{\omega\mu_0\delta_2}{2^{1/2}} = \frac{2\pi \cdot 4\pi \cdot 10^{-7} \cdot 2 \cdot 10^{-3}}{2^{1/2}} 10^4 \approx 10^{-4} \, \text{ohm}$$

Approximate Solution to the Boundary Value Problem

Following Eq. (10.62), the electric field in the surrounding medium can be presented as:

$$E_\phi^* = \frac{i\omega\mu}{\pi} Ir_k \left[\int_0^\infty I_1(rv) K_1(r_k v) \cos mz \, dm + \int_0^\infty D(v) K_1(rv) \cos mz \, dm \right]$$

(10.69)

where the first integral is the normal field; and D is an unknown function of the integrant, which determines the secondary field. At the same time, the magnetic field might be expressed directly from the Maxwell equation:

$$\frac{1}{r}\frac{\partial}{\partial r}rE_\phi^* = i\omega B_z^*$$

After substituting Eq. (10.69) into the latter, we receive

$$B_z^* = \frac{I\mu_0 r_k}{\pi}\left[\int_0^\infty \nu K_1(r_k\nu)I_0(r\nu)\cos mz\,dm - \int_0^\infty D\nu\cdot K_0(r\nu)\cos mz\,dm\right]$$

From Eq. (10.18), we have

$$i\omega\mu_0\frac{I\mu_0 r_k}{\pi}\left[\int_0^\infty I_1(a\nu)K_1(r_k\nu)\cos mz\,dm + \int_0^\infty D(\nu)K_1(a\nu)\cos mz\,dm\right]$$

$$= Z_2^*\cdot\frac{I\mu_0 r_k}{\pi}\left[\int_0^\infty \nu K_1(r_k\nu)I_0(a\nu)\cos mz\,dm - \int_0^\infty D\nu K_0(a\nu)\cos mz\,dm\right]$$

(10.70)

Let us introduce notation:

$$\xi = \frac{Z_2^*}{i\omega\mu_0}$$

Because $k^2 = -i\omega\mu\gamma$, for ξ we have

$$\xi = \frac{Z_2^*}{i\omega\mu_0} = \frac{1}{ik_2} = \frac{\delta_2}{(i-1)}$$

where δ_2 is a small number because it represents the skin depth of the highly conductive cylinder and the product $\xi\nu$ is dimensionless. Thus, for the secondary electric field, we have

$$E_\phi^{*s} = -\frac{i\omega\mu_0}{\pi}Ir_k\int_0^\infty\frac{I_1(a\nu) - \nu\xi I_0(a\nu)}{K_1(a\nu) + \nu\xi K_0(a\nu)}K_1(r_k\nu)K_1(r\nu)\cos mz\,dm \quad (10.71)$$

For instance, in the case of an ideal conductor $\xi = 0$, the electric field is

$$E_\phi^* = -\frac{i\omega\mu_0}{\pi}Ir_k\int_0^\infty\frac{I_1(a\nu)}{K_1(a\nu)}K_1(r_k\nu)K_1(r\nu)\cos mz\,dm$$

that coincides with Eq. (10.65) at $r = r_k$.

High-Frequency and Early Transient Stage Asymptote

The fraction in Eq. (10.71) can be presented as:

$$\frac{I_1(a\nu) - \nu\xi I_0(a\nu)}{K_1(a\nu) + \nu\xi K_0(a\nu)} - \frac{I_1(a\nu)}{K_1(a\nu)} + \frac{I_1(a\nu)}{K_1(a\nu)} \qquad (10.72)$$

where the last term corresponds to the case of the ideal conductor. Combining the first two terms in Eq. (10.22) and considering that

$$I_0 K_1 + I_1 K_0 = 1/(\nu a)$$

we obtain

$$\frac{I_1 K_1 - \nu\xi I_0 K_1 - I_1 K_1 - \nu\xi I_1 K_0}{K_1(K_1 + \nu\xi K_0)} = -\nu\xi \frac{I_0 K_1 + I_1 K_0}{K_1(K_1 + \nu\xi K_0)} = -\xi \frac{1}{a} \frac{1}{K_1[K_1 + \nu\xi K_0]}$$

Thus the total field comprises the three field components: the field E_ϕ^{*n} of the current ring in a uniform medium; the field E_ϕ^{*i}, caused by the presence of an ideally conductive cylinder; and the field E_ϕ^{*s}, characterizing diffusion, providing that the skin depth in the cylinder is sufficiently small. For the last component, we have

$$E_\phi^{*s} = -\frac{i\omega\mu_0}{\pi a}\xi Ir_k \int_0^\infty \frac{K_1(r_k\nu)K_1(r\nu)}{K_1(a\nu)[K_1(a\nu) + \nu\xi K_0(a\nu)]} \cos mz\, dm \qquad (10.73)$$

Suppose that the field is observed at a large distance from the current source. Then the integral is mainly defined by small values of m and κ. By neglecting the second term in the brackets of the denominator and replacing $K_1(r_k\nu)/K_1(a\nu)$ with an asymptotic value of a/r_k, we receive

$$E_\phi^{*s} \approx -\frac{i\omega\mu_0 I}{\pi a} r_k \xi \frac{a}{r_k} \int_0^\infty \frac{K_1(r\nu)}{K_1(a\nu)} \cos mz\, dm \approx -\frac{i\omega\mu_0 I}{\pi}\xi a \int_0^\infty \nu K_1(r\nu) \cos mz\, dm$$

or

$$E_\phi^{*s} \approx \frac{i\omega\mu_0 I}{\pi}\xi a \frac{\partial}{\partial r} \int_0^\infty K_0(r\nu) \cos mz\, dm = \frac{i\omega\mu_0 I}{2}\xi a \frac{\partial}{\partial r} R^{-1} \exp(ikR) \qquad (10.74)$$

$$E_\phi^{*s} \approx -\frac{i\omega\mu_0 I}{2R^2}\xi a \exp(ikR)(1 - ikR)\sin\theta$$

At first glance, the field E_ϕ^{*s} in Eq. (10.74) is small compared with the field of the ideal conductor. However, if the difference in radii of the ring and the cylinder is small, then the terms describing the normal field and the field from the ideal conductor almost cancel each other, and the diffusion term may dominate. Suppose that $|ka \ll 1|$. Then the frequency dependence of the electric field, caused by the diffusion in the cylinder, is

$$E_\phi^{*s} \approx \frac{1}{R^3} \frac{c}{(\omega \gamma)^{1/2}}$$

Correspondingly, the early stage of the transient process in the cylinder is defined as:

$$E_\phi(t) = \frac{c}{\gamma^{1/2} R^3} \frac{1}{\pi} \frac{2}{\pi} \int_0^\infty \frac{\cos \omega t}{\omega^{1/2}} d\omega = \frac{c}{R^3 \gamma^{1/2} t^{1/2}} \frac{1}{t^{1/2}} \quad (10.75)$$

This equation suggests that the finite conductivity of the cylinder causes a very slow decay $\propto 1/t^{1/2}$ of the transient field [2]. A large difference in conductivities leads to a very different decay of the transient field in the formation and the cylinder. Specifically, the late stage of the transient process (hundreds of microseconds) in the formation corresponds to the early transient stage in the cylinder. We may expect that in the presence of both cylinder and formation, the slowly decaying term $\propto 1/t^{1/2}$, corresponding to the cylinder, will completely dominate over the fast decaying signal $\propto 1/t^{5/2}$ from the formation.

10.4.4 Effect of Spacing on the Pipe Signal

Now we compare our qualitative analysis with numerical calculations using the advanced finite element numerical technique. In the model, the source is the current ring with the radius $r_k = 0.085$ m, which slightly exceeds a radius of the pipe, $a = 0.07$ m. The electromotive force is measured at the distance $z = 3$ m (Fig. 10.12A) and $z = 7$ m (Fig. 10.12B) between the transmitting and receiving coils. Resistivity of the entire space is set to $\rho = 100$, 10, and 1 ohm m.

As shown in Fig. 10.12, it is useful to distinguish three different time ranges:
1. The early time range $t < 10$ μs. In this case the signal is mainly defined by the resistivity of the formation.
2. The intermediate range $10 \leq t < 100$ μs, when the signal depends on properties of both the formation and the pipe.

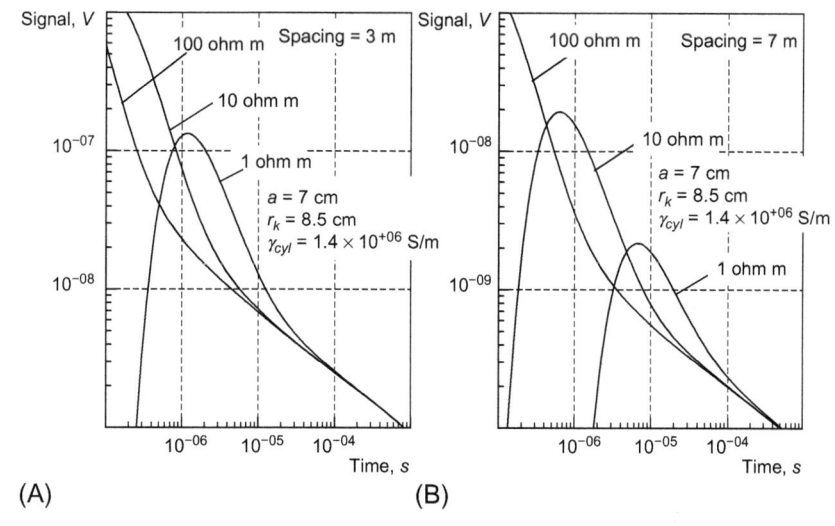

Induction Logging Based on Transient EM Measurements

Fig. 10.12 Transient response at $L=3$ m (A) and 7 m (B) spacing in the presence of conductive pipe in homogeneous whole space. Index of the curves is whole space resistivity.

3. At later times, the influence on currents in the pipe completely dominates, and obtaining measurements of formation resistivity becomes hardly possible.

At relatively early times, when the diffusion of currents in a pipe is insignificant, the influence of the surrounding medium becomes stronger. This fact is not occasional, because at such time range, the pipe behaves almost as an ideal conductor. Correspondingly, we focus our attention on conditions in which induced currents in the cylinder are located relatively close to the transmitter coil, but the diffusion in the surrounding medium is described by the intermediate and late stages. Those conditions correspond to the measurements at the relatively large spacing, when the early stage transient process can be measured at the expanded time window. However, this approach alone has a limited value, because it leads to undesirable increase in the tool length and does not preserve sensitivity to the deep parts of the formation (see Fig. 10.12, when at $t \geq 100$ μs signal from the pipe dominates).

The influence on the pipe signal is even more pronounced when the object of interest is located ahead of the two-coil probe. For example, let us consider the case of a conductive layer placed at two distances (distance to the boundary, or d2b) of 10 and 30 m ahead of the receiving coil R (Fig. 10.13). The resistivity around the tool and ahead-placed layer is

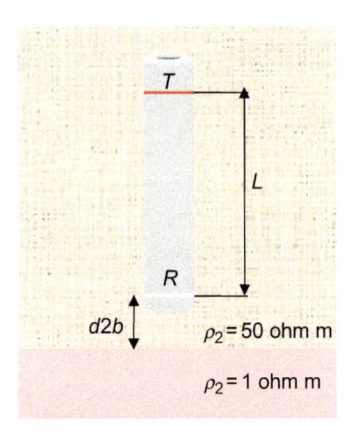

Fig. 10.13 Two-coil probe surrounded by resistive layer and conductive layer placed ahead of the probe.

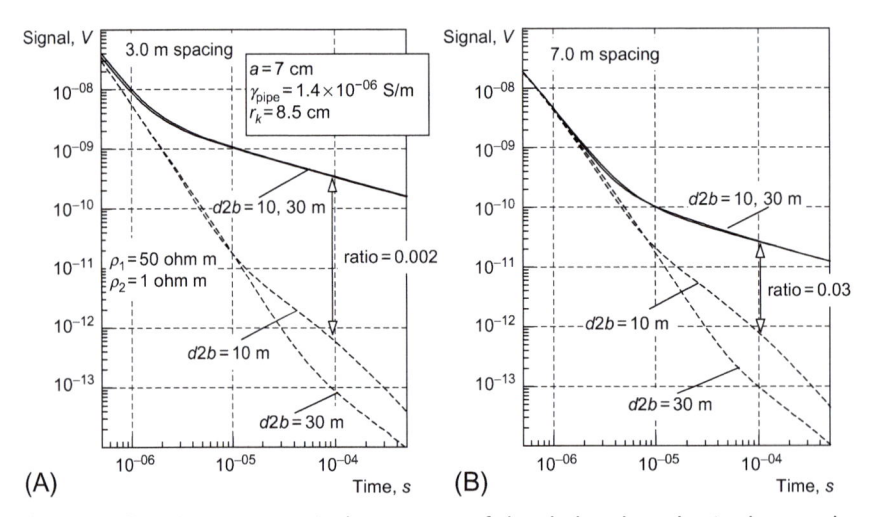

Fig. 10.14 Transient response in the presence of ahead-placed conductive layer at the spacing of 3 m (A) and (B) 7 m. Code of curves is d2b.

$\rho_1 = 50$ ohm m and $\rho_2 = 1$ ohm m, correspondingly. Our primary objective is to detect the ahead-placed boundary.

In Fig. 10.14, we show signals in the absence (dashed lines) and presence (solid lines) of the conductive pipe for two spacing of 3 m (Fig. 10.14A) and 7 m (Fig. 10.14B).

In the absence of the pipe, the signals (dashed lines) are very well distinguished and demonstrate high sensitivity to the distance to the boundary, while the presence of the pipe diminishes the sensitivity by making it impossible to resolve the target (overlapping solid lines). In other words, the transient process is entirely defined by the properties of the pipe.

Like in the previously considered case of a homogeneous formation (Fig. 10.12), increasing the spacing from $L_1 = 3$ m to $L_2 = 7$ m leads to the increase of the relative contribution of the formation into the total signal because the signal from the pipe drops as $\propto 1/z^3$, while signal from the formation, at least at the late stage, practically does not depend on the spacing. For example, at 3 m spacing and 100 μs, the ratio of signals from the pipe and formation is 0.002 (Fig. 10.14A), while at 7 m spacing it increases by a factor of $(7/3)^3 = 12.7$ to 0.03 (Fig. 10.14B). But even at the increased spacing, the ahead-placed boundary is still practically invisible.

10.4.5 Effect of the Increased Pipe Conductivity on the Transient Response

There is another approach that can also delay diffusion through the pipe, thus reducing its influence. Looking at the equation for the transient signal (Eq. 10.75) we may notice that the leading term is inversely proportional to the square root of pipe conductivity. This suggests a possible reduction of the signal from the pipe by covering it near the transmitter and receiver with a material (shield) that has higher conductivity than that of the steel.

First, this assumption is confirmed by the rigorous modeling for the pipe of different conductivity, changing from $\gamma = 1.4 \times 10^6$ S/m (Siemens per meter) for the steel to $\gamma = 0.6 \times 10^{12}$ S/m. The intermediate value of $\gamma = 0.6 \times 10^8$ S/m corresponds to the conductivity of the copper; while $\gamma = 0.6 \times 10^{10}$ S/m and $\gamma = 0.6 \times 10^{12}$ S/m to some hypothetical "superconductive" materials.

The modeling results for 3 m and 7 m spacing (Fig. 10.15) confirm reduction of the pipe signal with an increase of the conductivity. This signal reduction is in full agreement with Eq. (10.75), which indicates on $\infty \, 1/\sqrt{\gamma}$ dependence on the conductivity of the pipe and $\infty 1/z^3$ dependence on the spacing. It is also interesting to notice the fast, practically exponential, decay of the signal (Fig. 10.15, $\gamma = 1.4 \times 10^6$ S/m) at the very late stage, $t \geq 1$ ms, when thickness of the skin layer in the pipe becomes comparable with the pipe thickness.

Let us explore further the effect of the increased pipe conductivity by analyzing the transient response in the case of a highly conductive pipe surrounded by homogeneous formations of different resistivity. For illustration, a modeling is presented for the pipe with $\gamma = 0.6 \times 10^{12}$ S/m and a set of

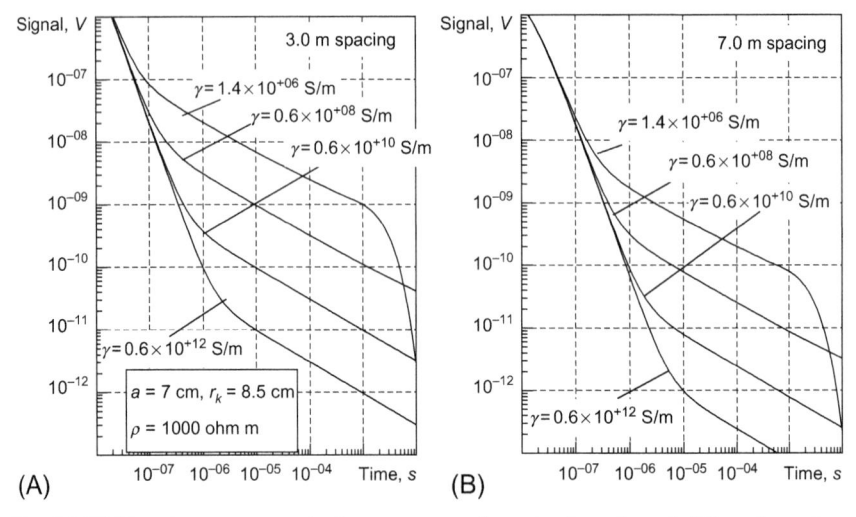

Fig. 10.15 Transient response in the presence of conductive pipe of different conductivities at (A) 3 m and (B) 7 m spacing. Index of the curves is conductivity of the pipe.

homogeneous formations with $\rho = 1$, 10, and 100 ohm m. The results (dashed lines) at spacing of 3 and 7 m are presented in Fig. 10.16A and B, correspondingly. In addition, the signals for the homogeneous medium in the absence of the pipe are also shown (solid lines).

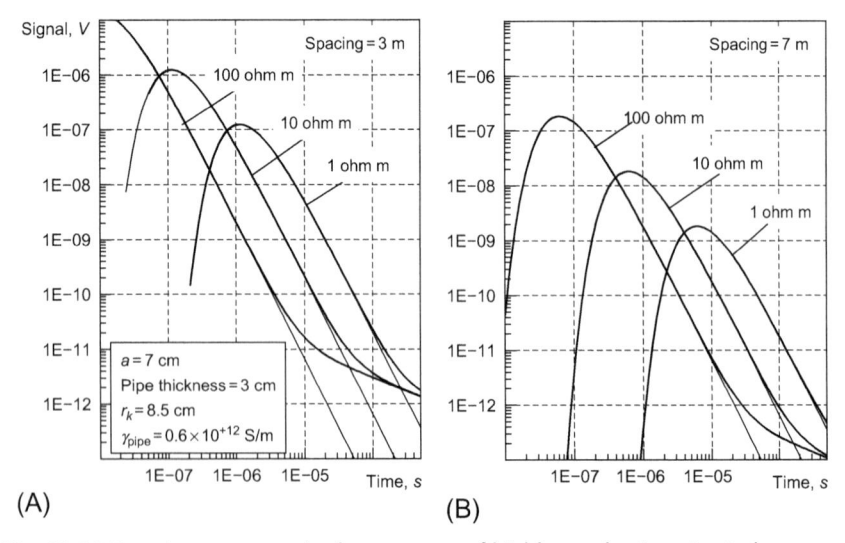

Fig. 10.16 Transient response in the presence of highly conductive pipe in homogeneous formation at (A) 3 m and (B) 7 m spacing. Index of the curves is whole space resistivity.

Analyzing these responses we notice the following:
- For the given formation resistivity and spacing there is a time range where behavior of the signal is slightly affected by the presence of the conductive pipe. In this time range the skin depth in the pipe is practically equal to zero and the pipe behaves almost as an ideal conductor (Eq. 10.65).
- The lower the resistivity of the formation, the greater the time range where the pipe behaves as an ideal conductor (overlapping dashed and solid lines). For example, in the case of the 100 ohm m formation and 3 m spacing, the overlap is observed up to $t \approx 4$ μs, while in 10- and 1-ohm m formations, it is extended to $t \approx 30$ μs and $t \approx 150$ μs, correspondingly.
- Increase in the spacing significantly extends the time range in which the response follows the response in the whole space. Comparison of the data in Fig. 10.16 demonstrates an extension by a factor of 4.

10.4.6 Reduction of the Pipe Signal Using Finite Size Copper Shield and Bucking

Unfortunately, there is no material with such high conductivity as in the example earlier ($\gamma = 0.6 \times 10^{12}$ S/m) to replace the steel. But we still may use available conductive materials, such as copper, $\gamma_c = 0.6 \times 10^{08}$ S/m, and partially reduce the signal from the pipe. According to Eq. (10.75) and the data in Fig. 10.15, the copper leads to reduction of the signal from the pipe by a factor of $\sqrt{\gamma_c/\gamma_s} = 6.55$. As we discussed before, the major effect from the conductive pipe comes from the region near the transmitting and receiving coils. For this reason, a thin and relatively short copper layer wrapped around a steel pipe (Fig. 10.17) serves as a shield. In the following numerical examples, transmitting and receiving coils are placed in the middle of the 0.75-m long copper shield. In addition, we can further suppress signal from the pipe by considering that the signal from the pipe decreases with the spacing as $\infty 1/L^3$, while the transient signal from the formation only slightly depends on the spacing, especially at the late stage. Thus, by combining two signals at two different spacings, we may substantially reduce contribution from the pipe into the total signal.

Obviously, the signals should be combined with the weights that are inversely proportional to L^3 or, more precisely, with the weights that provide no signal in the absence of the formation. For example, let us select two spacings, $L_1 = 5$ m and $L_2 = 7$ m, and estimate a coefficient $k(t)$, which provides the following condition:

$$S^{air}(t) = S^{air}_{L_2}(t) - k(t) \cdot S_{L_1}{}^{air}(t) = 0 \qquad (10.76)$$

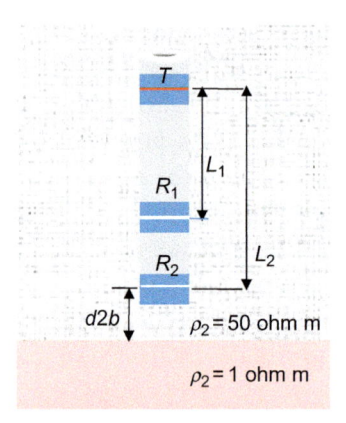

Fig. 10.17 Three-coil probe with copper shields surrounded by resistive layer and conductive layer placed ahead of the probe.

We call the coefficient $k(t)$ the bucking coefficient; the short-spaced receiving coil is the bucking coil; and the long-spaced coil is the main coil. In fact, the bucking coil in the transient regime plays a similar role to the bucking coil used in the three-coil induction system with harmonic excitation, compensating a large signal caused by eddy currents in the pipe.

As we mentioned earlier, the signal from the pipe (Fig. 10.18A) is proportional to $\infty 1/L^3$, thus the bucking coefficient should approach a constant value $k = (L_1/L_2)^3$ (Fig. 10.18B). However, with an increase of time, diffusion of currents in the pipe (see Fig. 10.15, $\gamma = 1.4 \times 10^6$ S/m) leads to a small deviation from the constant k: the higher the conductivity of the pipe (or conductivity of the shield), the later this deviation takes place.

Now let us show effectiveness in suppression of the pipe signal using both the copper shield and bucking technique when applied to the one of the most challenging tasks of geo-steering—detection of the target ahead of the drill bit. It is assumed that resistivity around the transient system is 50 ohm m, while the conductive layer ahead of the drill bit has resistivity of 1 ohm m. The distance from the transmitter to the bucking and main coils is 5 and 7 m, correspondingly. Modeling results are shown in Fig. 10.19A. First, we may notice the reduction of the signal amplitude caused by the copper shield. By comparing the upper dashed curve (representing steel) with the solid curve (representing the pipe with the copper shield) in Fig. 10.19A, we see that the signal drops by a factor of $\sqrt{\gamma_c/\gamma_s} = 6.55$.

Next, using the signals at 5 and 7 m spacing and applying Eq. (10.76), we obtain a family of the bucked curves corresponding to the different distances

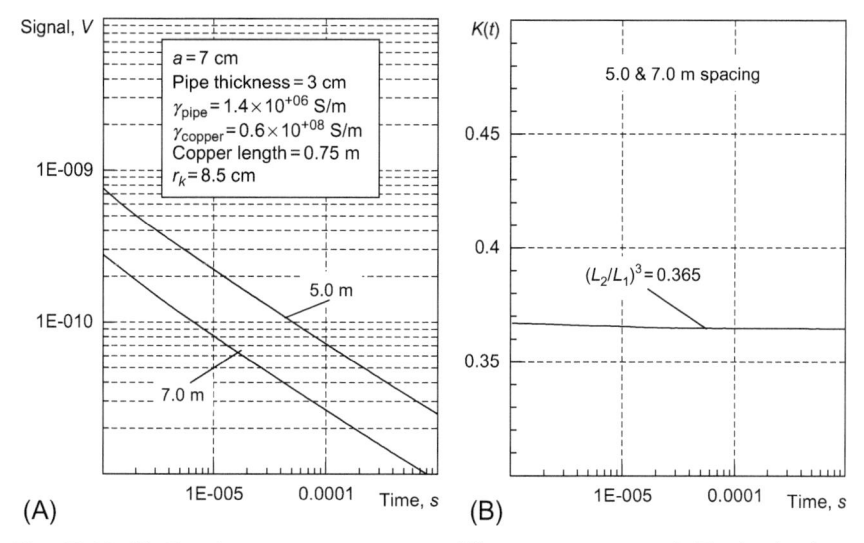

Induction Logging Based on Transient EM Measurements

Fig. 10.18 (A) Transient response at two different spacings and (B) the bucking coefficient.

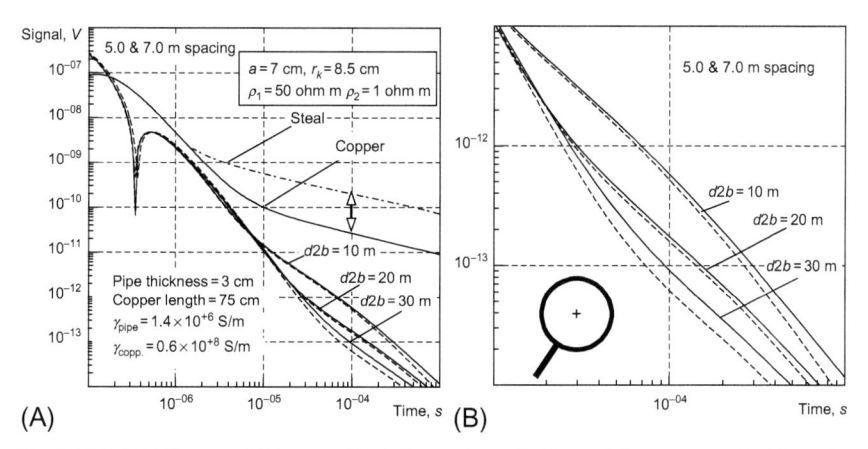

Fig. 10.19 (A) Effect and (B) magnified illustration of effect of the copper shield on the bucked response in the presence of ahead-placed boundary. Code of the curves is d2b.

to the boundary, $d2b = 10$, 20, and 30 m (solid curves). The curves show the effectiveness of the transformation defined by Eq. (10.76), and enabling us to essentially reduce the influence of the pipe and preserve sensitivity to the target. Moreover, the behavior of the bucked curves is very similar to the synthetic signals observed in the absence of the pipe (Fig. 10.19A, dashed lines): the shorter the distance to the target, the closer the bucked curves to the curves obtained in absence of the pipe. The difference between curves is most pronounced at the late stage when the target is located 30 m ahead of the main coil. The magnified mismatch is shown in Fig. 10.19B and it is caused by the increased influence of the skin effect in the pipe when the conductive target has limited contribution into the signal. Similar behavior was observed (Fig. 10.16) when the skin effect in the pipe was especially pronounced in the resistive formation of 100 ohm m.

10.4.7 Improving Formation/Pipe Signals Ratio Using Magnetic Shielding

To further reduce the influence of the pipe, one can use a magnetic shield in the form of a short nonconductive cylindrical ferrite with high magnetic permeability located between the coils and the pipe. The ferrite's high permeability causes the magnetic field lines to be concentrated in the core material, thus increasing the effective magnetic moment of the transmitting coil (Fig. 10.20). When ferrite is placed in the external magnetic field, it becomes magnetized, and magnetization currents of different directions arise both externally and internally of the ferrite's surface. Specifically, on the external surface, the currents produce a magnetic field of the same direction with an external field, while currents on the internal surface (close to the pipe surface) generate the magnetic field of the opposite direction. As a result, the total field in the vicinity of the pipe becomes smaller, thus reducing

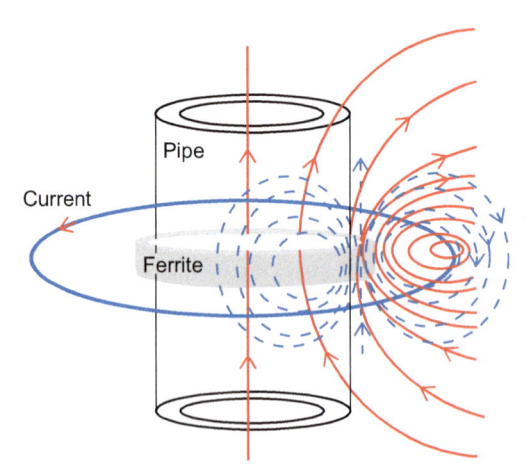

Fig. 10.20 Distribution of vector lines of the magnetic field in the presence of ferrite. *Solid lines* show the magnetic field of the primary source *(ring)*; *dashed lines* correspond to the magnetic field of the magnetized ferrite.

Induction Logging Based on Transient EM Measurements

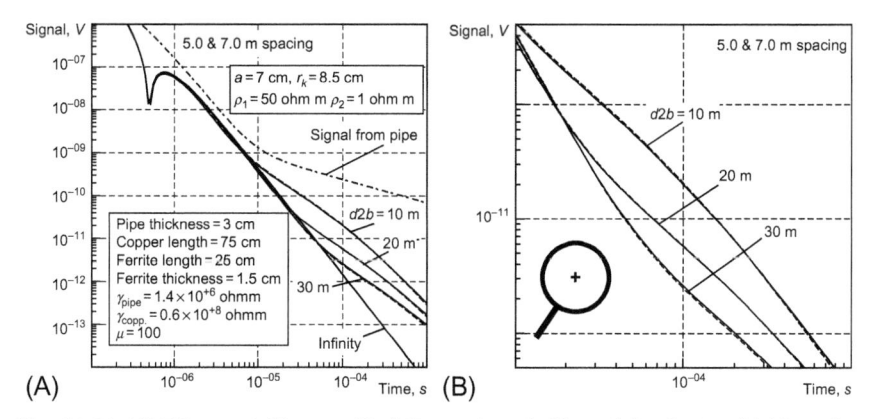

Fig. 10.21 (A) Effect and (B) magnified illustration of effect of the ferrite shield on the transient bucked response in the presence of ahead-placed boundary. Code of the curves is *d2b*.

the intensity of undesirable induced currents in the pipe. Of course, the shielding effect on the receiver side is similar to that on the transmitter side.

Let's see how the ferrite affects the transient response when it is added to the previously analyzed arrangement, based on the use of the copper shield and bucking technique. We assume that the ferrite inserts in the transmitting and each receiving coils are placed at 5 and 7 m, correspondingly. The length of the ferrite is 25 cm, its thickness is 1.5 cm, and relative permeability is 100. The position of each coil is centered with respect to the ferrite. Modeling results are shown in Fig. 10.21 and demonstrate increase of the signal level and improved resolution with respect to the ahead-placed boundary. Moreover, the bucked signals (solid lines) practically coincide with synthetic signals that are calculated in the absence of the pipe. This fact is desirable because it enables us to exclude pipe from the forward model, and eventually perform an inversion in a more reasonable amount of time.

Reducing pipe influence on the response was impossible when either shielding or bucking alone were used. On the other hand, by combining all the analyzed means we were able to reach a desirable level of pipe suppression.

10.5 INVERSION OF TRANSIENT DATA IN THE TASK OF GEO-STEERING

Inversion is a technique for determining the geo-electrical properties of a formation using induction logging data. In LWD, inversion constitutes a critical part of the technology because real-time data are used to determine the best course for the drilling operation. The measurements are used in geo-steering to determine dip angle and bed boundaries to keep the well in the sweet spot throughout long lateral sections. Typically, decisions are made by jointly interpreting data from gamma ray, acoustic, and resistivity tools. In the following section we illustrate the main aspects of inversion using a hypothetical induction probe operating in the time domain and measuring different field components.

10.5.1 Well- and Ill-Posed Problems

In the case of forward modeling, the problem is referred to as well-posed because the equations and coefficients, which are defined by properties of the formation, along with primary sources, are known. The corresponding boundary value problem has a unique solution, which continuously depends on parameters and the data- small changes in parameters result in small changes in the solution. With today's available computational power, solutions to almost any forward modeling problem can be found in a very reasonable amount of time.

In the case of the inverse problem, the coefficients are unknown and have to be found using a set of measurements taken with a logging device. It is called an inverse problem with respect to the forward problem because it uses the measurements and then calculates properties of the formation. The physics that relates the formation's parameters (i.e., the model parameters) to the observed data is governed by Maxwell equations. The vast majority of inverse problems are ill-posed because of the lack of uniqueness and continuity with respect to small changes in the data. Typically, solving an inverse problem requires solving systems of linear equations at some point. The main property of any system is so-called condition number, which is defined as ratio of the largest to smallest singular value in the singular value decomposition of a matrix of a system. If a system has linearly independent rows and columns, it is characterized by a small condition number; otherwise this number is large. As a matter of rule, all inverse problems lead to systems with a large condition number.

Example 1 Let us look at the system with linearly dependent rows (large condition number):

$$\begin{cases} x_1 + 5x_2 = 6 + \varepsilon \\ 10x_1 + 50.1x_2 = 60.1 \end{cases}$$

where ε is some small value. If ε is equal to zero, the solution is $x_1 = 1$, $x_2 = 1$, but the small perturbation of the right-hand side by $\varepsilon = 0.001$

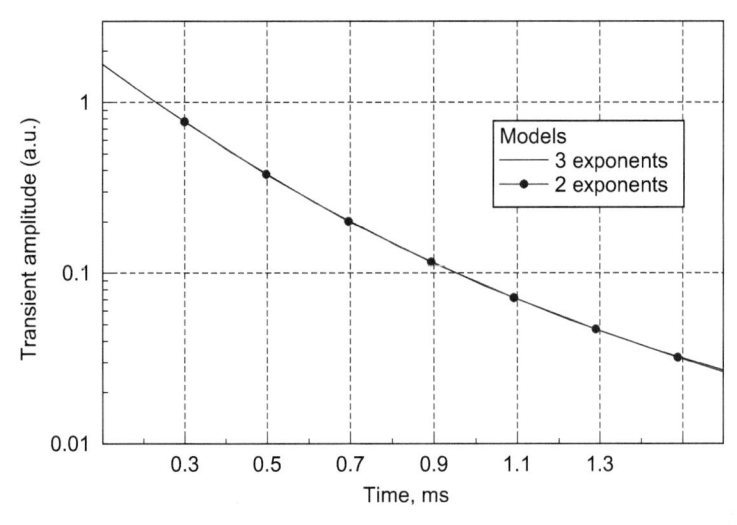

Fig. 10.22 The difference between two models is indistinguishable.

leads to the solution $x_1 = 2.0$, $x_2 = 0.8$, significantly deviating from the unperturbed case of $\varepsilon = 0$, and illustrating lack of continuity of the solution.

Example 2 This is the famous example of Lanczos, demonstrating the non-uniqueness or equivalence between different models. Lanczos fitted the same set of data using first, a set of two exponents and then a set of three exponents. The results are

$f_2(t) = 2.202 \exp(-4450t) + 0.305 \exp(-1580t)$,
$f_3(t) = 1.5576 \exp(-5000t) + 0.8607 \exp(-3000t) + 0.0951 \exp(-000t)$

The difference between $f_2(t)$ (Fig. 10.22, solid line) and $f_3(t)$ (Fig. 10.22, dotted line) is less than the line width used to plot the data. In the given time range it is impossible to establish the exact number of exponents in the model.

10.5.2 Main Elements of the Inversion Algorithm

Any inversion algorithm includes steps of comparison of measured and synthetic data, reproducing the probe response in the presence of the formation. The combination of formation parameters providing the minimum misfit is considered to be the solution of the inversion problem (see the following diagram). Normally, hundreds and thousands of comparisons between synthetic and measured data are needed to find the solution.

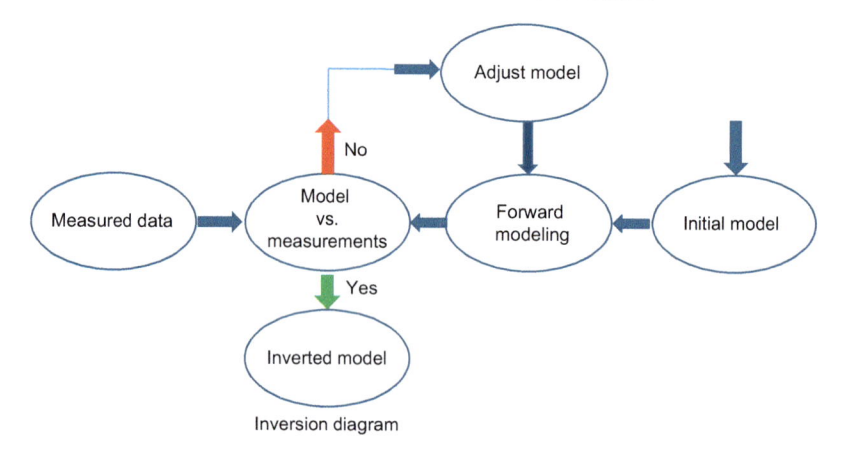

Inversion diagram

For this reason, forward modeling, which generates synthetic data for the different geo-electrical models, is the first critical element of the inversion algorithm. The next component is the strategy used to generate parameters of formation candidates to be tested for the best fit. Because of the ill-conditioned nature of the inversion problem, the means for stabilizing or regularizing the solution constitute another critical element of the inversion. In particular, regularization includes increase of the data set, optimal number of parameters subject to inversion, prior information about some of the parameters, for example, range of parameters variation, and so on. In fact, any additional information about the model or data constitutes regularized inversion.

Finally, when the inversion is complete, it is important to have an estimator of the error in inverted parameters. In other words, the interpreter needs to know error bars, indicating how an error in the measured data propagated into the error of estimated parameters.

10.5.3 Table-Based Inversion

When the number of parameters is limited, it is possible to invert data by applying an old-fashioned approach that is based on the precalculated master curves. Indeed, in the simplest geo-steering landing scenario, the typical formation model represents two layers and the borehole trajectory is tilted at some angle with respect to the boundary separating the layers (Fig. 10.23). Overall, we have only four parameters: resistivity of the upper layer ρ_1, resistivity of the lower layer ρ_2, dip angle α, and the most important parameter in this application, distance to bed $d2b$.

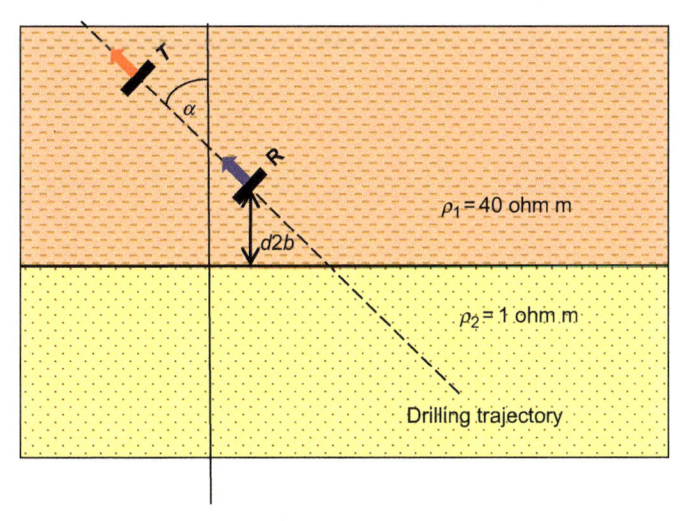

Induction Logging Based on Transient EM Measurements

Fig. 10.23 Formation model and drilling trajectory.

We use the simplest arrangement comprising one transmitting and one receiving coil placed in the upper layer. The task of geo-steering is to provide an optimal landing point, that is, a point where the well transition into the horizontal/lateral portion of the well occurs. Let us estimate the time needed to generate a table using, for example, 30 points to discretize each parameter. The total number of precalculated transient signals will be 30^4. Assuming that the typical time to calculate one master curve on an average dual-quad processor is approximately 0.1 s, the table can be generated in a matter of days using only one multicore processing unit. An important feature of the table-based inversion algorithm is that, for the given probe and formation model, the master curves are generated only once and used indefinitely.

The inversion algorithm is simple: we perform a global search by comparing the measured data with synthetic data from the table by calculating a least square deviation λ in each node of the four-parametric table:

$$\lambda = \frac{1}{N_T} \sum_{j=1}^{N_T} \left(\frac{S_e^j - S_s^j(\alpha, \rho_1, \rho_2, d2b)}{S_e^j} \right)^2$$

where S_e^j, S_s^j are measured and theoretical data, correspondingly; and N_T is the number of time discretes in the precalculated transient signals. Then the parameters corresponding to the node, providing a minimal value of λ, are accepted as a possible solution to the inverse problem.

The data may include some number of logging points along the drilling trajectory as well as the different measurements corresponding to the different orientation of transmitting and receiving coils. Another benefit of the table-based inversion is simplicity in implementing constraints on the invertible parameters: the regions of parameters specified by constraints are simply excluded from the scanning.

10.5.4 Stability of the Inverse Problem Solution

Stability of the inversion is mainly defined by the propagation of error in the measurements into errors in the inverted parameters.

The solution is stable when noise in the data is not amplified too drastically and the error in the reconstructed parameters is acceptable. Of course, such notion is subjective. For example, if 10% noise is translated into 10% error in the parameter, the solution is stable, whereas 100% error in the parameter would indicate lack of stability.

There are different approaches for studying the stability of the inversion scheme. One way would be to determine how the result is changing if input data have been perturbed by some small number. Another way is to examine how the result of inversion had been changing with respect to a different realization of the noise while the statistical properties of the noise are kept the same. The simplest example of such noise is a Gaussian noise having a probability density function equal to that of the normal distribution with a fixed standard deviation.

In the following examples we use model in Fig. 10.23 and the 30 nodes for each parameter to discretize the parameters with a geometric step. The discretized range for resistivity is from 1 to 200 ohm m, the distance from transmitter to boundary is in the range from 1 to 50 m, and the deviation angle is from 0 to 90 degrees. The transient signal is calculated with a geometric step in the time interval from 1 μs to 10 ms (100 points).

The probe is a simple two-coil system with axial transmitting and receiving coils separated by $L=5$ m spacing. The length of the perpendicular $d2b$ from the receiver to the boundary (Fig. 10.23) is called the distance to the boundary (or so-called true vertical depth), and the distance between the receiver and the boundary along the trajectory is called measured depth, $Rd2b=d2b/\cos(\alpha)$ (Fig. 10.23).

To study the stability of the table-based inversion algorithm, we constructed numerical experiments consisting of 100 consecutive inversion runs. In each run, a different noise realization of the same standard deviation is used to contaminate the data. The sets of inverted parameters comprise the results of our statistical inversion. One of the input parameters used in the experiments is the number of logging points along the drilling trajectory. By varying this number, we can see how additional logging points affect the stability of inverted parameters.

In the analysis we look at two sets of the data: in the first set, data from only one logging point closest to the boundary are included; in the second, data from two additional logging points from the previous toll positions are added. In the numerical experiments, there are three different deviation angles α of 0, 45, and 83 degrees are used, corresponding to the vertical, deviated, and near-horizontal borehole trajectory, correspondingly. The standard deviation of the relative Gaussian noise imposed on the data is either 10% or 20% of the signal level.

Statistical Inversion for the Case of a Vertical Well (0 Degree Deviation Angle)

In the first example, the boundary is placed at the distance of $d2b=23.5$ m from the receiver, and the resistivity values of the first and second layers are $\rho_1 = 40$ and $\rho_2 = 1$ ohm m, correspondingly. In the experiments 20% noise is added to the data. In Fig. 10.24, statistical inversion results are presented for the case when data are placed at either one logging point at the distance $Rd2b=23.5$ m from the boundary (left subplot) or three logging points at distances $Rd2b=23.5, 33, 45$ (right subplot). In each subplot, the x-axis represents the inverted deviation angle, and the left y-axis depicts inverted distance to boundary $d2b$ (dots). The right y-axis shows percentage P of

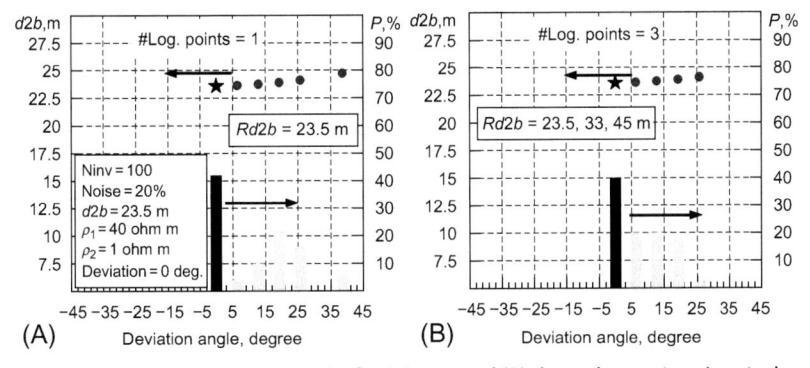

Fig. 10.24 Statistical inversion results for (A) one and (B) three data points (vertical well, 20% noise).

repetitions when inverted parameters *d2b* and deviation angle resulted in the same values (gray bars) during 100 sequential runs. The exact values of the parameters are represented by the black stars and bars.

As shown in Fig. 10.24, the percentage of runs when inversion accurately found both distance to bed *d2b* and deviation angle α practically does not depend on the number of logging points used in the inversion; in both cases it stays at approximately 40%. At the same time, the number of inversions resulted with a deviation angle larger than 20 degrees decreases from 17% to 3% when three logging points instead of one are used. Similarly, we can see reduction of outcomes with inverted *d2b* exceeding 24.0 m. In fact, by increasing the data set from one to three measurements, we effectively increase the signal-to-noise ratio, and it is eventually to the reduced number of outliers and increased stability of the inversion. The effect of reduced noise is further shown in Fig. 10.25 where statistical inversion was conducted for 10% noise. As seen from Figs. 10.24 and 10.25, the main consequence of reduced noise is reduced spread in the inverted parameters. Specifically, the results in Fig. 10.24 (right subplot, three logging points) are similar to those presented in Fig. 10.25 (left subplot, one logging point).

The stability of the inversion is further improved when three logging points are used (Fig. 10.25, right subplot). In particular, in the case of three logging points and 10% noise, there are no outcomes of inverted angles larger than 15 degrees. Interestingly, the inversion provided an accurate result for the resistivity of both layers in all analyzed cases (for this reason, inverted resistivity is not presented). These parameters are well defined

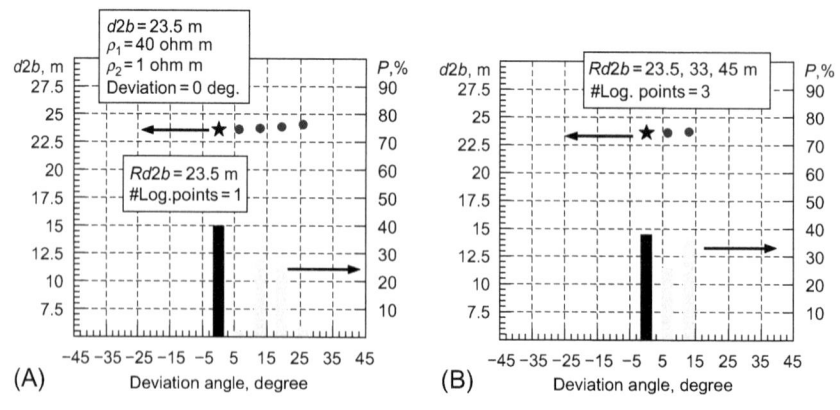

Fig. 10.25 Statistical inversion results for (A) one and (B) three data points (vertical well, 10% noise).

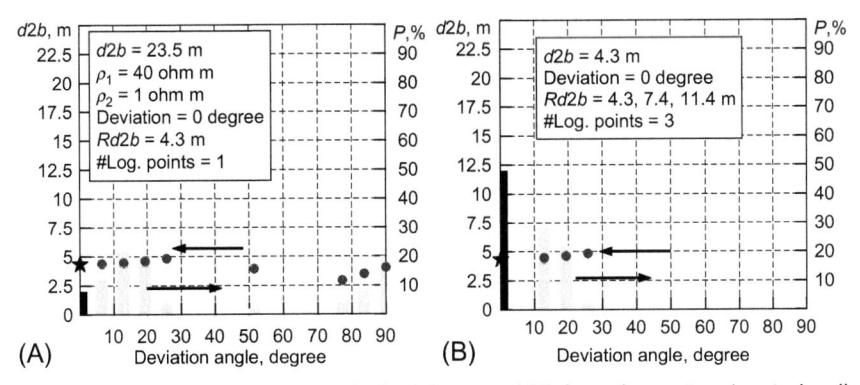

Induction Logging Based on Transient EM Measurements   373

Fig. 10.26 Statistical inversion results for (A) one and (B) three data points (vertical well, 20% noise).

because the data include both an early time range (first microseconds), when the transient process strongly depends on the resistivity of the upper layer, and the very late stage (hundreds of microseconds), which is mainly driven by the resistivity of the bottom layer.

Shallow Distance to the Boundary d2b = 4.3 m

When the ahead-placed boundary is located at the shallow distance to bed $d2b = 4.3$ m, it becomes comparable with the probe length ($L = 5.0$ m), and this leads to a significant uncertainty in the inverted parameters. This is shown in Fig. 10.26, in which results of statistical inversion are presented for the case of the data contaminated with 20% noise. In the left subplot, the spread in the deviation angle covers the entire range of 90 degrees. In other words, there is no stability in determining deviation. But the stability is drastically improved when data at two points $Rd2b = 7.4$ and 11.4 m are added. We can see (Fig. 10.26, right subplot) that the spread in the inverted deviation angle is reduced by a factor of approximately 4.

Also, the number of outcomes corresponding to the exact solution for $d2b$ increases from 8% to 48% (compare black bars in the left and right subplots), significantly reducing the number of outliers with deviation angle α larger than 20 degrees (gray bars, right subplot).

Statistical Inversion for Deviation Angle = 45 Degrees and Deep Distance to Boundary Rd2b = 25 m

The results of statistical inversion for the case of deep $d2b$ when $Rd2b = 25$ m are shown in Fig. 10.27 (noise = 20%). The major difference from the case

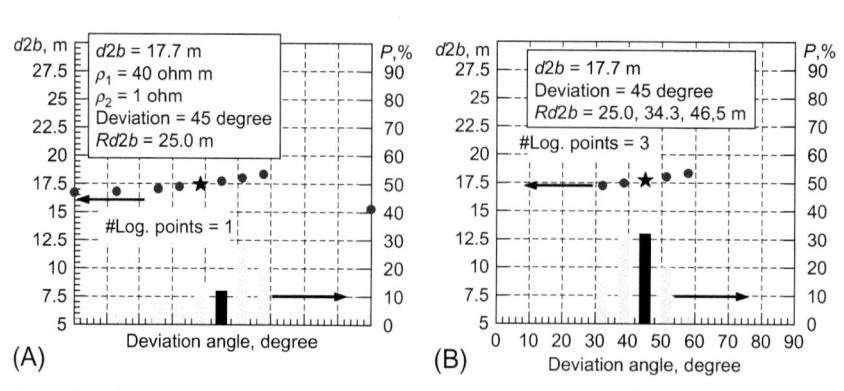

Fig. 10.27 Statistical inversion results for (A) one and (B) three data points (deviation angle = 45 degrees, deep d2b, 20% noise).

with a zero deviation angle is the significantly more pronounced effect caused by the increased number of logging points on the inversion result. Indeed, an increase to three logging points reduces the spread in $Rd2b$ from the range of 15.3–18.3 m to the range of 17.2–18.3 m.

Although additional logging points placed further from the boundary do not directly improve sensitivity to the boundary position, their combination does improve sensitivity to the deviation angle, and this eventually leads to the improved stability in inverted $d2b$.

Regularization Using Constraints

In the previous examples we assumed that all four parameters are unknown and we have no prior information about the parameters of the model. In many practical cases, however, we do have knowledge about the formation either in terms of approximate values of those parameters or, at least, the range of their possible variation. Incorporation of this knowledge into the inversion scheme reduces the size of the parameter's space and increases stability of the inversion. In application to the table-based inversion, incorporation of constraints assumes exclusion of some region of the constrained parameters from the global search or scanning, and there is no surprise that the reduced parameter space leads to the reduced uncertainty in the inverted parameters.

The effectiveness of the constraints in reducing uncertainties of parameters is illustrated in the following example in which we consider a two-layered formation with resistivity around the probe of 10 ohm m and resistivity of the ahead-placed second layer of 1 ohm m. The spacing between transmitter and receiver is 5 m and the level of the relative noise is 20%. In the inversion we use two logging points placed at distances of

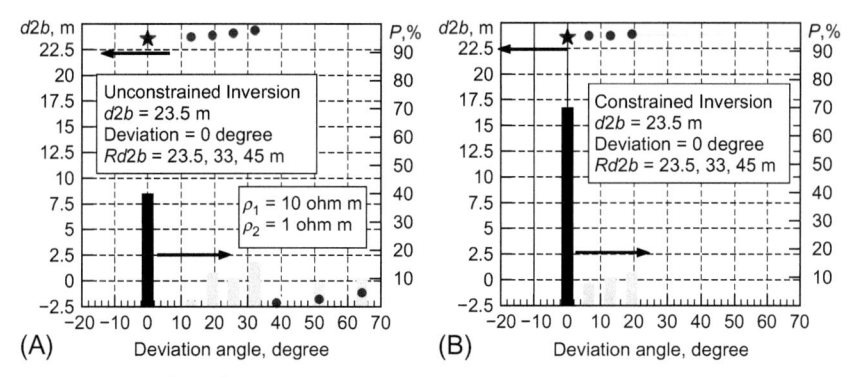

Fig. 10.28 Results of (A) unconstrained and (B) constrained inversion (deviation angle = 0 degree, 20% noise).

$Rd2b = 23.5$ m and $Rd2b = 33.5$ m. Results of the statistical inversion are shown in Fig. 10.28. The subplot on the left represents inversion results when no constraints are imposed on the parameters. We can see that there are some outliers in the region above 30 degrees (dots), which correspond to the scenarios of the receiver intersecting the boundary. The subplot on the right shows results of the inversion when a constraint on the boundary was applied ($d2b > 0$) by imposing the boundary to be below the receiver. The subplot on the right in Fig. 10.28 shows that the constraints removed all the erroneous values of parameter $d2b$ by eliminating outliers above 30 degrees.

Overall, table-based inversion is robust and provides satisfactory results in finding parameters of interest. Uncertainties in deviation angle and resistivities are in the acceptable range and permit accurate distance to the ahead-placed boundary. An increased number of logging points are the most beneficial in case of deviated trajectories and benchmarks with the distance to bed comparable with the transmitter-receiver spacing. Constraints help reduce uncertainties in inverted parameters and avoid erroneous inversion results.

10.5.5 Multiparametric Inversion

In the productive layer, it is important to navigate a horizontal well by detecting bed boundaries and keeping the well in the sweet spot throughout long lateral sections. The simplest model that describes this scenario consists of three layers. The trajectory is assumed to be parallel to the boundaries and there are at least five parameters of interest: resistivity of each layer, the thickness of the layer, and position of the probe with respect to the boundaries (Fig. 10.27B). The number of parameters can be even higher if additional layers or anisotropy are taken into account.

In this case, application of the table-based inversion is not effective because doing so requires large computation resources to generate the tables and significant time consumption to perform the scanning through each node in the table. For five or more invertible parameters, it is advisable to use advanced optimization techniques that do not rely on precalculated lookup tables but generate synthetic signal during the inversion.

Gauss-Newton Method

One of the most popular iterative methods for solving a least-squares problem is the Gauss-Newton (GN) method. This method requires calculation of the first derivatives of the minimized function $\Phi(\vec{x})$, called the Jacobian matrix $\hat{J}(\vec{x})$. The derivatives are calculated with respect to the parameters of interest (parameters of inversion), comprising a vector of unknowns $\vec{x}$. The function $\Phi(\vec{x})$ represents sum of squared residuals between measured $y_k (k=1,\ldots,n)^k$ and model-predicted values s_k.

In the least-squares formulation, the function $\Phi(\vec{x})$ is presented as:

$$f(\vec{x}) = \|\Phi(\vec{x})\|^2 = \sum_{k=1}^{n} \left(s_k(\vec{x}) - y_k \right)^2 \qquad (10.76)$$

where $\Phi(\vec{x})$ is the column-vector of n elements:

$$\Phi(\vec{x}) = \begin{bmatrix} (s_1(\vec{x}) - y_1) \\ (s_2(\vec{x}) - y_2) \\ \ldots \\ (s_n(\vec{x}) - y_n) \end{bmatrix} \qquad (10.77)$$

These kinds of nonlinear problems are quite popular and have many practical applications in finding sets of parameters satisfying the measurements. Let $\hat{J}(\vec{x})$ be a Jacobian matrix of function $\Phi(\vec{x})$, consisting of n rows and m columns:

$$\hat{J}(\vec{x}) = \begin{bmatrix} \dfrac{\partial s_1(\vec{x})}{\partial x_1} & \dfrac{\partial s_1(\vec{x})}{\partial x_2} & \cdots & \dfrac{\partial s_1(\vec{x})}{\partial x_m} \\ \dfrac{\partial s_2(\vec{x})}{\partial x_1} & \dfrac{\partial s_2(\vec{x})}{\partial x_2} & \cdots & \dfrac{\partial s_2(\vec{x})}{\partial x_m} \\ \ldots & \ldots & \ldots & \ldots \\ \dfrac{\partial s_n(\vec{x})}{\partial x_1} & \dfrac{\partial s_n(\vec{x})}{\partial x_2} & & \dfrac{\partial s_n(\vec{x})}{\partial x_m} \end{bmatrix} \qquad (10.78)$$

where n is the number of measurements and m is the number of parameters in the model.

Then, assuming some initial guess for the vector $\vec{x}_0$ the sequential approximations of $\vec{x}_{j+1}$ according to the GN method can be found as:

$$\vec{x}_{j+1} = \vec{x}_j - \left(J^T(\vec{x}_j)J(\vec{x}_j)\right)^{-1} J^T(\vec{x}_j)\Phi(\vec{x}_j) \qquad (10.79)$$

where $\hat{J}^T(\vec{x})$ is a transposed Jacobian matrix. The GN method relies on the fact that the second derivatives Hessian matrix, having a least-squares form of:

$$\hat{H}(\vec{x}) = \begin{bmatrix} \dfrac{\partial^2 f(\vec{x})}{\partial x_1 \partial x_1} & \dfrac{\partial^2 f(\vec{x})}{\partial x_1 \partial x_2} & \cdots & \dfrac{\partial^2 f(\vec{x})}{\partial x_1 \partial x_m} \\ \dfrac{\partial^2 f(\vec{x})}{\partial x_2 \partial x_1} & \dfrac{\partial^2 f(\vec{x})}{\partial x_2 \partial x_2} & \cdots & \dfrac{\partial^2 f(\vec{x})}{\partial x_2 \partial x_m} \\ \cdots & \cdots & \cdots & \cdots \\ \dfrac{\partial^2 f(\vec{x})}{\partial x_m \partial x_1} & \dfrac{\partial^2 f(\vec{x})}{\partial x_m \partial x_2} & \cdots & \dfrac{\partial^2 f(\vec{x})}{\partial x_m \partial x_m} \end{bmatrix} = \left[\dfrac{\partial f(\vec{x})}{\partial x_i \partial x_j}\right]_{i=1, j=1}^{m} \qquad (10.80)$$

$$= J^T(\vec{x})J(\vec{x}) + Q(\vec{x})$$

where

$$Q(\vec{x}) = \sum_{i=1}^{n} \Phi_i(\vec{x}) H_i(\vec{x}) \qquad (10.81)$$

can be approximated through the Jacobian matrix as $\hat{H}(\vec{x}) \approx J^T(\vec{x})J(\vec{x})$. The approximation is valid unless the residuals $\Phi_i(\vec{x})$ become large and the first term in Eq. (10.80) no longer dominates over the second term. The improved version of the GN method is the Levenberg-Marquardt (LM) algorithm, which is based on some heuristic ideas and allows improving stability and convergence of the iterations.

Levenberg-Marquardt Method

It happens that the GN method demonstrates oscillatory features during iterations, manifesting lack of robustness. To overcome this issue, Levenberg and Marquardt provided a damped least-squares algorithm, which adjusts some damping factor λ to control rate of convergence. The updated approximation in the LM for $\vec{x}_{j+1}$ is

$$\vec{x}_{j+1} = \vec{x}_j - \left(J^T(\vec{x}_j)J(\vec{x}_j) + \lambda_k \hat{I}\right)^{-1} J^T(\vec{x}_j)\Phi(\vec{x}_j) \qquad (10.82)$$

where $\hat{I}$ is the identity matrix and $\vec{x}_{j+1} - \vec{x}_j$ is the incremental update in the estimated vector of parameters. The (nonnegative) damping factor, λ, is adjusted at each iteration. If reduction of $\Phi(\vec{x})$ is rapid, a smaller value can be used, bringing the algorithm closer to the GN algorithm, whereas if an iteration gives insufficient reduction in the residual, λ can be increased. For large values of λ, the step will be taken approximately in the direction of the gradient. If either the length of the calculated step or the reduction of sum of squares from the latest parameter vector fall below predefined limits, iteration stops and the last parameter vector are considered to be the solution.

The drawback of the form Eq. (10.82) is that if the value of damping factor λ is large, inverting $\left(J^T(\vec{x}_j)J(\vec{x}_j) + \lambda_k \hat{I}\right)$ is not used at all [3]. Marquardt suggested scaling each component of the gradient according to the curvature so that there is larger movement along the directions where the gradient is smaller. This avoids slow convergence in the direction of small gradient. Therefore, Marquardt replaced the identity matrix, $\hat{I}$ with the diagonal matrix consisting of the diagonal elements of $\left(J^T(\vec{x}_j)J(\vec{x}_j)\right)$, resulting in the LM algorithm:

$$\vec{x}_{j+1} = \vec{x}_j - \left(J^T(\vec{x}_j)J(\vec{x}_j) + \lambda_k diag\left(J^T(\vec{x}_j)J(\vec{x}_j)\right)\right)^{-1} J^T(\vec{x}_j)\Phi(\vec{x}_j)$$

The choice of damping parameter λ is more or less heuristic and mainly depends on how well the initial problem is scaled. It is recommended to start from a large number λ_0, calculate the residual sum of squares, and then reduce λ in the next step by a factor of ν. If both of these are worse than the initial point, then the damping is increased by successive multiplication by ν until a better point is found with a new damping factor of $\lambda_0 \nu^k$ for some k.

If use of the damping factor λ_0/ν results in a reduction in squared residual then this is taken as the new value of and the process continues. If using λ/ν resulted in a worse residual, but using λ resulted in a better residual, then λ is left unchanged and the new optimum is taken as the value obtained with λ as damping factor.

The LM algorithm is a very popular curve-fitting algorithm used in many software applications for solving generic curve-fitting problems. However, this algorithm finds only a local minimum like all other iterative procedures, not a global minimum.

Combination of Iterative and Table-Based Inversion Algorithms

Both the GN and LM iterative methods of solving least-squares problems require some initial guess. In the case of geo-steering a horizontal well, it is a natural choice to use table-based inversion for finding three parameters of a two-layer model (Fig. 10.29A) and then, use them as an initial guess for the iterative procedure with five parameters corresponding to the three-layer model (Fig. 10.29B).

The model in Fig. 10.29B corresponds to a typical scenario of a probe that is located close to a reservoir roof at the distance $d2b_1$, and the task is to identify all the parameters, including a distance $d2b_2$ to the reservoir floor. Of course, the greater the thickness of the middle layer, for which the lower boundary of the second layer still can be detected, the better for the navigation. In the presented example the transient synthetic responses include XX, YY, and ZZ components affected by 10% random noise. In the notation the first index corresponds to the orientation of the transmitting dipole and the second to the receiving dipole. The models are selected to illustrate limitations of the inversion in resolving a distance to the lower boundary $d2b_2$. Therefore, we consider three cases with $d2b_2 = 2$, 5, and 10 m, correspondingly. The rest of parameters of the models are $\rho_1 = 1$ ohmm, $\rho_2 = 5$ ohmm, $\rho_3 = 1$ ohmm, and $d2b_1 = 1$ m.

Table 10.4 summarizes the results of inversion experiments. The cells contain the true value of the parameter (on top), result of the table-based search (second line), and result of the iterative inversion (at the bottom) for all three models.

The misfit F shows the relative difference between experimental and synthetic data normalized by the noise level, and N_F is the number of forward modeling performed to find the solution by iterative inversion.

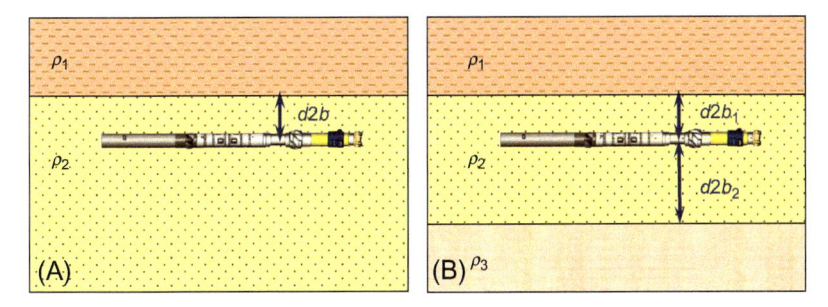

Fig. 10.29 (A) Three-parametric model used to find initial guess for the (B) five-parametric model in iterative inversion.

Table 10.4 Original Models and Inverted Parameters

Parameters	ρ_1 ohm m	ρ_2 ohm m	ρ_3 ohm m	$d2b_1$ m	$d2b_2$ m	Misfit F
Model 1	1	5	1	1	2	
Table-based inversion	1	5.46	5.46	0.7	n/a	63.0
Iterative inversion	0.998	5.05	1.01	0.98	2	0.39 (N_F=44)
Model 2	1	5	1	1	5	
Table-based inversion	1	5.46	5.46	0.7	n/a	55.0
Iterative inversion	0.98	5.0	1.03	0.97	5.2	0.47 (N_F=61)
Model 3	1	5	1	1	10	
Table-based inversion	1	5.46	5.46	0.7	n/a	35.1
Iterative inversion	0.97	5.1	1.3	0.95	7.5	0.35 (N_F=63)

In all three cases, the table-based inversion provides an accurate estimate of the resistivity $\rho_1 = 1$ ohm m of the upper layer located near the probe. The resistivities of the second and third layers are determined within the error of 10%, and this error leads to 30% error in $d2b_1$. The inverted $d2b_1$ values are further improved by the iterative inversion. Also, iterative inversion gives a very good estimate for the lower boundary $d2b_2$ when it is below 10 m (Models 2 and 3). But with increase of the thickness of the second layer, the data become less sensitive to the far-placed boundary, and the error in $d2b_2$ reaches 30% when $d2b_2 = 10$ m (Model 3).

10.5.6 Estimation of Parameter Uncertainties

In addition to estimation of parameters, inversion has to assess their uncertainties. This can be done by through linear approximation of the responses around the inverted model [4]. Let us denote f_i ($i = 1, \ldots, n$) measurements and ε_i ($i = 1, \ldots, n$) noise associated with these measurements, $\vec{p}^0$ — vector of parameters. This noise causes inverted parameters to be defined with some uncertainties $\delta p_j = p_j - p_{0j}$, $^{0j0j} j = 1, \ldots, m$, and our goal is to estimate those uncertainties (Fig. 10.30) assuming that the noise in the data is well known. The data set might comprise any combination of electromagnetic components taken at any subset of times and subarrays. The parameters are electrical and geometrical properties of the formation and resistivity of the layers.

Taking into account only the linear term of a Taylor series for the signal decomposition with respect to parameters, we have

$$\delta \vec{f} = \hat{Z} \cdot \delta \vec{p} \qquad (10.83)$$

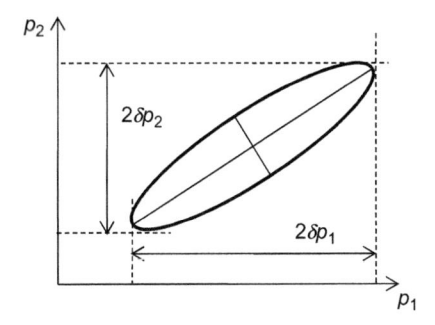

Fig. 10.30 Projection of equivalent ellipsoid onto parameter axis.

where the vector δf_i ($i = 1, \ldots, n$) describes the change of the signals corresponding to the change in the parameters $\delta \vec{p}_j$, while the matrix $\hat{Z}$ is a Jacobean matrix comprising partial derivatives of the signals with respect to the parameters of interest $\vec{p}^{\,0}$:

$$Z_{ij} = \frac{\partial f_i}{\partial p_j}(\vec{p}^{\,o}) \tag{10.84}$$

The matrix $\hat{Z}$ has n rows and m columns. Each measurement is characterized by the error ε_i comprising a vector:

$$\varepsilon_i = |f_i^0| Err_i + Aerr_i \tag{10.85}$$

where Err_i, $Aerr_i$ are relative and absolute measurements errors, correspondingly. The error might be described, for example, in terms of the standard deviation. In this case, for the covariance matrix we have

$$\hat{\Sigma} = \begin{pmatrix} \cdot & & & & 0 \\ & \cdot & & 0 & \\ & & \frac{1}{\varepsilon_i^2} & & \\ & 0 & & \cdot & \\ 0 & & & & \cdot \end{pmatrix} \tag{10.86}$$

Having defined matrixes $\hat{Z}$ and $\hat{\Sigma}$, we also introduce square matrix $\hat{A}$

$$\hat{A} = \hat{Z}^T \Sigma^2 Z \tag{10.87}$$

Normalizing the data by the noise we set up a data equivalence region:

$$\frac{1}{n}\vec{\delta f}^T \cdot \hat{\Sigma}^2 \cdot \vec{\delta f} = 1 \qquad (10.88)$$

and from Eqs. (10.83), (10.86) for the equivalence region for parameters we have

$$\frac{1}{n}\vec{\delta p}^T \cdot \hat{A} \cdot \vec{\delta p} = 1 \qquad (10.89)$$

Eq. (10.89) describes m-dimensional ellipsoid in the parameter space [4]. To derive an explicit expression for the variation in parameters $\vec{\delta p}$ we use eigenvectors $\vec{v}_i$ and eigenvalues λ_i of the matrix $\hat{A}$:

$$\hat{A}\hat{V} = \hat{\Lambda}\hat{V} \qquad (10.90)$$

where columns of $\hat{V}$ are the eigenvectors and the diagonal elements are eigenvalues λ_i. Then uncertainties in the parameters are expressed through projections of the ellipsoid to the parameter axis:

$$\vec{\delta p}_j = \sqrt{\sum_{i=1}^{m}(l_i \cdot v_{ji})^2} \qquad (10.91)$$

The described technique was applied to estimate uncertainties in the inverted parameters for Model 3 in Table 10.5. The uncertainties are expressed in % for resistivity and in meters (m) for distances.

It is seen that the better determined parameters (ρ_1, ρ_2) have a narrow range of uncertainty compared to the parameter $d2b_2$, which is determined less accurately.

The described approach does not guarantee that the inverted parameters are certainly placed in the range of estimated uncertainties; rather it estimates the range of uncertainties assuming that the inverted model is the one that fits the measurements.

To conclude, we must emphasize that the nonuniqueness in the parameters is an inherent part of the inversion and cannot be completely

Table 10.5 Uncertainties in the Inverted Parameters for Model 3

Model 3	$\rho_1 = 1.0$ (ohm m)	$\rho_2 = 5.0$ (ohm m)	$\rho_3 = 1.0$ (ohm m)	$d2b_1 = 1.0$ (m)	$d2b_2 = 10$ (m)
Inverted parameter	0.97	5.1	1.3	0.95	8.5
Range of variation	0.96–0.98	5.05–5.15	1.15–1.45	0.9–1.0	7.1–9.9

eliminated, but can be reduced. The means to reduce the nonuniqueness include: completeness of the measurements (spatial, orientational, and time sampling); decrease of the error in the data; optimal parameterization of the model, which takes into account only physically important parameters; good initial guess; and complementary prior information.

REFERENCES
[1] Kaufman A, Sokolov V. Theory of induction logging based on the use of transient field. Novosibirsk: Nauka; 1972.
[2] Bespalov A, Rabinovich M, Tabarovsky LA. Deep resistivity transient method for MWD applications using asymptotic filtering. Patent number US 7,027,922; 2006.
[3] Garcia-Cruz XM, Sergiyenko OY, Tyrsa V, Rivas-Lopez M, et al. Optimization of 3D laser scanning speed by use of combined variable step. Opt Lasers Eng 2014;54: 141–15. Elsevier.
[4] Nardi G, Martakov S, Nikitenko M, Rabinovich MB. Evaluation of parameter uncertainty utilizing resolution analysis in reservoir navigation increases the degree of accuracy and confidence in wellbore placement, In: SPWLA 51st annual logging symposium, Australia; 2010.

FURTHER READING
[1] Itskovich G. Method for measuring transient electromagnetic components to perform deep geosteering while drilling. Patent number US 7,046,009; 2006.
[2] Itskovich G. Method of eliminating conductive drill parasitic influence on the measurements of transient electromagnetic components in MWD tools. Patent number US 7,150,316; 2006.
[3] Itskovich G. Method for measuring transient electromagnetic components to perform deep geosteering while drilling. Patent number US 7,167,006; 2007.
[4] Itskovich G. Deep MWD resistivity measurements using EM shielding. Patent number US 8,278,930; 2012.
[5] Itskovich G, Reiderman A. Method and apparatus for while-drilling transient resistivity measurements. Patent number US 8,035,392; 2011.
[6] Itskovich G, Chemali R, Wang T. Electromagnetic and magneto-static shield to perform measurements ahead of the drill bit. Patent number US 7,994,790; 2011.

CHAPTER ELEVEN

Induction Logging Using Transversal Coils

Contents

11.1 Electromagnetic Field of the Magnetic Dipole in a Uniform Isotropic Medium	385
11.2 Boundary Value Problem for the Horizontal Magnetic Dipole in the Cylindrically Layered Formation	388
11.3 Magnetic Field in the Range of Small Parameter	396
11.4 Magnetic Field in the Far Zone	406
11.5 Magnetic Field in a Medium With Two Cylindrical Interfaces	417
11.6 Magnetic Field in Medium With a Thin Resistive Cylindrical Layer	421
11.7 Magnetic Field in Medium With One Horizontal Interface	426
11.8 Magnetic Field of the Horizontal Dipole in the Formation With Two Horizontal Interfaces	432
11.9 Profiling With a Two-Coil Induction Probe in a Medium With Horizontal Interfaces	441
Further Reading	445

In previous chapters, we have considered various aspects of induction logging when the source of the field is the vertical magnetic dipole and the induced currents are located in horizontal planes.

In these cases thin resistive layers, as well as caverns and fractures, that are perpendicular to the borehole, practically do not manifest themselves; even in an anisotropic medium only longitudinal conductivity defines a measured signal. To increase sensitivity to thin resistive and anisotropic layers and, possibly, improve the vertical response of the induction probe, we turn to modification of induction logging with horizontally oriented coils.

11.1 ELECTROMAGNETIC FIELD OF THE MAGNETIC DIPOLE IN A UNIFORM ISOTROPIC MEDIUM

We start with the simplest case of a uniform conducting and isotropic medium. By analogy with complex amplitudes caused by the vertical magnetic dipole (Chapter 5), for the x-oriented dipole, we have

$$E_\phi^* = \frac{i\omega\mu_0 M_x}{4\pi R^2} \exp(ikR)(1-ikR)\sin\theta$$

$$B_R^* = \frac{2\mu_0 M_x}{4\pi R^3} \exp(ikR)(1-ikR)\cos\theta \qquad (11.1)$$

$$B_\theta^* = \frac{\mu_0 M_x}{4\pi R^3} \exp(ikR)\left(1-ikR-k^2R^2\right)\sin\theta$$

where M_x is the transversal dipole moment, $k = (1+i)/\delta$ is the wave number, $\delta = \sqrt{(2/\omega\mu_0\gamma^0)}$ is the thickness of the skin layer, and θ is the inclination or polar angle of a spherical system of coordinates. Similar to the case of the vertical magnetic dipole, current lines are also circles, but they are located in planes perpendicular to the x-axis. When the field is excited by the transversal dipole, the main component, arising on the borehole axis, is also oriented along the x-axis.

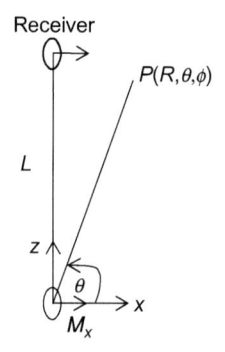

Transversal induction probe.

In accordance with Eq. (11.1), for the complex amplitude of this component, we have

$$B_x^* = B_0\left(1 - ikL - k^2L^2\right)\exp(ikL) \qquad (11.2)$$

where L is the length of the probe, and

$$B_0 = \frac{\mu_0 M_x}{4\pi L^3} \qquad (11.3)$$

is the field of the magnetic dipole in free space. Let us introduce function b_x^*, defined as

$$b_x^* = \frac{B_x^*}{B_0} = \exp(ikL)\left(1 - ikL - k^2L^2\right) \qquad (11.4)$$

Substituting $k = (1+i)/\delta$ into Eq. (11.4), we have the following expressions for the in-phase and quadrature components of b_x^*

$$Inb_x^* = [(1+p)\cos p + p(1+2p)\sin p]\exp(-p)$$
$$Qb_x^* = [(1+p)\sin p - p(1+2p)\cos p]\exp(-p) \quad (11.5)$$

where parameter $p = L/\delta$ is the distance from the dipole expressed in units of the skin depth. In accordance with Eq. (11.5), for the magnitude $A = \sqrt{(Inb_x^*)^2 + (Qb_x^*)^2}$ and the phase $\phi = \tan^{-1}(Qb_x^*/Inb_x^*)$ we have

$$A = \exp(-p)\left[(1+p)^2 + p^2(1+2p)^2\right]^{1/2}$$
$$\phi = p - \tan^{-1}[p(1+2p)/(1+p)] \quad (11.6)$$

First, consider a field in the near zone, when the parameter p is small. Expanding the exponent from Eq. (11.4) in a series and performing elementary transformations, we obtain

$$b_x^* = 1 + \sum_{n=0}^{\infty} \frac{(ikL)^{n+2}(n+1)}{n!(n+2)} \quad (11.7)$$

Restricting the sum in Eq. (11.7) to the first two terms we have

$$Inb_x^* \approx 1 + \frac{4}{3}p^3, \quad Qb_x^* \approx -p^2 + \frac{4}{3}p^3 \quad (11.8)$$

Thus, in the range of the small parameter, the quadrature component Qb_x^* prevails over the in-phase component $(Inb_x^* - 1)$ of the secondary field. The component Qb_x^* is directly proportional to the frequency and conductivity, and its magnitude is equal to that of the vertical magnetic dipole placed at the same distance along the z-axis (Chapter 5). In a wave zone at distances significantly exceeding the skin depth, the component B_θ^* is greater than B_R^*, and at an equatorial plane $B_{\theta=\pi/2}^* = B_x^*$, perpendicular to the x-axis:

$$B_x^* = -\frac{\mu_0 M_0}{4\pi L}k^2 \exp(ikL) \text{ if } |kL| \gg 1 \quad (11.9)$$

As follows from Eq. (11.1), the ratio of the electric field to the magnetic field at the wave zone does not depend on the distance, and it is equal to the impedance in a uniform medium:

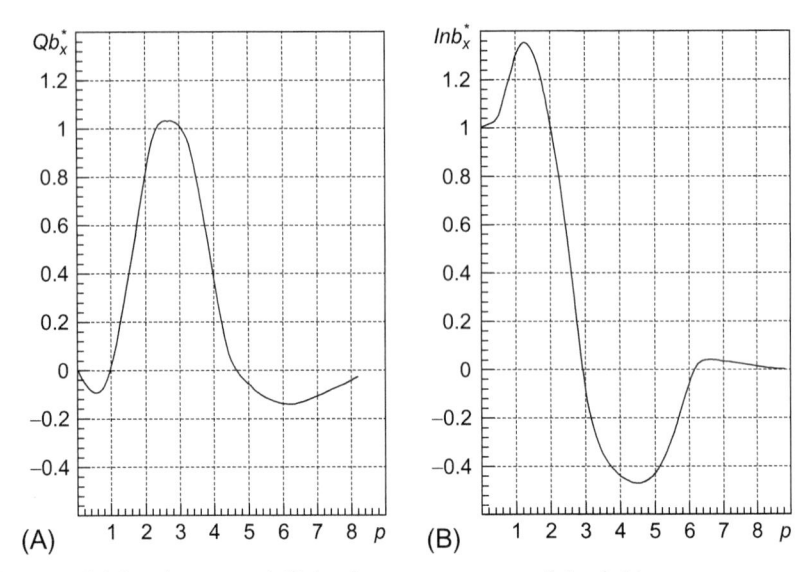

Fig. 11.1 (A) Quadrature and (B) in-phase components of the field.

$$\frac{E_\phi^*}{B_x^*} = -\frac{\omega}{k} \qquad (11.10)$$

Graphs of the quadrature and in-phase components of b_x^* are shown in Fig. 11.1, and an amplitude and phase of the secondary field $|b_x^* - 1|$ are shown in Fig. 11.2.

11.2 BOUNDARY VALUE PROBLEM FOR THE HORIZONTAL MAGNETIC DIPOLE IN THE CYLINDRICALLY LAYERED FORMATION

Next we consider a model consisting of formation, borehole, and horizontal magnetic dipole located on the axis. The radius and conductivity of the borehole are a and γ_1, respectively. The formation conductivity is γ_2 and the magnetic permeability of both regions coincides with that in free space. We introduce a cylindrical system of coordinates, and the magnetic dipole directed along the x-axis with moment $M = M_0 \exp(-\omega t)$ is placed at its origin (Fig. 11.3).

The system of equations for the quasistationary field is

$$\begin{array}{ll} curl E^* = i\omega B^* & div E^* = 0 \\ curl B^* = \gamma \mu_0 E^* & div B^* = 0 \end{array} \qquad (11.11)$$

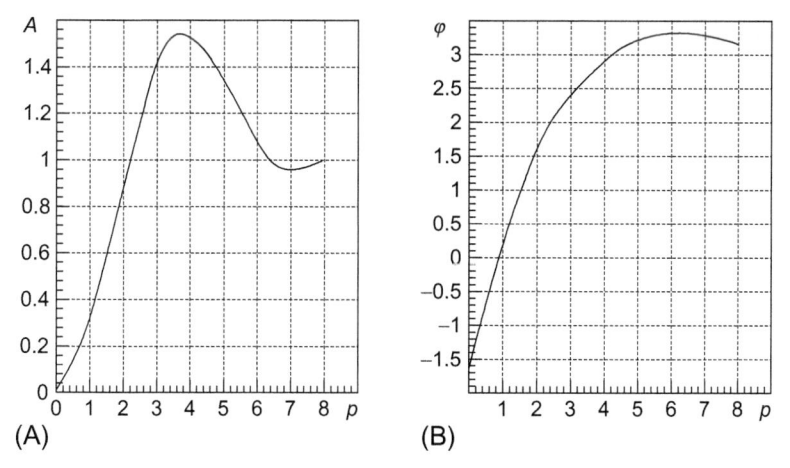

Fig. 11.2 (A) Amplitude and (B) phase of the secondary field.

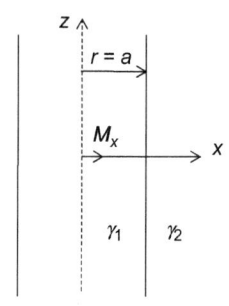

Fig. 11.3 Horizontal magnetic dipole on the borehole axis.

In the case of a horizontal dipole the primary vortex electric field, unlike that of a vertical dipole, intersects the boundary between media with different conductivities. For this reason electric charges arise on the borehole surface, and their density changes synchronously with the electric field at a given point. The charge density at each point depends on the conductivity contrast between the borehole and formation as well as on the coordinates of the point. In this case, when the sources of the secondary field are currents and charges, it is impossible to express the electromagnetic field using only one component of the vector potential. Solving the boundary value problem for the vector potential A^* leads to a system of differential equations of the second order. It is convenient to introduce two potentials, namely, an electric A_e type and a magnetic A_m type, and present a solution as a sum of the fields, corresponding to these potentials:

$$E^* = E^{(1)*} + E^{(2)*} \quad B^* = B^{(1)*} + B^{(2)*}$$

The relationships between the complex amplitudes of these fields and the amplitudes of the vector potentials are

$$E^{(1)*} = i\omega\, curl A_e^* \quad B^{(2)*} = curl A_m^* \tag{11.12}$$

Then, as follows from Eq. (11.11):

$$B^{(1)*} = k^2 A_e^* - grad\, U_e^* \quad E^{(2)*} = i\omega A_m^* - grad\, U_m^* \tag{11.13}$$

After introducing the following gage conditions:

$$\gamma U_e^* = -div A_e^* \quad \text{and} \quad U_m^* = -div A_m^* \tag{11.14}$$

we derive the following differential equations of the second order (Helmholtz equations):

$$\nabla^2 A_e^* + k^2 A_e^* = 0 \quad \nabla^2 A_m^* + k^2 A_m^* = 0 \tag{11.15}$$

In fact, the boundary value problem can be solved using only the vertical components of the vector potentials, that is:

$$A_e^* = (0, 0, A_{ez}^*), \quad A_m^* = (0, 0, A_{mz}^*) \tag{11.16}$$

In accordance with Eq. (11.12), the vertical component of the electric field is absent in magnetic potential, whereas the vertical component of the magnetic field is absent in the electric-type potential:

$$E_{ez}^* = 0 \quad \text{and} \quad B_{mz}^* = 0 \tag{11.17}$$

In the case of a uniform medium the fields of the magnetic dipole are fully described by a single vector potential of the magnetic type.

Connection between potential A_{mz}^* and corresponding electric and magnetic fields follows from Eqs. (11.12), (11.13):

$$E_{mr}^* = i\omega \frac{1}{r}\frac{\partial A_{mz}^*}{\partial \phi} \quad B_{mr}^*/\mu_0 = \frac{\partial^2 A_{mz}^*}{\partial r \partial z}$$

$$E_{m\phi}^* = -i\omega \frac{\partial A_{mz}^*}{\partial r} \quad B_{m\phi}^*/\mu_0 = \frac{1}{r}\frac{\partial^2 A_{mz}^*}{\partial \phi \partial z} \tag{11.18}$$

$$E_{mz}^* = 0 \quad B_{mz}^*/\mu_0 = k^2 A_{mz}^* + \frac{\partial^2 A_{mz}^*}{\partial z^2}$$

where potential A_{mz}^* satisfies the equation:

$$\frac{\partial^2 A_{mz}^*}{\partial r^2} + \frac{1}{r}\frac{\partial A_{mz}^*}{\partial r} + \frac{1}{r^2}\frac{\partial^2 A_{mz}^*}{\partial \phi^2} + \frac{\partial^2 A_{mz}^*}{\partial z^2} + k^2 A_{mz}^* = 0 \qquad (11.19)$$

Similarly, for potential A_{ez}^* and corresponding field components, we have

$$E_{er}^* = \frac{1}{\gamma}\frac{\partial^2 A_{ez}^*}{\partial r \partial z} \qquad B_{er}^*/\mu_0 = \frac{1}{r}\frac{\partial A_{ez}^*}{\partial \phi}$$

$$E_{e\phi}^* = \frac{1}{\gamma r}\frac{\partial^2 A_{ez}^*}{\partial \phi \partial r} \qquad B_{e\phi}^*/\mu_0 = -\frac{\partial A_{ez}^*}{\partial r} \qquad (11.20)$$

$$E_{ez}^* = \frac{1}{\gamma}\left(k^2 A_{ez}^* + \frac{\partial^2 A_{ez}^*}{\partial z^2}\right) \quad B_{ez}^* = 0$$

and

$$\frac{\partial^2 A_{ez}^*}{\partial r^2} + \frac{1}{r}\frac{\partial A_{ez}^*}{\partial r} + \frac{1}{r^2}\frac{\partial^2 A_{ez}^*}{\partial \phi^2} + \frac{\partial^2 A_{ez}^*}{\partial z^2} + k^2 A_{ez}^* = 0 \qquad (11.21)$$

Eqs. (11.19), (11.21) are not independent ones because only by mixing A_{ez}^* and A_{mz}^* it is possible to satisfy a continuity of tangential components at the boundary $(r=a)$. The continuity results in the following system of boundary conditions for potentials A_{ez}^* and A_{mz}^*:

$$\frac{1}{\gamma_1}\left(k_1^2 A_{e1}^* + \frac{\partial^2 A_{e1}^*}{\partial z^2}\right) = \frac{1}{\gamma_2}\left(k_2^2 A_{e2}^* + \frac{\partial^2 A_{e2}^*}{\partial z^2}\right)$$

$$\frac{1}{\gamma_1}\left(\frac{1}{a}\frac{\partial^2 A_{e1}^*}{\partial \phi \partial z} - k_1^2 \frac{\partial A_{m1}^*}{\partial r}\right) = \frac{1}{\gamma_2}\left(\frac{1}{a}\frac{\partial^2 A_{e2}^*}{\partial \phi \partial z} - k_2^2 \frac{\partial A_{m2}^*}{\partial r}\right)$$

$$k_1^2 A_{m1}^* + \frac{\partial^2 A_{m1}^*}{\partial z^2} = k_2^2 A_{m2}^* + \frac{\partial^2 A_{m2}^*}{\partial z^2} \qquad (11.22)$$

$$-\frac{\partial A_{e1}^*}{\partial r} + \frac{1}{a}\frac{\partial^2 A_{m1}^*}{\partial \phi \partial z} = -\frac{\partial A_{e2}^*}{\partial r} + \frac{1}{a}\frac{\partial^2 A_{m2}^*}{\partial \phi \partial z}$$

where k_1, A_{e1}^*, A_{m1}^* and k_2, A_{e2}^*, A_{m2}^* are the wave numbers and the complex amplitude of the z-component of the vector potentials in the borehole and formation, respectively.

Let us find expressions for potentials A_{e0}^* and A_{m0}^* in a uniform medium with conductivity γ_1. The latter is needed to formulate conditions in the vicinity of the dipole for A_{e1}^* and A_{m1}^*. As mentioned earlier, the field of the dipole in a uniform isotropic medium can be described by a single component of the magnetic-type potential:

$$\mathbf{A}_m^* = (A_{mx}^*, 0, 0)$$

or

$$A_{mx}^* = \frac{\mu_0 M_0}{4\pi R} \exp(ik_1 R) = \frac{\mu_0 M_0}{2\pi^2} \int_0^\infty K_0(m_1 r) \cos mz\, dm \qquad (11.23)$$

where $m_1 = (m^2 - k_1^2)^{1/2}$, and

$$\mathbf{E}^* = i\omega\,\mathrm{curl}\,\mathbf{A}_m^*, \quad \mathbf{B}^* = k^2 \mathbf{A}_m^* + \mathrm{grad}\,\mathrm{div}\,\mathbf{A}_m^*. \qquad (11.24)$$

Therefore, for vertical components of the field, we obtain

$$E_{z0}^* = i\omega\mu_0 \frac{M_0}{2\pi^2} \sin\phi \int_0^\infty m_1 K_1(m_1 r) \cos mz\, dm$$

$$B_{z0}^* = \frac{\mu_0 M_0}{2\pi^2} \cos\phi \int_0^\infty m m_1 K_1(m_1 r) \sin mz\, dm \qquad (11.25)$$

where

$$\cos\phi = x/r \quad \text{and} \quad r = (x^2 + y^2)^{1/2}$$

On the other hand, by analogy with Eqs. (11.18), (11.20), we have

$$E_{z0}^* = \frac{1}{\gamma_1}\left(k_1^2 A_{e0}^* + \frac{\partial^2 A_{e0}^*}{\partial z^2}\right) \quad B_{z0}^* = k_1^2 A_{m0}^* + \frac{\partial^2 A_{m0}^*}{\partial z^2}. \qquad (11.26)$$

The corresponding potentials for the fields in a uniform medium Eq. (11.26) are

$$A_{e0}^* = -k_1^2 \frac{\mu_0 M_0}{2\pi^2} \sin\phi \int_0^\infty \frac{1}{m_1} K_1(m_1 r) \cos mz\, dm$$

$$A_{m0}^* = -\frac{\mu_0 M_0}{2\pi^2} \cos\phi \int_0^\infty \frac{m}{m_1} K_1(m_1 r) \sin mz\, dm \qquad (11.27)$$

The set of two potentials A_{e0}^* and A_{m0}^* in Eq. (11.27) represents an alternative form to Eq. (11.23), which describes the field in a uniform medium using just one potential A_m^*. The potentials A_{e0}^* and A_{m0}^* are not independent and the connection is given by the following relationship:

$$\frac{\partial^2 A_{e0}^*}{\partial \phi \partial z} = -k_1^2 A_{m0}^*$$

Returning to the original problem, it is natural to assume that while approaching the dipole, potentials A_{e1}^* and A_{m1}^* tend to potentials in the whole space A_{e0}^* and A_{m0}^*, respectively. Taking into account the behavior of the field near the source and at infinity, we present the potentials inside and outside of the borehole as:

$$A_{e1}^* = A_{e0}^* + k_1^2 \frac{\mu_0 M_0}{2\pi^2} \sin\phi \int_0^\infty \frac{1}{m_1} C_1 I_1(m_1 r) \cos mz\, dm$$

$$A_{m1}^* = A_{m0}^* + \frac{\mu_0 M_0}{2\pi^2} \cos\phi \int_0^\infty \frac{m}{m_1} D_1 I_1(m_1 r) \sin mz\, dm$$

(11.28)

$$A_{e2}^* = -k_2^2 \frac{\mu_0 M_0}{2\pi^2} \sin\phi \int_0^\infty \frac{1}{m_2} C_2 K_1(m_2 r) \cos mz\, dm$$

$$A_{m2}^* = -\frac{\mu_0 M_0}{2\pi^2} \cos\phi \int_0^\infty \frac{m}{m_2} D_2 K_1(m_2 r) \sin mz\, dm$$

where $m_2 = (m^2 - k_2^2)^{1/2}$. From boundary conditions Eq. (11.22), we may further derive a system of equations for coefficients C, D:

$$K_1(m_1 a) - I_1(m_1 a)C_1 = \frac{m_2}{m_1} K_1(m_2 a)C_2 \frac{1}{m_1 a}[K_1(m_1 a) - I_1(m_1 a)C_1]$$

$$+ [K_1'(m_1 a) - I_1'(m_1 a)D_1] = \frac{1}{m_2 a} K_1(m_2 a)C_2$$

$$+ K_1'(m_2 a)D_2 K_1(m_1 a) - I_1(m_1 a)D_1$$

$$= \frac{m_2}{m_1} K_1(m_2 a)D_2 k_1^2 [K_1'(m_1 a) - I_1'(m_1 a)C_1]$$

$$+ \frac{m^2}{m_1 a}[K_1(m_1 a) - I_1(m_1 a)D_1]$$

$$= k_2^2 K_1'(m_2 a)C_2 + \frac{m^2}{m_2 a} K_1(m_2 a)D_2$$

(11.29)

Solving this system we find

$$C_1 = \frac{\Delta_c}{\Delta}, \quad D_1 = \frac{\Delta_d}{\Delta}, \quad C_2 = \frac{\overline{m}_1}{\overline{m}_2}\frac{K_1(\overline{m}_1) - I_1(\overline{m}_1)C}{K_1(\overline{m}_2)}, \quad D_2 = \frac{\overline{m}_1}{\overline{m}_2}\frac{K_1(\overline{m}_1) - I_1(\overline{m}_1)D}{K_1(\overline{m}_2)}$$

$$\Delta_c = I_0(\overline{m}_1)K_0(\overline{m}_1) + I_1(\overline{m}_1)K_1(\overline{m}_1)P_1 - I_1(\overline{m}_1)K_1(\overline{m}_1)P_2 - s\frac{K_0(\overline{m}_2)}{\overline{m}_2 K_1(\overline{m}_2)}$$

$$\Delta_d = I_0(\overline{m}_1)K_0(\overline{m}_1) + I_1(\overline{m}_1)K_1(\overline{m}_1)P_1 - I_1(\overline{m}_1)K_1(\overline{m}_1)P_2 - \frac{K_0(\overline{m}_2)}{\overline{m}_2 K_1(\overline{m}_2)}$$

$$\Delta = -I_0^2(\overline{m}_1) + I_1^2(\overline{m}_1)P_1 + I_0(\overline{m}_1)I_1(\overline{m}_1)P_2$$

(11.30)

where

$$P_1 = \frac{2\overline{m}^2 - \overline{m}_2^2}{\overline{m}_2^3}(1-s)\frac{K_0(\overline{m}_2)}{K_1(\overline{m}_2)} - \frac{\overline{m}_1^2}{\overline{m}_2^2}s\frac{K_0^2(\overline{m}_2)}{K_1^2(\overline{m}_2)}$$

$$P_2 = \frac{2\overline{m}^2 - \overline{m}_1^2}{\overline{m}_1 \overline{m}_2^2}(1-s) - \frac{\overline{m}_1}{\overline{m}_2}(1+s)\frac{K_0(\overline{m}_2)}{K_1(\overline{m}_2)}$$

(11.31)

In Eqs. (11.30), (11.31) we replaced variables: $ma \to \overline{m}$, $m_1 a \to \overline{m}_1$, $m_2 a \to \overline{m}_2$, and $s = \gamma_2/\gamma_1$.

The magnetic field on the axis of the borehole has only component B_x, which is parallel to the dipole moment. Using Eqs. (11.18), (11.20), (11.28), we obtain expressions for the magnetic field on the axis ($r = 0$):

$$B_x^* = B_{0x}^* + \frac{\mu_0 M_0}{2\pi^2}\frac{1}{a^3}\int_0^\infty \left(\frac{\overline{m}^2}{2}D_1 + \frac{k_1^2 a^2}{2}C_1\right)\cos\left(\frac{L}{a}\overline{m}\right)d\overline{m} \qquad (11.32)$$

where

$$B_{0x}^* = -\frac{\mu_0 M_0}{4\pi L^3}\left(1 - ik_1 L - k_1^2 L^2\right)\exp(ik_1 L)$$

is the complex amplitude of the field in a uniform medium, and L is the length of the probe. Correspondingly, the field, in units of the primary field, is

$$b_x^* = \left(1 - ik_1 L - k_1^2 L^2\right) \exp\left(ik_1 L\right)$$
$$-\frac{\alpha^3}{\pi} \int_0^\infty \left(\overline{m}^2 D_1 + k_1^2 a^2 C_1\right) \cos\alpha \overline{m} d\overline{m} \qquad (11.33)$$

Results of calculations of the secondary amplitude $\left|b_x^* - 1\right|$ and phase ϕ for different ratios of γ_2/γ_1 are shown in Figs. 11.4 and 11.5. Along the x-axis we depict the ratio of the borehole radius a and skin depth δ_1 in the homogeneous media with conductivity of the borehole γ_1. The left-hand side of the curves corresponds to the low-frequency part of the spectrum where the amplitude of the secondary field is small and mainly defined by the quadrature component.

The secondary field increases with the frequency and almost compensates for the primary field approaching unity (right-hand asymptote of curves in Fig. 11.4). The phase of the secondary field in Fig. 11.5 tends to $-\pi/2$ in the range of the small parameter due to the quadrature component being greater than the in-phase component and the component's opposite signs. At the high-frequencies, the secondary field tends to compensate the primary field, and, correspondingly, the phase approaches π.

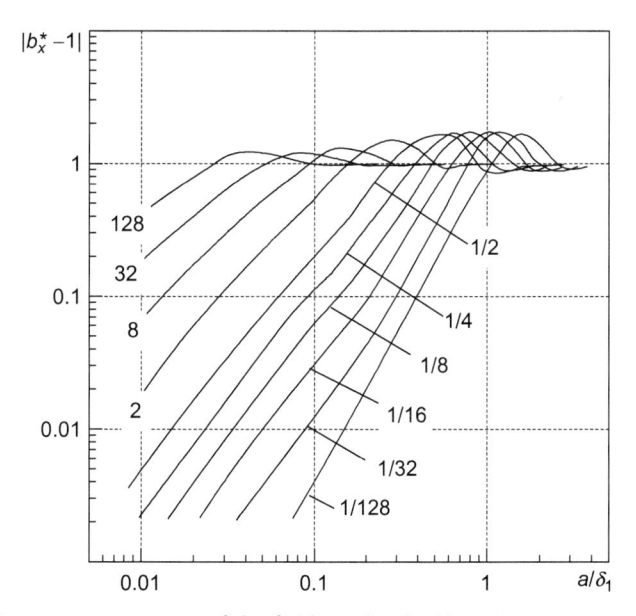

Fig. 11.4 Frequency responses of the field amplitude, $L/a = 10$.

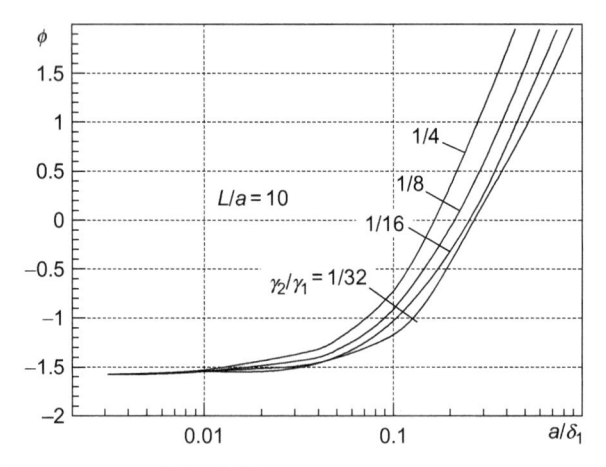

Fig. 11.5 Phase responses of the field.

11.3 MAGNETIC FIELD IN THE RANGE OF SMALL PARAMETER

A transversal dipole M_x generates eddy currents located in vertical planes. These currents intersect the borehole boundary, giving raise to surface charges. In the range of small parameter L/δ the intensity of these surface charges is proportional to the square of the wave numbers, i.e., $k_1^2 = i\gamma_1\mu_0\omega$ and $k_2^2 = i\gamma_2\mu_0\omega$. Naturally, surface charges affect the magnitude and direction of induced currents, but in the range of the small parameter, the phase of the currents is shifted by $\pi/2$ with respect to the primary current in the dipole. Thus both the magnetic field of induced currents and the secondary field of charges are proportional to ω. The secondary quadrature component of the electric field is relatively small and, therefore, it is not taken into account in the small parameter approximation. To obtain an asymptotic expression of the magnetic field, let us present an integrand of Eq. (11.33) in the form of a Maclaurin series expansion near $k^2 a^2 = 0$. Restricting the series to its first term, we obtain

$$b_x^* = b_{0x}^* - \frac{2\alpha^3}{\pi}\int_0^\infty \frac{\cos\alpha\overline{m}}{\Delta(0)}\left[\frac{k_1^2 a^2}{2}\Delta_c(0) - \frac{\overline{m}}{4}k_1^2 a^2 \frac{\partial \Delta_d(0)}{\partial \overline{m}_1} - \frac{\overline{m}}{4}k_2^2 a^2 \frac{\partial \Delta_d(0)}{\partial \overline{m}_2}\right] d\overline{m}$$

(11.34)

where

$$\Delta_c(0) = (1-s)\frac{K_0(\overline{m})}{\overline{m}K_1(\overline{m})}$$

$$\Delta = -I_0^2(\overline{m}_1) + I_1^2(\overline{m}_1)P_1 + I_0(\overline{m}_1)I_1(\overline{m}_1)P_2$$

$$\Delta(0) = -\frac{1}{\overline{m}K_1(\overline{m})}\left[I_0(\overline{m}) + sI_1(\overline{m})\frac{K_0(\overline{m})}{K_1(\overline{m})} - (1-s)\frac{I_1(\overline{m})}{\overline{m}}\right]$$

$$\frac{\partial \Delta_d(0)}{\partial \overline{m}_1} = (1-s)\left[I_1(\overline{m})K_0(\overline{m})\left(1+\frac{2}{\overline{m}^2}\right) + \frac{I_0(\overline{m})K_0(\overline{m})}{\overline{m}} + \frac{I_1(\overline{m})K_1(\overline{m})}{\overline{m}}\right.$$
$$\left. + \frac{I_0(\overline{m})K_0^2(\overline{m})}{K_1(\overline{m})} - \frac{K_0(\overline{m})}{\overline{m}^2 K_1(\overline{m})}\right] - \frac{1}{\overline{m}} + s\frac{K_0^2(\overline{m})}{\overline{m}K_1^2(\overline{m})} \quad (11.35)$$

$$\frac{\partial \Delta_d(0)}{\partial \overline{m}_2} = (1-s)\left[-\frac{2}{\overline{m}^2} + \frac{K_0(\overline{m})}{\overline{m}K_1(\overline{m})} - 1 + \frac{K_0^2(\overline{m})}{K_1^2(\overline{m})}\right]I_1(\overline{m})K_0(\overline{m})$$
$$- (1-s)\frac{I_1(\overline{m})K_1(\overline{m})}{\overline{m}} + \frac{1}{\overline{m}} - \frac{K_0^2(\overline{m})}{\overline{m}K_1^2(\overline{m})}$$

Also

$$b_{0x}^* = \frac{k_1^2 L^2}{2}$$

Thus, for the quadrature component of the field we have

$$Qb_x^* = -\left(\frac{L}{\delta_1}\right)^2 G_1(\alpha,s) - \left(\frac{L}{\delta_2}\right)^2 G_2(\alpha,s) \quad (11.36)$$

where $\delta_1 = (2/\gamma_1\mu_0\omega)^{1/2}$, $\delta_2 = (2/\gamma_2\mu_0\omega)^{1/2}$, and γ_1 and γ_2 are conductivities of medium of the borehole and formation. Also,

$$G_1(\alpha,s) = 1 + \frac{2\alpha}{\pi}\int_0^\infty \left[\Delta_c(0) - \frac{\overline{m}\partial\Delta_d(0)}{2\;\partial\overline{m}_1}\right]\frac{\cos\alpha\overline{m}}{\Delta(0)}d\overline{m}$$

$$(11.37)$$

$$G_2(\alpha,s) = -\frac{2\alpha}{\pi}\int_0^\infty \frac{\overline{m}\partial\Delta_d(0)}{2\;\partial\overline{m}_2}\frac{\cos\alpha\overline{m}\,d\overline{m}}{\Delta(0)}$$

As follows from Eq. (11.37), in a uniform medium:

$$G_1(\alpha,1) + G_2(\alpha,1) = 1 \quad (11.38)$$

Note that functions G_1 and G_2 are still called geometric factors regardless of their dependence on the ratio of conductivities. As was shown in Chapter 7, it is possible to obtain a more accurate expression for the low-frequency field. For instance, considering two terms of the expansion, we have

$$Inb_x^* = \frac{4}{3}(L/\delta_2)^3 \quad \text{and} \quad Qb_x^* \approx -a_1(L/\delta_1)^2 + \frac{4}{3}(L/\delta_2)^3 \qquad (11.39)$$

where the coefficient a_1 is defined by expression Eq. (11.37):

$$a_1 = G_1 + sG_2$$

Therefore, in the range of the small parameter, neither the in-phase component nor the second term of the quadrature component, Eq. (11.39), depends on the conductivity of the borehole. Similar behavior was already observed in the case of vertical magnetic dipole.

Let us consider functions G_1 and G_2 at the range of small parameter L/δ_1 and L/δ_2 at different values of α. If the probe length decreases, $\alpha \to 0$, then $G_2(\alpha, s) \to 0$, and $G_1(\alpha, s) \to 1$, approaching the geometric factor of a uniform medium with conductivity, γ_1. For large values of the parameter α due to rapid oscillations of the function $\cos(ma)$, the integral in Eq. (11.37) is defined by the integrand near $m=0$. For small values of m, we have

$$\Delta(0) \approx -\frac{1+s}{2} \quad \Delta_c(0) \approx (1-s)K_0(\overline{m})$$

$$\frac{\partial \Delta_d(0)}{\partial \overline{m}_1} \approx -\frac{\partial \Delta_d(0)}{\partial \overline{m}_2} = (1-s)\frac{K_0(\overline{m})}{m} \qquad (11.40)$$

By using the asymptotic presentation of the Summerfield integral:

$$\int_0^\infty K_0(\overline{m})\cos\alpha\overline{m}d\overline{m} = \frac{\pi}{2}\frac{1}{(1+\alpha^2)^{1/2}} \to \frac{\pi}{2\alpha} \quad \text{if } \alpha \to \infty$$

for $\alpha \gg 1$, we obtain

$$G_1(\alpha, s) \to 1 - \frac{1-s}{1+s} = \frac{2s}{1+s} = \frac{\gamma_2}{\gamma_{av}}$$

$$G_2(\alpha, s) = -\frac{1-s}{1+s} = \frac{\gamma_2 - \gamma_1}{\gamma_2 + \gamma_1} = -K_{12} \qquad (11.41)$$

where γ_{av} is the average value of conductivity and K_{12} is the contrast coefficient that characterizes the density of the charges on the surface of the borehole. Correspondingly, for the quadrature component, we have

$$Qb_x^* = -\left(\frac{L}{\delta_2}\right)^2 \quad \text{if } \alpha \gg 1 \qquad (11.42)$$

Thus, at the low-frequency limit with an increase of the length of the probe, the field tends toward that in a uniform medium with the conductivity of formation. In general, functions G_1 and G_2 depend on the resistivity of the medium regardless of the length of the probe. Now, let us introduce functions $G_1^*(\alpha, s)$ and $G_2^*(\alpha, s)$, which approach 0 and 1, respectively, when $\alpha \to \infty$:

$$G_1^*(\alpha, s) = G_1(\alpha, s) - \frac{2s}{1+s} \quad G_2^*(\alpha, s) = G_2(\alpha, s) + \frac{2}{1+s} \qquad (11.43)$$

Then, instead of Eq. (11.36), we can write

$$Qb_x^* = -\left(\frac{L}{\delta_1}\right)^2 \left[G_1^*(\alpha, s) + sG_2^*(\alpha, s)\right] \qquad (11.44)$$

First, consider the asymptotic behavior of function $G_1^*(\alpha, s)$ at large α. It is convenient to isolate singularity of the integrand in Eq. (11.37) for small values of m. For this purpose, we present $G_1^*(\alpha, s)$ as:

$$G_1^*(\alpha, s) = \frac{1-s}{1+s} + \frac{2\alpha}{\pi} \int_0^\infty \left\{ \frac{1}{\Delta(0)} \left[\Delta_c(0) - \frac{\overline{m}\partial\Delta_d(0)}{2 \partial \overline{m}_1}\right] + \frac{1-s}{1+s} K_0(\overline{m}) \right\}$$

$$\cos\alpha\overline{m}d\overline{m} - \frac{2\alpha}{\pi} \frac{1-s}{1+s} \int_0^\infty K_0(\overline{m}) \cos\alpha\overline{m}d\overline{m}$$

$$= \frac{1-s}{1+s}\left(1 - \frac{\alpha}{(1+\alpha^2)^{1/2}}\right) + \frac{2\alpha}{\pi}\int_0^\infty \phi(\overline{m}) \cos\alpha\overline{m}d\overline{m}$$

(11.45)

where

$$\phi(\overline{m}) = \frac{1}{\Delta(0)}\left[\Delta_c(\bar{k}) - \frac{\overline{m}\partial\Delta_d(0)}{2 \partial \overline{m}_1}\right] + \frac{1-s}{1+s}K_0(\overline{m})$$

Integrating Eq. (11.45) by parts and considering that the function $\phi(m)$ and its derivatives approach to zero when $m \to \infty$ we obtain

$$\int_0^\infty \phi(\overline{m}) \cos(\alpha \overline{m}) d\overline{m} = -\frac{1}{\alpha^2}\phi'(0) - \frac{1}{\alpha^2}\int_0^\infty \phi''(\overline{m}) \cos(\alpha \overline{m}) d\overline{m}$$

Using the known expressions for Bessel functions

$$K_0(\overline{m}) \approx -\left(1 + \frac{\overline{m}^2}{4}\right) \ln\frac{\overline{m}}{2} - C + \frac{\overline{m}^2}{4}(1-C)$$

$$K_1(\overline{m}) \approx \frac{1}{\overline{m}} + \frac{\overline{m}}{2}\ln\frac{\overline{m}}{2} - \frac{\overline{m}}{4}(1-2C)$$

$$I_0(\overline{m}) \approx 1 + \frac{\overline{m}^2}{4} \quad I_1(m) \approx \frac{\overline{m}}{2}\left(1 + \frac{\overline{m}^2}{8}\right)$$

we obtain

$$\phi(\overline{m}) = \frac{2s}{(1+s)^2}\overline{m}^2 \ln^2 \overline{m} - \left[\frac{3+3s+2s^2}{2(1+s)^2} + \frac{8s}{(1+s)^2}(\ln 2 - C)\right]\frac{\overline{m}^2}{2}\ln \overline{m} + const$$

$$\phi'(\overline{m}) = 0 \quad \text{and}$$

$$\phi''(\overline{m}) = \frac{4s}{(1+s)^2}\ln^2 \overline{m} - \left[\frac{3-21s+2s^2}{2(1+s)^2} + \frac{8s}{(1+s)^2}(\ln - C)\right]\ln \overline{m}$$

Here C is Euler constant. Inasmuch as the field at large distances from the dipole is defined by low-frequency spatial harmonics we can use an arbitrary number in the upper limit of the integral in Eq. (11.45). This gives

$$\int_0^\infty \phi(\overline{m})\cos\alpha\overline{m}d\overline{m} = -\frac{1}{\alpha^2}\left\{\frac{4s}{(1+s)^2}\int_0^1 \ln^2 \overline{m} \cos(\alpha \overline{m}) d\overline{m} \right.$$

$$\left. -\left[\frac{3-21s+2s^2}{2(1+s)^2} + \frac{8s}{(1+s)^2}(\ln 2 - C)\right]\int_0^1 \ln m \cos(\alpha m) d\overline{m}\right\}$$

Inasmuch as

$$\int_0^1 \ln^2 \overline{m} \cos(\alpha \overline{m}) d\overline{m} \approx \frac{1}{\alpha} \ln^2 \overline{m} \sin(\alpha \overline{m})\Big|_0^1 - \frac{2}{\alpha} \int_0^\infty \frac{\ln \overline{m}}{\overline{m}} \sin(\alpha \overline{m}) d\overline{m} = \frac{\pi}{2}(\ln \alpha + C)$$

then

$$\int_0^\infty \phi(\overline{m}) \cos(\alpha \overline{m}) d\overline{m} = -\frac{\pi}{\alpha^3}\left[\frac{4s}{(1+s)^2} \ln 2\alpha + \frac{3 - 21s + 2s^2}{4(1+s)^2}\right]$$

whence

$$G_1^*(\alpha, s) = -\frac{1}{\alpha^2(1+s)^2}\left(8s \ln 2\alpha + \frac{2 - 21s + 3s^2}{2}\right) \quad \text{if } \alpha \gg 1 \quad (11.46)$$

Similar transformations lead to:

$$G_2^*(\alpha, s) = 1 + \frac{1}{2\alpha^2}(1 + 5s - 22s^2 + 16s^2 \ln 2\alpha)\frac{1}{(1+s)^2}, \quad \alpha \gg 1 \quad (11.47)$$

If the formation resistivity is significantly higher than that of the borehole, then Eqs. (11.46), (11.47) can be further simplified:

$$G_1^*(\alpha, s) \approx -\frac{1}{\alpha^2} - \frac{s}{\alpha^2}(8 \ln 2\alpha - 12.5)$$

$$G_2^*(\alpha, s) \approx 1 + \frac{1}{2\alpha^2}, \quad \text{if } (\alpha \gg 1, s < 1)$$

$$(11.48)$$

Then, using Eq. (11.44), we have the following expression for the magnetic field:

$$Qb_x^* = \left(\frac{L}{\delta_1}\right)^2 \left\{\frac{1}{\alpha^2} - s\left[1 - \frac{1}{\alpha^2}(8 \ln 2\alpha - 13)\right]\right\} \quad \text{if } \gamma_2/\gamma_1 \ll 1 \text{ and } \alpha \gg 1$$

$$(11.49)$$

Table 11.1 contains $G_1^*(\alpha, s)$, $G_2^*(\alpha, s)$, and $G_1^* + sG_2^*$ for different values of s and α.

Suppose that the formation resistivity exceeds that of the borehole ($s < 1$). Then the function $G_1^*(\alpha, s) + sG_2^*(\alpha, s)$ and, the quadrature component, respectively, may change signs twice, because surface charges create a

Table 11.1 Functions G_1^*, G_2^*, $G_1^* + sG_2^*$

s		1/128			1/64	
α	G_1^*	G_2^*	$G_1^* + s \cdot G_2^*$	G_1^*	G_2^*	$G_1^* + s \cdot G_2^*$
2	−0.135	0.1240E + 1	−0.1256	−0.1357	0.1204E + 1	−0.1169
4	−0.7001E − 1	0.1037E + 1	−0.6191E − 1	−0.7163E − 1	0.1038E + 1	−0.5540E − 1
6	−0.3020E − 1	0.1014E + 1	−0.2228E − 1	−0.3162E − 1	0.1014E + 1	−0.1577E − 1
8	−0.1672E − 1	0.1007E + 1	−0.8851E − 2	−0.1778E − 1	0.1008E + 1	−0.2039E − 2
10	−0.1074E − 1	0.1005E + 1	−0.2896E − 2	−0.1156E − 1	0.1005E + 1	0.4144E − 2
12	−0.7528E − 2	0.1003E + 1	0.3104E − 3	−0.8169E − 2	0.1003E + 1	0.7511E − 2
16	−0.4310E − 2	0.1002E + 1	0.3518E − 2	−0.4739E − 2	0.1002E + 1	0.1092E − 1
20	−0.2798E − 2	0.1001E + 1	0.5024E − 2	−0.3176E − 2	0.1001E + 1	0.1254E − 1

s		1/32			1/16	
α	G_1^*	G_2^*	$G_1^* + s \cdot G_2^*$	G_1^*	G_2^*	$G_1^* + s \cdot G_2^*$
2	−0.1370E − 1	0.124E + 1	−0.9942E − 1	−0.1395	0.1203E + 1	−0.6433E − 1
4	−0.7469E − 1	0.1040E + 1	−0.4220E − 1	−0.8024E − 1	0.1043E + 1	−0.1504E − 1
6	−0.3432E − 1	0.1015E + 1	−0.2600E − 2	−0.3924E − 1	0.1017E + 1	0.2431E − 1
8	−0.1982E − 1	0.1008E + 1	0.1169E − 1	−0.2354E − 1	0.1009E + 1	0.3954E − 1
10	−0.1311E − 1	0.1005E + 1	0.1830E − 1	−0.1596E − 1	0.1006E + 1	0.4692E − 1
12	−0.9394E − 2	0.1004E + 1	0.2197E − 1	−0.1164E − 1	0.1004E + 1	0.5112E − 1
16	−0.5559E − 2	0.1002E + 1	0.2576E − 1	−0.7062E − 2	0.1002E + 1	0.5559E − 1
20	−0.3699E − 2	0.1001E + 1	0.2759E − 1	−0.4781E − 2	0.1002E + 1	0.5782E − 1

Induction Logging Using Transversal Coils

s		1/8				1/4	
α	G_1^*	G_2^*	$G_1^* + s \cdot G_2^*$		G_1^*	G_2^*	$G_1^* + s \cdot G_2^*$
2	−0.1440	0.1201E + 1	0.6s155E − 2		−0.1516	0.1198E + 1	0.1479
4	−0.8936E − 1	0.1049E + 1	0.4185E − 1		−0.1019	0.1064E + 1	0.1637
6	−0.4746E − 1	0.1021E + 1	0.8013E − 1		−0.5910E − 1	0.1030E + 1	0.1983
8	−0.2979E − 1	0.1012E + 1	0.9669E − 1		−0.3874E − 1	0.1018E + 1	0.2157
10	−0.2076E − 1	0.1008E + 1	0.1052		−0.2764E − 1	0.1012E + 1	0.2254
12	−0.1542E − 1	0.1006E + 1	0.1103		−0.2085E − 1	0.1009E + 1	0.2313
16	−0.9591E − 2	0.1003E + 1	0.1158		−0.1322E − 1	0.1005E + 1	0.2381
20	−0.6600E − 2	0.1002E + 1	0.1187		−0.9202E − 2	0.1004E + 1	0.2417

s		1/2				2	
α	G_1^*	G_2^*	$G_1^* + s \cdot G_2^*$		G_1^*	G_2^*	$G_1^* + s \cdot G_2^*$
2	−0.1630	0.1190E + 1	0.4322		−0.1944	0.1162E + 1	0.2129
4	−0.1144	0.1086E + 1	0.4285		−0.1177	0.1160E + 1	0.2203
6	−0.7119E − 1	0.1048E + 1	0.4526		−0.7506E − 1	0.1115E + 1	0.2155
8	−0.4820E − 1	0.1031E + 1	0.4672		−0.5133E − 1	0.1083E + 1	0.2115
10	−0.3495E − 1	0.1022E + 1	0.4759		−0.3723E − 1	0.1062E + 1	0.2087
12	−0.2660E − 1	0.1016E + 1	0.4815		−0.2827E − 1	0.1048E + 1	0.2068
16	−0.1704E − 1	0.1010E + 1	0.4880		−0.1799E − 1	0.1031E + 1	0.2044
20	−0.1194E − 1	0.1007E + 1	0.4916		−0.1252E − 1	0.1022E + 1	0.2032

Continued

Table 11.1 Functions G_1^*, G_2^*, $G_1^* + sG_2^*$—cont'd

		8			16	
α	G_1^*	G_2^*	$G_1^* + s \cdot G_2^*$	G_1^*	G_2^*	$G_1^* + s \cdot G_2^*$
2	−0.2194	0.1134E + 1	0.8855E + 1	−0.2259	0.1127E + 1	0.1780E + 2
4	−0.1020	0.1220E + 1	0.9656E + 1	−0.9664E − 1	0.1235E + 1	0.1966E + 2
6	−0.5858E − 1	0.1176E + 1	0.9350E + 1	−0.5267E − 1	0.1192E + 1	0.1902E + 2
8	−0.3758E − 1	0.1133E + 1	0.9026E + 1	−0.3256E − 1	0.1146E + 1	0.1831E + 2
10	−0.2606E − 1	0.1102E + 1	0.8787E + 1	−0.2197E − 1	0.1112E + 1	0.1778E + 2
12	−0.1914E − 1	0.1080E + 1	0.8619E + 1	−0.1579E − 1	0.1089E + 1	0.1740E + 2
16	−0.1163E − 1	0.1052E + 1	0.8409E + 1	−0.9305E − 2	0.1058E + 1	0.1693E + 2
20	−0.7854E − 2	0.1037E + 1	0.8291E + 1	−0.6156E − 2	0.1041E + 1	0.1666E + 2

field in the direction opposite to that of the primary electric field. Near the source, $(L/a < 1)$ the influence of the charges is small, and the field coincides with that in a uniform medium with the conductivity of the borehole γ_1: $Qb_x \approx -(L/\delta_1)^2$.

At the range of large distances, when $L/a \gg (\gamma_1/\gamma_2)^{1/2}$ as follows from Eq. (11.49), the effect caused by the charges is also small and $Qb_x \approx -(L/\delta_2)^2$. For intermediate values of probe lengths, the field caused by the charges is comparable with the vortex field, and it is oriented in the opposite direction, causing a zero crossing of the total field. At the vicinity of these α and s, the conditions of the small parameter are met only for very low frequencies, such that we can disregard terms smaller than k^2. Table 11.2 shows intervals within which the quadrature component vanishes to zero.

When $\gamma_2/\gamma_1 \gg 1$, the function $G_1^* + sG_2^*$ does not change sign, and the expression for the quadrature component is

$$Qb_x^* = -\left(\frac{L}{\delta_1}\right)^2 [G_1^*(\alpha, s) + sG_2^*(\alpha, s)] + \frac{4}{3}\left(\frac{L}{\delta_2}\right)^3$$

$$\text{or } Qb_x^* \approx -\left(\frac{L}{\delta_1}\right)^2 \beta + \frac{4}{3}\left(\frac{L}{\delta_2}\right)^3 \quad (11.50)$$

Returning to Eq. (11.39), we may notice one interesting feature of the quadrature component. The value

$$\beta = G_1^* + sG_2^*$$

is an oscillating function of α and s, hence the magnetic field Qb_x^* may increase with frequency faster than linearly when $a_1 \ll 0$. This feature is not observed in the media excited by a vertical magnetic dipole. If $\alpha \gg 1$ and $s \ll 1$, we have

$$Qb_x^* \approx \left(\frac{L}{\delta_1}\right)^2 \frac{1}{\alpha^2} - \left(\frac{L}{\delta_2}\right)^2 + \frac{4}{3}\left(\frac{L}{\delta_2}\right)^3 \quad (11.51)$$

In Eqs. (11.49), (11.51), the magnetic field is presented as a sum of two terms, each depending on either the conductivity of the borehole or the

Table 11.2 Intervals Within Which the Quadrature Component Vanishes to Zero

s	1/128	1/64	1/32	1/16	1/8
α	$1 \div 2$	$1 \div 2$	$1 \div 2$	$1 \div 2$	
	$11 \div 12$	$8 \div 9$	$6 \div 7$	$4 \div 5$	$1 \div 3$

formation. This feature is favorable for application of the previously described focusing probes, which permit a significant decrease of the borehole influence. The simplest of these is a three-coil probe. The signal of the three-coil probe becomes a borehole insensitive (provided that the electromotive force in each receiver coil is the same) because one part of the field Qb_x^* is proportional to γ_1 and does not depend on the probe length, L_1

In Table 11.3 we present calculated values of $\Delta b_x^* = Qb_x^*(L_1) - Qb_x^*(L_2)$, using exact Eq. (11.33), Δb_x and approximate Eq. (11.49), Δb^{apr} formulas. In addition, we present a difference in the magnitudes in the short- and long-spaced receivers.

These data show that the signal Δb_x^* of the probe does not practically depend on the resistivity of the borehole even beyond the range of the small parameter. As in the case of the vertical magnetic dipole, there are conditions when induced currents in the borehole and surface charges have no influence on the skin effect in the formation. Rather, the skin effect occurs in the same manner as in a uniform medium with resistivity of the formation. For this reason, instead of Eq. (11.51) for the quadrature component of the field b_x^*, we have

$$Qb_x^* = -\left(\frac{L}{\delta_1}\right)^2 G_1^*(\alpha, s) - \left(\frac{L}{\delta_2}\right)^2 G_1^{**}(\alpha, s) + Qb_{x0}^*\left(\frac{L}{\delta_2}\right), \qquad (11.52)$$

where

$$G_2^{**}(\alpha, s) = G_2^*(\alpha, s) - 1 \qquad (11.53)$$

The expression for Qb_x^* is valid in a broad range of α and s. The maximum values of parameter a/δ_1, for which the results of calculations by exact and approximate formulas Eq. (11.52) do not differ by more than 5%, are given in Table 11.4.

11.4 MAGNETIC FIELD IN THE FAR ZONE

Now we derive asymptotic formulas for the field B_x in the far zone, $\alpha \gg 1$. To proceed, we deform the contour of integration in Eq. (11.33) on the complex plane of m in the same manner as was done for the case of the vertical magnetic dipole. However, such a procedure requires either the proof of the absence of poles of the integrand or evaluation of their contribution to the integral. Complexity of the integrand makes it extremely difficult determination of poles.

Table 11.3 Calculated Values Δb_x^*

$L_1/a = 10; L_2/a = 8$

s		1/32	1/16	1/8	1/4	1/2
$\rho_2 = 2.5\,\text{ohm m}; \Delta b^{apr} \approx -0.88 \times 10^{-2}$	a/δ_1	0.1	0.07	0.05	0.035	0.025
	$\Delta b_x^* \times 10^2$	—	−0.73	−0.73	−0.74	−0.75
	$(A_1 - A_2) \times 10^2$	0.76	0.77	0.77	0.79	0.80
$\rho_2 = 5.0\,\text{ohm m}; \Delta b^{apr} \approx -0.44 \times 10^{-2}$	a/δ_1	0.07	0.05	0.035	0.025	0.018
	$\Delta b_x^* \times 10^2$	−0.41	−0.41	−0.41	−0.42	−0.43
	$(A_1 - A_2) \times 10^2$	0.2	0.43	0.43	0.43	0.44
$\rho_2 = 2.5\,\text{ohm m}; \Delta b^{apr} \approx -0.22 \times 10^{-2}$	a/δ_1	0.05	0.035	0.025	0.018	0.012
	$\Delta b_x^* \times 10^2$	−0.22	−0.22	−0.22	−0.23	−0.23
	$(A_1 - A_2) \times 10^2$	0.23	0.23	0.23	0.23	0.24
$\rho_2 = 2.5\,\text{ohm m}; \Delta b^{apr} \approx -0.12 \times 10^{-2}$	a/δ_1	0.035	0.025	0.018	0.012	0.0088
	$\Delta b_x^* \times 10^2$	−0.12	−0.12	−0.12	−0.12	−0.12
	$(A_1 - A_2) \times 10^2$	0.12	0.12	0.12	0.12	0.12

Table 11.4 The Maximum Values of Parameter a/δ_1

s		1/128	1/64	1/32	1/16	1/8	½	8
$\alpha = 4$	a/δ_1	0.6	0.7	0.8	0.9	0.2	0.2	0.1
$\alpha = 8$	a/δ_1	0.15	0.2	0.2	0.2	0.13	0.13	0.05

At the same time, sufficient agreement between the results of calculations by asymptotic and exact formulae allows us to think that contribution of unaccounted poles from the upper half-plane of m in a considered part of the spectrum is sufficiently small. Let us present the integral in Eq. (11.33) in the following form:

$$\frac{\alpha^3}{\pi}\int_0^\infty \left(\overline{m}^2 D_1 + k_1^2 a^2 C_1\right) \cos\alpha\overline{m}\, d\overline{m} = \frac{\alpha^3}{2\pi}\int_{-\infty}^\infty \left(\overline{m}^2 D_1 + k_1^2 a^2 C_1\right) \exp(i\alpha\overline{m})\, d\overline{m}$$

(11.54)

We suppose that in the upper half-plane of complex variable m, there are no singularities except the branch points $m_1 = k_1 a$ and $m_2 = k_2 a$. Choosing crosscuts along lines $\mathrm{Re}\, m_1 = 0$ and $\mathrm{Re}\, m_2 = 0$, it is assumed that the real parts of radicals $(m^2 - k_1^2)^{1/2}$ and $(m^2 - k_2^2)^{1/2}$ are positive on the complex plane of m. As follows from the asymptotic behavior of the Bessel functions, the integrand in Eq. (11.54) increases with $m \to \infty$, but not faster than $\exp(2|m|)$. For this reason, convergence of the integral in Eq. (11.54) in the upper half-plane for $\alpha > 2$ is provided by the multiplier $\exp(i\alpha m)$ irrespective of the sign of the real part of radicals m_1 and m_2. We draw crosscuts from branch points $k_1 a$ and $k_2 a$ parallel to the imaginary axis and deform the contour of integration in Γ (Fig. 11.6).

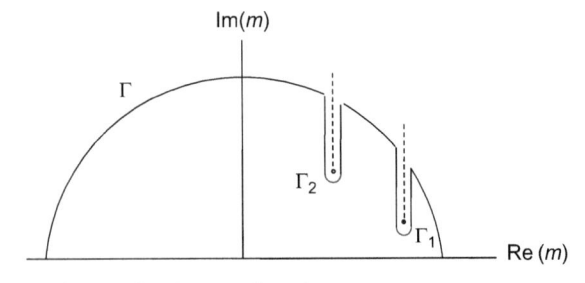

Fig. 11.6 Contour integration in complex plane.

The integral along arcs with an infinite radius, which is due to the presence of the term $\exp(i\alpha m)$, vanishes owing to $\operatorname{Im} m > 0$, $\alpha > 2$. For this reason the integral along the real axis Eq. (11.54) is equal to the sum of the integrals along the sides of crosscuts Γ_1 and Γ_2. First, let us evaluate the integral along crosscut Γ_1. In passing from the left side of the crosscut to the right side, the value of m_1 changes sign. Thus the integral along crosscut Γ_1 is equal to:

$$\frac{\alpha^3}{\pi}\int_{\Gamma_1} \{\overline{m}^2[D_1(\overline{m}_1) - D_1(-\overline{m}_1)] + k_1^2 a^2[C_1(\overline{m}_1) - C_1(-\overline{m}_1)]\} \exp(i\alpha\overline{m})d\overline{m} \tag{11.55}$$

Using properties of Bessel functions:

$$\begin{aligned} I_0(-z) &= I_0(z) & K_0(-z) &= K_0(z) + i\pi I_0(z) \\ I_1(-z) &= -I_1(z) & K_1(-z) &= -K_1(z) + i\pi I_1(z) \end{aligned} \tag{11.56}$$

it is fairly straightforward to show that for the functions D and C we have

$$D_1(-\overline{m}_1) = D_1(\overline{m}_1) - i\pi \quad C_1(-\overline{m}_1) = C_1(\overline{m}_1) - i\pi \tag{11.57}$$

Thus the integral in Eq. (11.55) has the form:

$$\frac{\alpha^3}{\pi} i \int_{\Gamma_1} (\overline{m}^2 + k_1^2 a^2) \exp(i\alpha\overline{m}) d\overline{m} \tag{11.58}$$

Letting $m = t + k_1 a$, we obtain

$$\exp(ik_1 L)\frac{\alpha^3}{2}\int_0^\infty (t^2 - 2itk_1 a - 2k_1^2 a^2)\exp(-\alpha t)dt = (1 - ik_1 L - k_1^2 L^2)$$

$$\exp(ik_1 L) = b_{x0}^*(L/\delta_1)$$

where $b_{x0}^*(L/\delta_1)$ is the x-component of the magnetic field in a uniform medium with the resistivity of the borehole. Correspondingly, as follows from Eq. (11.33), the magnetic field is expressed, as in the case of the vertical magnetic dipole, only through the integral along the crosscut Γ_2:

$$b_x^* = -\frac{\alpha^3}{2\pi}\int_{\Gamma_2} (\overline{m}^2 D_1 + k_1^2 a^2 C_1) \exp(i\alpha\overline{m}) d\overline{m} \tag{11.59}$$

To transform the integrand in Eq. (11.59), we use relationships that follow from Eq. (11.56):

$$\frac{K_0(z)}{K_1(z)} + \frac{K_0(-z)}{K_1(-z)} = \frac{i\pi}{zK_1(z)K_1(-z)}$$

$$\frac{K_0^2(z)}{K_1^2(z)} - \frac{K_0^2(-z)}{K_1^2(-z)} = \frac{i\pi}{zK_1(z)K_1(-z)}\left[\frac{K_0(z)}{K_1(z)} - \frac{K_0(-z)}{K_1(-z)}\right]$$

$$\frac{K_0^2(z)K_0(-z)}{K_1^2(z)K_1(-z)} + \frac{K_0^2(-z)K_0(z)}{K_1^2(-z)K_1(z)} = \frac{i\pi}{zK_1(z)K_1(-z)}\frac{K_0(z)K_0(-z)}{K_1(z)K_1(-z)}$$

After relatively simple transformations, we have the following expression for the difference between the values of function C in both sides of the crosscut:

$$C_1(\overline{m}_2) - C_1(-\overline{m}_2) = \frac{i\pi}{\overline{m}_2^2 K_1(\overline{m}_2)K_1(-\overline{m}_2)\Delta(\overline{m}_2)\Delta(-\overline{m}_2)}$$

$$\times \left\{\frac{I_1^2(\overline{m}_1)}{\overline{m}_1}(1-s)^2\frac{\overline{m}^2 + k_1^2a^2}{\overline{m}_2^2}\frac{\overline{m}^2 + k_2^2a^2}{\overline{m}_2^2} + sI_0^2(\overline{m}_1) - I_0(\overline{m}_1)\frac{I_1(\overline{m}_1)}{\overline{m}_1}(1-s)\right.$$

$$\times \left[s\frac{\overline{m}^2 + k_1^2a^2}{\overline{m}_2^2} + \frac{\overline{m}^2 + k_2^2a^2}{\overline{m}_2^2}\right] - s\overline{m}_1\left[\frac{\overline{m}^2 + k_1^2a^2}{\overline{m}_2^2}\frac{I_1^2(\overline{m}_1)}{\overline{m}_1^2}(1-s) - I_0(\overline{m})\frac{I_1(m_1)}{\overline{m}_1}\right]$$

$$\times \left[\frac{K_0(\overline{m}_2)}{\overline{m}_2 K_1(\overline{m}_2)} - \frac{K_0(-\overline{m}_2)}{\overline{m}_2 K_1(-\overline{m}_2)}\right] + s(1+2s)I_1^2(\overline{m}_1)\overline{m}_1^2\frac{K_0(\overline{m}_2)K_0(-\overline{m}_2)}{\overline{m}_2^2 K_1(\overline{m}_2)K_1(-\overline{m}_2)}\right\}$$

$$(11.60)$$

where

$$\Delta(\pm\overline{m}_2) = -I_0^2(\overline{m}_1) + I_1^2(\overline{m}_1)\left[\frac{\overline{m}^2 + k_2^2a^2}{\overline{m}_2^2}(1-s)\frac{K_0(\pm\overline{m}_2)}{\pm\overline{m}_2 K_1(\pm\overline{m}_2)}\right.$$

$$\left. - s\overline{m}_1^2\frac{K_0^2(\pm\overline{m}_2)}{\overline{m}_2^2 K_1^2(\pm\overline{m}_2)}\right] + I_0(\overline{m}_1)\frac{I_1(\overline{m}_1)}{\overline{m}_1}\left[\frac{\overline{m}^2 + k_1^2a^2}{\overline{m}_2^2}(1-s)\right.$$

$$\left. -(1+s)\overline{m}_1^2\frac{K_0(\pm\overline{m}_2)}{\pm\overline{m}_2 K_1(\pm\overline{m}_2)}\right]$$

Inasmuch as the function D_1 can be presented in the form:

$$D_1 = C_1 + (s-1)\frac{K_0(\overline{m}_2)}{\overline{m}_2 K_1(\overline{m}_2)\Delta(\overline{m}_2)}$$

for the discontinuity of the function D_1, we have

$$D_1(\overline{m}_2) - D_1(-\overline{m}_2) = C_1(\overline{m}_2) - C_1(-\overline{m}_2) + A(\overline{m}_2)$$

where

$$A(\overline{m}_2) = \frac{i\pi(s-1)}{\overline{m}_2^2 K_1(\overline{m}_2) K_1(-\overline{m}_2) \Delta(\overline{m}_2) \Delta(-\overline{m}_2)}$$
$$\times \left[I_0(\overline{m}_1) \frac{I_1(\overline{m}_1)\overline{m}^2 + k_1^2 a^2}{\overline{m}_1 \overline{m}_2^2}(1-s) - I_0^2(\overline{m}_1) - s\overline{m}_1^2 I_1^2(\overline{m}_1) \frac{K_0(\overline{m}_2) K_0(-\overline{m}_2)}{\overline{m}_2^2 K_1(\overline{m}_2) K_1(-\overline{m}_2)} \right]$$

(11.61)

Thus, instead of Eq. (11.59), we have

$$b_x^* = -\frac{\alpha^3}{2\pi} \int_{k_2 a}^{-\infty + k_2 a} \{(\overline{m}^2 + k_1^2 a^2)[C_1(\overline{m}_2) - C_1(-\overline{m}_2)]$$
$$+ \overline{m}^2 A(\overline{m}_2)\} \exp(i\alpha\overline{m}) d\overline{m}$$

(11.62)

Now we introduce a new variable, letting $\overline{m} = it + k_2 a$. Along the cross-cut, the variable t changes from zero to infinity, and

$$\overline{m}_1 = \left[-t^2 + 2ik_2 at + (k_2^2 - k_1^2)a^2 \right]^{1/2}, \quad \overline{m}_2 = \left(-t^2 + 2ik_2 at \right)^{1/2}$$

Correspondingly, the expression for the magnetic field has the form:

$$b_x^* = \exp(-ik_2 L) \frac{i\alpha^3}{2\pi} \int_0^\infty \{(\overline{m}^2 + k_1^2 a^2)[C_1(\overline{m}_2) - C_1(-\overline{m}_2)]$$
$$+ \overline{m}^2 A(\overline{m}_2)\} \exp(-\alpha t) dt$$

(11.63)

In spite of the cumbersome character of the integrand, presentation Eq. (11.63) turns out to be useful for the calculations when ($\alpha \gg 1$), because unlike Eq. (11.33), the integral in Eq. (11.63) does not contain the oscillating function cos ma. Moreover, in the wave zone, $|k_2 L| > 1$, when the value of the field is exponentially small, it is very difficult to provide the smallness of the integral Eq. (11.33) for ($\alpha \gg 1$) by summation of large oscillating values of the integrand. By contrast, the form Eq. (11.63) essentially facilitates the calculations since the small value of the integral is provided by the multiplier $\exp(-ik_2 L)$, which stands in front of the integral of the nonoscillatory function. Now proceeding from Eq. (11.63), we obtain the asymptotic formula that describes the field in the far zone ($\alpha \gg 1$). In this case the value of the integral is defined by the range $t \leq 1/\alpha < 1$. Generally, the integrand in Eq. (11.63) depends on m_1 in a rather complicated manner. However, if conditions

$$\frac{1}{\alpha^2} \ll |k_1^2 a^2| \quad \text{i.e.,} \quad |k_1 L| > 1, \quad s < 1$$

are met, we can approximate $\overline{m}_1$ as:

$$\overline{m}_1 \approx \left(k_2^2 a^2 - k_1^2 a^2\right)^{1/2}$$

and think of $\overline{m}_1$ and functions of $\overline{m}_1$ as being independent of the variable of integration, t. For the radical $\overline{m}_2$, we have

$$\overline{m}_2 \approx \left(-\frac{1}{\alpha^2} + 2i\frac{k_2 a}{\alpha}\right)^{1/2}, \quad \text{i.e.,} \quad |\overline{m}_2| \ll 1$$

By keeping the terms of orders $s/\overline{m}_2^4, s^2/\overline{m}_2^4, 1/\overline{m}_2^2, \left(s \ln(\overline{m}_2)/\overline{m}_2^2\right)$ and omitting the terms $s/\overline{m}_2^2, \ldots$, we can present the expression Eq. (11.60) in the form:

$$C_1(\overline{m}_2) - C_1(-\overline{m}_2) = \frac{i\pi}{\overline{m}_2^2 K_1(\overline{m}_2) K_1(-\overline{m}_2) \Delta(\overline{m}_2) \Delta(-\overline{m}_2)}$$

$$\times \left[\frac{I_1^2(\overline{m}_1)}{\overline{m}_1^2}(1 - 2s) \frac{\left(\overline{m}^2 + k_1^2 a^2\right)\left(\overline{m}^2 + k_2^2 a^2\right)}{\overline{m}_2^2} - 2s I_1^2(\overline{m}_1) \frac{\overline{m}^2 + k_1^2 a^2}{\overline{m}_2^2} K_0(\overline{m}_2) \right].$$

(11.64)

By analogy, we have

$$\overline{m}_2^2 K_1(\overline{m}_2) K_1(-\overline{m}_2) \Delta(\overline{m}_2) \Delta(-\overline{m}_2) \approx -\left\{ I_0(\overline{m}_1) \frac{I_1(\overline{m}_1)}{\overline{m}_1} \frac{\overline{m}^2 + k_1^2 a^2}{\overline{m}_2^2} \right.$$

$$\left. + K_0(\overline{m}_2) \times \left[I_1^2(\overline{m}_1) \frac{\overline{m}^2 + k_2^2 a^2}{\overline{m}_2^2}(1 - s) - I_0(\overline{m}_1) \frac{I_1(\overline{m}_1)}{\overline{m}_1} \frac{\left(3\overline{m}^2 - k_1^2 a^2\right)}{2} \right] \right\}$$

(11.65)

Substituting expression Eq. (11.65) into Eq. (11.64) and after simple algebra, we obtain

$$C_1(\overline{m}_2) - C_1(-\overline{m}_2) = -\frac{i\pi}{I_0^2(\overline{m}_1)} \frac{\overline{m}^2 K_0(\overline{m}_2) K_0(\overline{m}_2)}{\left(\overline{m}^2 + k_1^2 a^2\right)} \left(\overline{m}^2 + k_2^2 a^2 - 2s\right)$$

$$\times \left[\overline{m}_1^2 + 2\overline{m}_1 \frac{I_1(\overline{m}_1)}{I_0(\overline{m}_1)}\right] - 2\overline{m}_1 \frac{I_1(\overline{m}_1)}{I_0(\overline{m}_1)}$$

$$\times \left[\frac{\overline{m}_2^2 - 2s\overline{m}^2}{\overline{m}^2 + k_1^2 a^2} - \overline{m}_2^2 \frac{I_0(\overline{m}_1)}{I_1(m_1)\overline{m}_1} \frac{3\overline{m}^2 - k_1^2 a^2}{2\overline{m}^2 + k_1^2 a^2}\right] \left(\overline{m}^2 + k_2^2 a^2\right) K_0(\overline{m}_2)$$

(11.66)

For function $A(m_2)$ we have

$$A(\overline{m}_2) \approx \frac{i\pi(s-1)}{\overline{m}_2^2 K_1(\overline{m}_2) K_1(-\overline{m}_2) \Delta(\overline{m}_2) \Delta(-\overline{m}_2)} I_0(\overline{m}_1) \frac{I_1(\overline{m}_1)}{\overline{m}_1} \frac{\overline{m}^2 + k_1^2 a^2}{\overline{m}_2^2}$$

$$\approx -i\pi(s-1) \frac{\overline{m}_1}{I_0(\overline{m}_1) I_1(\overline{m}_1)} \frac{\overline{m}_2^2}{\overline{m}^2 + k_1^2 a^2} \left\{1 - 2\overline{m}_1 \frac{I_1(\overline{m}_1)}{I_0(\overline{m}_1)}\right.$$

$$\times \left[\frac{\overline{m}^2 + k_2^2 a^2}{\overline{m}_2^2} - \frac{I_0(\overline{m}_1)}{\overline{m}_1 I_1(\overline{m}_1)} \frac{3\overline{m}^2 - k_1^2 a^2}{2}\right] \left.\frac{\overline{m}_2^2 K_0(\overline{m}_2)}{\overline{m}^2 + k_1^2 a^2}\right\}$$

(11.67)

Substituting expressions Eqs. (11.66), (11.67) into Eq. (11.63) and discarding terms, giving after integration values of the order of $1/\alpha^4$ we obtain

$$b_x^* \approx -\frac{\exp(ik_2 L) \alpha^3}{I_0^2(\overline{m}_1)} \frac{1}{2} \left\{ \int_0^\infty (\overline{m}^2 + k_2^2 a^2) \exp(-\alpha t) dt \right.$$

(11.68)

$$\left. - 2s \left[\overline{m}^2 + 2\overline{m}_1 \frac{I_1(\overline{m}_1)}{I_0(\overline{m}_1)}\right] \int_0^\infty \overline{m}_2^2 K_0(\overline{m}_2) \exp(-\alpha t) dt \right\}$$

where

$$\overline{m}^2 = -t^2 + 2ik_2\alpha t + k_2^2 a^2, \quad \overline{m}_2 = \left(-t^2 + 2ik_2\alpha t\right)^{1/2}, \text{ and } \overline{m}_1 = \left(k_2^2 a^2 - k_1^2 a^2\right)^{1/2}$$

The first integral is expressed in terms of elementary functions:

$$\int_0^\infty (\overline{m}^2 + k_2^2 a^2) \exp(-\alpha t) dt = -\frac{2}{\alpha^3} \left(1 - ik_2 L - k_2^2 L^2\right).$$

The second integral can be presented in the form:

$$\int_0^\infty \overline{m}_2^2 K_0(\overline{m}_2) \exp(-\alpha t) dt \approx - \int_0^\infty \overline{m}_2^2 \ln \overline{m}_2 \exp(-\alpha t) dt$$

$$= \frac{1}{2}\left(\frac{\partial^2}{\partial \alpha^2} + 2ik_2 a \frac{\partial}{\partial \alpha}\right) \times \int_0^\infty \ln(-t^2 + 2ik_2 \alpha t) \exp(-\alpha t) dt \qquad (11.69)$$

$$= \frac{1}{2}\left(\frac{\partial^2}{\partial \alpha^2} + 2ik_2 a \frac{\partial}{\partial \alpha}\right) \int_0^\infty [\ln(-t) + \ln(t - 2iak_2)] \exp(-\alpha t) dt$$

We have

$$\int_0^\infty \ln(-t) \exp(-\alpha t) dt = -\frac{1}{\alpha}(\ln \alpha + C) + \frac{i\pi}{\alpha} \approx -\frac{\ln \alpha}{\alpha}$$

The second integral in Eq. (11.69) is expressed through the integral exponential function:

$$\int_0^\infty \ln(t - 2ik_2 a) \exp(-\alpha t) dt = \frac{1}{\alpha}[\ln(-2ik_2 a) - \exp(-2ik_2 L) Ei(2ik_2 L)]$$

Correspondingly, for the magnetic field we have

$$b_x^* \approx \frac{1}{I_0^2(\overline{m}_1)} b_{0x}(L/\delta_2) + \frac{\exp(ik_2 L)}{I_0^2(\overline{m}_1)} 2s \left[\overline{m}_1^2 + 2\overline{m}_1 \frac{I_1(\overline{m}_1)}{I_0(\overline{m}_1)}\right] P(k_2 a, \alpha) \quad (11.70)$$

where b_{0x}^* is the complex amplitude of the field in a uniform medium with resistivity of formation and

$$P(k_2 a, \alpha) = \frac{a^3}{4}\left(\frac{\partial^2}{\partial \alpha^2} + 2ik_2 a \frac{\partial}{\partial \alpha}\right)\left[-\frac{\ln \alpha}{\alpha} + \frac{\ln(-2ik_2 a)}{\alpha}\right.$$

$$\left. - \frac{\exp(-2ik_2 L)}{\alpha} Ei(2ik_2 L)\right]$$

Next, we consider several cases. Assuming that $|k_2 L \ll 1|$, we have

$Ei(2ik_2 L) \approx \ln(-2ik_2 a) = \ln \alpha + \ln(-2ik_2 a)$ and $P(k_2, a, \alpha) \approx -\ln \alpha$

Therefore, Eq. (11.70) becomes

Induction Logging Using Transversal Coils

$$b_x^* = \frac{1}{I_0^2(\overline{m}_1)}\left[1-\left(\frac{L}{\delta_2}\right)^2\right] - \frac{2s\ln\alpha}{I_0^2(\overline{m}_1)}\left[\overline{m}_1^2 + 2\overline{m}_1\frac{I_1(\overline{m}_1)}{I_0(\overline{m}_1)}\right] \quad (11.71)$$

In particular, if the skin depth in the borehole is greater than its radius, then Bessel's functions $I_0(m_1)$ and $I_1(m_1)$ can be expanded in a series, thus instead of Eq. (11.71) we obtain

$$b_x^* = 1 - \frac{\overline{m}_1^2}{2} - \left(\frac{L}{\delta_2}\right)^2 - 4s\overline{m}_1^2\ln\alpha$$

Inasmuch as $s \ll 1$ and

$$\overline{m}_1^2 \approx -\frac{k_1^2 L^2}{\alpha^2}$$

for the quadrature component of the field, we have

$$Qb_x^* \approx \frac{1}{\alpha^2}\left(\frac{L}{\delta_1}\right)^2 - \left(1 - \frac{8\ln\alpha}{\alpha^2}\right)\left(\frac{L}{\delta_2}\right)^2$$

which up to the term s/α^2, coincides with Eq. (11.49), derived for the range of small parameters. In the wave zone when $|k_2 L| > 1$, by using the following asymptotic expression:

$$Ei(2ik_2 L) \approx \frac{\exp(2ik_2 L)}{2ik_2 L}$$

we obtain

$$P(k_2 a, \alpha) \approx \frac{\alpha^3}{4}\left(\frac{\partial^2}{\partial\alpha^2} + 2ik_2\alpha\frac{\partial}{\partial\alpha}\right)\left[-\frac{\ln\alpha}{\alpha} + \frac{\ln(-2ik_2 a)}{\alpha}\right]$$

$$\approx \frac{ik_2 L}{2}[\ln\alpha - \ln|k_2 a|]$$

Table 11.5 provides a comparison between $A = |b_x^*|$, calculated using exact solution Eq. (11.33) and the asymptotic expression Eq. (11.70).

It is natural to distinguish three frequency ranges of the amplitude spectrum; i.e., the range of small parameters, the intermediate zone, and the wave zone. As follows from Table 11.5, the asymptotic expression Eq. (11.70) is sufficiently accurate at the range of the small parameter and the intermediate zone when $a/\delta_1 < 1$. If parameter a/δ_1 exceeds 1 and

Table 11.5 A Comparison Between Exact and Asymptotic Values of A

		s	1/64		1/16		1/4	
a/δ_1	α	A	A^{apr}	A	A^{apr}	A	A^{apr}	
0.1	4	1.00	1.00	1.00	1.00	1.00	1.00
	10	1.00	1.00	1.02	1.01	1.09	1.08
	12	1.00	1.00	1.03	1.02	1.13	1.12
	20	1.01	1.01	1.09	1.09	1.32	1.31
	24	1.03	1.03	1.12	1.13	1.39	1.38
	30	1.06	1.05	1.21	1.21	1.44	1.44
0.2	4	1.00	1.00	1.00	1.00	1.05	1.03
	10	1.01	1.01	1.09	1.08	1.31	1.28
	12	1.03	1.02	1.13	1.12	1.38	1.36
	20	1.09	1.09	1.32	1.31	1.39	1.39
	24	1.14	1.13	1.39	1.38	1.28	1.28
	30	1.21	1.21	1.44	1.44	1.05	1.05
0.4	4	1.00	1.00	1.05	1.02	1.21	1.12
	10	1.09	1.07	1.30	1.27	1.34	1.34
	12	1.13	1.12	1.36	1.34	1.23	1.24
	20	1.31	1.30	1.37	1.38	0.62	0.65
	24	1.38	1.37	1.26	1.27	0.39	0.41
	30	1.43	1.43	1.03	1.04	0.18	0.19
0.8	4	1.99	1.97	1.13	1.05	1.25	1.17
	10	1.23	1.21	1.22	1.26	0.51	0.61
	12	1.29	1.28	1.11	1.18	0.32	0.39
	20	1.30	1.31	0.56	0.60	0.033	0.043
	24	1.21	1.22	0.35	0.38	0.0094	0.012
	30	0.91	1.00	0.16	0.18	0.0013	0.0017

$s > 1/4$, the accuracy of the approximate solution rapidly drops due to the unaccounted poles in deriving Eq. (11.70).

In the far zone, $|k_1 L| \gg 1$ and $(\alpha \gg 1)$; Eq. (11.70) is reduced to:

$$b_x^* \approx \frac{1}{I_0^2(\overline{m}_1)} b_{0x}^* \left(\frac{L}{\delta_2}\right)$$

where $\overline{m}_1 = \left(k_2^2 a^2 - k_1^2 a^2\right)^{1/2}$.

The last expression suggests that the ratio of amplitudes or phase difference of two probes of the length L_1 and L_2 does not depend on the radius and conductivity of the borehole, but rather is defined by the formation parameters only. This feature of the ratio of amplitudes and phase difference, previously observed in b_z^* of a vertical magnetic dipole, is shown in Fig. 11.7, where we show modeling results of both ratio of amplitudes and phase difference for different contrast between resistivity of the formation and the borehole.

In fact, similar to the case of the vertical magnetic dipole, instead of the ratio of amplitudes A_1 and A_2 we use the attenuation $At = 20\log|A_2/A_1|$, assuming that receiving moments $M_1/M_2 = (L_2/L_1)^3$ are selected to provide zero attenuation of the field in the air. The calculated attenuation and phase difference are normalized by the corresponding values in a uniform medium with resistivity of the formation:

$$P_A(\alpha) = \left|\frac{At(B_x^*)}{At(B_x^{*un})}\right| \quad \text{and} \quad P_{\Delta\phi}(\alpha) = \frac{\Delta\phi(B_x^*)}{\Delta\phi(B_x^{*un})}$$

Here, $At(B_x^*un)$ and $\Delta\phi(B_x^*un)$ are attenuation and phase difference in a uniform medium with resistivity of the formation. Calculations are performed for two three-coil probes with the longest two-coil probes at $L = 0.75$ and $L = 1.5$ m, correspondingly. The length's ratio between the short- and long-spaced coils is equal to 0.75 for each probe. The frequency range is between 10 kHz and 10 MHz. The radius of the borehole is 10 cm, the resistivity of the formation is fixed at $\rho_2 = 10$ ohmm, and the resistivity of the borehole is equal $\rho_1 = 1.0, 0.1,$ and 0.01 ohmm. Index of curves is γ_2/γ_1. The data (two upper subplots in Fig. 11.7) are in full agreement with the theoretically predicted behavior: both attenuation and phase difference are practically insensitive to the properties of the borehole, especially when the resistivity contrast is less than 100. (The reader should not be confused with the "horn effect" on the curves caused by zero-crossings of the corresponding functions.) When the resistivity contrast reaches 1000 (bottom subplots, Fig. 11.7), the short 0.75 m probe fails to remove signal from the borehole, but the long 1.5 m probe is still practically insensitive to it. We may notice only a slight advantage of attenuation in removing borehole signal compared with the phase difference.

11.5 MAGNETIC FIELD IN A MEDIUM WITH TWO CYLINDRICAL INTERFACES

Let us consider the field of the transversal magnetic dipole in a medium with two cylindrical interfaces. Analysis of the solution is helpful in analyzing influence of the invasion zone on the radial response of transversal probes. Using results derived for the case of one cylindrical boundary, the potentials in the presence of two interfaces might be presented as:

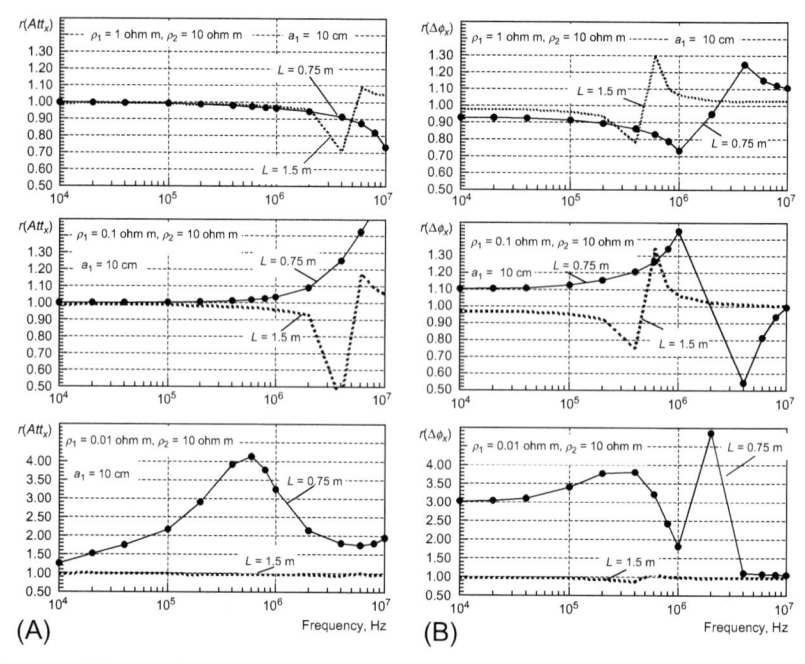

Fig. 11.7 (A) Normalized attenuations and (B) phase differences for the three-coil probe at different contrasts between resistivity of the formation and the borehole.

$$A_{e1}^* = A_{e0}^* + k_1^2 \frac{M}{2\pi^2} \sin\phi \int_0^\infty \frac{1}{m_1} C_1 I_1(m_1 r) \cos mz\, dm$$

$$A_{m1}^* = A_{m0}^* + \frac{M}{2\pi^2} \cos\phi \int_0^\infty \frac{m}{m_1} D_1 I_1(m_1 r) \sin mz\, dm$$

$$A_{e2}^* = k_2^2 \frac{M}{2\pi^2} \sin\phi \int_0^\infty \frac{1}{m_2} [-C_2 K_1(m_2 r) + C_3 I_1(m_2 r)] \cos mz\, dm$$

$$A_{m2}^* = \frac{M}{2\pi^2} \cos\phi \int_0^\infty \frac{m}{m_2} [-D_2 K_1(m_2 r) + D_3 I_1(m_2 r)] \sin mz\, dm \quad (11.72)$$

$$A_{3e}^* = -k_3^2 \frac{M}{2\pi^2} \sin\phi \int_0^\infty \frac{1}{m_3} C_4 K_1(m_3 r) \cos mz\, dm$$

$$A_{m3}^* = -\frac{M}{2\pi^2} \cos\phi \int_0^\infty \frac{m}{m_3} D_4 K_1(m_3 r) \sin mz\, dm$$

Here, A_{0e}^* and A_{0m}^* are complex amplitudes of vector potentials in a uniform medium with resistivity ρ_1; and indexes 1, 2, 3 correspond to the potentials in the borehole, invasion zone, and formation, correspondingly. Similar to the case of one cylindrical boundary, Eq. (11.72) account for behavior of the field near the source and at infinity. The continuity of tangential components of the electric and magnetic fields at boundaries $r = a$ and $r = b$ provides a system of eight equations with respect to eight unknown coefficients C_i and D_i ($i = 1, ..., 4$):

$$m_1[K_1(m_1a) - C_1 I_1(m_1a)] = m_2[C_2 K_1(m_2 a) - C_3 I_1(m_2 a)]$$

$$\frac{1}{m_1 a}[K_1(m_1 a) - C_1 I_1(m_1 a)] + K_1'(m_1 a) - D_1 I_1'(m_1 a)$$

$$= \frac{1}{m_2 a}[C_2 K_1(m_2 a) - C_3 I_1(m_2 a)] + D_2 K_1'(m_2 a) - D_3 I_1'(m_2 a)$$

$$m_1[K_1(m_1 a) - D_1 I_1(m)] = m_2[D_2 K_1(m_2 a) - D_3 I_1(m_2 a)]$$

$$k_1^2[K_1'(m_1 a) - C_1 I_1'(m_1 a)] + \frac{m^2}{m_1 a}[K_1(m_1 a) - D_1 I_1(m_1 a)]$$

$$= k_2^2[C_2 K_1'(m_2 a) - C_3 I_1'(m_2 a)] + \frac{m^2}{m_2 a}[D_2 K_1(m_2 a) - D_3 I_1(m_2 a)]$$

$$m_2[C_2 K_1(m_2 b) - C_3 I_1(m_2 b)] = m_3 C_4 K_1(m_3 b)$$

$$\frac{1}{m_2 b}[C_2 K_1(m_2 b) - C_3 I_1(m_2 b)] + D_2 K_1'(m_2 b) - D_3 I_1'(m_2 b)$$

$$= \frac{1}{m_3 b} C_4 K_1(m_3 b) + D_4 K_1'(m_3 b)$$

$$m_2[D_2 K_1(m_2 b) - D_3 I_1(m_2 b)] = m_3 D_4 K_1(m_3 b)$$

$$k_2^2[C_2 K_1'(m_2 b) - C_3 I_1'(m_2 b)] + \frac{m^2}{m_3 b}[D_2 K_1(m_2 b) - D_3 I_1(m_2 b)]$$

$$= k_3^2 C_4 K_1'(m_3 b) + \frac{m^2}{m_3 b} D_4 K_1(m_3 b)$$

By numerically solving this system of linear equations, we find all the coefficients of integrands in Eq. (11.72) that are needed to calculate potentials and, correspondingly, the complex amplitudes of the field components.

The expression for the magnetic field on the axis of the borehole has the form:

$$b_x^* = (1 - ik_1 L - k_1^2 L^2) \exp(ik_2 L) - \frac{\alpha^3}{\pi} \int_0^\infty [m^2 D_1 + k_1^2 a C_1] \cos\frac{L}{a_1} dm \quad (11.73)$$

where a_1 is the radius of the borehole. The derived solution can be used to confirm the efficiency of attenuation and phase difference of the three-coil probe in reducing the effect of the invasion zone. To proceed, we use the same parameters for probes, borehole, and formations as in the case of one cylindrical boundary (Fig. 11.7). In addition, the formation model includes an invasion zone with the radius $b = 2a$. The calculated attenuation and phase difference, shown in Fig. 11.8, are normalized by the corresponding values in a uniform medium with resistivity of the formation.

The data in the left subplots (Fig. 11.8A) indicate the ability of the attenuation of the long probe $(L = 1.5\,\text{m})$ to remove the influence of both borehole and invasion. The advantage of the long probe over the short one $(L = 0.75\,\text{m})$ is especially pronounced in the case of very conductive invasion with $\rho_2 = 0.01\,\text{ohm m}$. Also, by comparing attenuation and phase difference we see that attenuation has clear advantage and enables us to reduce influence of the borehole at shorter spacings.

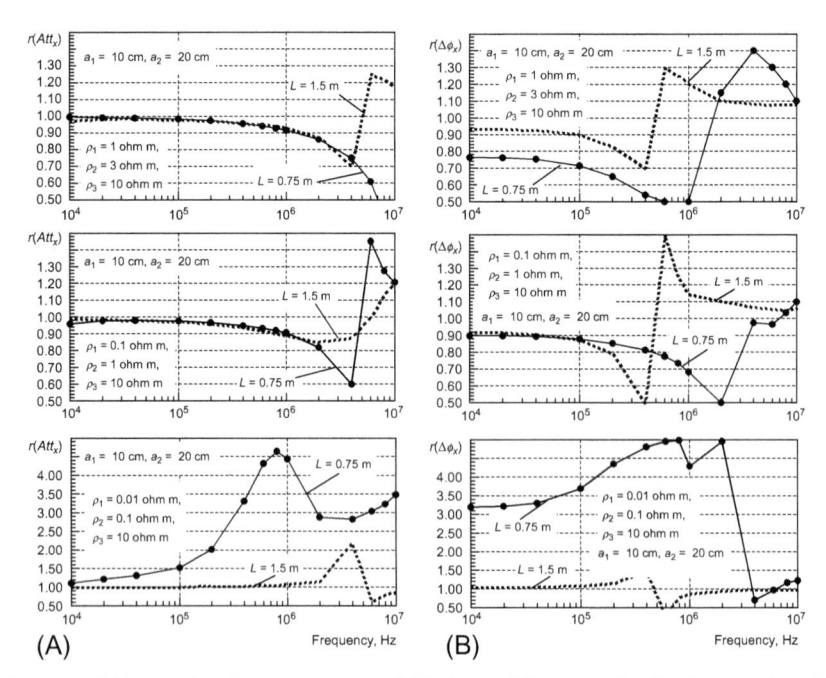

Fig. 11.8 (A) Normalized attenuations and (B) phase differences for the three-coil probe in the presence of invasion zone.

11.6 MAGNETIC FIELD IN MEDIUM WITH A THIN RESISTIVE CYLINDRICAL LAYER

We suppose that a relatively thin and resistive invasion zone is formed due to penetration of the very resistive oil-based mud in the formation. Although in the analysis we assume that the conductivity of the borehole and the formation are equal, generalization for the case of the different conductivities does not require special effort. Thus, in a uniform medium with conductivity γ_1, there is a thin cylindrical layer with radius a and thickness h that has conductivity γ_2. These parameters satisfy conditions: $h/a \ll 1$, $\gamma_1/\gamma_2 \gg 1$. The electric properties of the layer are characterized by the transversal resistance, $T = h/\gamma_2$. At the surface, $r = a$ tangential components of the magnetic field are continuous:

$$B_{1z} = B_{2z} \quad B_{1\phi} = B_{2\phi} \tag{11.75}$$

where B_1 and B_2 are fields in the borehole and the formation, respectively. The tangential components of the electric field are discontinuous owing to the presence of the double layer, so we have

$$E_{2z} = E_{1z} + T\gamma_1 \frac{\partial E_{1z}}{\partial z} \quad \text{and} \quad E_{2\phi} = E_{1\phi} + T\frac{\gamma_1}{a} \frac{\partial E_{1z}}{\partial \phi} \tag{11.76}$$

Substituting expressions for field components through potentials Eq. (11.28) into Eqs. (11.75), (11.76), we obtain

$$\begin{aligned}
K_1(m_1) - I_1(m_1)D &= K_1(m_1)Ga^2k_1^2\left[K_1'(m_1) - I_1'(m_1)C\right] \\
&+ \frac{m^2}{m_1}[K_1(m_1) - I_1(m_1)D] = ak_1^2 K_1'(m_1)E \\
&+ \frac{m^2}{m_1}K_1(m_1)GK_1(m_1) - I_1(m_1)C \\
&- \tau\left[\frac{m^2}{m_1}I_1(m_1)D + \frac{m^2}{m_1}K_0(m_1) + \frac{m^2}{m_1}I_1'(m_1)C\right] \\
&= K_1(m_1)E - K_0(m_1) - \frac{I_1(m_1)}{m_1}C - I_1'(m_1)D \\
&- \tau\left[\frac{I_1(m_1)}{m_1}D + K_0(m_1) + I_1'(m_1)C\right] \\
&= \frac{K_1(m_1)}{m_1}E + K_1'(m_1)G
\end{aligned} \tag{11.77}$$

where

$$m_1 = (m^2 - k_1^2 a^2)^{1/2} \quad \text{and} \quad \tau = \frac{T\gamma_1}{a} = \frac{T}{T_0}$$

By analogy, T_0 can be called the transversal resistance of the borehole. Solving system Eq. (11.77) gives

$$C = \frac{m^2 K_0(m_1) K_1'(m_1)}{1 - \tau \left[\frac{k_1^2 a^2}{m_1^2} I_1(m_1) K_1(m_1) + m^2 I_1'(m_1) K_1'(m_1) \right]} \tau$$

$$D = \frac{\frac{k_1^2 a^2}{m_1} K_0(m_1) K_1(m_1)}{1 - \tau \left[\frac{k_1^2 a^2}{m_1^2} I_1(m_1) K_1(m_1) + m^2 I_1'(m_1) K_1'(m_1) \right]} \tau \quad (11.78)$$

Thus the magnetic field on the z-axis of the borehole is

$$b_x^* = b_{0x}^*(\gamma_1) + k_1^2 a^2 \frac{a^3}{\pi} \tau \int_0^\infty \frac{m^2 K_0^2(m_1) \cos \alpha m}{1 - \tau \left[\frac{k_1^2 a^2}{m_1^2} I_1(m_1) K_1(m_1) + m^2 I_1'(m_1) K_1'(m_1) \right]} dm$$

(11.79)

For $\tau \to 0$ we obtain $b_x^* \to b_{0x}^*(\gamma_1)$, whereas in the opposite case, as $\tau \to \infty$:

$$b_x^* = b_{0x}^*(\gamma_1) - k_1^2 a^2 \frac{a^3}{\pi} \int_0^\infty \frac{m^2 K_0(m_1) \cos \alpha m \, dm}{\frac{k_1^2 a^2}{m_1^2} I_1(m_1) K_1(m_1) + m^2 I_1'(m_1) K_1'(m_1)} \quad (11.80)$$

The calculations show that Eq. (11.80) describes field b_x^* with sufficient accuracy when $\tau > 10$. The amplitudes of the secondary field as a function of the parameter, L/a, are shown in Fig. 11.9. The index is a/δ_1.

In the presence of the thin resistive layer, the primary field is practically compensated by the secondary field if $L/a > 10$ and $a/\delta_1 > 0.8$.

At the range of the small parameter, the quadrature component of the magnetic field can be presented as:

$$Q b_x^* = -\left(\frac{L}{\delta_1}\right)^2 (1 + G_\tau)$$

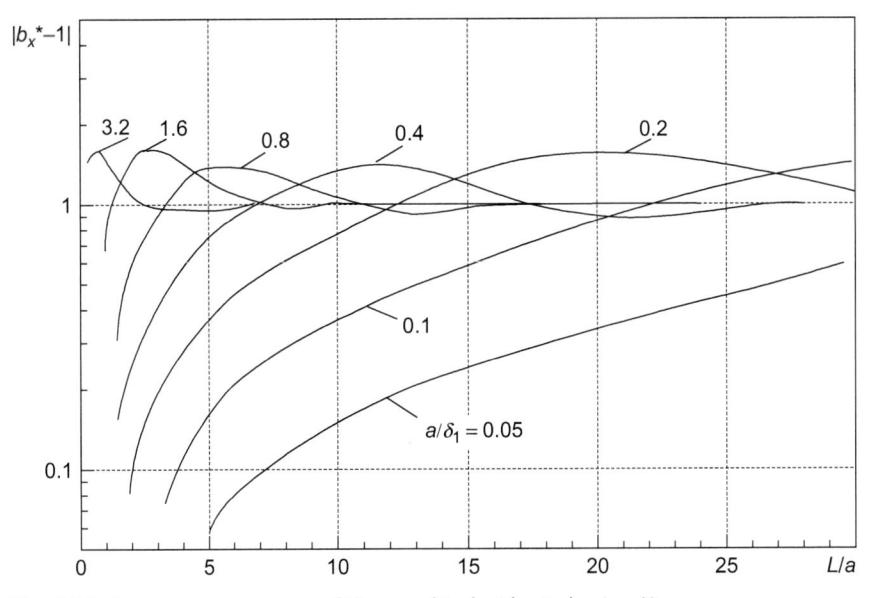

Induction Logging Using Transversal Coils 423

Fig. 11.9 Frequency responses of the amplitude. The index is a/δ_1.

where

$$G_\tau = -\frac{2\tau\alpha}{\pi}\int_0^\infty \frac{m^2 K_0^2(m)\cos\alpha m}{1-\tau m^2 I_1'(m)K_1'(m)}dm$$

If the length of the probe is much larger than the radius of the borehole ($\alpha \gg 1$), then, by performing integration by parts in the integral above we obtain

$$G_\tau \approx -\frac{2\tau}{1+\tau/2\pi}\alpha\int_0^\infty m^2 K_0^2(m)\cos\alpha m\, dm \approx \frac{4\tau}{1+(\tau/2)}\frac{\ln\alpha}{\alpha^2}$$

Thus

$$Qb_x^* \approx -\left(\frac{L}{\delta_1}\right)^2\left(1+\frac{4\tau}{1+\tau/2}\frac{\ln\alpha}{\alpha^2}\right)$$

and for large values of τ, it gives

$$Qb_x^* \approx -\left(\frac{L}{\delta_1}\right)^2\left(1+8\frac{\ln\alpha}{\alpha^2}\right) \qquad (11.81)$$

Next, we derive an asymptotic expression for the field in the far zone ($\alpha \gg 1$) for arbitrary parameter L/δ_1. For simplicity, we set parameter τ to be much greater than unity (i.e., $\tau \gg 1$). The integrand in Eq. (11.80) has a branch point on the complex plane of the variable of integration m, when $m = k_1 a$, and, by deforming the contour of the integration along the crosscut and expanding the integrand by powers of m_1, we obtain

$$b_x^* = b_{0x}^* - k_1^2 a^2 \frac{\alpha^3}{\pi} \int_{k_1 a}^{i\infty + k_1 a} m \left[K_0^2(m_1) - K_0^2(-m_1) \right] \exp(i\alpha m) dm$$

If $|m_1| \ll 1$, then

$$K_0^2(m_1) - K_0^2(-m_1) = 2i\pi K_0(m_1) I_0(m_1) + \pi^2 I_0^2(m_1) = -2i\pi K_0(m_1)$$

and

$$b_x^* = b_{0x}(\gamma_1) - 2ia^2 k_1^2 \alpha^3 \int_{k_1 a}^{i\infty + k_1 a} m^2 K_0(m_1) \exp(i\alpha m) dm$$

Letting $m_1 = it + k_1 a$, where variable t changes as

$$0 \leq t < \infty$$

we have

$$b_x^* = b_{0x}^*(\gamma_1) + 2a^2 k_1^2 \alpha^3 \exp(ik_1 L) \int_0^\infty (it + k_1 a)^2 K_0(m_1) \exp(-\alpha t) dt \quad (11.82)$$

where

$$m_1 = \left(-t^2 + 2itk_1 a\right)^{1/2}$$

For $\alpha \gg 1$ the integral in Eq. (11.82) is expressed through the integral exponential function:

$$\int_0^\infty (t+k_1 a)^2 K_0(m_1)\exp(-\alpha t)dt = \frac{1}{2}\left(\frac{\partial^2}{\partial \alpha^2} + 2ik_1 a\frac{\partial}{\partial \alpha} - k_1^2 a^2\right)$$

$$\times \int_0^\infty [\ln(-t) + \ln(t-2ik_1 a)]\exp(-\alpha t)dt = \frac{1}{2}\left(\frac{\partial^2}{\partial \alpha^2} + 2ik_1 a\frac{\partial}{\partial \alpha} - k_1^2 a^2\right)$$

$$\times \left[-\frac{\ln\alpha}{\alpha} + \frac{\ln(-2ik_1 a)}{\alpha} - \frac{\exp(-2ik_1 L)}{\alpha}Ei(2ik_1 L)\right] \tag{11.83}$$

If $|k_1 L| \ll 1$, then

$$Ei(2ik_1 L) \approx \ln(-2ik_1 L)$$

and Eq. (11.83) becomes

$$\left(\frac{\partial^2}{\partial \alpha^2} + 2ik_1 a\frac{\partial}{\partial \alpha} - k_1^2 a^2\right)\left(\frac{-\ln\alpha}{\alpha}\right) - \frac{2\ln\alpha}{\alpha}$$

Therefore, for the magnetic field we have

$$b_x^* = b_{0x}^*(\gamma_1) - 4k_1^2 a^2 \exp(ik_1 L)\ln\alpha = -\frac{k_1^2 L^2}{2}\left(1 + 8\frac{\ln\alpha}{\alpha^2}\right)$$

This expression coincides with Eq. (11.81), which is valid for the range of small parameters. In the opposite case, i.e., when $k_1 L \gg 1$:

$$Ei(2ik_1 L) \approx \frac{\exp(2ik_1 L)}{2ik_1 L}$$

and, instead of Eq. (11.83), we obtain

$$\frac{1}{2}\left(\frac{\partial^2}{\partial \alpha^2} + 2ik_1 a\frac{\partial}{\partial \alpha} - k_1^2 a^2\right)\left[-\frac{\ln\alpha}{\alpha} + \frac{\ln(-2ik_1 a)}{\alpha}\right] = \frac{k_1^2 L^2}{2}\frac{\ln\alpha - \ln|k_1 a|}{\alpha^3}$$

and

$$b_x^* = b_{0x}^* + k_1^2 a^2 k_1^2 L^2(\ln\alpha - \ln|k_1 a|)\exp(ik_1 L) \tag{11.84}$$

For the range of large parameters $|k_1 L| \gg 1$, we have

$$b_{0x}^*(\gamma_1) \approx -k_1^2 L^2 \exp(ik_1 L)$$

therefore, Eq. (11.84) can be presented in the form:

$$b_x^* = b_{ox}^*(\gamma_1)\left(1 - k_1^2 L^2 \frac{\ln\alpha - \ln|k_1 a|}{\alpha^2}\right) \quad (11.85)$$

11.7 MAGNETIC FIELD IN MEDIUM WITH ONE HORIZONTAL INTERFACE

Now we begin to study the field of the transversal magnetic dipole in the presence of horizontal boundaries. Let us place the dipole at the origin of the coordinates and direct the dipole moment along the x-axis:

$$\mathbf{M} = \operatorname{Re} M_0 \exp(-i\omega t)\mathbf{x}_0 \quad (11.86)$$

where $M_0 = I_0 n S$.

As before, we proceed from the field equations

$$\begin{array}{ll} \operatorname{curl}\mathbf{E} = i\omega\mathbf{B} & \operatorname{curl}\mathbf{B} = \gamma\mu_0\mathbf{E} \\ \operatorname{div}\mathbf{E} = 0 & \operatorname{div}\mathbf{B} = 0 \end{array} \quad (11.87)$$

Introduction of the vector potential of the magnetic type

$$\mathbf{E} = i\omega \operatorname{curl}\mathbf{A} \quad (11.88)$$

gives

$$\nabla^2 \mathbf{A} + k^2 \mathbf{A} = 0 \quad (11.89)$$

and the relationships between the vector potential and the field are

$$\mathbf{E} = i\omega \operatorname{curl}\mathbf{A} \quad \text{and} \quad \mathbf{B} = k^2\mathbf{A} + \operatorname{grad}\operatorname{div}\mathbf{A} \quad (11.90)$$

We look for a solution, assuming that the y-component of the vector potential is equal to zero:

$$A_y = 0$$

Then, in accordance with Eq. (11.90), we have

$$E_x = i\omega \frac{\partial A_z}{\partial y} \quad E_y = i\omega\left(\frac{\partial A_x}{\partial z} - \frac{\partial A_z}{\partial x}\right) \quad E_z = -i\omega \frac{\partial A_x}{\partial y}$$

$$B_x = k^2 A_x + \frac{\partial}{\partial x}\operatorname{div}\mathbf{A} \quad B_y = \frac{\partial}{\partial y}\operatorname{div}\mathbf{A} \quad B_z = k^2 A_z + \frac{\partial}{\partial z}\operatorname{div}\mathbf{A}$$

$$(11.91)$$

Induction Logging Using Transversal Coils

As follows from Eqs. (11.90), (11.91), for continuity of the tangential components of the field at the interface $z=h$ it is sufficient to provide the continuity of values

$$A_z, \quad \frac{\partial A_x}{\partial z}, \quad k^2 A_x, \quad \text{div}\,\mathbf{A}$$

Thus, for components of the vector potential we obtain two groups of conditions, i.e.,

$$k_1^2 A_{1x} = k_2^2 A_{2x}, \quad \frac{\partial A_{1x}}{\partial z} = \frac{\partial A_{2x}}{\partial z} \tag{11.92}$$

and

$$A_{1z} = A_{2z}, \quad \text{div}\,\mathbf{A}_1 = \text{div}\,\mathbf{A}_2 \tag{11.93}$$

In a uniform medium the field is described by one component of the vector potential, which has the form:

$$A_{0x}^* = \frac{\mu_0 M_0}{4\pi} \frac{\exp(ik_1 R)}{R}$$

or

$$A_{0x}^* = \frac{\mu_0 M_0}{4\pi} \int_0^\infty \frac{m}{m_1} \exp m_1(-|z|) J_0(mr)\,dm$$

where

$$m_1 = \left(m^2 - k_1^2\right)^{1/2}$$

By analogy with the case of the vertical magnetic dipole (Chapter 9), we represent the component A_x^* in both parts of the medium as

$$A_{1x}^* = \frac{\mu_0 M_0}{4\pi} \int_0^\infty \left[\frac{m}{m_1}\exp(-m_1|z|) + A_m \exp(m_1 z)\right] J_0(mr)\,dm \quad \text{if } z < h$$

$$A_{2x}^* = \frac{\mu_0 M_0}{4\pi} \int_0^\infty B_m \exp(-m_2 z) J_0(mr)\,dm, \quad \text{if } z > h$$

$$\tag{11.94}$$

where $m_2 = (m^2 - k_2^2)^{1/2}$. The boundary conditions at $z = h$ give

$$\frac{m}{m_1}e^{-m_1 h} + A_m e^{m_1 h} = sB_m e^{-m_2 h}$$
$$-me^{-m_1 h} + m_1 A_m e^{m_1 h} = -m_2 B_m e^{-m_2 h}$$

Whence

$$A_m = \frac{m}{m_1}\frac{sm_1 - m_2}{sm_1 + m_2}e^{-2m_1 h}$$

and

$$B_m = \frac{2m}{sm_1 + m_2}e^{-(m_1 - m_2)h} \tag{11.95}$$

where $s = \gamma_2/\gamma_1$.

Thus

$$A_{1x}^* = A_{0x}^*(\gamma_1) + \frac{\mu_0 M_0}{4\pi}\int_0^\infty \frac{m}{m_1}\frac{sm_1 - m_2}{sm_1 + m_2}e^{m_1(z-2h)}J_0(mr)\,dm$$

$$A_{2x}^* = \frac{\mu_0 M_0}{4\pi}\int_0^\infty \frac{2m}{sm_1 + m_2}e^{-(m_1 - m_2)h}e^{-m_2 z}J_0(mr)\,dm$$

(11.96)

To determine the component A_z^*, we use condition of continuity of $div\mathbf{A}^*$:

$$\frac{\partial}{\partial x}\left(A_{1x}^* - A_{2x}^*\right) = \frac{\partial}{\partial z}\left(A_{2z}^* - A_{1z}^*\right)$$

Inasmuch as

$$\frac{\partial A_x}{\partial x} = \frac{\partial A_x}{\partial r}\frac{\partial r}{\partial x} = \cos\phi \int_0^\infty F(m)e^{\pm m_i z}J_1(mr)\,dm$$

To provide continuity of $div A^*$, it is appropriate to present the solution for A_z^* in the following form:

$$A_{1z}^* = \frac{\mu_0 M_0}{4\pi} \cos\phi \int_0^\infty C_m e^{m_1 z} J_1(mr) dm \quad \text{and} \quad A_{2z}^*$$

$$= \frac{\mu_0 M_0}{4\pi} \cos\phi \int_0^\infty D_m e^{-m_2 z} J_1(mr) dm$$

In accordance with Eq. (11.93), we have

$$\begin{aligned} C_m e^{m_1 h} &= D_m e^{-m_2 h} \\ (s-1) m B_m e^{-m_2 h} &= m_2 D_m e^{m_2 h} + m_1 C_m e^{m_1 h} \end{aligned} \quad (11.97)$$

Solving this system we obtain

$$C_m = \frac{(s-1)mB_m}{m_1 + m_2} e^{-(m_1+m_2)h} \quad D_m = \frac{(s-1)mB_m}{m_1 + m_2} \quad (11.98)$$

Thus

$$A_{1z}^* = \frac{\mu_0 M_0}{4\pi} \cos\phi \int_0^\infty \frac{(s-1)mB_m}{m_1 + m_2} e^{-(m_1+m_2)h} e^{m_1 z} J_1(mr) dm$$

and

$$A_{2z}^* = \frac{\mu_0 M_0}{4\pi} \cos\phi \int_0^\infty \frac{(s-1)mB_m}{m_1 + m_2} e^{-m_2 z} J_1(mr) dm \quad (11.99)$$

The magnetic field on the z-axis has component B_x only, and, in accordance with Eqs. (11.91), (11.99), we have

$$\begin{aligned} b_{1x}^* &= b_{0x}^* - L \int_0^\infty \phi_1(m) e^{m_1 L} dm \\ b_{2x}^* &= -L \int_0^\infty \phi_2(m) e^{-m_2 L} dm \end{aligned} \quad (11.100)$$

where b_x is the magnetic field expressed in units of the field in free space:

$$b_x = \frac{B_x}{B_0} \quad B_0 = -\frac{\mu_0 M_0}{4\pi L^3}$$

and L is the length of the probe. Also,

$$b_{0x}^* = e^{ik_1 L}\left(1 - ik_1 L - k_1^2 L^2\right)\phi_1(m)$$

$$= \left(k_1^2 L^2 - \frac{m^2 L^2}{2}\right)\frac{m}{m_1}\frac{sm_1 - m_2}{m_1 + m_2}e^{-2m_1 h}$$

$$+ m^2 L^2 \frac{m^3(s-1)e^{-2m_1 h}}{m_1(sm_1 + m_2)(m_1 + m_2)}\phi_2(m) \qquad (11.101)$$

$$= \left(k_2^2 L^2 - \frac{m^2 L^2}{2}\frac{m_1 + sm_1}{m_1 + m_2}\right)\frac{2m}{sm_1 + m_2}e^{-(m_1 - m_2)h}$$

First, consider the field at the low-frequency limit when the skin depth in both media exceeds the distance from the dipole to the interface as well as the length of the probe. In deriving the asymptotic formulas, we use the approach described in Chapter 8, namely the interval of integration is presented as the sum of two parts, i.e., the internal part where $0 < mL < m_0 L \ll 1$ and the external part where $m > m_0$. Within the external interval, radicals m_1 and m_2 can be expanded in a series by powers of k_1^2/m^2 and k_2^2/m^2. For this reason, the integral at the external interval is presented as a series of terms that have even powers of k. Within the internal interval, the exponents can be expanded in series ($mL < 1$), and the integral is reduced to the sum of tabular integrals, which in its turn can be presented as a series with respect to the wave number k. Unlike the integral at the external interval, these series contain odd powers of k and logarithmic terms. For example, in a medium, where the dipole is located, at the low-frequency limit, we have

$$Inb_x^* = 1 + a_1\left(\frac{L}{\delta_1}\right)^3 \quad \text{and} \quad Qb_x^* \approx d\left(\frac{L}{\delta_1}\right)^2 + a_1\left(\frac{L}{\delta_1}\right)^3 \qquad (11.102)$$

where

$$a_1 = \frac{2}{s^2 - 1}\left[\frac{4}{3}s^{3/2}\left(s^{1/2} - 1\right) - \frac{1}{5}s\left(s^{3/2} - 1\right) + \frac{2}{15}\left(s^{7/2} - 1\right)\right.$$

$$\left. + \frac{s^2}{2(s+1)^{1/2}}\ln\frac{\sqrt{s+1} - 1}{\sqrt{s+1} + 1}\frac{\sqrt{s+1} + \sqrt{s}}{\sqrt{s+1} - \sqrt{s}}\right]$$

and

$$d = -1 - \frac{1}{4}\frac{(s+5)(s-1)}{(s+1)}\frac{L}{2h - L} \qquad (11.103)$$

where

$$\delta_1 = \left(\frac{2}{\gamma_1 \mu_0 \omega}\right)^{1/2}, \quad \delta_2 = \left(\frac{2}{\gamma_2 \mu_0 \omega}\right)^{1/2}, \quad L/\delta_1 \ll 1, \quad L/\delta_2 \ll 1, \quad s = \gamma_2/\gamma_1$$

If the interface is located at a sufficient distance from the source and the observation point $(L/h \ll 1)$, coefficient d tends to -1, corresponding to a uniform medium. At the same time, coefficient a_1 does not depend on the position of the probe with respect to the boundary, and it is a function of the resistivity of both media. The second terms in Eq. (11.102) are proportional to $\omega^{3/2}$ and sensitive only to the deepest part of the formation. (In fact, by measuring these terms we can reach the same depth of investigation as that achievable at the late stage of the transient field). It is obvious that, as $s \to 1$, coefficients a_1 and d correspond to a uniform medium:

$$a_1 = \frac{4}{3} \quad \text{and} \quad d = -1$$

Deriving asymptotic expression at the high-frequency limit, we use the following relationship:

$$I_n = \int_0^\infty m^n \exp\left(m^2 + k_1^2 L^2\right)^{1/2} dm \approx a_n (k_1 L)^{(n+1)/2} \exp(-k_1 L) \quad (11.104)$$

where $|k_1 L \gg 1|$ and a_n are functions of the number n. In particular, for the first three values of n, they are equal to $1, (\pi/2)^{1/2}$, and 2, respectively. Note that integrals of type Eq. (11.104) for odd values of n are reduced to elementary functions, but, for even values they are expressed through modified Bessel functions $K_n(k_1 L)$. After elementary transformations, by presenting the field through integrals of type I_n, and taking into account exponential decay at $|k_1 L| \gg 1$, we obtain

$$b_{1x}^* = b_{0x}^*(\gamma_1) - k_1^2 L^2 \frac{\sqrt{s}-1}{\sqrt{s}+1} \frac{\exp[ik_1 L(2\alpha - 1)]}{2\alpha - 1} \approx b_{0x}^*(\gamma_1) \quad (11.105)$$

where $\alpha = h/L > 1$. The field becomes the same as that in a uniform medium with conductivity, γ_1 due to the skin effect. However, if the dipole or the observation point is located at the interface $\alpha = 1$, then the field is a function of the conductivities of both media regardless of the frequency. In accordance with Eq. (11.105), we have

$$b_{1x}^* \approx -k_1^2 L^2 \frac{2\sqrt{s}}{\sqrt{s}+1} \exp(ik_1 L) \qquad (11.106)$$

It is proper to note one specific feature of the current distribution when the conductivity of the medium, surrounding the dipole, is equal to zero ($s \to \infty$). As seen in Eq. (11.96), the component A_{2x} vanishes. For this reason, in the conducting part of the medium, the electric field and induced currents do not have a vertical component, and the distribution of currents is symmetrical with respect to the plane yoz, which is not intersected by current lines.

11.8 MAGNETIC FIELD OF THE HORIZONTAL DIPOLE IN THE FORMATION WITH TWO HORIZONTAL INTERFACES

Suppose that the magnetic dipole is located within the formation. Then, according to the results obtained in the previous section, the expressions for the vector potential have the following forms:

$$\begin{cases} A_{1x}^* = \frac{\mu_0 M_0}{4\pi} \int_0^\infty D_1 e^{m_1 z} J_0(mr) \, dm \\ \\ A_{1z}^* = \frac{\mu_0 M_0}{4\pi} \cos\phi \int_0^\infty F_1 e^{m_1 z} J_1(mr) \, dm \end{cases} \quad \text{if } z < -h_2$$

$$\begin{cases} A_{2x}^* = \frac{\mu_0 M_0}{4\pi} \int_0^\infty \left[\frac{m}{m_2} e^{-m_2|z|} + D_2 e^{m_2 z} + D_3 e^{-m_2 z} \right] J_0(mr) \, dm \\ \\ A_{2z}^* = \frac{\mu_0 M_0}{4\pi} \cos\phi \int_0^\infty [F_2 e^{m_2 z} + F_3 e^{-m_2 z}] J_1(mrdm) \end{cases} \quad \text{if } -h_2 < z < h_1$$

$$\qquad (11.107)$$

$$\begin{cases} A_{3x}^* = \frac{\mu_0 M_0}{4\pi} \int_0^\infty D_4 e^{-m_1 z} J_0(mr) \, dm \\ \\ A_{3z}^* = \frac{\mu_0 M_0}{4\pi} \cos\phi \int_0^\infty F_4 e^{-m_1 z} J_1(mr) \, dm \end{cases} \quad \text{if } z > h_1$$

From the system of equations, following from the boundary conditions at $z = h_1$ and $z = -h_2$, we can determine the coefficients D_1, D_2, D_3, D_4 and F_1, F_2, F_3, F_4. In particular, for the horizontal component of the magnetic field on the z-axis, when the two-coil probe is located symmetrically with respect to the horizontal boundaries, we obtain

$$b_x^* = b_{0x}^*(\gamma_2) - \int_0^\infty \left\{ \left(\frac{m}{2} - k_2^2 L^2\right) 2q_{12}(1 - q_{12}e^{-am_2}\cosh m_2) + \frac{(1-s)(1-q_{12})m^2 m_2}{(m_1+m_2)d_2} \right.$$
$$\left. \times [1 - (q_{12} - K_{12})e^{-am_2}\cosh m_2 - K_{12}q_{12}e^{-am_2}] \right\} \frac{m}{m_2 d_1} e^{-am_2} dm, \text{ if } \alpha = H/L \geq 1$$

(11.108)

where

$$d_1 = 1 - q_{12}^2 e^{-2m_2}, \quad d_2 = 1 - K_{12}^2 e^{-2m_2}, \quad q_{12} = \frac{sm_1 - m_2}{sm_1 + m_2}, \quad s = \frac{\gamma_2}{\gamma_1}, \quad K_{12} = \frac{m_1 - m_2}{m_1 + m_2},$$

γ_2 and γ_1 are the conductivities of the layer and shoulders, respectively, H is the thickness of the formation, and L is the length of the probe.

By analogy, when the length of the probe exceeds the thickness of the layer, and the transmitter and receiver coils are located symmetrically with respect to the layer, an expression for the field is

$$b_x^* = \int_0^\infty \left[\frac{m^2}{2} s - k_2^2 L^2 + \frac{m^2 m_1^2}{2(m_1+m_2)^2} \frac{(s-1)^2}{d_2} \left(1 - e^{-2am_2}\right) \right]$$
$$\times \frac{4mm_2 e^{-[am_2 + (1-\alpha)m_1]}}{(sm_1+m_2)^2 d_1} dm \quad \alpha < 1$$

(11.109)

Owing to the symmetrical position of the coils, the field is defined by three parameters:

$$p = \frac{L}{\delta} \quad s = \frac{\gamma_2}{\gamma_1} \quad \text{and} \quad \alpha = \frac{H}{L}$$

First, consider the field at the low-frequency limit, when parameter $p = L/\delta_2 \to 0$ and the probe is located within the bed γ_2. Proceeding from the approach, described in Chapter 7, we present Eq. (11.109) as a sum of two integrals: first integral, corresponding to small values of integration variable m and the second integral, representing the residual part. The first integral gives an asymptotic expression for Inb_x^*:

$$Inb_x^* \approx \frac{4}{3}\left(\frac{L}{\delta_1}\right)^3$$

and for the quadrature component, we have

$$Qb_x^* \approx -\left(\frac{L}{\delta_2}\right)^2\left[1 - 2\frac{s-1}{s+1}\int_0^\infty \frac{1 - (s-1)/(s+1)e^{-\alpha m}\cosh\alpha m}{((s-1)/(s+1))^2 e^{-2\alpha m}}dm + \frac{1-s}{2\alpha s}\right]$$
$$+ \frac{4}{3}\left(\frac{L}{\delta_1}\right)^3$$

(11.110)

It is essential that the in-phase component of the field at the low-frequency limit coincides with the in-phase component of the field in a uniform medium with conductivity of the shoulders, γ_1. A similar result is obtained when the source is the vertical magnetic dipole. This indicates that surface charges, occurring at the interfaces between layer and the surrounding medium affect only the quadrature component of the magnetic field. Now we present the quadrature component Qb_x^* as the sum of two terms:

$$Qb_x^* = Qb_{1x}^* + Qb_{2x}^*$$

where

$$Qb_{1x}^* = -\left(\frac{L}{\delta_2}\right)^2\left(1 - \frac{1}{2\alpha}\right) - \left(\frac{L}{\delta_1}\right)^2\frac{1}{2\alpha}$$
$$Qb_{2x}^* = \left(\frac{L}{\delta_2}\right)^2 2F(\beta, \alpha)$$

(11.111)

where

$$F(\beta, \alpha) = \beta\int_0^\infty \frac{1 - \beta e^{-\alpha m}\cosh m}{1 - \beta^2\exp(-2\alpha m)}dm$$

(11.112)

and

$$\beta = \frac{s-1}{s+1}, \quad -1 < \beta < 1$$

At the range of the small parameter, the component Qb_{1x}^* coincides with the vertical component Qb_z^* of the vertical magnetic dipole and it consists of two terms, each depending on either conductivity shoulder γ_1 or the bed γ_2. Correspondingly, we can introduce a geometric factor for each term. In accordance with Eq. (11.111), we define

$$G_1 = 1 - 1/2\alpha, \quad G_2 = 1/2\alpha, \quad G_1 + G_2 = 1$$

and

$$Qb_{1x}^* = -\frac{\mu_0 \omega L^2}{2}[\gamma_1 G_1(\alpha) + \gamma_2 G_2(\alpha)] \quad (11.113)$$

The expressions for the geometric factors are the same as in the case of the excitation of the field by a vertical magnetic dipole. The second term Qb_{2x}^* includes the function $F(\beta, \alpha)$, which depends on the ratio of the conductivities, or more precisely, on parameter β. The appearance of this term can be explained in the following way. The primary electric field gives rise to the surface charges, whose density is

$$\sigma(a) = \frac{1}{2\pi}\frac{s-1}{s+1}E_n^{av}(a) \quad (11.114)$$

where $E_n^{av}(a)$ is the magnitude of the normal component of the field, created by the vortex field of currents and all charges, except those, located at the point a. In this approximation, the field of electric charges and the primary field are directly proportional to frequency. Let us present Eq. (11.112) as:

$$Qb_x^* = -\frac{\omega \mu_0 L^2}{2}[\gamma_2 G_2^*(\alpha, s) + \gamma_1 G_1(\alpha, s)] \quad (11.115)$$

where

$$G_2^*(\alpha, s) = 1 - \frac{1}{2\alpha} - 2F(\beta, s)$$

If the layer resistivity exceeds that of the shoulders $(s < 1)$, then the electric charges increase the field within the layer, and function G_2^* becomes larger. In a more conductive layer the electric field of the charges reduces the primary field, and, under certain conditions, the function G_2^* crosses zero and changes sign.

Table 11.6 contains the values of the functions $G_1^* + (1/s) \cdot G_2$ and $F(\beta, \alpha)$ for some values of α. It is possible to show that function $F(\beta, \alpha)$ is expressed through hypergeometric series $_2F_1$

Table 11.6 Maximum Values of (L/δ_1), for Which the Difference Between Exact and Approximate Values Below 5%

$\alpha = 4$		$\alpha = 8$		$\alpha = 16$	
$G_2^* + \dfrac{G_1^*}{s}$	$F(\beta, \alpha)$	$G_2^* + \dfrac{G_1^*}{s}$	$F(\beta, \alpha)$	$G_2^* + \dfrac{G_1^*}{s}$	$F(\beta, \alpha)$
−1.03	18.9	−0.520	9.998	−0.26	5.49
−0.703	6.28	−0.351	3.64	−0.175	2.32
−0.377	2.63	−0.188	1.81	−0.094	1.41
−0.102	1.33	−0.0507	1.16	−0.0253	1.08
0.0717	0.794	0.0359	0.897	0.018	0.948
0.142	0.606	0.0718	0.802	0.0359	0.901
0.164	0.552	0.0825	0.774	0.0414	0.887
0.169	0.538	0.0854	0.767	0.0428	0.883

$$F(\beta, \alpha) = \frac{\beta}{2}\left[\frac{1}{\beta\alpha}\ln\frac{1+\beta}{1-\beta} - \frac{\beta}{2\alpha+1}\cdot {}_2F_1\left(1, 1+\frac{1}{2\alpha}, 2+\frac{1}{2\alpha}, \beta^2\right)\right.$$
$$\left. -\frac{\beta}{2\alpha-1}\cdot {}_2F_1\left(1, 1-\frac{1}{2\alpha}, 2-\frac{1}{2\alpha}, \beta^2\right)\right] \quad (11.116)$$

When the length of the probe is equal to the formation thickness ($\alpha = 1$), $F(\beta, \alpha)$ is expressed through elementary function:

$$F(\beta, \alpha) = \frac{1}{2} - \frac{1}{s^2-1}\ln s \quad (11.117)$$

and for the quadrature component, we have

$$Qb_x^* = -\left(\frac{L}{\delta_2}\right)^2\left(-\frac{1}{2} + \frac{2}{s^2-1}\ln s\right) - \frac{1}{2}\left(\frac{L}{\delta_1}\right)^2 \quad (11.118)$$

For the large α, function $F(\beta, \alpha)$ decreases inversely proportional to α:

$$F(\beta, \alpha) \approx \frac{1}{\alpha}\ln\frac{2s}{s+1} \quad (11.119)$$

and the function $G_1^*(\alpha, s)$ remains positive for all values of s. With increase of resistivity of the layer ($s \to 0$), the term Qb_{2x}^* trends to zero.

The asymptotic presentation for the field, when the formation is located within the probe, is derived in a similar manner, and we obtain

$$Inb_x^* = \frac{4}{3}\left(\frac{L}{\delta_1}\right)^3$$

Induction Logging Using Transversal Coils

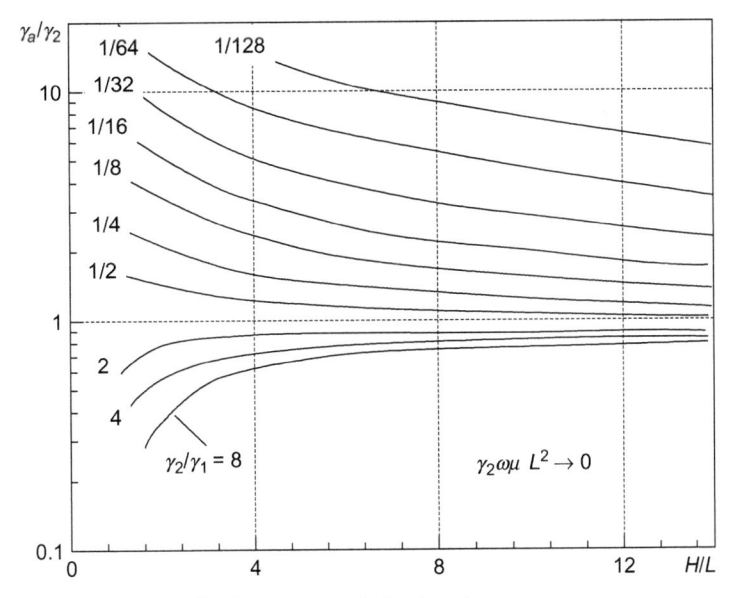

Fig. 11.10 Apparent conductivity curves. Index is γ_2/γ_1.

and

$$Qb_x^* = -\left(\frac{L}{\delta_2}\right)^2 \left[\frac{4}{(s+1)^2}\int_0^\infty \frac{e^{-m}\,dm}{1-((s-1)/(s+1))^2\exp(-2am)} - \frac{\alpha}{2}\right] - \frac{\alpha}{2}\left(\frac{L}{\delta_1}\right)^3$$

(11.120)

The integral in this expression can be also presented using a hyperbolic function. Now consider the responses of a two-coil probe placed in the middle of the bed at the range of small parameters, if $\alpha \geq 2$ (Fig. 11.10). The apparent conductivity is introduced as:

$$\frac{\gamma_a}{\gamma_2} = \frac{Qb_z^*}{Qb_{0z}(\gamma_2)}$$

where Qb_z^* is the quadrature component of the vertical component of the field.

It is natural that the influence of the surrounding medium increases with an increase of its conductivity γ_1 and a decrease of the thickness of the layer.

Comparison of responses caused by the vertical and horizontal dipoles shows that the influence of a more conductive surrounding medium on the fields is practically the same. If the layer is more conductive, the influence of the shoulder is more pronounced in the case of transversal dipoles and is caused by the influence of the electrical charges. This can be seen in Fig. 11.10 where apparent conductivity curve does not reach an asymptotic value of 1 even for the thick layers and conductivity contrast $\gamma_2/\gamma_1 = 128$. As the frequency increases, the skin effect becomes more pronounced, causing reduced influence of the surrounding medium.

Consider the frequency responses of the field (Fig. 11.11). We can see that at the low-frequency spectrum in Fig. 11.11A the secondary field is relatively small. Then it increases and in the limit when the skin depth in the layer is small, the amplitude of the secondary field approaches to that of the primary field.

Like in the case of the vertical magnetic dipole, the phase shift of the secondary field at the range of small parameter is $-\pi/2$ (Fig. 11.11B).

Now let us consider the influence of relatively thin layers ($\alpha < 1$). At the low-frequency limit we present a field as the sum of two terms: a field in a uniform medium with the conductivity of the surrounding medium and the part of the field that takes into account the influence of the bed:

$$Qb_x^* = Qb_{0x}^* \left(\frac{L}{\delta_1}\right) + \left(\frac{L}{\delta_1}\right)^2 G_2(\alpha, s) \tag{11.121}$$

Here

$$G_2(\alpha, s) = -\frac{4s}{(1+s)^2} \int_0^\infty \frac{e^{-m}}{1 + \left(\frac{s-1}{s+1}\right)^2 e^{-2\alpha m}} dm + \frac{\alpha(s-1)}{2} + 1$$

The latter coincides with Eq. (11.120) in the range of small parameters ($L/\delta < 1$), and for certain combinations of α and s, it is valid for a wider range of parameters (L/δ). Table 11.7 provides the maximum values of (L/δ_1), for which the difference of the quadrature components obtained from the exact solution and the approximate formula Eq. (11.121) does not exceed 5%.

These data demonstrate how the maximal value of parameter L/δ_1 increases as parameter α decreases. If a thin layer has a relatively high resistivity or conductivity, the range of application of Eq. (11.121) is restricted to

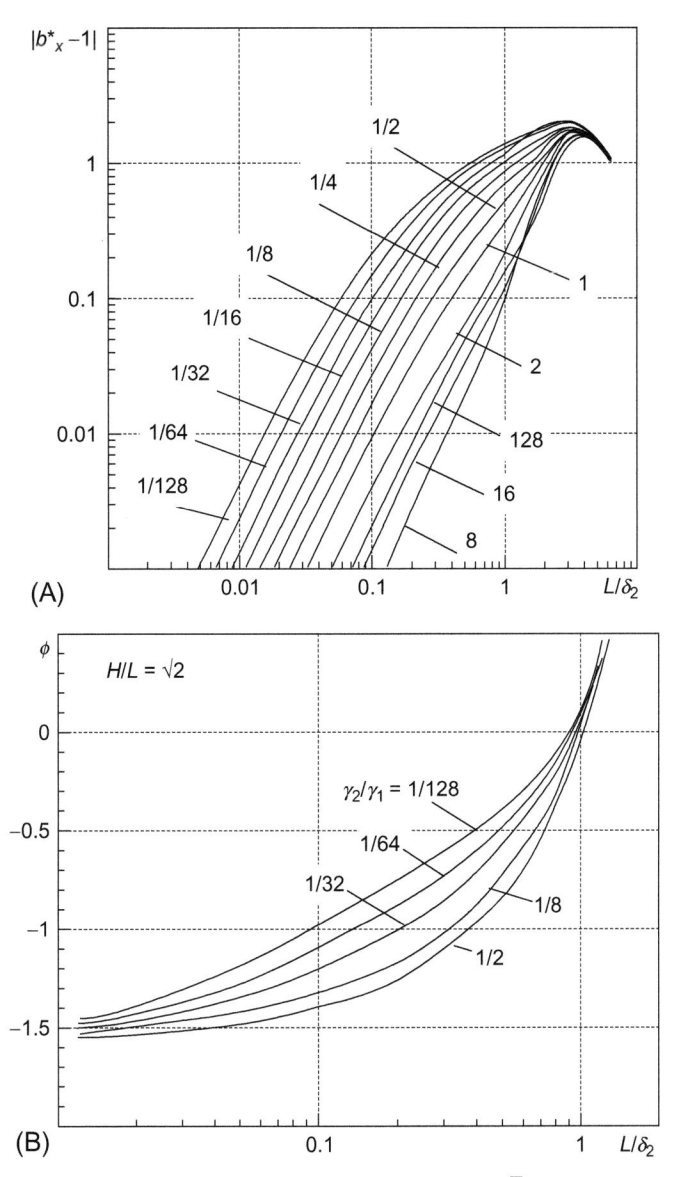

Fig.11.11 (A) Frequency responses of amplitude $H/L = \sqrt{2}$. Curve index is γ_2/γ_1. (B) Frequency responses of the phase. Curve index is γ_2/γ_1.

smaller values of L/δ_1. When s approaches unity, the maximum value of parameter L/δ_1 increases. Let us analyze the field at the low-frequency limit when the thickness of the layer is sufficiently small. By expanding the denominator of the integrand in Eq. (11.121) by powers of α, we obtain

Table 11.7 Maximum of Parameter (L/δ_1), for Which the Difference Between Exact and Approximate Quadrature Components Below 5%

α	$s=\dfrac{1}{128}$	$\dfrac{1}{64}$	$\dfrac{1}{32}$	$\dfrac{1}{16}$	$\dfrac{1}{4}$	$\dfrac{1}{2}$	2	8	16	32	64
$\alpha=\dfrac{1}{16}$	0.05	0.1	0.15	0.3	0.4	0.6	0.8	0.3	0.2	0.2	0.15
$\alpha=\dfrac{1}{8}$	0.03	0.07	0.1	0.2	0.4	0.6	0.6	0.2	0.1	0.1	0.07

$$Qb_x^* = -\left(\frac{L}{\delta_2}\right)^2 \frac{4}{(s+1)^2} \int_0^\infty \frac{\exp(-m)\,dm}{1-((s-1)/(s+1))^2(1-2\alpha m)}$$

$$+ \left(\frac{L}{\delta_1}\right)^2 \frac{\alpha(s-1)}{2} = \left(\frac{L}{\delta_1}\right)^2 \left[t\exp t \cdot Ei(-t) + \frac{\alpha(s-1)}{2}\right] \quad (11.122)$$

where

$$t = \frac{2s}{\alpha(s-1)^2}, \quad Ei(-t) = -e^{-t}\int_0^\infty \frac{\exp(-x)}{x+t}dx$$

is the integral exponential function. As is known:

$$Ei(-t) \to \ln t \text{ if } t \to 0 \quad \text{and} \quad Ei(-t) \to -e^{-t}\left(\frac{1}{t}-\frac{1}{t^2}\right) \text{ if } t \to \infty$$

For illustration, consider two extreme cases: $s \ll 1$ and $s \gg 1$, which correspond to either a very conductive or a very resistive thin layer, respectively.

Case 1: Very conductive thin layer ($s \gg 1$)

If parameter $s \gg 1$, then $t \approx 2/\alpha s$. Using the asymptotic value of function $Ei(-t)$ for $t \ll 1$, provided that $s \gg 2/\alpha$, instead of Eq. (11.122), we obtain

$$Qb_x^* = \left(-\frac{L}{\delta_1}\right)^2 \left(\frac{\alpha s}{2} - \frac{\alpha}{2} - \frac{2}{\alpha s}\ln\frac{2}{\alpha s}\right) \approx \frac{\alpha}{2}\left(\frac{L}{\delta_2}\right)^2, \quad \text{if } \frac{\alpha s}{2} \ll 1 \quad (11.123)$$

But, if $1 \ll s < {}^2/\alpha$, then

$$Qb_x^* = -\left(\frac{L}{\delta_1}\right)^2\left(1-\alpha s + \frac{\alpha}{2}\right) = -\left(\frac{L}{\delta_1}\right)^2 + \alpha\left(\frac{L}{\delta_2}\right)^2 \text{ as } \frac{\alpha s}{2} \ll 1 \quad (11.124)$$

Thus the field Qb_x^* can be presented as the sum of the field in a uniform medium with conductivity γ_1, and the field due to the presence of a thin conducting layer with conductivity γ_2:

$$Qb_x^* = Qb_{0x}^*(\gamma_1) + \alpha\left(\frac{L}{\delta_2}\right)^2 \qquad (11.125)$$

Case 2: Very resistive thin layer ($s \ll 1$)
For the parameter t we have $t = 2s/\alpha$. If $s < \alpha/2$, then $t \ll 1$ and correspondingly:

$$Qb_x^* = \left(\frac{L}{\delta_1}\right)^2 \left(\frac{2s}{\alpha}\ln\frac{2s}{\alpha} + \frac{s\alpha}{2} - \frac{\alpha}{2}\right) \approx \left(\frac{L}{\delta}\right)^2 \left(\frac{2s}{\alpha} - \frac{\alpha}{2}\right) \qquad (11.126)$$

In the case of ($s \gg \alpha/2$), we have

$$Qb_x^* = -\left(\frac{L}{\delta_1}\right)^2 \left(1 - \frac{\alpha}{2s} + \frac{\alpha}{s} - \frac{\alpha s}{2}\right) \approx -\left(\frac{L}{\delta_1}\right)^2 + \frac{\alpha}{2s}\left(\frac{L}{\delta_1}\right)^2 \qquad (11.127)$$

Generalizing this expression for higher frequencies, we obtain

$$Qb_x^* = Qb_{0x}^*(\gamma_1) + \frac{\alpha}{2s}\left(\frac{L}{\delta_1}\right)^2 \qquad (11.128)$$

Thus the smaller parameters s and α/s, for higher frequencies, Eq. (11.121) is applied.

11.9 PROFILING WITH A TWO-COIL INDUCTION PROBE IN A MEDIUM WITH HORIZONTAL INTERFACES

Considering the profiling curves, it is appropriate to distinguish four specific positions of the probe with respect to the interfaces (Fig. 11.12).

Case 1: The probe is located outside the formation (Fig. 11.12A). In accordance with the results obtained in the previous section, we have

$$b_x^* = b_{0x}^*(\gamma_1) - \int_0^\infty \left[\left(k_1^2 L^2 - \frac{m^2}{2}\right)D_1 e^{m_1} + \frac{mm_1}{2}F_1 e^{m_1}\right] dm \qquad (11.129)$$

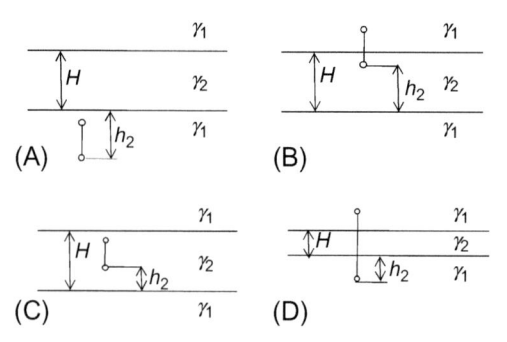

Fig. 11.12 (A–D) Different positions of two-coil probe.

where

$$D_1 = \frac{m}{m_1} \frac{q_{12}}{d_1} \left(1 - e^{-2\beta m_1}\right)$$

$$F_1 = -F\left(1 - e^{-2\alpha m_2}\right)\left(1 - q_{12}K_{12}e^{-2\alpha m_2}\right)e^{-\beta m_1}$$

$$F = \frac{2(1-s)m^2}{(m_1 + m_2)(sm_1 + m_2)d_1 d_2}$$

$$K_{12} = \frac{m_1 - m_2}{m_1 + m_2}, \quad q_{12} = \frac{sm_1 - m_2}{sm_1 + m_2}, \quad d_1 = 1 - q_{12}e^{-2\alpha m_2}, \quad d_2 = 1 - K_{12}^2 e^{-2\alpha m_2}$$

$$\alpha = \frac{H}{L}, \quad 0 \leq \alpha < \infty, \quad \beta = \frac{h_2}{L}, \quad \beta \geq 1$$

Case 2: The coils of the probe are located on both sides of interface. In this case (Fig. 11.12B), we have

$$b_x^* = -\int_0^\infty \left[\left(k_1^2 L^2 - \frac{m^2}{2}\right) D_4 e^{-m_1} + \frac{mm_1}{2} F_4 e^{-m_1}\right] dm$$

$$D_4 = s \frac{m}{m_2} \frac{e^{(\alpha-\beta)(m_1-m_2)}}{d_1}(1 - q_{12})\left(1 - q_{12}e^{-2\beta m_2}\right) \tag{11.130}$$

$$F_4 = Fe^{(\alpha-\beta)(m_1-m_2)}\left[(K_{12} - q_{12})e^{-2\beta m_2}\left(1 - e^{-2(\alpha-\beta)m_2}\right)\right.$$
$$\left. + \left(1 - K_{12}q_{12}e^{-2\alpha m_2}\right)\left(1 - e^{-2\beta m_2}\right)\right]$$

$$0 \leq \alpha < \infty, \quad 0 \leq \beta \leq \alpha, \quad \beta \leq 1$$

Case 3: The probe is located within the layer.
For this probe location, we have (Fig. 11.12C)

$$b_x^* = b_{0x}^*(\gamma_2) - \int_0^\infty \left[\left(k_2^2 L^2 - \frac{m^2}{2}\right)(D_2 e^{m_2} + D_3 e^{-m_2}) + \frac{mm_2}{2}(F_2 e^{m_2} - F_3 e^{-m_2})\right] dm$$

$$D_2 = -\frac{m}{m_2}\frac{q_{12}}{d_1}e^{-(\alpha-\beta)m_2}\left(1 - q_{12}e^{-2\beta m_2}\right) \quad D_3 = -\frac{m}{m_2}\frac{q_{12}}{d_1}e^{-2\beta m_2}\left(1 - q_{12}e^{-2(\alpha-\beta)m_2}\right)$$

$$F_2 = Fe^{-2(\alpha-\beta)m_2}\left[(K_{12} - q_{12})e^{-2\beta m_2} + 1 - K_{12}q_{12}e^{-2\alpha m_2}\right]$$

$$F_3 = -Fe^{-2\beta m_2}\left[(K_{12} - q_{12})e^{-2(\alpha-\beta)m_2} + 1 - K_{12}q_{12}e^{-2\alpha m_2}\right]$$

$$1 \leq \alpha \leq \infty, \ 0 \leq \beta \leq \alpha - 1$$

(11.131)

In the case of the probe in a symmetrical position with respect to the boundaries $\beta = \frac{\alpha-1}{2}$ and Eq. (11.131) coincides with Eq. (11.108).

Case 4: The layer is located between the coils of the probe.
For this location of the probe (Fig. 11.12D), we have

$$b_x^* = -\int_0^\infty \left[\left(k_1^2 L^2 - \frac{m^2}{2}\right)D_4 e^{-m_1} - \frac{mm_1}{2}F_4 e^{-m_1}\right] dm \qquad (11.132)$$

where

$$D_4 = \frac{4smm_2}{(sm_1 - m_2)^2 d_1}e^{\alpha(m_1 - m_2)} \quad F_4 = 2F(1-s)\frac{m_1 m_2 e^{\alpha(m_1 - m_2)}\left(1 - e^{-2\alpha m_2}\right)}{(m_1 + m_2)(sm_1 + m_2)}$$

The shoulders on both sides of the layer have the same conductivity; thus the field does not depend on position of the layer with respect to the coils.

Eqs. (11.129)–(11.132) permit calculation of the field along the trajectory, which crosses the layer. Some results corresponding to the case of the thick bed with $\alpha = H/L = 4$ are shown in Fig. 11.13. The apparent conductivity is introduced as

$$\frac{\gamma_a}{\gamma_2} = \frac{|b_x^* - 1|}{|b_{0x}^*(\gamma_2) - 1|}$$

where $b_{0x}^*(\gamma_2)$ is the field amplitude in the whole space with conductivity of the bed γ_2. The profiling curves are plotted for the fixed values of

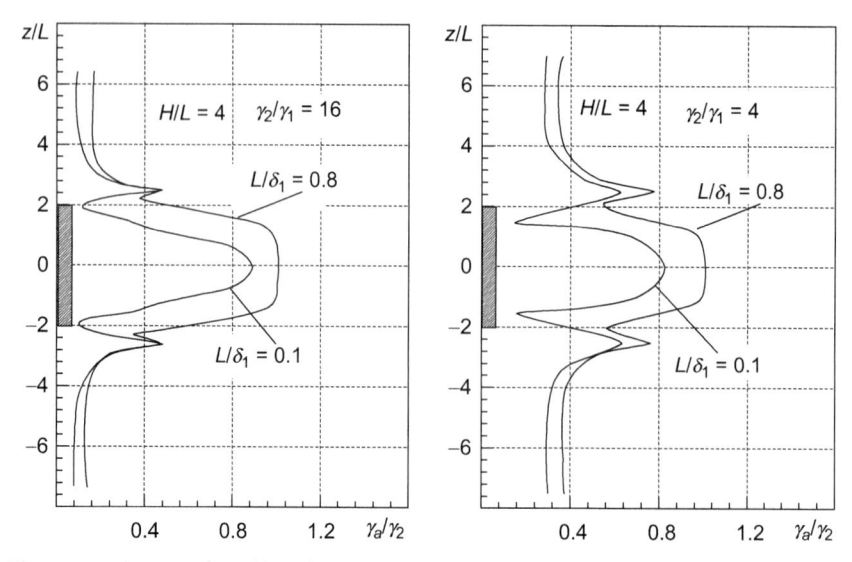

Fig. 11.13 Curves of profiling for two values of parameter L/δ_1.

$\alpha = H/L = 4$ and $s = \gamma_2/\gamma_1 = 16;4$. The index of curves is parameter L/δ_1. The horizontal axis depicts the value of γ_a/γ_2, and the vertical axis indicates the distance from the center of the bed to the middle of the probe, expressed in units of the layer thickness. In the middle of the bed, the apparent conductivity is approaching a true conductivity of the bed. When either the transmitter or the receiver is near the boundary of the layer, surface charges lead to a rapid change of the field. If the distance between the layer and the probe, located outside of the bed increases and slightly exceeds the layer thickness, the value of the apparent conductivity γ_a/γ_2 asymptotically approaches the following limit:

$$\frac{\gamma_a}{\gamma_2} = \frac{|b^*_{ox}(\gamma_1) - 1|}{|b^*_{0x}(\gamma_2) - 1|}$$

In the range of the small parameter, this limit is equal to γ_1/γ_2. The distance d between "horns" on the curves of profiling is related to the thickness of the formation (i.e., $d \approx H + L$).

If conductivity of the thick layer is lower than that of the shoulders ($\gamma_2 < \gamma_1, H \geq L$), profiling curves are still indicative of the layer thickness. But if thickness of the layer is several times less than the probe length ($H \ll L$), the influence of the shoulder makes determining the thickness H practically impossible, regardless of the conductivity contrast γ_1/γ_2.

FURTHER READING

[1] Kaufman AA, Kaganskiy AM. Induction logging with transversal magnetic dipoles. USSR, Science; 1972.
[2] Eydman IE. Method of induction logging. Patent no 272448, USSR; 1970.

CHAPTER TWELVE

The Influence of Anisotropy on the Field

Contents

12.1 Anisotropy of a Layered Medium	448
12.2 Electromagnetic Field of Magnetic Dipole in a Uniform and Anisotropic Medium	452
12.3 Magnetic Field in an Anisotropic Formation of Finite Thickness	459
Further Reading	465

Detecting and evaluating low-resistivity pay zones using conventional induction logging tools is a major challenge in hydrocarbon exploration. Traditional induction tools comprise transmitter and receiver sensors whose axes are aligned parallel to the borehole. If the formation dip is small, the induced currents flow mainly parallel to the bedding planes, thus measuring the horizontal resistivity of the formation. However, many geological formations exhibit resistivity anisotropy, i.e., the resistivity varies with direction. For example, in thinly laminated sand/shale sequences, where the sand is hydrocarbon bearing, the vertical resistivity measured perpendicular to the bedding is higher than the horizontal resistivity. The low-resistivity shales dominate the horizontal resistivity, whereas the vertical resistivity is more sensitive to the more resistive sand layers. Induction tools with vertically oriented coils cannot accurately detect and delineate this type of low-resistivity reservoir because the measured resistivity will be biased toward the low-resistivity shales. To resolve formation parameters in electrically anisotropic reservoirs, transversal coils should be used. Baker Hughes Incorporated was the first service company to build such a tool, and successfully used it to resolve an anisotropic formation and find the relative dip of the tool with respect to the formation. Today, all major service companies offer similar services. In this chapter, we consider the electromagnetic field of a magnetic dipole in the presence of uniform and horizontally layered anisotropic media.

12.1 ANISOTROPY OF A LAYERED MEDIUM

First, suppose that a medium is an alternation of elementary isotropic layers of two types: one has conductivity γ_1 and dielectric constant ε_1; the other has conductivity and dielectric constant γ_2 and ε_2, respectively (Fig. 12.1).

Let us assume that in such an elementary layer, which is denoted by index (1), a uniform electric field $\mathbf{E}_1^* = \mathbf{E}\exp(-i\omega t)$ is given, and is located at the xz plane. The current density in this layer is:

$$\mathbf{j}_1^* = \gamma_1 \mathbf{E}_1^* \tag{12.1}$$

Thicknesses of the skin depths, δ_1 and δ_2, are assumed to be sufficiently large such that they significantly exceed the thickness of an elementary layer and the skin effect within these layers can be disregarded. Now, we express $\mathbf{E}^*$ and $\mathbf{j}^*$ in every isotropic layer through current $\mathbf{j}_1^*$. Maxwell's equations result in the following conditions at the interface of the first and second layers:

$$E_{2x}^* = E_{1x}^*, \quad \varepsilon_2 E_{2z}^* - \varepsilon_1 E_{1z}^* = \sigma_0^* \tag{12.2}$$

where σ_0^* is the complex amplitude of free surface charges. From the principle of charge conservation, at the boundary between the first and the second layer we have:

$$j_{2z}^* - j_{1z}^* = i\omega \sigma_0^* \tag{12.3}$$

By eliminating σ_0^* from Eqs. (12.2), (12.3) and applying Ohm's law, we obtain the following expressions for the current and the field in the second layer:

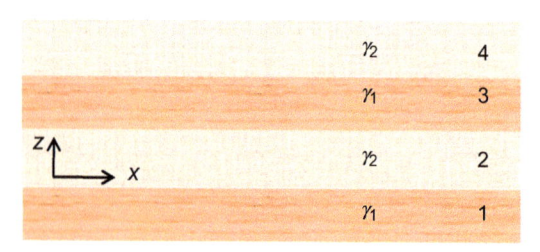

Fig. 12.1 Anisotropic layered medium.

The Influence of Anisotropy on the Field

$$j_{2x}^* = \frac{\gamma_2}{\gamma_1} j_{1x}^*, \quad j_{2z}^* = \frac{1 - \dfrac{i\omega\varepsilon_1}{\gamma_1}}{1 - \dfrac{i\omega\varepsilon_2}{\gamma_2}} j_{1z}^*$$

$$E_{2x}^* = \frac{j_{2x}^*}{\gamma_2} = \frac{j_{1x}^*}{\gamma_1}, \quad E_{2z}^* = \frac{j_{2z}^*}{\gamma_2} = \frac{1 - \dfrac{i\omega\varepsilon_1}{\gamma_1}}{1 - \dfrac{i\omega\varepsilon_2}{\gamma_2}} \frac{j_{1z}^*}{\gamma_2} \quad (12.4)$$

By analogy, from conditions on the surface between the second and third layers, we have:

$$j_{3x}^* = \frac{\gamma_3}{\gamma_2} j_{2x}^*, \quad j_{3z}^* = \frac{1 - \dfrac{i\omega\varepsilon_2}{\gamma_2}}{1 - \dfrac{i\omega\varepsilon_3}{\gamma_3}} j_{2z}^* \quad (12.5)$$

Owing to $\gamma_3 = \gamma_1$, $\varepsilon_3 = \varepsilon_1$, Eq. (12.5) becomes:

$$\mathbf{j}_3^* = \mathbf{j}_1^*, \quad \mathbf{E}_3^* = \mathbf{E}_1^* \quad (12.6)$$

Thus, in the formation consisting of alternating thin layers of both types, the field and current density have paired values, that is, $\mathbf{E}_1^*, \mathbf{j}_1^*$, and $\mathbf{E}_2^*, \mathbf{j}_2^*$, corresponding to the first and second layers. Let us consider an arbitrary layer with the thickness D, in which the relative contribution into conductivity of layers with conductivity γ_2 is equal to n. Then, for average values of current and the field, we have:

$$\langle j_x^{*av} \rangle = \left(1 - n + n \cdot \frac{\gamma_2}{\gamma_1}\right) j_{1x}^*, \quad \langle j_z^{*av} \rangle = \left(1 - n + n \cdot \frac{1 - \dfrac{i\omega\varepsilon_1}{\gamma_1}}{1 - \dfrac{i\omega\varepsilon_2}{\gamma_2}}\right) j_{1z}^* \quad (12.7)$$

and

$$\langle E_x^{*av} \rangle = \frac{j_{1x}^*}{\gamma_1}, \quad \langle E_z^{*av} \rangle = \left(1 - n + n \cdot \frac{\gamma_1}{\gamma_2} \frac{1 - \dfrac{i\omega\varepsilon_1}{\gamma_1}}{1 - \dfrac{i\omega\varepsilon_2}{\gamma_2}}\right) \frac{j_{1z}^*}{\gamma_1}$$

Defining the longitudinal γ_t and transversal γ_n conductivities as:

we obtain:

$$\gamma_t = \frac{j_x^{*av}}{E_x^{*av}}, \quad \gamma_n = \frac{j_z^{*av}}{E_z^{*av}}$$

$$\gamma_t = \gamma_1\left(1 - n + n\cdot\frac{\gamma_2}{\gamma_1}\right) \text{ and } \gamma_n = \gamma_1 \frac{1 - n[1 - p(\omega)]}{1 - n\left[1 - \frac{\gamma_1}{\gamma_2}p(\omega)\right]} \quad (12.8)$$

where

$$p(\omega) = \frac{1 - \dfrac{i\omega\varepsilon_1}{\gamma_1}}{1 - \dfrac{i\omega\varepsilon_2}{\gamma_2}} \quad (12.9)$$

In the quasistationary approximation, dependence on the dielectric constant is absent, and, consequently, expressions for transversal conductivity and coefficient of anisotropy λ have the form:

$$\gamma_n = \frac{\gamma_1}{1 - n + n\cdot\dfrac{\gamma_1}{\gamma_2}} \quad (12.10)$$

and

$$\lambda = \left(\frac{\rho_n}{\rho_t}\right)^{1/2} = \left[\left(1 - n + n\frac{\gamma_2}{\gamma_1}\right)\left(1 - n + n\frac{\gamma_1}{\gamma_2}\right)\right]^{1/2} \quad (12.11)$$

Fig. 12.2 illustrates the dependence of anisotropy coefficient λ on parameters γ_2/γ_1 and n.

In general, when the influence of displacement currents is essential, the transversal resistivity depends on frequency. If the electric field is not uniform and changes along the layer, we can assume that the longitudinal conductance also is a function of frequency. Fig. 12.3 shows the influence of displacement currents on the anisotropy coefficient λ.

If n remains constant within interval D, and the probe length is much greater than the layer thickness, this part of a medium can be considered as a uniform anisotropic layer with coefficient of anisotropy λ.

The Influence of Anisotropy on the Field 451

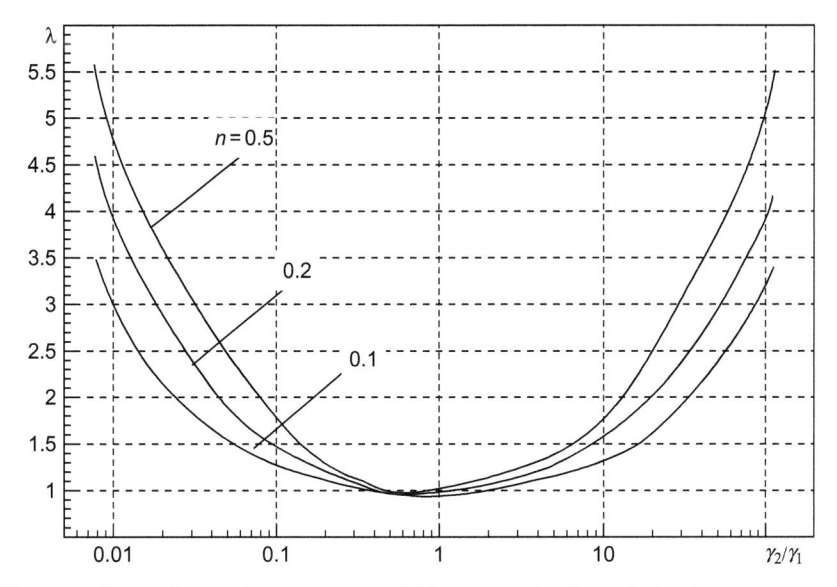

Fig. 12.2 Dependence of anisotropy coefficient on γ_2/γ_1. Curve index is n.

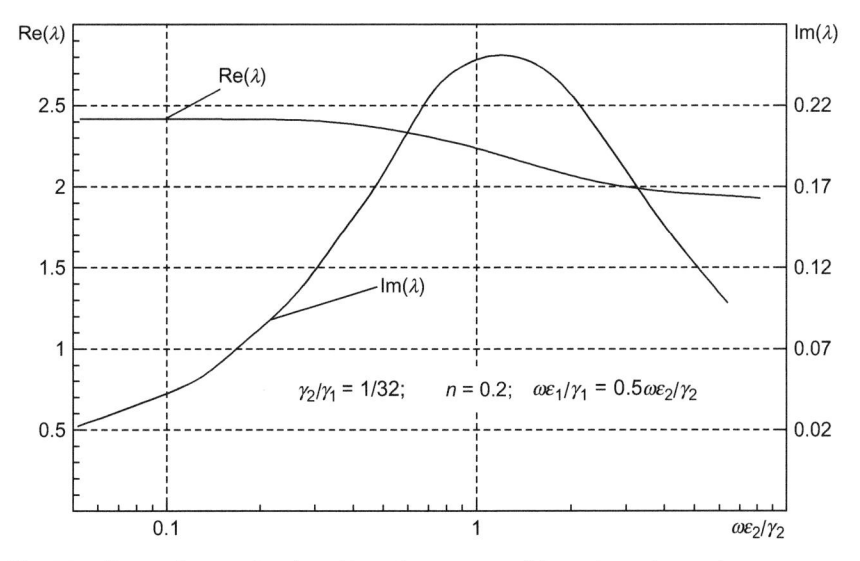

Fig. 12.3 Dependence of real and imaginary parts of λ on the ratio $\omega\varepsilon_2/\gamma_2$.

12.2 ELECTROMAGNETIC FIELD OF MAGNETIC DIPOLE IN A UNIFORM AND ANISOTROPIC MEDIUM

Let us consider a uniform anisotropic medium with the tensor of conductivity:

$$\gamma_{ik} = \begin{pmatrix} \gamma_t & 0 & 0 \\ 0 & \gamma_t & 0 \\ 0 & 0 & \gamma_n \end{pmatrix} \qquad (12.12)$$

An arbitrarily oriented magnetic dipole can be presented as the sum of two dipoles, oriented vertically and horizontally. A vertical magnetic dipole induces currents in horizontal planes, and they do not depend on the transversal conductivity γ_n. Features of the field in a uniform medium, caused by the vertical dipole, were discussed in detail earlier.

Now we explore the case when the moment of the dipole is oriented horizontally. Under such type of excitation, volume charges occur in the anisotropic medium. In fact, by presenting the equation of the quasistationary field $div\,\mathbf{j} = 0$ in the form:

$$\gamma_t\, div\,\mathbf{E} + (\gamma_n - \gamma_t)\frac{\partial E_z}{\partial z} = 0$$

and using the equation:

$$div\,\mathbf{E} = \delta/\varepsilon_0$$

we obtain an expression for the volume density of the charges at an arbitrary point in a medium:

$$\delta = \varepsilon_0\left(1 - \frac{\gamma_n}{\gamma_t}\right)\frac{\partial E_z}{\partial z} \quad \text{or} \quad \delta = \varepsilon_0\left(1 - \frac{1}{\lambda^2}\right)\frac{\partial E_z}{\partial z} \qquad (12.13)$$

To describe the field, we use Maxwell equations in the following form:

$$curl\,\mathbf{E} = i\omega\mathbf{B}, \quad div\,\mathbf{E} = \delta/\varepsilon_0$$
$$curl_x\mathbf{B} = \gamma_t\mu_0 E_x, \quad curl_y\mathbf{B} = \gamma_t\mu_0 E_y \qquad (12.14)$$
$$curl_z\mathbf{B} = \gamma_n\mu_0 E_z, \quad div\,\mathbf{B} = 0$$

Inasmuch as the volume density δ is not zero, it is impossible to introduce the vector potential of the electric type $\mathbf{E} = curl\,\mathbf{A}_m$. Thus, we let:

$$\mathbf{B} = curl\,\mathbf{A} \qquad (12.15)$$

The Influence of Anisotropy on the Field

Then, from the first equation of the set (12.14), it follows that:
$$\mathbf{E} = i\omega \mathbf{A} - \text{grad } U \tag{12.16}$$

Thus, we have for the potential A, the following equations:

$$\frac{\partial}{\partial x} \text{div} \mathbf{A} - \nabla^2 A_x = \gamma_t \left(i\omega\mu_0 A_x - \frac{\partial U}{\partial x} \right)$$

$$\frac{\partial}{\partial y} \text{div} \mathbf{A} - \nabla^2 A_y = \gamma_t \left(i\omega\mu_0 A_y - \frac{\partial U}{\partial y} \right)$$

$$\frac{\partial}{\partial z} \text{div} \mathbf{A} - \nabla^2 A_z = \gamma_n \left(i\omega\mu_0 A_z - \frac{\partial U}{\partial z} \right)$$

By choosing the gauge condition in the form:
$$\text{div} \mathbf{A} = -\gamma_t U$$

we have:

$$\nabla^2 A_x + k_t^2 A_x = 0$$
$$\nabla^2 A_y + k_t^2 A_y = 0 \tag{12.17}$$
$$\nabla^2 A_z + k_n^2 A_z = \left(1 - \frac{1}{\lambda^2}\right) \frac{\partial}{\partial z} \text{div} \mathbf{A}$$

where

$$k_t^2 = i\gamma_t \mu_0 \omega, \quad k_n^2 = i\gamma_n \mu_0 \omega, \quad \lambda^2 = \frac{\gamma_t}{\gamma_n}$$

The behavior of the vector potential of electrical type $\mathbf{A}$ near the magnetic dipole is not known beforehand; therefore, it is appropriate to present the magnetic dipole as a sum of two vertical and two horizontal electric dipoles (Fig. 12.4) and find a solution for each of them.

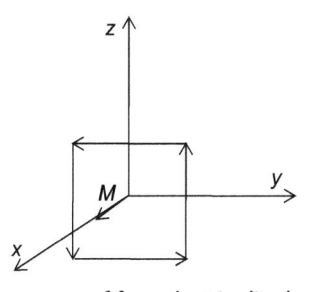

Fig. 12.4 Magnetic dipole as a sum of four electric dipoles.

The vector potential of the vertical electric dipole can be described by only one component, A_z^v, because, due to axial symmetry, the magnetic field has only one component B_ϕ. In accordance with Eq. (12.17), equations for component A_z^v have the form:

$$\frac{\partial^2 A_z^v}{\partial x^2} + \frac{\partial^2 A_z^v}{\partial y^2} + \frac{1}{\lambda^2}\frac{\partial^2 A_z^v}{\partial z^2} + k_n^2 A_z^v = 0 \qquad (12.18)$$

After replacing variable z with $z_1 = \lambda z$, Eq. (12.18) coincides with the equation for a uniform isotropic medium, therefore:

$$A_z^v = C \frac{\exp(ik_n R_*)}{R_*} \qquad (12.19)$$

where

$$R_* = \left(x^2 + y^2 + \lambda^2 z^2\right)^{1/2}$$

To determine the constant C, we use the expression for the potential of the electrode in a uniform anisotropic medium:

$$\phi = \frac{I}{4\pi(\gamma_t \gamma_n)^{1/2} R_*} \qquad (12.20)$$

where I is the value of the direct current.

Assuming small size of the electrode and differentiating Eq. (12.20) with respect to z, we obtain an expression for the potential of the vertical electric dipole when the distance between the electrodes Δz is equal to a:

$$U = -\frac{\partial \phi}{\partial z}\Delta z = \frac{Ia}{4\pi(\gamma_t \gamma_n)^{1/2}} \frac{\lambda^2 z}{R_*^3} \qquad (12.21)$$

At the same time, taking into account the gauge condition:

$$U = -\frac{1}{\gamma_t}\frac{\partial A_z}{\partial z}$$

we have:

$$A_z = \frac{C \lambda^2 z}{\gamma_t R_*^3} \qquad (12.22)$$

Comparing Eqs. (12.19), (12.21), (12.22), we obtain the following expression for the constant C:

$$C = \frac{Ia}{4\pi}\left(\frac{\gamma_t}{\gamma_n}\right)^{1/2} = \frac{Ia}{4\pi}\lambda$$

Thus,

$$A_z^v = \frac{Ia}{4\pi}\lambda\frac{\exp(ik_n R_*)}{R_*} \tag{12.23}$$

Now consider the field of a horizontal electric dipole with the moment directed along the y-axis. We look for the solution of Eq. (12.17) assuming that $A_x^h = 0$. Then, for components A_y^h and A_z^h, we have:

$$\nabla^2 A_y^h + k_t^2 A_y^h = 0$$

and

$$\frac{\partial^2 A_z^h}{\partial x^2} + \frac{\partial^2 A_z^h}{\partial y^2} + \frac{1}{\lambda^2}\frac{\partial^2 A_z^h}{\partial z^2} + k_n^2 A_z^h = \left(1 - \frac{1}{\lambda^2}\right)\frac{\partial^2 A_y^h}{\partial y \partial z} \tag{12.24}$$

Let us present A_y^h as:

$$A_y^h = C_1 \frac{\exp(ik_t R)}{R} = C_1 \int_0^\infty \frac{m}{m_t} e^{-m_t|z|} J_0(mr)\,dm \tag{12.25}$$

where $m_t = (m^2 - k_t^2)^{1/2}$.

It is convenient to present component A_z^h as:

$$A_z^h = \frac{y}{r}\int_0^\infty F_m(m,z) J_1(mr)\,dm = -\frac{\partial}{\partial y}\int_0^\infty \frac{F_m(m,z)}{m} J_0(mr)\,dm \tag{12.26}$$

The expression for A_z^h is defined by conditions of excitation and the relationship between the scalar and vector potentials. Substituting Eqs. (12.25), (12.26) into Eq. (12.24), we obtain an equation for function $F_m(m,z)$

$$\frac{d^2 F_m}{dz^2} - \lambda^2 m_n^2 F_m = \text{sign}(z) C_1 (\lambda^2 - 1) m^2 e^{-m_t|z|} \tag{12.27}$$

Here,

$$m_n = (m^2 - k_n^2)^{1/2}$$

The solution of Eq. (12.27) is:

$$F_m = \text{sign}(z) C_1 \left(e^{-\lambda m_n|z|} - e^{-m_t|z|}\right) \tag{12.28}$$

For $z=0$, function $F(m)$ is continuous together with its first derivative. Thus,

$$A_z^h = C_1 \, sign(z) \frac{y}{r} \int_0^\infty \left[e^{-\lambda m_n |z|} - e^{-m_t |z|} \right] J_0(mr) \, dm \qquad (12.29)$$

or

$$A_z^h = C_1 \frac{y}{r^2} \left[\frac{z}{R} e^{ik_t R} - \frac{\lambda z}{R_*} e^{ik_n R_*} \right] \qquad (12.30)$$

Constant C_1 is determined from the gauge condition and the behavior of the field at infinity, and is equal to:

$$C_1 = \frac{Ia}{4\pi}$$

Thus, for a horizontal electric dipole, we have:

$$\mathbf{A}^h = \left(0, A_y^h, A_z^h \right)$$

Here,

$$A_y^h = \frac{Ia}{4\pi} \frac{e^{ik_t R}}{R}$$

$$A_z^h = \frac{Ia}{4\pi} \frac{y}{r^2} \left[\frac{z}{R} e^{ik_t R} - \frac{\lambda z}{R_*} e^{ik_n R_*} \right] \qquad (12.31)$$

The components of the magnetic dipole, A_y and A_z, are determined by summation of the corresponding components of the electric dipoles:

$$A_y^* = \lim \left[A_y^{(1)} + A_y^{(2)} + A_y^{(3)} + A_y^{(4)} \right] = -\frac{M_0}{4\pi} \frac{\partial}{\partial z} \frac{e^{ik_t R}}{R}$$

$$A_z^* = \lim \left[A_z^{(1)} + A_z^{(2)} + A_z^{(3)} + A_z^{(4)} \right] =$$

$$\frac{M_0}{4\pi} \lambda \frac{\partial}{\partial y} \frac{e^{ik_n R_*}}{R_*} - \frac{M_0}{4\pi} sign(z) \frac{\partial}{\partial z} \frac{y}{r^2} \cdot \left(\frac{z}{R} e^{ik_t R} - \frac{\lambda z}{R_*} e^{ik_n R_n} \right), \quad (a \to 0)$$

$$(12.32)$$

According to Eqs. (12.15), (12.32), for the magnetic field along the z-axis, we have:

$$B_y = B_z = 0$$

and

$$b_x^* = \frac{B_x^*}{B_0} = \left(1 - ik_tL - -k_t^2L^2\frac{1+\lambda^2}{2\lambda^2}\right)e^{ik_tL} \quad (12.33)$$

where

$$B_0 = -\mu_0 M_0/4\pi L^3$$

From Eq. (12.33) in the near zone, $|kL \ll 1|$, we have:

$$b_x^* \approx 1 - \frac{k_n^2L^2}{2} + \left(\frac{1}{2\lambda^2} + \frac{1}{6}\right)k_t^3L^3$$

or

$$Inb_x^* \approx 1 + \left(\frac{1}{\lambda^2} + \frac{1}{3}\right)\left(\frac{L}{\delta_t}\right)^3 \text{ and } Qb_x^* \approx -\left(\frac{L}{\delta_n}\right)^2 + \left(\frac{1}{\lambda^2} + \frac{1}{3}\right)\left(\frac{L}{\delta_t}\right)^3 \quad (12.34)$$

where

$$\delta_t = 2/(\sigma_t\omega\mu)^{1/2} \text{ and } \delta_n = 2/(\sigma_n\omega\mu)^{1/2}$$

Thus, at the range of small parameters L/δ, the quadrature component of the field is directly proportional to the transversal conductivity γ_n, and the ratio of quadrature components, corresponding to the vertical and horizontal dipoles, permits determination of the anisotropy coefficient:

$$\frac{Qb_z^{*v}}{Qb_x^{*h}} \approx \lambda^2 \quad (12.35)$$

Inasmuch as $\lambda \geq 1$, the in-phase component in the anisotropic medium (12.34) is smaller than that in the isotropic medium with conductivity, γ_t. For large values of the anisotropy coefficient $(\gamma_n \to 0)$, both components of the secondary field become the same at the range of small parameter:

$$Qb_x^* = Inb_x^* - 1 = \frac{1}{3}\left(\frac{L}{\delta_t}\right)^3 \text{ if } \lambda \gg 1 \quad (12.36)$$

In the wave zone, when $|kL \gg 1|$, we have:

$$b_x^* \approx -\frac{k_t^2L^2}{2}\left(1 + \frac{1}{\lambda^2}\right)\exp(ik_tL) \quad (12.37)$$

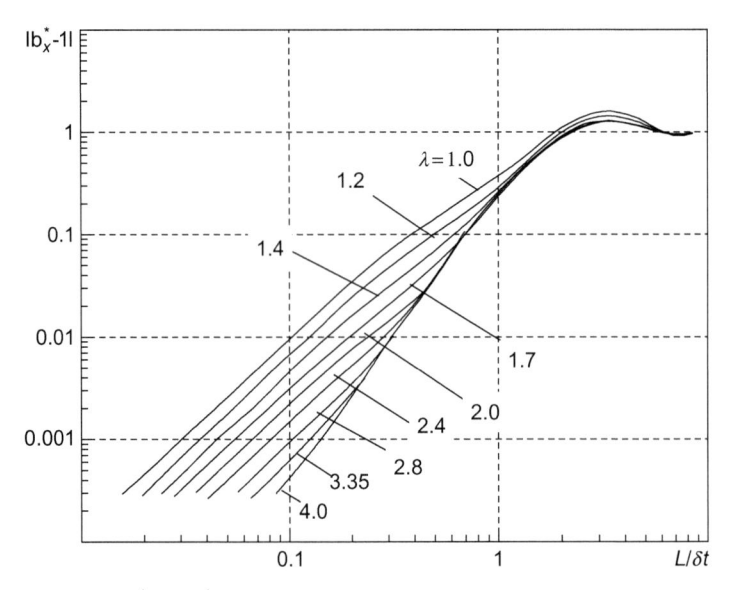

Fig. 12.5 Function $|b_x^* - 1|$ in anisotropic whole space. Index of curves is anisotropy coefficient.

and the influence of anisotropy decreases as λ increases. Typical frequency responses for the function $|b_x^* - 1|$, are presented in Fig. 12.5.

The curves show increased sensitivity of responses to the anisotropy coefficient in the range of small parameter (low frequency, $L/\delta_t < 0.5$) and $\lambda < 2$. Generally, when condition $\dfrac{1}{\lambda^2} \geq \dfrac{1}{3}$ is met, the anisotropy coefficient can be reliably determined.

Applying the Fourier transform to Eq. (12.33), we find the transient response of field b_x when the current in the dipole is turned off:

$$b_x(t) = \Phi(u) - \left(\frac{2}{\pi}\right)^{1/2}\left(1 + \frac{1+\lambda^2}{2\lambda^2}u^2\right)u e^{-u^2/2}$$

where

$$\Phi(u) = \left(\frac{2}{\pi}\right)^{1/2}\int_0^u e^{-t^2/2}dt$$

is the probability integral, and $u = L(\mu_0 \gamma_t / 2t)^{1/2}$.

Table 12.1 contains the values of b_x as a function of λ and $1/u$.
In the limited cases of $t \to 0$ and $t \to \infty$, we obtain:

Table 12.1 Dependence of b_x on Anisotropy Coefficient λ

u^{-1}	$\lambda = 1$	$\lambda = 1.2$	$\lambda = 1.4$	$\lambda = 1.6$	$\lambda = 1.8$	$\lambda = 2.0$
0.1	−1.0000	−1.0000	−1.0000	−1.0000	−1.0000	−1.0000
	−1.0000	−1.0000	−1.0000	−1.0000	−1.0000	−1.0000
0.2	−0.9996	−0.9997	−0.9997	−0.9997	−0.9997	−0.9998
	−0.99265	−0.9369	−0.9432	−0.9472	−0.9500	−0.9520
0.4	−0.3642	−0.4479	−0.4983	−0.5311	−0.5535	−0.5696
	0.2320	0.09083	0.005718	−0.04952	−0.08739	−0.1145
0.8	0.2472	0.01382	0.07247	0.02981	0.00056	−0.02036
	0.06129	0.004334	−0.03001	−0.05230	−0.06758	−0.07851
1.6	−0.05280	−0.07728	−0.09204	−0.1016	−0.1082	−0.1129
	−0.08575	−0.09530	−0.1010	−0.1048	−0.1073	−0.1092
3.2	−0.08082	−0.08436	−0.08650	−0.08789	−0.08884	−0.08952
	−0.06488	−0.06616	−0.06694E	−0.06744	−0.06779	−0.06803
6.4	−0.04872	−0.04918	−0.04946	−0.04964	−0.04976	−0.04985
	−0.03547	0.03564	−0.03574	−0.0358	−0.03585	−0.03588

$$b_x(t) \approx -\left(\frac{2}{\pi}\right)^{1/2}\left(\frac{1}{2} + \frac{1}{2\lambda^2}\right)u^3 e^{-u^2/2} \quad \text{if} \quad u \to \infty \text{ or } t \to 0$$

$$b_x \approx -\left(\frac{2}{\pi}\right)^{1/2}\left(\frac{1}{6} + \frac{1}{2\lambda^2}\right)u^3 \quad \text{if} \quad u \to 0 \text{ or } t \to \infty$$

(12.38)

Therefore, at the late stage and relatively small anisotropy coefficient, the field is inversely proportional to λ^2.

12.3 MAGNETIC FIELD IN AN ANISOTROPIC FORMATION OF FINITE THICKNESS

Using results obtained in the previous section, we can define the magnetic field in a formation with finite thickness when the medium is anisotropic. The main axes of the tensor of conductivity in all three layers coincide with the coordinate lines. Equation of interfaces: $z = h_1$ and $z = -h_2$ (Fig. 12.6).

All quantities that characterize the layer are denoted by the index (2) and quantities characterizing the medium above and beneath, by the index (1). We assume that $\gamma_{ik}^{(1)} = \gamma_{ik}^{(3)}$. In medium (2), the magnetic dipole is located at the origin of the coordinates, and its moment is oriented along the x-axis. In accordance with Eq. (12.32), near the source, the field can be described by

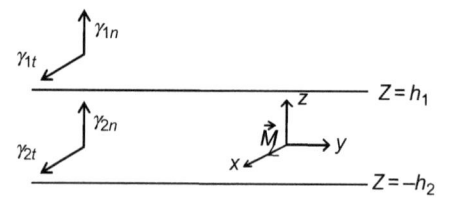

Fig. 12.6 An anisotropic layer model.

the vector potential of electric type $\mathbf{A}^{(0)}$, which has two components: $A_y^{(0)}$ and $A_z^{(0)}$.

Here,

$$A_y^{(0)*} = -\frac{\mu_0 M_0}{4\pi} \frac{\partial}{\partial z} \frac{e^{ik_{2t}R}}{R} = -\frac{\mu_0 M_0}{4\pi} \frac{\partial}{\partial z} \int_0^\infty \frac{m}{m_{2t}} e^{-m_{2t}|z|} J_0(mr)\,dm$$

$$A_z^{(0)*} = \frac{\mu_0 M_0}{4\pi} \lambda_2 \frac{\partial}{\partial y} \frac{e^{ik_{2n}R_*}}{R_*} - \frac{\mu_0 M_0}{4\pi} \frac{y}{r^2} \frac{\partial}{\partial z} \left(\frac{z}{R} e^{ik_{2t}R} - \frac{\lambda_2 z}{R_*} e^{ik_{2n}R_*} \right) \quad (12.39)$$

$$= -\frac{\mu_0 M_0}{4\pi} \frac{y}{r} \int_0^\infty \left[\frac{k_{2t} k_{2n}}{m_{2n}} e^{-m_{2n}\lambda_{2n}|z|} + m_{2t} e^{ik_{2t}|z|} \right] J_1(mr)\,dm$$

where

$$k_{2t}^2 = i\omega\mu_0\gamma_{2t}, \quad k_{2n}^2 = i\omega\mu_0\gamma_{2n}$$

$$m_{2n} = (m^2 - k_{2n}^2)^{1/2}, \quad m_{2t} = (m^2 - k_{2t}^2)^{1/2}$$

$$\lambda_2^2 = \gamma_{2t}/\gamma_{2n}, \quad R_*^2 = x^2 + y^2 + \lambda_2^2 z^2$$

Potentials $A_y^{(0)*}$ and $A_z^{(0)*}$ satisfy the equations:

$$(\nabla^2 + k_{2t}^2) A_y^{(0)*} = 0, \quad (\nabla^2 + k_{2n}^2) A_z^{(0)*} = \left(1 - \frac{1}{\lambda_2^2}\right) \frac{\partial^2 A_y^{(0)*}}{\partial z \partial y},$$

$$\nabla^2 = \frac{\partial^2}{\partial x^2} + \frac{\partial^2}{\partial y^2} + \frac{1}{\lambda_2^2} \frac{\partial^2}{\partial z^2} \quad (12.40)$$

For potentials in a layered medium, we have:

$$A_{1y}^* = \frac{\mu_0 M_0}{4\pi} \int_0^\infty D_1 J_0(mr) e^{m_{1t}z}\,dm$$

$$A_{2y}^* = A_y^{(0)*} + \frac{\mu_0 M_0}{4\pi} \int_0^\infty (D_2 e^{m_{2t}z} + D_3 e^{-m_{2t}z}) J_0(mr)\,dm \quad (12.41)$$

$$A_{3y}^* = \frac{\mu_0 M_0}{4\pi} \int_0^\infty D_3 e^{-m_{3t}z} J_0(mr)\,dm$$

and

$$(\nabla^2 + k_{it}^2)A_{iy}^* = 0$$

Also,

$$A_{1z}^* = \frac{\gamma\mu_0 M_0}{r}\frac{1}{4\pi}\int_0^\infty F_{1z}(z)J_1(mr)\,dm$$

$$A_{2z}^* = A_z^{(0)*} + \frac{\gamma\mu_0 M_0}{r}\frac{1}{4\pi}\int_0^\infty F_2(z)J_1(mr)\,dm \qquad (12.42)$$

$$A_{3z}^* = \frac{\gamma\mu_0 M_0}{r}\frac{1}{4\pi}\int_0^\infty F_3(z)J_1(mr)\,dm$$

Using Eqs. (12.40)–(12.42), we obtain equations determining functions $F_i(z)$:

$$\frac{d^2 F_1(z)}{dz^2} - \lambda_1^2 m_{1n}^2 F_1(z) = -mm_{1t}(\lambda_1^2 - 1)D_1 e^{m_{1t}z}$$

$$\frac{d^2 F_2(z)}{dz^2} - \lambda_2^2 m_{2n}^2 F_2(z) = -mm_{2t}(\lambda_2^2 - 1)(D_2 e^{m_{2t}z} - D_3 e^{-m_{2t}z}) \qquad (12.43)$$

$$\frac{d^2 F_3(z)}{dz^2} - \lambda_1^2 m_{1n}^2 F_3(z) = -mm_{1t}(\lambda_1^2 - 1)D_4 e^{m_{1t}z}$$

Taking into account the behavior of the field at infinity, we can write the solution to the Eq. (12.43) as:

$$F_1(z) = A_1 e^{\lambda_1 m_{1n}z} + \frac{m_{1t}}{m}D_1 e^{m_{1t}z}$$

$$F_2(z) = A_2 e^{\lambda_2 m_{2n}z} + B_2 e^{-\lambda_2 m_{2n}z} + \frac{m_{2t}}{m}(D_2 e^{m_{2t}z} - D_3 e^{-m_{2t}z}) \qquad (12.44)$$

$$F_3(z) = B_3 e^{-\lambda_2 m_{1n}z} - \frac{m_{1t}}{m}D_4 e^{-m_{1t}z}$$

Substituting Eq. (12.44) into Eq. (12.42), we have:

$$A_{1z}^* = \frac{\mu_0 M_0 \gamma}{4\pi}\frac{1}{r}\int_0^\infty \left(A_1 e^{\lambda_1 m_{1n}z} + \frac{m_{1t}}{m}D_1 e^{m_{1t}z}\right)J_1(mr)\,dm$$

$$A_{2z}^* = A_z^{(0)*} + \frac{\mu_0 M_0 \gamma}{4\pi}\frac{1}{r}\int_0^\infty \Big[A_2 e^{\lambda_2 m_{2n}z} + B_2 e^{-\lambda_2 m_{2n}z}$$

$$+ \frac{m_{2t}}{m}(D_2 e^{m_{2t}z} - D_3 e^{-m_{2t}z})\Big]J_1(mr)\,dm \qquad (12.45)$$

$$A_{3z}^* = \frac{\mu_0 M_0 \gamma}{4\pi}\frac{1}{r}\int_0^\infty \left(B_3 e^{-\lambda_1 m_{1n}z} - \frac{m_{1t}}{m}D_4 e^{-m_{1t}z}\right)J_1(mr)\,dm$$

To determine the coefficients $D_1, D_2, D_3, D_4, A_1, A_2, B_2, B_3$, we use the boundary conditions at $z = h_1$ and $z = -h_2$. The continuity of the tangential components of the electric and magnetic fields results in the following relations for A_y^*:

$$A_{1y}^* = A_{2y}^*, \quad \frac{\partial A_{1y}^*}{\partial z} = \frac{\partial A_{2y}^*}{\partial z}, \quad \text{if } z = -h_2$$
$$A_{2y}^* = A_{3y}^*, \quad \frac{\partial A_{2y}^*}{\partial z} = \frac{\partial A_{3y}^*}{\partial z}, \quad \text{if } z = h_1 \qquad (12.46)$$

and more complicated relations for A_z^*

$$A_{1z}^* = A_{2z}^*, \quad \frac{1}{\gamma_{1t}} \operatorname{div} \mathbf{A}_1^* = \frac{1}{\gamma_{2t}} \operatorname{div} \mathbf{A}_2^*, \quad \text{if } z = -h_2$$
$$A_{2z}^* = A_{3z}^*, \quad \frac{1}{\gamma_{2t}} \operatorname{div} \mathbf{A}_2^* = \frac{1}{\gamma_{3t}} \operatorname{div} \mathbf{A}_3^*, \quad \text{if } z = h_1 \qquad (12.47)$$

Substituting Eq. (12.41) into Eq. (12.46), we obtain a system of equations for the coefficients, D_i:

$$D_1 e^{-m_{1t}h_2} = -m e^{-m_{2t}h_2} + D_2 e^{-m_{2t}h_2} + D_3 e^{m_{2t}h_2}$$
$$m_{1t} D_1 e^{-m_{1t}h_2} = -m m_{2t} e^{-m_{2t}h_2} + m_{2t} D_2 e^{-m_{2t}h_2} - m_{2t} D_3 e^{m_{2t}h_2}$$
$$D_4 e^{-m_{1t}h_1} = m e^{-m_{2t}h_1} + D_2 e^{m_{2t}h_1} + D_3 e^{-m_{2t}h_1} \qquad (12.48)$$
$$m_{1t} D_4 e^{-m_{1t}h_1} = m m_{2t} e^{-m_{2t}h_1} - m_{2t} D_2 e^{m_{2t}h_1} + m_{2t} D_3 e^{-m_{2t}h_1}$$

Solving the system, we find:

$$D_2 = -m l_{12} e^{-2m_{2t}h_1} \frac{1 + l_{12} e^{-2m_{2t}h_2}}{1 - l_{12}^2 e^{-2m_{2t}H}} \quad \text{and}$$

$$D_3 = m l_{12} e^{-2m_{2t}h_2} \frac{1 + l_{12} e^{-2m_{2t}h_1}}{1 - l_{12}^2 e^{-2m_{2t}H}} \qquad (12.49)$$

where

$$l_{12} = \frac{m_{1t} - m_{2t}}{m_{1t} + m_{2t}}$$

Now we define coefficients A and B. At $z = -h_2$, component A_y^* is a continuous function, and correspondingly

$$\partial A_{1y}^* / \partial y = \partial A_{2y}^* / \partial y$$

and from Eq. (12.47), we have:

$$A^*_{1z} = A^*_{2z} \text{ and } S_t \frac{\partial A^*_{1z}}{\partial z} - \frac{\partial A^*_{2z}}{\partial z} = (1-S_t)\frac{\partial A^*_{2y}}{\partial y}, \text{ at } z = -h_2 \quad (12.50)$$

where $S_t = \dfrac{\gamma_{2t}}{\gamma_{1t}}$.

By analogy, at $z = h_1$, we have:

$$A^*_{2z} = A^*_{3z} \text{ and } S_t \frac{\partial A^*_{3z}}{\partial z} - \frac{\partial A^*_{2z}}{\partial z} = (1-S_t)\frac{\partial A^*_{2y}}{\partial y} \quad (12.51)$$

Substituting Eq. (12.45) in Eqs. (12.50), (12.51), we obtain the system of equations for coefficients A_1, A_2, B_2, and B_3:

$$A_1 e^{-\lambda_1 m_{1n} h_2} + \frac{m_{1t}}{m} D_1 e^{-m_{1t} h_2} = -\frac{k_{2t} k_{2n}}{m_{2n}} e^{-\lambda_2 m_{2n} h_2} - m_{2t} e^{-m_{2t} h_2} + A_2 e^{-\lambda_2 m_{2n} h_2}$$

$$+ B_2 e^{\lambda_2 m_{2n} h_2} + \frac{m_{2t}}{m}\left(D_2 e^{-m_{2t} h_2} - D_3 e^{m_{2t} h_2}\right)$$

$$S_t\left(\lambda_1 m_{1n} A_1 e^{-\lambda_1 m_{1n} h_2} + \frac{m_{1t}^2}{m} D_1 e^{-m_{1t} h_2}\right) + k_{2t}^2 e^{-m_{2n}\lambda_2 h_2} + m_{2t}^2 e^{-m_{2t} h_2} - \lambda_2 m_{2n}$$

$$A_2 e^{-\lambda_2 m_{2n} h_2} + \lambda_2 m_{2n} B_2 e^{\lambda_2 m_{2n} h_2} - \frac{m_{2t}^2}{m}\left(D_2 e^{-m_{2t} h_2} + D_2 e^{m_{2t} h_2}\right) = -m(1-S_t) D_1 e^{-m_{1t} h_2}$$

$$B_3 e^{-\lambda_1 m_{1n} h_1} - \frac{m_{1t}}{m} D_4 e^{-m_{1t} h_1} = -\frac{k_{2t} k_{2n}}{m_{2n}} e^{-\lambda_2 m_{2n} h_1} - m_{2t} e^{-m_{2t} h_1} + A_2 e^{\lambda_2 m_{2n} h_1}$$

$$+ B_2 e^{-\lambda_2 m_{2n} h_1} + \frac{m_{2t}}{m}\left(D_2 e^{m_{2t} h_1} - D_3 e^{-m_{2t} h_1}\right)$$

$$S_t\left(-\lambda_1 m_{1n} B_3 e^{-\lambda_1 m_{1n} h_1} + \frac{m_{1t}^2}{m} D_4 e^{-m_{1t} h_1}\right) - k_{2t}^2 e^{-\lambda_2 m_{2n} h_1} - m_{2t}^2 e^{-m_{2t} h_1}$$

$$-\lambda_2 m_{2n} A_2 e^{\lambda_2 m_{2n} h_1} + \lambda_2 m_{2n} B_2 e^{-\lambda_2 m_{2n} h_1} - \frac{m_{2t}^2}{m}\left(D_2 e^{m_{2t} h_1} + D_3 e^{-m_{2t} h_1}\right)$$

$$= -m(1-S_t) D_4 e^{m_{1t} h_1}$$

$$(12.52)$$

Using Eq. (12.48), establishing connections between coefficients D_i, the system (12.52) can be easily reduced to the form:

$$e^{-\lambda_1 m_{1n} h_2} A_1 - e^{\lambda_2 m_{2n} h_2} A_2 - e^{-\lambda_2 m_{2n} h_2} B_2 = \frac{k_{2t} k_{2n}}{m_{2n}} e^{-\lambda_2 m_{2n} h_2} + S_t \lambda_1 \frac{m_{1n}}{m} e^{-\lambda_1 m_{1n} h_2} A_1$$

$$-\lambda_2 \frac{m_{2n}}{m} e^{-\lambda_2 m_{2n} h_2} A_2 + \frac{\lambda_2 m_{2n}}{m} e^{-\lambda_2 m_{2n} h_2} = -\frac{k_{2t}^2}{m} e^{-\lambda_2 m_{2n} h_2}$$

$$e^{\lambda_1 m_{1n} h_1} B_3 - e^{-\lambda_2 m_{2n} h_1} A_2 - e^{-\lambda_2 m_{2n} h_1} B_2 = -\frac{k_{2t} k_{2n}}{m_{2n}} e^{-\lambda_2 m_{2n} h_1}$$

$$S_t \lambda_1 \frac{m_{1n}}{m} e^{-\lambda_1 m_{2n} h_1} B_3 + \lambda_2 \frac{m_{2n}}{m} e^{\lambda_2 m_{2n} h_1} A_2 - \lambda_2 \frac{m_{2n}}{m} e^{-\lambda_2 m_2 h_1} B_2 = -\frac{k_{2t}^2}{m} e^{-\lambda_2 m_{2n} h_1}$$

(12.53)

From these equations, we derive:

$$A_2 = \frac{k_{2t} k_{2n}}{m_{2n}} W \frac{e^{-2\lambda_2 m_{2n} h_1}}{1 - L^2 e^{-2\lambda_2 m_{2n} H}} \left(1 - W e^{-2\lambda_1 m_{2n} h_2}\right)$$

(12.54)

$$B_2 = \frac{k_{2t} k_{2n}}{m_{2n}} W \frac{e^{-2\lambda_2 m_{2n} h_2}}{1 - L^2 e^{-2\lambda_2 m_{2n} H}} \left(1 - W e^{-2\lambda_1 m_{2n} h_1}\right)$$

where:

$$W = \frac{S_t \lambda_1 m_{1n} - \lambda_2 m_{2n}}{S_t \lambda_1 m_{1n} + \lambda_2 m_{2n}}$$

The expression for the horizontal component of magnetic field B_x on the z-axis within the layer has the form:

$$B_x^* = \frac{\partial A_z^*}{\partial y} - \frac{\partial A_y^*}{\partial z} = B_{0x}^* + \frac{\mu_0 M_0}{4\pi} \int_0^\infty \left[\frac{m_{2t}}{2}(D_2 e^{m_{2t} z} - D_3 e^{-m_{2t} z})\right.$$
$$\left. - \frac{m}{2}(A_2 e^{\lambda_2 m_{2n} z} + B_2 e^{-\lambda_2 m_{2n} z})\right] dm \quad \text{if} \; -h_2 < z < h_1 \quad (12.55)$$

The derived solution can be used to study all features of the field excited and measured by transversal dipoles. Let us consider sensitivity of the low-frequency component $Q(b_x^*)$ to the anisotropy of the bed, surrounded by isotropic shoulders. To proceed, we introduce a function r_{Qx}, which corresponds to the ratio of the quadrature component of the field $Q(b_x^*)$ in an anisotropic bed to the corresponding component in the isotropic bed. The ratio is always equal to one when $\lambda^2 = \gamma_{2t}/\gamma_{2n} = 1$, and decreases with increase of γ_{2t}/γ_{2n}. To illustrate, we consider two scenarios: a conductive layer with $\gamma_{2t} = 0.2 \, (\text{ohmm})^{-1}$ surrounded by less conductive shoulders with $\gamma_{1t} = 0.1 \, (\text{ohmm})^{-1}$, Fig. 12.7A, and a resistive layer with $\gamma_{2t} = 0.1 \, (\text{ohmm})^{-1}$ surrounded by the more conductive shoulders,

The Influence of Anisotropy on the Field

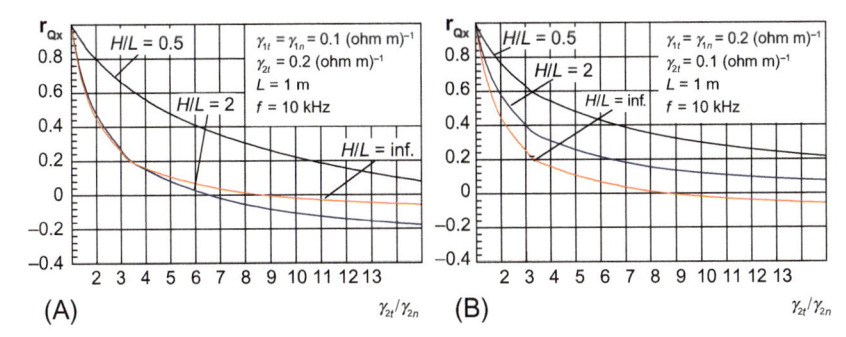

Fig. 12.7 The ratio of the quadrature component of the field $Q(b_x)$ in an anisotropic bed to the corresponding component in the isotropic bed for the case of a conductive (A) and resistive (B) bed.

$\gamma_{1t} = 0.2\,(\text{ohm m})^{-1}$, Fig. 12.7B. The two-coil probe with $L = 1\,\text{m}$ is placed symmetrically with respect to the boundaries and operating frequency is 10 kHz. Modeling results are presented for infinitely thick ($H/L = \text{inf}$), thin ($H/L = 0.5$), and moderately thick $H/L = 2.0$ layers. The x-axis depicts the ratio $\lambda^2 = \gamma_{2t}/\gamma_{2n}$.

In the case of $H/L = \text{inf}$ or $H/L = 2.0$, there is a region of $\gamma_{2t}/\gamma_{2n} < 2$ in which the function r_{Qx}, in accordance with Eq. (12.34), practically linearly drops with γ_{2t}/γ_{2n}. With further increase of the ratio γ_{2t}/γ_{2n}, the function r_{Qx} changes the sign and reaches an asymptote, exhibiting no sensitivity to the anisotropy when $\gamma_{2t}/\gamma_{2n} > 10$. Of course, the thicker the layer, the higher the sensitivity to the ratio γ_{2t}/γ_{2n}. Minimal sensitivity to the anisotropy is observed in the thin layer, $H/L = 0.5$, surrounded by the more conductive shoulders, Fig. 12.7B.

FURTHER READING

[1] Kaufman AA, Kaganski AM. Induction logging with transversal magnetic dipoles. Novosibirsk: Nauka; 1972.
[2] Kriegshauser B, Fanini O, Forgang S, Itskovich G, Rabinovich M, Tabarovsky L, et al. A new multi-component induction logging tool to resolve anisotropic formations. SPWLA; 2000.

APPENDIX

ELECTROMAGNETIC RESPONSE OF ECCENTRED MAGNETIC DIPOLE IN CYLINDRICALLY LAYERED MEDIA

M. Nikitenko, G.B. Itskovich

A.1 INTRODUCTION

Electromagnetic induction logging with coaxially oriented coils is the primary method for evaluating water and hydrocarbon saturation in reservoirs. Standard array-induction tools have dramatically improved induction logging by increasing the depth of investigation up to several feet while still maintaining high vertical resolutions down to 1 ft in smooth wellbores [1,2].

In addition, newly developed multifrequency dielectric array tools permit valuable information about formation petro-physical properties by applying assumed mixing lows to derive water saturation, water salinity, and hydrocarbon volume. Interpretation of induction and dielectric logs rely on sophisticated processing techniques [3], which require a tool eccentricity [4] and radial distribution of the near-borehole geo-electrical properties of the media to be taken into account [5,6]. This requirement motivated us to develop a fast modeling algorithm capable of simulating tool response in cylindrically layered media excited by an eccentred magnetic dipole.

The corresponding code must be fast enough to serve the needs of the on-site radial inversion and permit modeling in the presence of several radial zones with piecewise changing conductivity and dielectric constant. In general, for this type of boundary value problems with no symmetry, either finite difference [7–9] or a finite element method [10,11] is employed. In [4] the integral equation and finite difference methods were combined to find amplitudes of the azimuthal Fourier harmonics of the quasistationary field resulting from an off-axis source exciting the 2D axially symmetric media. In Nam et al.[12] the authors introduced an algorithm to simulate triaxial induction measurements that combines a Fourier series expansion in nonorthogonal system of coordinates with a 2D goal-oriented finite element method. In Wang et al. [13] Fourier series expansion was used to reduce the original 3D problem to a series of independent 2D problems that were solved semianalytically using normalized Bessel and Hankel functions. In addition, an analytical low-frequency formulation based on the generalized reflection and transmission coefficient matrices was used to simulate and study the effect of eccentricity on induction tool responses [14].

The potential of analytical approaches is not exhausted yet. In particular, a semianalytical treatment, presented below, leads to the ultra-fast and accurate simulation of electromagnetic responses in a wide frequency range, thereby serving the needs of induction and dielectric logging.

A.2 SOLUTION TO THE BOUNDARY VALUE PROBLEM

A.2.1 Problem Definition

Let us consider a boundary value problem of an electromagnetic field excited by a vertical magnetic dipole in cylindrically layered medium (Fig. A.1). The dipole is located in the first cylindrical layer (borehole) and its current is a simple harmonic function of time $I(t) = I_0 e^{-i\omega t}$.

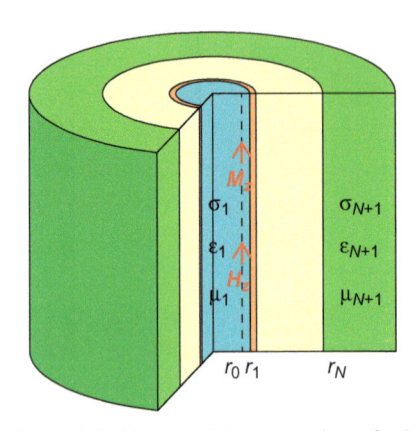

Fig. A.1 The formation is modeled by an arbitrary number of cylindrical layers with prescribed radii r_j, conductivity σ_j, dielectric permittivity ε_j, and magnetic permeability μ_j. These layers may describe the borehole, mud cake, invasion zone, and the uninvaded formation. The tool is comprised of the eccentric magnetic dipole M_z and the eccentric dipole receivers H_z located in the borehole.

A separation of variables method is used to solve the assigned problem [15,16]. We present electric and magnetic fields in the first layer as a sum of a normal and anomalous field:

$$\begin{cases} \vec{E} = \vec{E}_1 + \vec{E}_0 \\ \vec{H} = \vec{H}_1 + \vec{H}_0 \end{cases} \quad (A.1)$$

The fields $\vec{E}_1$, $\vec{H}_1$, $\vec{E}_0$, $\vec{H}_0$ as well as fields in the jth layer $\vec{E}_j$, $\vec{H}_j$ satisfy Maxwell equations:

$$\begin{cases} \text{rot } \vec{H}_0 = \gamma_1 \vec{E}_0 \\ \text{rot } \vec{E}_0 = i\omega\mu_1 \vec{H}_0 - \vec{j}^{\mu} \end{cases} \quad (A.2)$$

$$\begin{cases} \text{rot } \vec{H}_j = \gamma_j \vec{E}_j \\ \text{rot } \vec{E}_j = i\omega\mu_j \vec{H}_j \end{cases} \quad (A.3)$$

Here $\gamma_j = \sigma_j - i\omega\varepsilon_j$ is the complex conductivity; $\vec{j}^{\mu} = \left(0, 0, j_z^{\mu}\right)$ is the magnetic current; $j_z^{\mu} = -i\omega\mu_1 \cdot M_z \cdot U(P, P_0)$, $M_z = I_0 \cdot S \cdot n_t$ are the dipole moment; S is the coil square; n_t is the number of turns; $U(P, P_0)$ is the source function; P_0 is the coordinates of the source (transmitter); and P is the coordinates of the observation point (receiver).

At the boundaries, tangential components of the electric and magnetic fields are continuous and satisfy the following conditions:

$$\begin{cases} \left[E_{t1}^*\right]_{r=r_1} = -E_{t0}(r_1) \\ \left[H_{t1}\right]_{r=r_1} = -H_{t0}(r_1) \\ \left[E_{tj}^*\right]_{r=r_j} = 0 \\ \left[H_{tj}\right]_{r=r_j} = 0, \ j = 2, N \end{cases} \quad (A.4)$$

where the subscript "t" refers to the ϕ- or z-component of electromagnetic field. Square brackets in Eq. (A.4) denote the jump in a quantity across the boundary. Eqs. (A.2)–(A.4) uniquely determine an electromagnetic field at any point of the medium.

A.2.2 Fourier Transform

Let us find fields $\vec{E}_j$, $\vec{H}_j$, assuming that normal fields $\vec{E}_0$, $\vec{H}_0$ are known. Forward and inverse Fourier transform are determined as:

$$\begin{cases} A^*(r,\phi,\lambda) = \int_{-\infty}^{\infty} A(r,\phi,z) e^{-i\lambda z} dz \\ A(r,\phi,z) = \dfrac{1}{2\pi} \int_{-\infty}^{\infty} A^*(r,\phi,\lambda) e^{i\lambda z} d\lambda. \end{cases} \quad (A.5)$$

Then, by using Eqs. (A.3)–(A.5) Fourier transforms of tangential components can be expressed through Fourier transforms of vertical components as:

$$\begin{cases} E^*_{rj} = \dfrac{1}{p_j^2}\left(-i\lambda \dfrac{dE^*_{zj}}{dr} - \dfrac{i\omega\mu_j}{r}\dfrac{dH^*_{zj}}{d\phi}\right) \\ E^*_{\phi j} = \dfrac{1}{p_j^2}\left(i\omega\mu_j \dfrac{dH^*_{zj}}{dr} - \dfrac{i\lambda}{r}\dfrac{dE^*_{zj}}{d\phi}\right) \\ H^*_{rj} = \dfrac{1}{p_j^2}\left(-i\lambda \dfrac{dH^*_{zj}}{dr} - \dfrac{\gamma_j}{r}\dfrac{dE^*_{zj}}{d\phi}\right) \\ H^*_{\phi j} = \dfrac{1}{p_j^2}\left(\gamma_j \dfrac{dE^*_{zj}}{dr} - \dfrac{i\lambda}{r}\dfrac{dH^*_{zj}}{d\phi}\right) \end{cases} \quad (A.6)$$

$p_j^2 = \lambda^2 + k_j^2$, $k_j^2 = -i\omega\mu_j\gamma_j$. Fourier transforms of vertical components of electric and magnetic fields follow equations:

$$\begin{cases} \dfrac{1}{r}\dfrac{d}{dr}\left(r\dfrac{dE^*_{zj}}{dr}\right) + \dfrac{1}{r^2}\dfrac{d^2 E^*_{zj}}{d\phi^2} - p_j^2 E^*_{zj} = 0 \\ \dfrac{1}{r}\dfrac{d}{dr}\left(r\dfrac{dH^*_{zj}}{dr}\right) + \dfrac{1}{r^2}\dfrac{d^2 H^*_{zj}}{d\phi^2} - p_j^2 H^*_{zj} = 0 \end{cases} \quad (A.7)$$

By applying Fourier transform (A.5) to the boundary conditions (A.4), we receive the following conditions for tangential components at the first boundary $r = r_1$:

$$\begin{cases} \left[E^*_{zj}\right]_{r=r_1} = -E^*_{z0} \\ \left[H^*_{zj}\right]_{r=r_1} = -H^*_{z0} \\ \left[\dfrac{1}{p_j^2}\left(i\omega\mu_j \dfrac{dH^*_{zj}}{dr} - \dfrac{i\lambda}{r}\dfrac{dE^*_{zj}}{d\phi}\right)\right]_{r=r_1} = -\dfrac{1}{p_1^2}\left(i\omega\mu_1 \dfrac{dH^*_{z0}}{dr} - \dfrac{i\lambda}{r}\dfrac{dE^*_{z0}}{d\phi}\right) \\ \left[\dfrac{1}{p_j^2}\left(\gamma_j \dfrac{dE^*_{zj}}{dr} - \dfrac{i\lambda}{r}\dfrac{dH^*_{zj}}{d\phi}\right)\right]_{r=r_1} = -\dfrac{1}{p_1^2}\left(\gamma_1 \dfrac{dE^*_{z0}}{dr} - \dfrac{i\lambda}{r}\dfrac{dH^*_{z0}}{d\phi}\right) \end{cases} \quad (A.8)$$

At the boundaries $r = r_j$, $j = 2, N$ we have

$$\begin{cases} \left[E^*_{zj}\right]_{r=r_j} = 0 \\ \left[H^*_{zj}\right]_{r=r_j} = 0 \\ \left[\dfrac{1}{p_j^2}\left(i\omega\mu_j \dfrac{dH^*_{zj}}{dr} - \dfrac{i\lambda}{r}\dfrac{dE^*_{zj}}{d\phi}\right)\right]_{r=r_j} = 0 \\ \left[\dfrac{1}{p_j^2}\left(\gamma_j \dfrac{dE^*_{zj}}{dr} - \dfrac{i\lambda}{r}\dfrac{dH^*_{zj}}{d\phi}\right)\right]_{r=r_j} = 0 \end{cases} \quad (A.9)$$

A.2.3 Expansion in Series

Let us expand E^*_{zj}, E^*_{zj} and E^*_{z0}, E^*_{z0} into a series:

$$\begin{cases} E^*_{zj} = \sum_{n=0}^{\infty}\left(e^c_{nj}\cos n\overline{\phi} + e^s_{nj}\sin n\overline{\phi}\right) \\ H^*_{zj} = \sum_{n=0}^{\infty}\left(h^c_{nj}\cos n\overline{\phi} + h^s_{nj}\sin n\overline{\phi}\right) \\ E^*_{z0} = \sum_{n=0}^{\infty}\left(e^c_{n0}\cos n\overline{\phi} + e^s_{n0}\sin n\overline{\phi}\right) \\ H^*_{z0} = \sum_{n=0}^{\infty}\left(h^c_{n0}\cos n\overline{\phi} + h^s_{n0}\sin n\overline{\phi}\right) \end{cases} \quad (A.10)$$

Here $\overline{\phi} = \phi - \phi_0$ is the difference between receiver and transmitter angular coordinates (Fig. A.2).

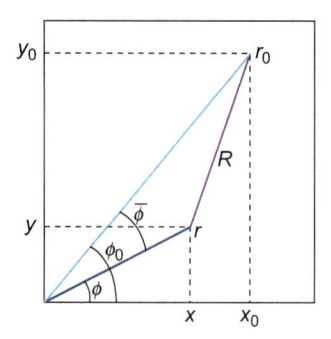

Fig. A.2 The polar coordinate system.

Substituting Eq. (A.10) into Eqs. (A.7)–(A.9), we obtain Eqs. (A.11)–(A.13) determining unknown functions $e_{nj}^{c,s}$, $h_{nj}^{c,s}$:

$$\begin{cases} \dfrac{1}{r}\dfrac{d}{dr}\left(r\dfrac{de_{nj}^{c,s}}{dr}\right) - \left(\dfrac{n^2}{r^2} + p_j^2\right)e_{nj}^{c,s} = 0 \\ \\ \dfrac{1}{r}\dfrac{d}{dr}\left(r\dfrac{dh_{nj}^{c,s}}{dr}\right) - \left(\dfrac{n^2}{r^2} + p_j^2\right)h_{nj}^{c,s} = 0 \end{cases} \quad (A.11)$$

$$\begin{cases} \left[e_{nj}^{c,s}\right]_{r=r_1} = -e_{n0}^{c,s} \\ \\ \left[h_{nj}^{c,s}\right]_{r=r_1} = -h_{n0}^{c,s} \\ \\ \left[\dfrac{1}{p_j^2}\left(i\omega\mu_j\dfrac{dh_{nj}^{c,s}}{dr} \mp \dfrac{i\lambda n}{r}e_{nj}^{s,c}\right)\right]_{r=r_1} = -\dfrac{1}{p_1^2}\left(i\omega\mu_1\dfrac{dh_{n0}^{c,s}}{dr} \mp \dfrac{i\lambda n}{r}e_{n0}^{s,c}\right) \\ \\ \left[\dfrac{1}{p_j^2}\left(\gamma_j\dfrac{de_{nj}^{c,s}}{dr} \mp \dfrac{i\lambda n}{r}h_{nj}^{s,c}\right)\right]_{r=r_1} = -\dfrac{1}{p_1^2}\left(\gamma_1\dfrac{de_{n0}^{c,s}}{dr} \mp \dfrac{i\lambda n}{r}h_{n0}^{s,c}\right) \end{cases} \quad (A.12)$$

$$\begin{cases} \left[e_{nj}^{c,s}\right]_{r=r_j} = 0 \\ \\ \left[h_{nj}^{c,s}\right]_{r=r_j} = 0 \\ \\ \left[\dfrac{1}{p_j^2}\left(i\omega\mu_j\dfrac{dh_{nj}^{c,s}}{dr} \mp \dfrac{i\lambda n}{r}e_{nj}^{s,c}\right)\right]_{r=r_j} = 0 \\ \\ \left[\dfrac{1}{p_j^2}\left(\gamma_j\dfrac{de_{nj}^{c,s}}{dr} \mp \dfrac{i\lambda n}{r}h_{nj}^{s,c}\right)\right]_{r=r_j} = 0, \quad j=2, N \end{cases} \quad (A.13)$$

A.2.4 Solution for Angular Harmonics

Solution to Eq. (A.11) is a linear combination of modified Bessel functions[17]:

$$\begin{cases} e_{nj}^{c,s} = C_{nj}^{c,s} I_n(p_j r) + P_{nj}^{c,s} K_n(p_j r) \\ h_{nj}^{c,s} = D_{nj}^{c,s} I_n(p_j r) + Q_{nj}^{c,s} K_n(p_j r), \quad \text{Re} p_j > 0 \end{cases} \quad (A.14)$$

Applying boundary conditions $E_{z1}^*, H_{z1}^*\big|_{r \to 0} \to 0$ and $E_{zN+1}^*, H_{zN+1}^*\big|_{r \to \infty} \to 0$, we find $P_1^{c,s} = Q_1^{c,s} = C_{N+1}^{c,s} = D_{N+1}^{c,s} = 0$. The remaining unknown coefficients $C_{nj}^{c,s}, D_{nj}^{c,s}, j = 1, N$ and $P_{nj}^{c,s}, Q_{nj}^{c,s}, j = 2, N+1$ can be determined through boundary conditions (A.12), (A.13). For this purpose it is necessary to define harmonics of the normal fields $e_{n0}^{c,s}, h_{n0}^{c,s}$.

A.2.5 Determination of the Normal Field

To simplify derivation of harmonics of the normal field, the Cartesian coordinate system is used. The point source function $U(P, P_0)$ (Eq. A.2) is presented as $U(P, P_0) = \delta(x - x_0) \delta(y - y_0) \delta(z - z_0)$, $P_0 = (x_0, y_0, z_0)$, $P = (x, y, z)$. The Fourier transform with respect to all coordinates (x, y, z) is determined as:

$$\begin{cases} A^+(\xi, \eta, \lambda) = \int_{-\infty}^{\infty}\int_{-\infty}^{\infty}\int_{-\infty}^{\infty} A(x, y, z) e^{-i\xi x} e^{-i\eta y} e^{-i\lambda z} dx dy dz, \\ A(x, y, z) = \frac{1}{(2\pi)^3} \int_{-\infty}^{\infty} A^+(\xi, \eta, \lambda) e^{i\xi x} e^{i\eta y} e^{i\lambda z} d\xi d\eta d\lambda \end{cases} \quad (A.15)$$

Applying transformation Eq. (A.15) to Eq. (A.2) and using properties of Fourier transform and delta-function [18,19], we derive simple algebraic expression determining the normal field:

$$H_{z0}^+ = -\frac{e^{-i\xi x_0} e^{-i\eta y_0} e^{-i\lambda z_0}}{\xi^2 + \eta^2 + p_1^2} \cdot p_1^2 \cdot M_z \quad (A.16)$$

while the z-component of the electrical normal field is equal to zero:

$$E_{z0}^+ = 0$$

Applying inverse Fourier transform with respect to coordinates (x,y) we obtain Fourier transformants of magnetic field with respect to the coordinate z:

$$H_{z0}^* = -\frac{e^{-i\lambda z_0}}{2\pi} \cdot p_1^2 \cdot M_z \cdot K_0(p_1 R) \quad (A.17)$$

$$R = \sqrt{(x-x_0)^2 + (y-y_0)^2}$$

In cylindrical coordinate system (Fig. A.2)

$$x = r\cos\phi, \quad y = r\sin\phi, \quad x_0 = r_0\cos\phi_0, \quad y_0 = r_0\sin\phi_0,$$

$$R = \sqrt{r^2 + r_0^2 - 2rr_0\cos\overline{\phi}}$$

Applying an addition theorem for Bessel functions [20]:

$$K_0(p_1 R) = I_0(p_1 r_0)K_0(p_1 r) + 2\sum_{n=1}^{\infty} I_n(p_1 r_0)K_n(p_1 r)\cos n\overline{\phi}, \quad r_0 < r \quad (A.18)$$

we derive the following expressions for the Fourier harmonics of the normal fields:

$$\begin{cases} h_{n0}^c = -\dfrac{e^{-i\lambda z_0}}{2\pi} \cdot l \cdot p_1^2 \cdot M_z \cdot I_n(p_1 r_0)K_n(p_1 r) \\ h_{n0}^s = 0 \\ e_{n0}^{c,s} = 0 \end{cases} \quad (A.19)$$

where

$$l = \begin{cases} 2, & n \neq 0 \\ 1, & n = 0 \end{cases}$$

A.2.6 Final Representation of the Magnetic Field

After taking into account Eq. (A.19), we determine $C_{nj}^s = D_{nj}^s = 0$, $j = 1, N$ and $P_{nj}^c = Q_{nj}^s = 0$, $j = 2, N+1$. To find nonzero coefficients we use Eq. (A.14) and rewrite Eqs. (A.12), (A.13) in the matrix form:

$$\begin{cases} \hat{W}_1[3,4](r_1)\vec{\psi}_1 + \hat{W}_1[1,2](r_1)\vec{\Pi} = \hat{W}_2(r_1)\vec{\psi}_2 \\ \hat{W}_j(r_j)\vec{\psi}_j = \hat{W}_{j+1}(r_j)\vec{\psi}_{j+1}, \quad j = 2, \; N-1 \\ \hat{W}_N(r_N)\vec{\psi}_N = \hat{W}_{N+1}[1,2](r_N)\vec{\psi}_{N+1} \end{cases} \quad (A.20)$$

The matrix $\hat{W}_j(r)$ is defined as:

$$\hat{W}_j(r) = \begin{bmatrix} K_n(p_j r) & 0 & I_n(p_j r) & 0 \\ 0 & K_n(p_j r) & 0 & I_n(p_j r) \\ -\alpha_j K_n(p_j r) & \beta_j K'_n(p_j r) & -\alpha_j I_n(p_j r) & \beta_j I'_n(p_j r) \\ \varsigma_j K'_n(p_j r) & \alpha_j K_n(p_j r) & \varsigma_j I'_n(p_j r) & \alpha_j I_n(p_j r) \end{bmatrix} \quad (A.21)$$

$$\alpha_j = \frac{i\lambda n}{p_j^2 r}, \quad \beta_j = \frac{i\omega\mu_j}{p_j}, \quad \varsigma_j = \frac{\gamma_j}{p_j}.$$

The following notations for the matrixes $\hat{W}$ and vectors $\vec{\psi}$ and $\vec{\Pi}$ are introduced:

$\hat{W}_1[3,4]$: 3rd and 4th columns of $\hat{W}_1$;
$\hat{W}_1[1,2]$: 1st and 2nd columns of $\hat{W}_1$; and
$\hat{W}_{N+1}[1,2]$: 1st and 2nd columns of $\hat{W}_{N+1}$.

$$\vec{\psi}_j = \begin{bmatrix} P^s_{nj} \\ Q^c_{nj} \\ C^s_{nj} \\ D^c_{nj} \end{bmatrix}, j = 2, N, \quad \vec{\psi}_1 = \begin{bmatrix} C^s_{n1} \\ D^c_{n1} \end{bmatrix}, \quad \vec{\psi}_{N+1} = \begin{bmatrix} P^s_{nN+1} \\ Q^c_{nN+1} \end{bmatrix}, \quad (A.22)$$

$$\vec{\Pi} = \begin{bmatrix} 0 \\ -M_z \dfrac{p_1^2}{\pi} I_n(p_1 r_0) \end{bmatrix}$$

In the expression for the right part $\vec{\Pi}$ the factor $e^{-i\lambda z_0}$ is intentionally omitted and is taken into account later in the inverse Fourier transform (A.5). After introducing vector $\vec{X} = \begin{bmatrix} 0 \\ \Pi_2 \\ C^s_{n1} \\ D^c_{n1} \end{bmatrix}$, Eq. (A.20) can be rewritten as:

$$\begin{cases} \hat{W}_1(r_1) \vec{X} = \hat{W}_2(r_1) \vec{\psi}_2 \\ \hat{W}_j(r_j) \vec{\psi}_j = \hat{W}_{j+1}(r_j) \vec{\psi}_{j+1}, \quad j = 2, N-1 \\ \hat{W}_N(r_N) \vec{\psi}_N = \hat{W}_{N+1}[1,2](r_N) \vec{\psi}_{N+1} \end{cases} \quad (A.23)$$

From Eq. (A.23) we obtain

$$\begin{aligned} \vec{X} &= \hat{V} \vec{\psi}_{N+1} \\ \hat{V} &= \hat{W}_1^{-1}(r_1) \hat{W}_2(r_1) \hat{W}_2^{-1}(r_2) \hat{W}_3(r_2) \cdots \hat{W}_N^{-1}(r_N) \hat{W}_{N+1}[1,2](r_N) \end{aligned} \quad (A.24)$$

Using equation for the Wronskian, the inverse matrix $\hat{W}_j^{-1}(r)$ is expressed explicitly:

$$\hat{W}_j^{-1}(r) = p_j r \cdot \begin{bmatrix} I_n'(p_j r) & \dfrac{\alpha_j}{\varsigma_j} I_n(p_j r) & 0 & -\dfrac{1}{\varsigma_j} I_n(p_j r) \\ -\dfrac{\alpha_j}{\beta_j} I_n(p_j r) & I_n'(p_j r) & -\dfrac{1}{\beta_j} I_n(p_j r) & 0 \\ -K_n'(p_j r) & -\dfrac{\alpha_j}{\varsigma_j} K_n(p_j r) & 0 & \dfrac{1}{\varsigma_j} K_n(p_j r) \\ \dfrac{\alpha_j}{\beta_j} K_n(p_j r) & -K_n'(p_j r) & \dfrac{1}{\beta_j} K_n(p_j r) & 0 \end{bmatrix}$$

(A.25)

From Eqs. (A.24), (A.21), (A.25) all the unknown coefficients D_{n1}^c are determined and Eqs. (A.10), (A.14) permit the Fourier transform of the magnetic field in the 1st layer (borehole):

$$H_{z1}^* = \sum_{n=0}^{\infty} D_{n1}^c I_n(p_1 r) \cos n\bar{\phi} \qquad (A.26)$$

Finally, we apply the inverse Fourier transform (A.5) to Eqs. (A.26), (A.17) to derive the total H_z and normal field H_{z0}, correspondingly:

$$H_z = H_{z0} + \dfrac{M_z}{\pi} \int_0^{\infty} \left(\sum_{n=0}^{\infty} D_{n1}^c I_n(p_1 r) \cos n\bar{\phi} \right) \cos \lambda(z - z_0) d\lambda$$

$$H_{z0} = \dfrac{M_z e^{-k_1 \bar{R}}}{2\pi \bar{R}^3} \left[1 + k_1 \bar{R} - \dfrac{R^2}{\bar{R}^2} \left(3 + 3 k_1 \bar{R} + (k_1 \bar{R})^2 \right) \right], \quad \bar{R}^2 = R^2 + (z - z_0)^2, \quad \operatorname{Re} k_j > 0.$$

(A.27)

Here we used the evenness of the integrand to reduce an integration pass from $(-\infty, +\infty)$ to $(0, +\infty)$:

$$\int_{-\infty}^{\infty} F(\lambda) \cdot e^{i\lambda(z-z_0)} d\lambda = 2 \int_0^{\infty} F(\lambda) \cos \lambda(z - z_0) d\lambda$$

A.3 NUMERICAL IMPLEMENTATION
A.3.1 Normalization

To prevent the exponential growth of modified Bessel functions $I_n(z)$, $K_n(z)$ and thereby improve numerical stability at large arguments, we use the following normalization:

$$\begin{cases} I_n(z) = \overline{I}_n(z) \cdot e^z \\ K_n(z) = \overline{K}_n(z) \cdot e^{-z} \end{cases} \quad (A.28)$$

Matrixes from Eq. (A.24) are presented in the normalized form:

$$\begin{cases} \hat{W}_j^{-1}(r_j) = \begin{bmatrix} e^{p_j r_j} & 0 & 0 & 0 \\ 0 & e^{p_j r_j} & 0 & 0 \\ 0 & 0 & e^{-p_j r_j} & 0 \\ 0 & 0 & 0 & e^{-p_j r_j} \end{bmatrix} \cdot \overline{\hat{W}}_j^{-1}(r_j), \quad j = 1, N \\ \hat{W}_j(r_{j-1}) = \overline{\hat{W}}_j(r_{j-1}) \cdot \begin{bmatrix} e^{-p_j r_{j-1}} & 0 & 0 & 0 \\ 0 & e^{-p_j r_{j-1}} & 0 & 0 \\ 0 & 0 & e^{p_j r_{j-1}} & 0 \\ 0 & 0 & 0 & e^{p_j r_{j-1}} \end{bmatrix}, \quad j = 2, N+1 \end{cases}$$

(A.29)

Matrixes $\overline{\hat{W}}_j$, $\overline{\hat{W}}_j^{-1}$ contain normalized functions $\overline{I}_n(z)$, $\overline{K}_n(z)$ instead of $I_n(z)$, $K_n(z)$. Then, for the product $\hat{W}_j(r_{j-1}) \cdot \hat{W}_j^{-1}(r_j)$ and matrix $\hat{D}_j$ we get

$$\hat{W}_j(r_{j-1}) \cdot \hat{W}_j^{-1}(r_j) = \overline{\hat{W}}_j(r_{j-1}) \cdot \hat{D}_j \cdot \overline{\hat{W}}_j^{-1}(r_j) \cdot e^{p_j(r_j - r_{j-1})}$$

$$\hat{D}_j = \begin{bmatrix} 1 & 0 & 0 & 0 \\ 0 & 1 & 0 & 0 \\ 0 & 0 & e^{-2p_j(r_j - r_{j-1})} & 0 \\ 0 & 0 & 0 & e^{-2p_j(r_j - r_{j-1})} \end{bmatrix}, \quad j = 2, N$$

Denoting:

$$\hat{D}_1 = \begin{bmatrix} e^{p_1 r_1} & 0 & 0 & 0 \\ 0 & e^{p_1 r_1} & 0 & 0 \\ 0 & 0 & e^{-p_1 r_1} & 0 \\ 0 & 0 & 0 & e^{-p_1 r_1} \end{bmatrix}, \quad \hat{D}_{N+1} = \begin{bmatrix} e^{-2p_{N+1} r_N} & 0 & 0 & 0 \\ 0 & e^{-2p_{N+1} r_N} & 0 & 0 \\ 0 & 0 & 1 & 0 \\ 0 & 0 & 0 & 1 \end{bmatrix}$$

we rewrite Eq. (A.24) as:

$$\vec{X} = \hat{V} \cdot \vec{\psi}_{N+1} \tag{A.30}$$

where

$$\begin{cases} \vec{X} = \hat{D}_1^{-1} \cdot \vec{X} \\ \hat{V} = \hat{W}_1^{-1}(r_1) \cdot \hat{W}_2(r_1) \cdot \hat{D}_2 \cdot \hat{W}_2^{-1}(r_2) \cdot \hat{W}_3(r_2) \cdot \hat{D}_3 \cdots \hat{W}_N^{-1}(r_N) \cdot \hat{W}_{N+1}(r_N) \cdot \hat{D}_{N+1} \\ \vec{\psi}_{N+1} = e^{p_2(r_2 - r_1)} \cdots e^{p_N(r_N - r_{N-1})} e^{p_{N+1} r_N} \cdot \vec{\psi}_{N+1} \end{cases} \tag{A.31}$$

Similarly, presenting vector $\vec{\psi}_{N+1}$ as $\vec{\psi}_{N+1} = \begin{bmatrix} y_1 \\ y_2 \end{bmatrix}$, we derive the following system of linear equations:

$$\begin{cases} 0 = \overline{V}_{11} y_1 + \overline{V}_{12} y_2 \\ A = \overline{V}_{21} y_1 + \overline{V}_{22} y_2 \\ e^{p_1 r_1} C_{n1}^s = \overline{V}_{31} y_1 + \overline{V}_{32} y_2 \\ e^{p_1 r_1} D_{n1}^c = \overline{V}_{41} y_1 + \overline{V}_{42} y_2 \end{cases} \tag{A.32}$$

where

$$A = e^{-p_1 r_1} \Pi_{n2} = -M_z \frac{p_1^2}{\pi} \overline{I}_n(p_1 r_0) e^{-p_1(r_1 - r_0)}$$

From the 1st and 2nd equations of the system (A.23) we find y_1, y_2:

$$\begin{cases} y_1 = -A \dfrac{\overline{V}_{12}}{\overline{V}_{11} \overline{V}_{22} - \overline{V}_{12} \overline{V}_{21}} \\ y_2 = A \dfrac{\overline{V}_{11}}{\overline{V}_{11} \overline{V}_{22} - \overline{V}_{12} \overline{V}_{21}} \end{cases}$$

and the coefficient D_{n1}^c is determined from the 4th equation as $D_{n1}^c = e^{p_1 r_1} A \dfrac{\overline{V}_{42} \overline{V}_{11} - \overline{V}_{41} \overline{V}_{12}}{\overline{V}_{11} \overline{V}_{22} - \overline{V}_{12} \overline{V}_{21}}$. Now, in Eq. (A.27) the factor $D_{n1}^c I_n(p_1 r)$ is of the form:

$$D_{n1}^c I_n(p_1 r) = A \frac{\overline{V}_{42} \overline{V}_{11} - \overline{V}_{41} \overline{V}_{12}}{\overline{V}_{11} \overline{V}_{22} - \overline{V}_{12} \overline{V}_{21}} \overline{I}_n(p_1 r) e^{-p_1(r_1 - r_0)} \tag{A.33}$$

Derived presentations for the matrix $\widehat{\overline{V}}$, factor $D_{n1}^{c} I_{n}(p_{1}r)$, and coefficient A are very convenient because they do not contain exponentially growing factors, which represent a significant obstacle for the numerical implementation.

A.3.2 Convergence of Series

Convergence is reached when each of the following terms in the expansion changes the sum by less than some predefined small number. The number of terms varies from 1, when eccentricity is equal to zero, to a large number, when the eccentricity approaches the borehole radius. When the transmitter and receiver radii approach the borehole radius, convergence of the series is very slow. In that case it is advisable to transpose integration and summation.

A.3.3 Integration

To ensure fast decay of the integrand, it is necessary to transform oscillating factor $\cos\lambda(z-z_0)$ into a decaying factor. This can be accomplished by integration in the plane of complex numbers λ [16]. Integrand $D_{n1}^{c} I_{n}(p_{1}r)$ depends on the radicals p_j ($p_j^2 = \lambda^2 + k_j^2$, $k_j^2 = -i\omega\mu_j\gamma_j = -i\omega\mu_j\sigma_j - \omega^2\mu_j\varepsilon_j$, $\mathrm{Re}\, p_j > 0$, $\mathrm{Re}\, k_j > 0$), which have branch points $p_j = 0$ in the complex plane of λ. The branch points $\lambda_j = (\lambda_{xj}, \lambda_{yj})$ are determined by the following relationships:

$$\begin{cases} \lambda_{xj} = \mathrm{Re}\,\lambda_j = -\mathrm{Im}\,k_j = \sqrt{\dfrac{\omega^2\mu_j\varepsilon_j + \sqrt{\left(\omega^2\mu_j\varepsilon_j\right)^2 + \left(\omega\mu_j\sigma_j\right)^2}}{2}} \\ \\ \lambda_{yj} = \mathrm{Im}\,\lambda_j = \mathrm{Re}\,k_j = \sqrt{\dfrac{-\omega^2\mu_j\varepsilon_j + \sqrt{\left(\omega^2\mu_j\varepsilon_j\right)^2 + \left(\omega\mu_j\sigma_j\right)^2}}{2}} \end{cases} \quad (A.34)$$

The cut, which separates Riemann surface sheets, is a part of the hyperbola in the complex plane λ with the origin in the branch point:

$$\begin{cases} 2\lambda_{xj}\lambda_{yj} = \omega\mu_j\sigma_j \\ \lambda_{xj} \leq \sqrt{\lambda_{yj}^2 + \omega^2\mu_j\varepsilon_j} \end{cases} \quad (A.35)$$

The larger dielectric permittivity ε_j the closer branch point to the real axis of integration λ_x. For illustration purpose, the branch points and cuts are shown in Fig. A.3.

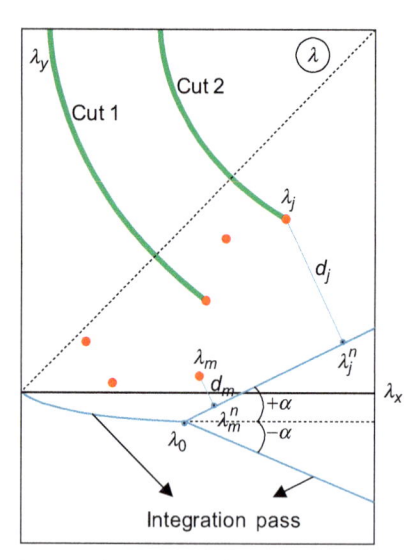

Fig. A.3 Plane of complex numbers λ. Branch points λ_j, cuts on a Riemann surface for two branch points, and the integration path.

The appropriate path of integration in the complex plane λ avoids an intersection with the cuts. Because the integrand in the vicinity of the branch points is irregular, it also must avoid approaching the branch points closely. Having that in mind, we set the following integration path.

The first part is parabola from the point $\lambda = (0, 0)$ to the point $\lambda_0 = (\lambda_{x0}, \lambda_{y0})$, having zero derivative at λ_0. The path then falls into the two rays along the angles $\pm \alpha$ with respect to the axis λ_x (Fig. A.3). The coordinate $\lambda_{y0} = d_0$ is set to be a fixed small number, while λ_{x0} is determined by the minimal distance from the branch points to the upper ray. Let d_j be the distances from the branch points to the upper ray, λ_j^n is the projections of the branch points to the ray, and d_m is the minimal distance for which the corresponding projection λ_j^n is higher than λ_0: $d_m = \min \left\{ d_j | \lambda_{yj}^n > \lambda_{y0} \right\}$. Then λ_{x0} is chosen such that $d_m = d_{0m}$, where d_{0m} is fixed small number. The constructed integration path neither intersects the cuts nor closely approaches the branch points.

Let us present the integral in Eq. (A.27) as a sum:

$$I = \int_0^\infty F(\lambda) \cos \lambda (z - z_0) d\lambda = I_1 + I_2 + I_3 \qquad (A.36)$$

For the terms earlier we have

$$I_1 = \int_0^{\lambda_0} F(\lambda)\cos\lambda(z-z_0)d\lambda$$

$$-\int_0^{\lambda_{x0}} F\left[\overline{\lambda}+i\lambda_{y0}\left(1-\left(\frac{\overline{\lambda}}{\lambda_{x0}}-1\right)^2\right)\right]\left(1-i\lambda_{y0}\frac{2}{\lambda_{x0}}\left(\frac{\overline{\lambda}}{\lambda_{x0}}-1\right)\right)\cos\lambda(z-z_0)d\overline{\lambda}$$

(A.37)

$$I_2 + I_3 = \int_{\lambda_0}^{\infty+\lambda_0} F(\lambda)\frac{e^{i\lambda(z-z_0)}+e^{-i\lambda(z-z_0)}}{2}d\lambda$$

$$I_2 = \frac{1}{2}\int_{\lambda_0}^{\infty+\lambda_0} F(\lambda)e^{i\lambda(z-z_0)}d\lambda \qquad (A.38)$$

$$I_3 = \frac{1}{2}\int_{\lambda_0}^{\infty+\lambda_0} F(\lambda)e^{-i\lambda(z-z_0)}d\lambda$$

Integration in I_2 is performed along the upper ray, while in I_3 it is performed along the lower ray. In that case the oscillating factor $\cos\lambda(z-z_0)$ becomes decaying one $e^{-tg\alpha(\overline{\lambda}-\lambda_{x0})}$:

$$\begin{cases} I_2 = \dfrac{1+i\cdot tg\alpha}{2}e^{-\lambda_{y0}(z-z_0)}\int_{\lambda_{x0}}^{\infty} F\left[\overline{\lambda}+i\left(tg\alpha(\overline{\lambda}-\lambda_{x0})+\lambda_{y0}\right)\right]e^{i\overline{\lambda}(z-z_0)}e^{-tg\alpha(\overline{\lambda}-\lambda_{x0})}d\overline{\lambda}, \\[2ex] I_3 = \dfrac{1-i\cdot tg\alpha}{2}e^{\lambda_{y0}(z-z_0)}\int_{\lambda_{x0}}^{\infty} F\left[\overline{\lambda}+i\left(-tg\alpha(\overline{\lambda}-\lambda_{x0})+\lambda_{y0}\right)\right]e^{-i\overline{\lambda}(z-z_0)}e^{-tg\alpha(\overline{\lambda}-\lambda_{x0})}d\overline{\lambda} \end{cases}$$

(A.39)

Eqs. (A.36), (A.37), (A.39) determine advanced integration in the complex plane λ with a rapidly decaying integrand. In the case of low frequencies when $\sigma \gg \omega\varepsilon$, the branch points are close to the ray $\lambda_y = \lambda_x$, and positioned far from the real axis λ_x. In that case the integration in I_1 (Eq. A.37) may be simplified and performed along the real axis λ_x:

$$I_1 = \int_0^{\lambda_0} F(\lambda)\cos\lambda(z-z_0)d\lambda \qquad (A.40)$$

where λ_0 is a prescribed small real number or zero. Correspondingly, the integrals I_2, I_3 (Eq. A.39) are transformed into

$$\begin{cases} I_2 = \dfrac{1}{2}\int_{\lambda_0}^{\infty} F(\lambda)e^{i\lambda(z-z_0)}d\lambda = \dfrac{1+i\cdot tg\alpha}{2}\int_{\lambda_0}^{\infty} F\left[\overline{\lambda}+i\cdot tg\alpha(\overline{\lambda}-\lambda_0)\right]e^{i\overline{\lambda}(z-z_0)}e^{-tg\alpha(\overline{\lambda}-\lambda_0)}d\overline{\lambda} \\ I_3 = \dfrac{1}{2}\int_{\lambda_0}^{\infty} F(\lambda)e^{-i\lambda(z-z_0)}d\lambda = \dfrac{1-i\cdot tg\alpha}{2}\int_{\lambda_0}^{\infty} F\left[\overline{\lambda}-i\cdot tg\alpha(\overline{\lambda}-\lambda_0)\right]e^{-i\overline{\lambda}(z-z_0)}e^{-tg\alpha(\overline{\lambda}-\lambda_0)}d\overline{\lambda} \end{cases}$$

$$(A.41)$$

A.3.4 Code Performance

Based on the described algorithm a computer 1.5D code for numerical calculations was developed. The program permits arbitrary number of layers with prescribed conductivity and dielectric constant in each layer. The performance strongly depends on the value of eccentricity and the number of cylindrical layers in the model—the greater the eccentricity and the number of layers the more processing time is required. For the majority of practical cases with several boundaries and dipole in the middle of the borehole, the processing time on a 3.2-GHz processor is less than 0.1 c, providing results with an error less than 0.01%. This kind of performance is essential while solving an inversion problem, where a large number of repetitive modeling calls are required before matching synthetic and measured data.

We verified our code versus 2.5D finite element (FE) code developed by [10] and Commercial CST FE 3D code. A comparison example is shown in Fig. A.4, where the green line corresponds to the 2.5D, the blue line to the CST, and the red line to our code. For the calculations we used Model 3 (Table A.1) and selected the following arrangements: transmitter-receivers spacings 0.2 and 0.14 m, frequency 100 MHz. The mismatch between 1.5D and 2.5D codes is less than 2%, while our code is, at least, 1000 times faster. Commercial CSTFE 3D code requires about 10 min per frequency, and the mismatch with 1.5D grows up to 4% at the extreme eccentricity value.

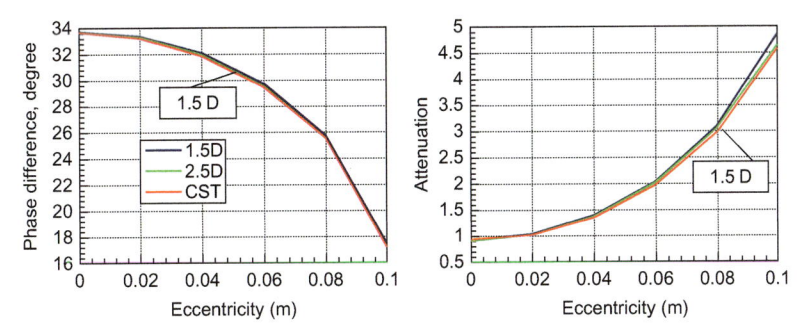

Fig. A.4 Comparison between 1.5D, 2.5D, and CST codes.

Table A.1 Parameters of the Models

Model	Mud	Formation	Borehole Radius
1	$\rho_1 = 0.02$ ohm m $\varepsilon_1^* = 50$	$\rho_2 = 10$ ohm m $\varepsilon_2^* = 40$	$r_1 = 0.108$ m
2	$\rho_1 = 2$ ohm m $\varepsilon_1^* = 70$		
3	$\rho_1 = 200$ ohm m $\varepsilon_1^* = 10$		

A.4 EFFECT OF ECCENTRICITY

In this section we present numerical results showing the usefulness of the algorithm for studying the influence of eccentricity on the induction responses. Eccentricity is defined as a displacement of the tool from the borehole axis, and it is equal to the transmitter/receiver radius $r = r_0$. While induction well logging operates in the frequency range from tens of KHz to tens of MHz, the frequency of the dielectric logging varies between tens and hundreds of MHz.

Following we show how the eccentricity and the dielectric permittivity affect the responses at different frequencies and conductivities of the borehole mud.

Three benchmarks of the borehole are considered:
1. high-conductivity mud (salty or biopolymer);
2. medium-conductivity mud (fresh); and
3. low-conductivity mud (oil-based).

The selected parameters of the models are presented in Table A.1: resistivities $\rho_j = \dfrac{1}{\sigma_j}$, relative dielectric permittivities ε_j^*, and borehole radius r_1.

Relative dielectric permittivity is determined by the relationship: $\varepsilon_j = \varepsilon_0 \varepsilon_j^*$, where $\varepsilon_0 = \dfrac{10^{-9}}{36\pi}$ F/m is the permittivity of free space. Magnetic permeability is equal to the permeability of free space: $\mu_j = \mu_0 = 4\pi 10^{-7}$ H/m.

Following we present the dependence of the signals on the eccentricity to determine cases when it significantly affects the measurements and cannot be ignored in the processing. Variation of the signals is presented in the normalized form as a ratio between signals of the eccentred and noneccentred probes.

A.4.1 Low-Frequency Induction Logging

The measured signal is the imaginary part of magnetic field in a three-coil array:

$$S = \mathrm{Im}\left(H_z(L_2) - \left(\frac{L_1}{L_2}\right)^3 H_z(L_1)\right), \text{ where } L = z - z_0, L_1 \text{ is the distance}$$

between transmitter and short-spaced coil (bucking coil) and L_2 is the distance between transmitter and long-spaced coil (main coil); $L_2 = 0.25, 0.5,$ and 1 m, $L_1 = 0.8\, L_2$, frequencies 10 and 200 kHz. At these frequencies, even for the case of low-conductive mud (Model 3) $\sigma \gg \omega\varepsilon$, the influence of the dielectric permittivity is small and the field is mainly defined by the conductivity of the media. For example, for $\rho_1 = 200$ ohm m, $\varepsilon_j^* = 10$, and at 200 kHz, the estimate for $\omega\varepsilon = 0.0001$ S/m ($\sigma = 0.005$ S/m) and the condition $\sigma \gg \omega\varepsilon$ is held.

The eccentricity must be taken into account when mud is highly conductive (Fig. A.5) or when subarrays are short (Figs. A.6 and A.7). In those cases the eccentricity should become a subject of inversion.

When the influence of the eccentricity is relatively small (10% and less), well-known correction algorithms can be used. For example, a specific combination of responses at two [21] or more [22] different frequencies is less sensitive to the near-borehole zone and, in particularly, to the eccentricity of the probe.

This is shown in A.8, where dual-frequency responses $\Delta S = S(\omega_1) - \dfrac{\omega_1}{\omega_2} S(\omega_2)$ for the Model 1 and different arrangements are presented. The combination diminishes the influence of the eccentricity.

In Table A.2 we present the signal dynamic range, corresponding to the single-frequency (third column) and dual-frequency responses (fourth column). The dynamic range is defined as a ratio of the signals at 0.09 m and

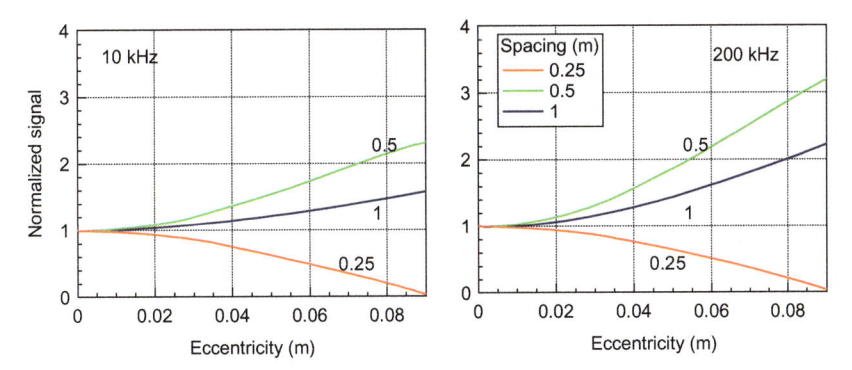

Fig. A.5 Conductive mud, Model 1. Relative variation of the low-frequency induction signals with eccentricity at 10 and 200 kHz.

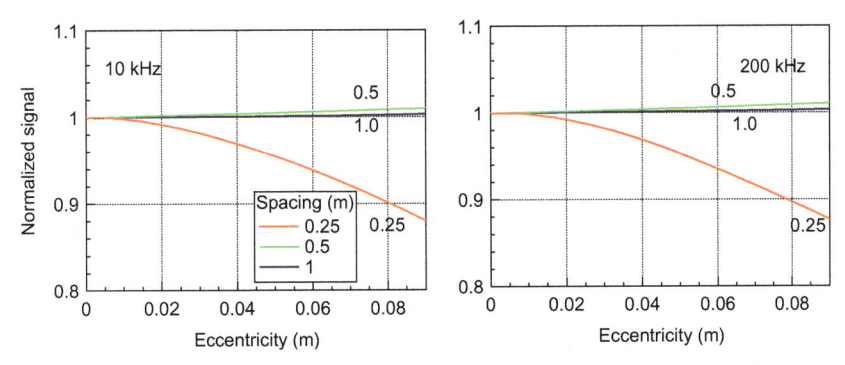

Fig. A.6 Moderately conductive mud, Model 2. Relative variation of the low-frequency induction signals with eccentricity at 10 and 200 kHz.

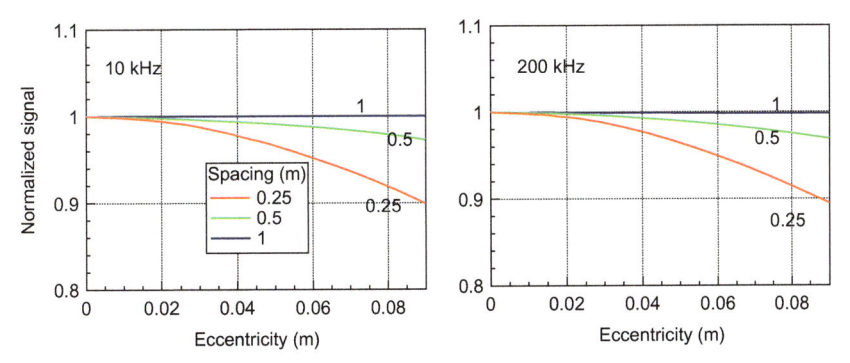

Fig. A.7 Resistive mud, Model 3. Relative variation of the low-frequency induction signals with eccentricity at 10 and 200 kHz.

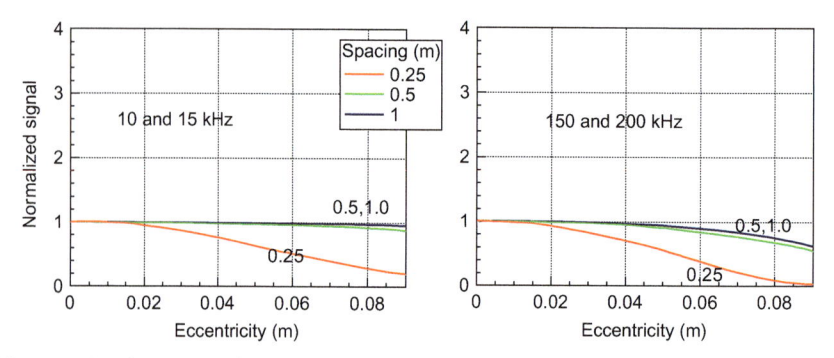

Fig. A.8 Conductive mud, Model 1. Eccentricity correction by the dual-frequency combination at 10 and 15 kHz and at 150 and 200 kHz.

Table A.2 Dynamic Range of Single-frequency and Dual-frequency Signals

Frequencies (kHz)	Spacing (m)	Single-Frequency Range	Dual-Frequency Range	Range Decrease
10	0.25	22	5.2	4.2
15	0.5	2.3	1.1	2.1
	1.0	1.6	1.1	1.5
150	0.25	–	–	–
200	0.5	3.2	1.8	1.8
	1.0	2.2	1.6	1.4

0 value of eccentricity. In the last column the ratio between the third and fourth columns demonstrates the benefit of the dual frequency.

Dual frequency effectively reduces sensitivity to the eccentricity in all cases, except the case of a large eccentricity when the signal rapidly changes due to zero crossing (combination of 150 and 200 kHz, 0.25-m spacing).

A.4.2 High-Frequency Induction Logging

In high-frequency induction logging, the measured values are phase difference $\Delta\phi$ and attenuation dA in three-coil array:

$$\begin{cases} \Delta\phi = arctg \dfrac{\mathrm{Im}\, S}{\mathrm{Re}\, S} \cdot \dfrac{180°}{\pi} \\ dA = -20 \lg |S| \end{cases}$$

where $S = \dfrac{H_z(L_2)}{H_z(L_1)} \Big/ \dfrac{H_z^{air}(L_2)}{H_z^{air}(L_1)}$, and $H_z^{air}\left(k_1 = -i\omega\sqrt{\mu_0 \varepsilon_0}\right)$ is the magnetic field in the air used for the calibration. It is calculated using expression for the normal field H_{z0} (Eq. A.27). At the low-frequency limit the calibration factor is defined only by the spacings L_1 and L_2: $\dfrac{H_z^{air}(L_2)}{H_z^{air}(L_1)} = \left(\dfrac{L_1}{L_2}\right)^3$. We use the following arrangements: $L_2 = 0.5$, 1, and 2 m; $L_1 = 0.8\ L_2$; frequencies 1 and 15 MHz. At the frequency of 15 MHz the parameter of formation $\omega\varepsilon = 0.03$ S/m is comparable with $\sigma = 0.1$ S/m and the permittivity has to be taken into account. For the low-conductive mud (Model 3) when $\sigma = 0.005$ S/m and the parameter $\omega\varepsilon = 0.0075$ S/m, the influence of the dielectric term is even more pronounced. This is especially true for shallow subarrays, which are severally affected by the conductivity of the mud. At the frequency of 1 MHz the influence of permittivity is negligible.

The effect of the eccentricity on the high-frequency logging responses is demonstrated in Figs. A.9–A.11. The long 2 m subarray is practically unsusceptible to the eccentricity. It is interesting to see how a very conductive mud hugely attenuates the signal at the frequency of 15 MHz (Fig. A.9, subplots on the right), making the response insensitive to all other parameters including standoff. At the same time the eccentricity has to be taken into account in the processing of the short- and medium-spacing responses (<1 m) at relatively low frequencies (Fig. A.9, subplots on the left). When

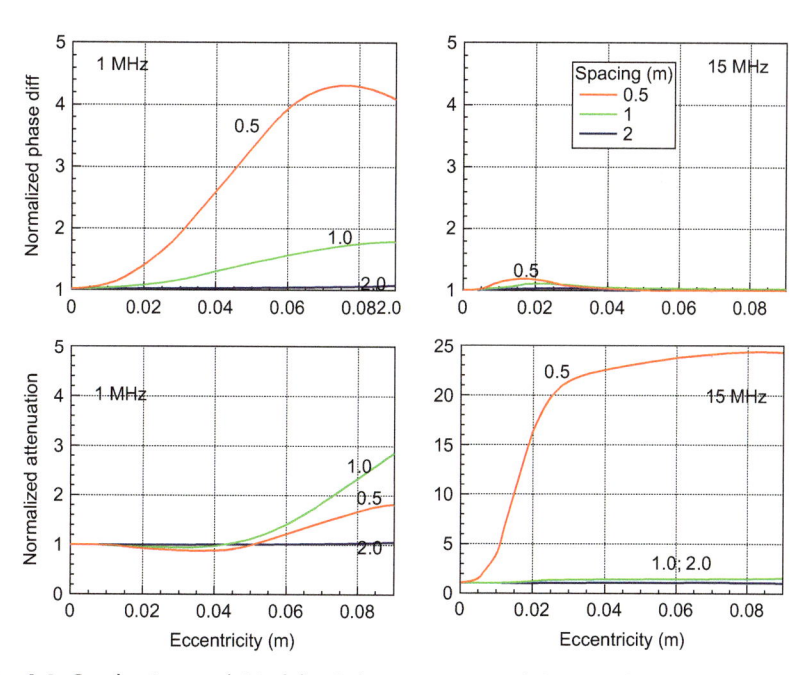

Fig. A.9 Conductive mud, Model 1. Relative variation of the high-frequency phase difference (upper part) and attenuation (lower part) with eccentricity at 1 and 15 MHz.

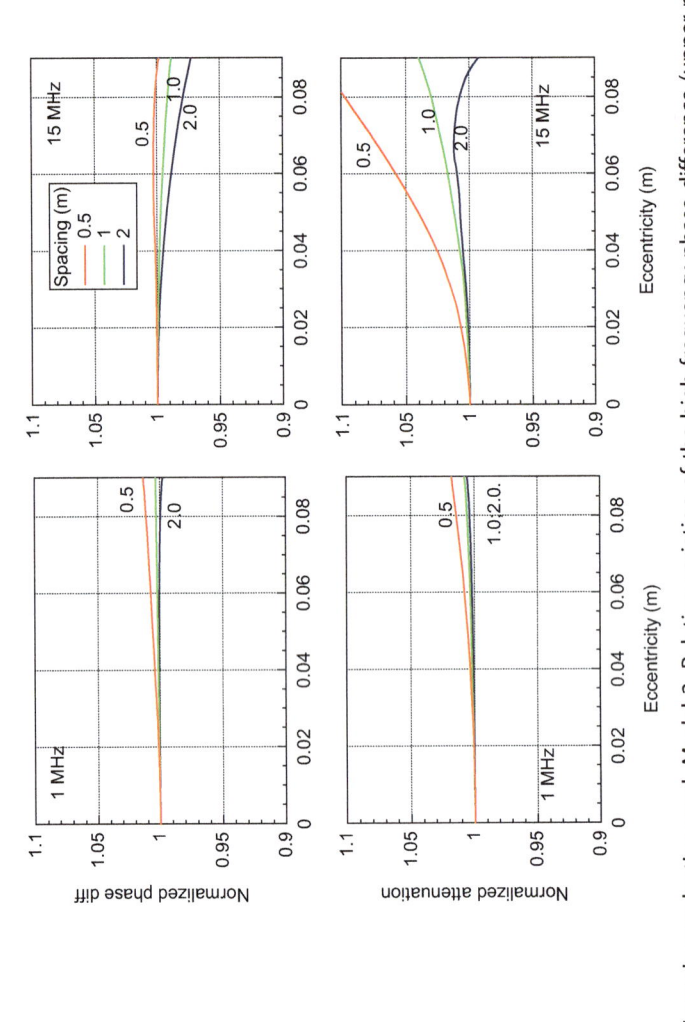

Fig. A.10 Moderately conductive mud, Model 2. Relative variation of the high-frequency phase difference (upper part) and attenuation (lower part) with eccentricity at 1 and 15 MHz.

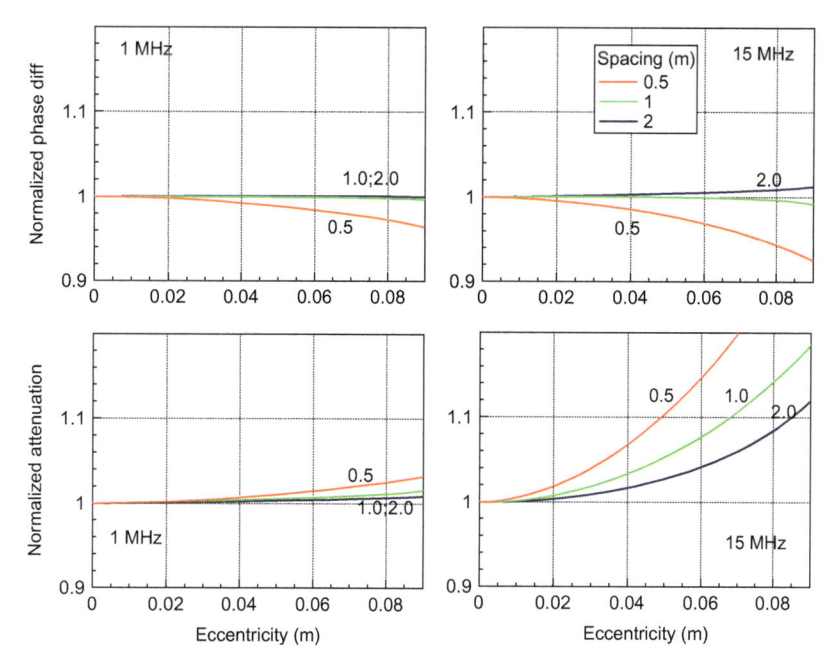

Fig. A.11 Resistive mud, Model 3. Relative variation of the high-frequency phase difference (upper part) and attenuation (lower part) with eccentricity at 1 and 15 MHz.

the mud is medium or low conductive, the eccentricity can be neglected regardless of the spacing (Figs. A.10 and A.11). This conclusion is in agreement with observation made by [23] about the negligible effect of the eccentricity in the analyzed range of resistivity contrasts. They also pointed out that the effect increases dramatically when the contrast between mud and formation approaches values of 10,000 and above.

A.5 CONCLUSIONS

We believe that the potential of analytical approaches is not yet exhausted; in this study a semianalytical approach enabled us to develop very fast code for simulation of electromagnetic responses of eccentred dipole located in cylindrically layered media. The code permits calculations in a wide range of frequencies, serving the needs of induction and dielectric logging. For effective implementation of the algorithm we stress the importance of a proper normalization of the integrands and the means increasing convergence of the infinite series. Also, when performing integration in the complex plane, cautions are needed in order to avoid intersection with the cuts located on Riemann surface.

Validation of our 1.5D code against two finite element codes, applicable for more complex formation structures, showed superior performance of the code in a simple cylindrically layered media.

In most cases of the low-frequency induction measurements, the influence eccentricity either can be neglected or corrected. In case of high-frequency logging the eccentricity is the most pronounced when a short probe is placed in a salty mud.

ACKNOWLEDGMENTS

The authors thank to our colleagues Fei Le and Alex Seryakov for conducting comparison of the code versus CST and 2.5D modeling codes.

REFERENCES

[1] Barber TD, Rosthal RA. Using a multiarray induction tool to achieve high-resolution logs with minimum environmental effects. In: SPE 22725; 1991.
[2] Beard D, Zhou Q, Bigelow E. Practical applications of a new multichannel and fully digital spectrum induction system. In: SPE36504. Annual technical conference and exhibition; 1996.
[3] Zhdanov MS. Geophysical electromagnetic theory and methods. Amsterdam: Elsevier; 2009. p.846.
[4] Tamarchenko T, Tabarovsky L. Fast frequency domain electromagnetic modeling in axially symmetric layered media. Radio Sci 1994;29.
[5] Hizem M, Budan H, Deville B, Faivre O, Mosse L, Simon M. Dielectric dispersion: a new wireline petrophysical measurement. In: SPE 116130; 2008.
[6] Seleznev N, Habashy T, Boyd A, Hizem M. Formation properties derived from a multi-frequency dielectric measurement. In: SPWLA 47th annual logging symposium, Veracruz, Mexico; 2006.
[7] Abubakar A, Habashy TM, Druskin VL, Knizhnerman L, Alumbaugh D. 2.5D forward and inverse modeling for interpreting low-frequency electromagnetic measurements. Geophysics 2008;73(4):F165–77.
[8] Davydycheva S, Druskin V, Habashy T. An efficient finite-difference scheme for electromagnetic logging in 3D anisotropic inhomogeneous media. Geophysics 2003;68(5):1525–36.
[9] Wang T, Fang S. 3-D electromagnetic anisotropy modeling using finite differences. Geophysics 2001;66(6):1386–98.
[10] Bespalov A, Kuznetsov YA, Tabarovsky L. Mathematical modeling and numerical methods for induction logging applications. In: Mathematical and numerical aspects of wave propagation WAVES, Berlin; 2013. p. 32–8.
[11] Everett ME, Badea EA, Shen LC, Merchant GA, Weiss CJ. 3-D finite element analysis of induction logging in a dipping formation. IEEE Trans Geosci Remote Sens 2003;39:2244–52.
[12] Nam MJ, Pardo D, Torres-Verdin C. Simulation of tri-axial induction measurements in dipping, invaded, and anisotropic formations using a Fourier series expansion in a

Non-orthogonal system of coordinates and a self-adaptive hp finite-element method. Geophysics 2010;75(3):F83–95.
[13] Wang GL, Torres-Verdín C, Gianzero S. Fast simulation of triaxial borehole induction measurements acquired in axially symmetrical and transversely isotropic media. Geophysics 2009;74(6):E233–49.
[14] Sun X, Nie Z, Li A, Luo X. Analysis and correction of borehole effect on the responses of multicomponent induction tools. Prog Electromagn Res 2008;85:211–26.
[15] Kaufman AA, Kagansky AM, Krivoputsky VS. Radial investigation characteristics of the induction tools displaced relatively the borehole axis. Geol Geophys 1974;7:102–16.
[16] Tabarovsky LA. Application of integral equations method to geoelectric problems. Novosibirsk: USSR, Nauka; 1975. p.144.
[17] Gradshtein IS, Ryzhik IM. Tables of integrals, sums, series and products. Moscow: USSR, Fizmatgiz; 1963. p.1100.
[18] Korn G, Korn T. Mathematics handbook. Moscow: USSR, Nauka; 1984. 832 p.
[19] Vladimirov VS. Equations of mathematical physics. Moscow: USSR, Nauka; 1971. 512 p.
[20] Watson GN. A treatise on the theory of Bessel functions. Cambridge: Cambridge University Press; 1922.
[21] Kaufman AA, Keller GV. Induction logging (methods in geochemistry and geophysics). New York: Elsevier; 1989.
[22] Tabarovsky LA, Rabinovich MB. Real time 2D inversion of induction logging data. J Appl Geophys 1998;38.
[23] Wu J, Wisler M. Effects of eccentering MWD tools on electromagnetic resistivity measurements. In: SPWLA 31st Annual Logging Symposium, Lafayette, Louisiana; 1990.

FURTHER READING
[1] Nikitenko M, Itskovich G, Seryakov A. Fast electromagnetic modeling in cylindrically-layered media excited by eccentred magnetic dipole. Radio Sci 2016;51:573–88.

INDEX

Note: Page numbers followed by f indicate figures, and t indicate tables.

A

Ampere's law, 4–5, 6f
Anisotropy
  coefficient, 451f, 458f
  conductivity, 448–449, 452, 459
  electromagnetic field of magnetic dipole, 452–459
  formation with finite thickness, 459–465
  Fourier transform, 458
  gauge condition, 453–454, 456
  horizontal electric dipole, 456
  in-phase component, 457
  of layered medium, 448–451
  Maxwell equations, 452
  Ohm's law, 448–449
  quadrature component of field, 465f
  quasistationary approximation, 450
Apparent conductivity, 181–182, 229
  corrections
    borehole, 225–226
    skin effect, 222–225, 223–224f
  curves, 301–305, 305f
  electromotive force, 181
  elementary geometric factor, 181–182
  in finite thickness bed, 188–194, 189f, 192f
  horizontal magnetic dipole, 437, 437f
  for 6FF40 probe, 222–223, 224f
  for two-coil probe, 222–223, 223f
Apparent resistivity curves, 181, 202f
  behavior of function, 336f
  on borehole axis, 334f
  invasion zone, 335–337, 336f
  of transient signals, 333–337
Attenuation, 275–277, 276f
  displacement effects, 287–288, 288f

B

Bessel functions, 254–256, 258, 289–295
  borehole, 195–196

boundary value problem, 254
  in internal integral, 265–266
  known expressions for, 400
  properties of, 409
Biot–Savart law, 21, 36, 59, 67, 164–166, 169, 239
  charge conservation principle, 62–63
  for direct currents, 19
  induction current in conducting ring, 105, 112–113
  magnetic field and, 5–9, 13, 22, 30–31
  Maxwell's equations, 67, 69
  quadrature and in-phase components, 122
  sinusoidal plane wave, 136
Borehole
  corrections, 225–226
  electrical methods, 8
  geophysics, 20–21
  Green's formula, 243
  in-phase component, 233
  integral equations, 233
  quadrature component, 233
Borehole axis
  apparent resistivity curves on, 334f
  approximate solution validity, 274–277, 276f
  Cauchy's formula, 267–274
  eccentricity, 483
  horizontal magnetic dipole, 389f
  integration contour, deformation of, 267–274, 268f
  magnetic field, 266–284, 325–333
  probe displacement, 284–288, 286–288f
  three-coil probe, 277–279, 277–278f
  two-coil probe, 279–284, 280f
Born approximation, 246–247
Boundary value problem, 136
  angular harmonics, 473
  Bessel functions, 254

493

Boundary value problem *(Continued)*
cylindrical boundaries, 250–251, 251f
deep-reading measurements while drilling, 347–354
expansion in series, 471–472
final representation of magnetic field, 474–476
formulation, 89–90, 249–250
Fourier transform, 470–471
Helmholtz's equation, 252–254
for horizontal magnetic dipole, 388–395
Maxwell's equations, 251–252
normal field, determination, 473–474
polar coordinate system, 472f
problem definition, 468–469
vector potential, 250
vertical magnetic dipole, 250, 251f

C

Cauchy formula, 267
Charge conservation principle, 46–48
displacement currents, 61–63
quasi-stationary electromagnetic field, 48
stationary field, 48
Charge density, 53–54
Code performance, 482
Conductivity distribution, 179–180, 180f
Conservation of energy. *See* Energy conservation
Convergence, 479
Coulomb's law, 4, 9–10, 42, 64, 67
distribution of electric charges, 48
Maxwell's equations, 48, 67–69
Current density, 8, 14, 21, 32–33
distribution, 170f
flux of, 40f, 46–48, 58
magnetic field, 169
quadrature component, 170–171
uniform medium, 168–169
Cylindrical conductor, magnetic field due to current in, 30–32, 31f

D

D'Alembert's method, 92–93, 134
Delta function, 125–126
Differential equation
for current, 113
first order, 50–51, 55–56, 73–74
for volume density, 49
Diffusion equations, 137–138, 148, 160
Diffusion process, 75, 171–172
Dipole moment, 141
magnitude of, 109
time-variable, 106f
varies with time, 110
Dirac delta function, 150
Displacement currents, 8, 61f, 62, 69, 104–105
behavior, 96–97
charge conservation principle, 61–63
in circuit with capacitor, 63–67, 64f
density, 60, 63
flux of, 99, 102
magnetic field, second source of, 59–61
total current, 61–63
Divergence of vector potential, 16–18
Doll's range, 281
Doll's theory, 173–174, 300. *See also* Hybrid method
apparent conductivity curves, 301–305, 305f
profiling curves, 306
quadrature component, 299
Duhamel's integral, 153–156

E

Eccentricity
borehole, 483
conductive mud, 485–487f
effect of, 483–489
high-frequency induction logging, 486–489
low-frequency induction logging, 484–486
moderately conductive mud, 485f, 488f
resistive mud, 485f, 489f
single-frequency and dual-frequency signals, 486t
Electric charges, 48
charge density δ_{02}, behavior of, 53–54
nonuniform medium, 50–52
quasi-stationary field, 52–53
slowly varying field, 57–58
surface distribution, 54–56

uniform medium, 49–50
voltage of, 65
volume, 49
Electric current circuit, 4–5, 85–86, 85f
Electric field
  along receiving loop, 129f
  complex amplitude, 144
  in conducting and polarizable medium, 2–4
  induced measurements, 128–131
  on plane wave (see Plane wave)
  surface integral equation for
    Born approximation, 246–247
    cylindrically layered formation, 238–244
    horizontally layered formation, 244–246
  transient responses of, 158–159, 159f
Electromagnetic field, 25
  anisotropic medium, 452–459
  expressions for, 109, 149–153
  of plane wave, 99
  propagation, 63
  quasi-stationary, 48
  scalar potentials, 76
  sources, 67–68, 67t
  theory of, 72
  uniform isotropic medium, 385–388, 452–459
Electromagnetic induction, 41, 43–44, 63
Electromagnetic potentials, 75–78
Electromagnetic wave, propagation, 69, 96–97, 148
Electromotive force, 40–41, 101, 129–130, 176–177, 272–274
  on moving circuit, 9–13
Elementary ring, 174–175, 175–176f
Energy conservation, 81–83, 87
Energy density, 83–85
Energy field
  charges and change of, 41f
  near an interface, 40f
Euler's formula, 78–79

F

Faraday's law, 13, 40–46, 40f, 68, 101, 130
  induction current in conducting ring, 111

Focusing probes, 205–206
Fourier's transform, 81
Fredholm integral equation, 242–243
Frequency asymptotics, 166–168
Frequency domain, 52, 141–148, 312, 318, 349–350

G

Gauss–Newton method, 376–377
Gauss's theorem, 17–18, 46–47, 61–62
Geometrical factors
  of borehole, 195–199, 199f
  of elementary layer, 182–184, 185f
  of elementary ring, 174–178, 175–176f
  of layer with finite thickness, 184–188
  magnetic dipole, 398
Geo-steering, transient field inversion
  elements of, 367–368
  Gauss–Newton method, 376–377
  inverse problem solution, 370–375
  iterative and table-based inversion algorithms, 379–380
  Levenberg–Marquardt method, 377–378
  multiparametric inversion, 375–380
  parameter uncertainties, estimation, 380–383
  statistical inversion for case of vertical well, 371–373
  table-based inversion, 368–370
  well-and ill-posed problems, 366–367
Green's formula, 238, 240

H

Heaviside step-function, 153–154
Helmholtz equation, 238, 289–291
  one-dimensional, 135
  solution, 141–145
Hessian matrix, 377
High-frequency induction logging, 486–489
Hybrid method, 227–228

I

Induced currents, 173–175
  expression for, 168–172
Induction current in conducting ring
  equation of, 111–113, 111f

Induction current in conducting ring
    (Continued)
  expressions, 114–115
  harmonic excitation, 122–125
  in-phase and quadrature component, 118, 119f, 120
  measurements of, 128–131
  sinusoidal primary magnetic field, 117–122
  step-function excitation, 125–128
  step-function varying primary magnetic field, 115–117
  strong interaction, 127, 128f
  transient responses, 113–115
  weak interaction, 126, 127f
Induction logging
  apparent conductivity corrections
    borehole, 225–226
    skin effect, 222–225, 223–224f
  electromagnetic, 467
  high-frequency, 486–489
  low-frequency, 484–486
  multicoil/focusing induction probe
    application, 205–206
    evolution, 221–222
    6FF40 probe, 215–221, 215f, 218f
    three coil, 208–215, 210f, 212f
  standard array-induction tools, 467
  two-coil probe
    apparent conductivity, 181–182
    elementary ring, geometrical factor of, 174–178, 175–176f
    forward problem solutions, 179–181
    radial characteristics of, 195–204
    vertical responses of, 182–194
In-phase components, 165, 167–168, 168f, 171–172, 228–229, 231
  amplitude of, 165–166
  Biot–Savart law, 122
  of field approach zero, 167
  Helmholtz's equation, 145
  horizontal bed with invasion, 233–234
  induction current in conducting ring, 118, 119f, 120–121
  magnetic field, 260–265
  vertical magnetic dipole, 297–298, 298f, 301

Integral equations
  for electric field
    Born approximation, 246–247
    cylindrically layered formation, 238–244
    horizontally layered formation, 244–246
    hybrid method, 227–228
    linear approximation, 235–238
  for magnetic field
    borehole and layer of finite thickness, 233
    cylindrical boundaries, 231–233, 232f
    horizontal bed with invasion, 233–235
Integration, 479–482
Integration contour, 268
Invasion zone, 233–234

J

Jacobian matrix, 376–377
Joule's law, 81–83

K

Kirchhoff law, 122–123

L

Laplace's equation, 16–18, 33
Leontovich boundary condition, 352–353
Levenberg–Marquardt method, 377–378
Linear approximation
  electric field, 235–236
  Faradey's law, 235–236
  Fredholm integral equation, 236–237
  Ohm's law, 236
Logging-while-drilling (LWD) measurements, 344
Lorentz force, 9–10
  on moving circuit, 9–13
Low-frequency induction logging, 484–486

M

Magnetic dipole
  in anisotropic formation of finite thickness, 459–465
  in anisotropic medium, 452–459
  asymptotic expression, 396–397, 415–416, 424, 431–432
  Bessel functions, 400

borehole and formation, 421–422
contour integration, 408f
electromagnetic field in uniform isotropic medium, 385–388
in far zone, 406–417
frequency responses of amplitude, 423f
geometric factors, 398
horizontal boundaries, 290f
horizontal magnetic dipole
  amplitude, 389f
  apparent conductivity curves, 437, 437f
  borehole axis, 389f
  boundary value problem, 388–395
  conductive thin layer, 440
  formation with two horizontal interfaces, 432–441
  frequency responses of amplitude/phase, 438, 439f
  frequency responses of field amplitude, 395f
  phase of secondary field, 389f
  phase responses of field, 396f
  probe between the coils, 443
  probe on both sides of interface, 442
  probe on outside the formation, 441–442
  probe within the layer, 443
  resistive thin layer, 441
  two-coil induction probe, profiling with, 441–444, 442f
in-phase and quadrature components, 387, 388f
integral exponential function, 414
magnetic field, 27–30
in medium with one horizontal interface, 426–432
in medium with thin layer, 290f
in medium with thin resistive cylindrical layer, 421–426
in medium with two cylindrical interfaces, 417–420
nonconducting medium, 106f, 108–109
normalized attenuations for three-coil probe, 418f, 420
phase differences for three-coil probe, 418f, 420
in range of small parameter, 396–406
with sinusoidal current, 163–164

Summerfield integral, 398–399
transient field in medium with cylindrical interfaces
  apparent resistivity curves, 333–337
  borehole axis, early and late stage, 325–333
  Fourier integral and calculation, 321–324
  harmonic amplitudes, 322
transient field in uniform medium, 312–321
  Biot–Savart's law, 318
  electromotive force, 321t
  expressions for field, 313–314
  Faraday's law, 314
  features, 314–321
  field components, 317f
  graphs of function, 318
  in-phase component, 318
  transmitter-receiver moment, 321
transient field in vertical magnetic dipole
  Fourier integral, 341
  layer thickness, 340–343
  medium with one horizontal boundary, 337–340
transversal dipole moment, 385–386
transversal induction probe, 386f
two-coil induction probe, 297
  apparent conductivity curves, 301–305, 305f
  conductive bed, 306
  dependence of field on parameter $p=L/\delta_1$, 297–301
  in-phase and quadrature component, 298f
  profiling curves for, 306–310, 307f
  thick resistive bed, 308, 309f
  thin conductive bed, 308, 309f
  thin resistive bed, 308, 310f
in uniform medium, 141–153, 163–172, 452–459
vertical component
  in the bed, 294
  Bessel functions, 289–295
  Helmholtz equation, 291
  magnetic field, 289–295
  outside the bed, 292
  thin conducting plane, 295–297

Index

Magnetic field, 228–229
  Biot–Savart law, 5–9
  on borehole axis, 266–284
  change with time, 42–43
  circulation, 20
  complex amplitude, 144
  current-carrying objects, 21–37
  current filament, 22–23, 22f
  due to current in cylindrical conductor, 30–32, 31f
  features of, 171
  induced measurements, 128–131
  of infinitely long solenoid, 32–35, 32f
  in-phase components, 168f
  intensity, 25
  in linear magnetic moment, 156f
  magnetic dipole, 27–30
  quadrature component of, 168f, 194
  second source, 59–61
  in small induction number, 257–266
  surface current, 7f
  system of equations, 18–21, 37
  tangential component, 61
  toroid, 32–33, 35–37
  vector potential, 13–16, 24–27
Magnetic flux, 41–42, 105, 112, 116–117
Magnetic moment of loop, 25
Maxwell's equation, 19–21, 67–70, 135
  electromagnetic potentials, 75–78
  fields E and B, 73–75
  inductive electric field, 108
  in integral form, 43
  magnetic field, 151–152
  in piecewise uniform medium, 72–73
  plane wave, 95, 101
  second form, 70–72, 102
  sinusoidal fields, 78–81
  stationary electric field, 2
  third of, 4
Maxwell's equations, piecewise uniform medium, 72–73
Multicoil induction probes, 204–206
  application, 205–206
  evolution, 221–222
  geometric factor of, 206–207
  nonsymmetrical, 207
  6FF40 probe, 215–221, 215f, 218f
  symmetrical, 207, 207f
  three coil, 208–215, 210f, 212f

N

Neumann series, 246
Nonconducting medium
  field in, 153–156
  magnetic dipole, 106f, 108–109
  quasistationary field in, 104–110
Nonuniform medium, electric charges, 50–52
Normalization, 477–479
Numerical implementation
  code performance, 482
  convergence of series, 479
  integration, 479–482
  normalization, 477–479

O

Ohm's law, 2, 49, 86, 175
  induction current in conducting ring, 111–112, 121
Ordinary differential equation, 113

P

Phase difference, 275–277, 276f
  displacement effects, 287–288, 288f
Plane wave, 94
  as function of time and distance, 136–138
  models, 97–104, 98f
  phase of, 100
  propagation, 95–97
  sinusoidal (see Sinusoidal waves)
  in uniform medium, 91–104
  velocity of propagation, 93–94, 94f
Poisson's equation
  vector potential, 17–18, 33
Poynting vector, 83–85, 85f
  directional energy flux density, 96
Primary magnetic field, 110, 112
  sinusoidal, 117–122
  step-function varying, 115–117
  varies with time, 113
Probe position, 185, 186–187f, 187, 190
Propagation of plane wave, 93–97, 94f

Q

Quadrature components, 165, 167–168, 168f, 171–172, 228–230
  amplitude of, 165–166
  Biot–Savart law, 122

of field approach zero, 167
Helmholtz's equation, 145
horizontal bed with invasion, 233–234
induction current in conducting ring, 118, 119f, 120–121
magnetic field, 258–259
vertical magnetic dipole, 297, 298f, 299, 301
Quasistationary approximation, 45–46, 105, 107, 109, 111, 130, 160
Quasistationary fields, 3–4, 137–138, 164–165
  amplitudes, 163–164
  electric charges, 52–53
  magnetic dipole, 106f, 108–109
  in nonconducting medium, 104–110
  slowly varying, 57–58
  in uniform medium, 163–172

R
Radius of convergence, 120
Ramp time, 114–116
Resistivity
  horizontal resistivity, 447
  low-resistivity pay zones, 447
Ring inductance, 112
Runge–Kutta method, 125–126

S
Self-inductance, 112–113
Sinusoidal fields, Maxwell's equations, 78–81
Sinusoidal primary magnetic field, 117–122
Sinusoidal waves, 103, 165
  as function of time and distance, 136–138, 137f
  high frequency limit, 138–141
  low frequency limit, 138–141, 141f
  in uniform medium, 133–141
6FF40 focusing probe, 215–221, 218–220f
  apparent conductivity, 218, 219f
  borehole geometrical factor, 218, 218f
  configuration, 215–216, 215f
  dual induction probe, 221, 221f
  electromotive force, 216
  elementary layer for, 219, 220f
  parameters, 216, 216t
  profiling curves, 219–221, 220f

Skin effect, 228, 281
  corrections, 222–225, 223–224f
Solenoids
  inductive electric field of, 105–107
  magnetic field, 32–35, 32f
  vortex field of, 106f
Sommerfeld integral, 292
Step-functions
  arbitrary and, 154f
  excitation, 125–128, 160
  Heaviside step-function, 153–154
  spectrum of, 149
  of time, 149
  varying primary magnetic field, 115–117
Stoke's theorem, 20, 43, 60–61
Superposition of waves, 147–148
Superposition principle, 5, 7, 22

T
Taylor's expansion, 258
Three-coil probe, 203–204, 203f, 208
  amplitude ratio, 277, 277f
  approximate *vs.* exact solution, 275–276, 276f
  attenuation, 275–277, 276f, 278f
  borehole geometrical factor for, 210–211, 210f
  displacement effect, 286–287, 286f
  elementary layer for, 212–213, 212f
  normalized apparent conductivity, 211–212, 211f
  phase difference, 275–277, 276–278f
  wave path, 272, 273f
Toroid, magnetic field, 32–33, 35–37
Transient field
  in conducting medium, 157–160, 159f
  deep-reading measurements while drilling
    boundary condition, 352–353
    boundary value problem, 347–354
    finite conductivity of cylinder, 352–356
    formation/pipe signals ratio using magnetic shielding, improving, 364–365
    high-frequency and early transient stage asymptote, 355–356
    homogeneous formations, 359–360, 360f
    modified Bessel functions, 346

Transient field *(Continued)*
　　pipe conductivity, increasing, 359–361
　　pipe signal reduction using finite size copper shield and bucking, 361–364
　　spacing effect on pipe signal, 356–359
　　two-coil probe, 358f
　　uniform conducting medium, 344–347
　inversion in task of geo-steering
　　elements of, 367–368
　　Gauss–Newton method, 376–377
　　inverse problem solution, 370–375
　　iterative and table-based inversion algorithms, 379–380
　　Levenberg–Marquardt method, 377–378
　　multiparametric inversion, 375–380
　　parameter uncertainties, estimation, 380–383
　　statistical inversion for case of vertical well, 371–373
　　table-based inversion, 368–370
　　well-and ill-posed problems, 366–367
　magnetic dipole in medium with cylindrical interfaces
　　apparent resistivity curves, 333–337
　　borehole axis, early and late stage, 325–333
　　Fourier integral and calculation, 321–324
　　harmonic amplitudes, 322
　magnetic dipole in uniform medium
　　Biot–Savart's law, 318
　　electromotive force, 321t
　　expressions for field, 313–314
　　Faraday's law, 314
　　features, 314–321
　　field components, 317f
　　graphs of function, 318
　　in-phase component, 318
　　transmitter-receiver moment, 321
　vertical magnetic dipole
　　Fourier integral, 341
　　layer thickness, 340–343
　　medium with one horizontal boundary, 337–340
Transmission line, 85–86, 85f
Transverse plane wave, 96

Trial and error method, 92
Two-coil probe
　apparent conductivity, 181–182, 211f
　Biot–Savart law, 279
　displacement effect, 286–287, 286–287f
　elementary layer for, 212–213, 213f
　forward problem solutions, 179–181
　frequency responses
　　in three-layered formation, 279, 280f
　　in two-layered formation, 279, 280f
　geometrical factor
　　borehole, 195–199, 199f
　　elementary layer, 182–184, 185f
　　of elementary ring, 174–178, 175–176f
　　layer with finite thickness, 184–188
　induced currents density, 279
　in-phase component, 279
　with one interface, 188–189, 189f
　quadrature component, 279
　radial characteristics, 199–204
　vertical magnetic dipole, 279
　wave path, 272, 273f

U
Uniform medium
　anisotropic medium, 452–459
　electric charges, 49–50
　magnetic dipole in, 141–153, 163–172, 452–459
　plane wave in, 91–104
　sinusoidal plane wave in, 133–141
Uniqueness theorem, 86–89

V
Vector potential
　within borehole, 254–255
　boundary value problem, 250
　complex amplitude, 144
　components of, 15
　divergence and Laplacian of, 16–18
　expression for, 149–153
　Laplace's equation, 16–18, 33
　magnetic field, 13–16, 24–27
　z-component of, 142–143
Velocity of propagation, 134, 140
　frequency range, 139
　plane wave, 93–94, 94f
　sinusoidal wave, 136

Vertical magnetic dipole
　borehole axis
　　approximate solution validity, 274–277, 276f
　　Cauchy's formula, 267–274
　　integration contour, deformation of, 267–274, 268f
　　probe displacement, 284–288, 286–288f
　　three-coil probe, 277–279, 277–278f
　　two-coil probe, 279–284, 280f
　boundary value problem, 249–254, 251f
　field components, 254–257
　small induction number
　　asymptotic expressions, of field, 265–266
　　in-phase component, 260–265
　　quadrature component, 258–259
Volume electric charges, 49

W

Wave equation, 94
　solution to, 92–93
Wave path
　in three-coil probe, 272, 273f
　in two-coil probe, 272, 273f